Design and Analysis of Experiments for Statistical Selection, Screening, and Multiple Comparisons

WILEY SERIES IN PROBABILITY AND STATISTICS

Established by WALTER A. SHEWHART and SAMUEL S. WILKS

Editors: *Vic Barnett, Ralph A. Bradley, Nicholas I. Fisher, J. Stuart Hunter, J. B. Kadane, David G. Kendall, David W. Scott, Adrian F. M. Smith, Jozef L. Teugels, Geoffrey S. Watson*

A complete list of the titles in this series appears at the end of this volume

Design and Analysis of Experiments for Statistical Selection, Screening, and Multiple Comparisons

ROBERT E. BECHHOFER
School of Operations Research and Industrial Engineering
Cornell University

THOMAS J. SANTNER
Department of Statistics
The Ohio State University

DAVID M. GOLDSMAN
School of Industrial and Systems Engineering
Georgia Institute of Technology

A Wiley-Interscience Publication
JOHN WILEY & SONS, INC.
New York • Chichester • Brisbane • Toronto • Singapore

QA
279
.B43
1995

This text is printed on acid-free paper.

Copyright © 1995 by John Wiley & Sons, Inc.

All rights reserved. Published simultaneously in Canada.

Reproduction or translation of any part of this work beyond
that permitted by section 107 or 108 of the 1976 United
States Copyright Act without the permission of the copyright
owner is unlawful. Requests for permission or further
information should be addressed to the Permissions Department,
John Wiley & Sons, Inc., 605 Third Avenue, New York,
NY 10158-0012.

Library of Congress Cataloging-in-Publication Data:
Bechhofer, Robert E. (Robert Eric), 1919–
　　Design and analysis of experiments for statistical selection,
　screening, and multiple comparisons / Robert E. Bechhofer, Thomas J.
　Santner, David Goldsman.
　　　　p.　cm. — (Wiley series in probability and mathematical
statistics. Applied probability and statistics)
　　Includes bibliographical references and indexes.
　　ISBN 0-471-57427-9 (acid-free)
　　1. Experimental design.　2. Analysis of variance.　I. Santner,
Thomas J., 1947– .　II. Goldsman, David Morris, 1958– .
III. Title.　IV. Series.
QA279.B43　1995
519.5′38—dc20　　　　　　　　　　　　　94-41843

Printed in the United States of America

10　9　8　7　6　5　4　3　2　1

To Joan, Gail,
and our families
for their patience and encouragement
during the writing of this book

Contents

Preface   xi

1   The Rationale of Selection, Screening and Multiple Comparisons   1

    1.1   Introduction   1
    1.2   Basic Concepts of Classical Experimental Design   4
           1.2.1   Qualitative Versus Quantitative Factors   4
           1.2.2   Principles of Experimental Design   5
    1.3   Formulations Considered in This Book   7
    1.4   Organization of the Book   12

2   Selecting the Best Treatment in a Single-Factor Normal Response Experiment Using the Indifference-Zone Approach   15

    2.1   Introduction   15
    2.2   A Single-Stage Procedure for Common Known Variance   18
           2.2.1   The Completely Randomized Design   18
           2.2.2   Designs with Blocking   24
           2.2.3   Robustness of Normal Theory Procedure   28
    2.3   Multi-Stage Procedures for Common Known Variance   31
           2.3.1   A Closed Two-Stage Procedure with Elimination   32
           2.3.2   A Closed Multi-Stage Procedure without Elimination   35
           2.3.3   A Closed Multi-Stage Procedure with Elimination   38
    2.4   Comparison of Common Known Variance Procedures   41
    2.5   A Single-Stage Procedure for Selecting the s Best of t Treatments Having a Common Known Variance   43
           2.5.1   Selection without Regard to Order   44
           2.5.2   Selection with Regard to Order   46
    2.6   A Single-Stage Procedure for Unequal Known Variances   48
    2.7   Multi-Stage Procedures for Common Unknown Variance   51
           2.7.1   An Open Two-Stage Procedure without Elimination   52
           2.7.2   An Open Multi-Stage Procedure with Elimination   54
           2.7.3   Comparison of Procedures   56

	2.8	Procedures for Unequal Unknown Variances	60
		2.8.1 Variances Bounded	60
		2.8.2 Variances Unbounded	60
	2.9	Chapter Notes	65

3 Selecting a Subset Containing the Best Treatment in a Normal Response Experiment   69

	3.1	Introduction	69
	3.2	Single-Stage Procedures	71
		3.2.1 Balanced Experiments	71
		3.2.2 Unbalanced Experiments	76
		3.2.3 Robustness of Procedure $\mathcal{N}_G$	79
	3.3	Experiments with Blocking	80
		3.3.1 Randomized Complete Block Designs	80
		3.3.2 Balanced Incomplete Block Designs	81
		3.3.3 Latin Square Designs	84
	3.4	Alternative Goals	86
		3.4.1 Selection of the s Best Treatments	86
		3.4.2 Selection of δ^*-Near-Best Treatments	91
		3.4.3 A Bounded Procedure for δ^*-Near-Best Treatments	94
	3.5	Chapter Notes	96

4 Multiple Comparison Approaches for Normal Response Experiments   100

	4.1	Introduction	100
	4.2	Simultaneous Confidence Intervals for Orthogonal Contrasts	101
	4.3	Simultaneous Confidence Intervals for All Pairwise Differences	109
	4.4	Simultaneous Confidence Intervals for Comparing All Treatments with the Best	112
	4.5	Chapter Notes	114

5 Problems Involving a Standard or Control Treatment in Normal Response Experiments   116

	5.1	Introduction	116
	5.2	Selecting the Best Treatment Using the IZ Approach	117
		5.2.1 Selection Involving a Standard (Common Known σ^2)	118
		5.2.2 Selection Involving a Standard (Common Unknown σ^2)	123
		5.2.3 Selection Involving a Control (Common Known σ^2)	125
	5.3	Selecting a Subset of Treatments	126

		5.3.1 Screening Involving a Standard	126
		5.3.2 Screening Involving a Control Treatment	129
	5.4	Simultaneous Confidence Intervals	131
		5.4.1 Comparison with a Standard	131
		5.4.2 Comparison with a Control Treatment	133
		5.4.3 Comparisons with Respect to Two Control Treatments . . .	138
	5.5	Chapter Notes	141

6 Selection Problems in Two-Factor Normal Response Experiments — 142

	6.1	Introduction	142
	6.2	Indifference-Zone Selection Using Completely Randomized Designs (Common Known Variance)	144
		6.2.1 Single-Stage Procedure	146
		6.2.2 A Closed Sequential Procedure without Elimination	151
		6.2.3 A Closed Sequential Procedure with Elimination	152
		6.2.4 Comparison of the Performance Characteristics of the Procedures	154
	6.3	Indifference-Zone Selection Using Split-Plot Designs (Common Known Variance)	159
		6.3.1 Introduction	159
		6.3.2 IZ Selection When $(\sigma_w^2, \sigma_\epsilon^2)$ Is Known	161
	6.4	IZ Selection Using Completely Randomized Designs (Common Unknown Variance)	163
	6.5	IZ Selection Using Split-Plot Designs (Common Unknown Variance)	166
	6.6	Subset Selection Using CR Designs	167
	6.7	Subset Selection Using Split-Plot Designs	169
		6.7.1 Subset Selection When γ_w Is Known	170
		6.7.2 Subset Selection When σ_w^2 and σ_ϵ^2 Are Unknown	173
	6.8	Chapter Notes	175

7 Selecting Best Treatments in Single-Factor Bernoulli Response Experiments — 177

	7.1	Introduction	177
	7.2	A Single-Stage Procedure for the Indifference-Zone $p_{[t]} - p_{[t-1]} \geq \Delta^\star$	182
		7.2.1 Completely Unknown p-Values	182
		7.2.2 An Alternative Design Specification	184
	7.3	A Closed Adaptive Sequential Procedure	186
		7.3.1 Introduction	186

		7.3.2	The Procedure	187
		7.3.3	Optimality Properties of Procedure $\mathcal{B}_{BK}$	189
	7.4	Open Sequential Procedures for the Odds Ratio Indifference-Zone		192
		7.4.1	An Open Sequential Procedure without Elimination	192
		7.4.2	An Open Sequential Procedure with Elimination	195
		7.4.3	Comparison of Procedures $\mathcal{B}_{BKS}$ and $\mathcal{B}_P$	196
	7.5	A Single-Stage Subset Selection Procedure		200
		7.5.1	Screening in Balanced Experiments	201
		7.5.2	Screening in Unbalanced Experiments	209
		7.5.3	An Alternative Design Specification	211
	7.6	Chapter Notes		211
8	**Selection Problems for Categorical Response Experiments**			**214**
	8.1	Introduction		214
	8.2	Indifference-Zone Procedures for Multinomial Data		216
		8.2.1	A Single-Stage Procedure	217
		8.2.2	Use of Curtailment When the Maximum Number of Observations Is Specified	227
		8.2.3	A Closed Sequential Procedure	229
		8.2.4	Applications	233
	8.3	Subset Procedures for Multinomial Data		236
		8.3.1	A Single-Stage Procedure	236
		8.3.2	Curtailment of Procedure $\mathcal{M}_{GN}$	237
		8.3.3	A Curtailed Sequential Procedure	240
	8.4	Indifference-Zone Procedures for Cross-Classified Data		241
	8.5	Chapter Notes		244
Appendix A	**Relationships Among Critical Points and Notation**			**247**
Appendix B	**Tables**			**253**
Appendix C	**FORTRAN Programs**			**277**
References				**303**
Author Index				**319**
Subject Index				**323**

Preface

This book was written for experimenters in applied areas. It is self-contained and can be read profitably by practitioners who have knowledge of statistical methods through classical experimental design. It can also be used to augment traditional courses in the Analysis of Variance and the Design of Experiments.

Arguably, the early articles by Paulson (1949, 1952) and Bahadur (1950) feature the type of alternative to the then-popular hypothesis testing formulations that might be considered the starting point of these studies. A steady stream of contributions to "ranking and selection" theory began in earnest about 1954 with publication of "A single-sample multiple decision procedure for ranking means of normal populations with known variances" by the first author, together with work by Milton Sobel and related ones by Shanti S. Gupta. Since then more than 1000 articles and several books on the general theme of ranking and selection and related methodologies have appeared in the statistical literature. Researchers throughout the world have made many important contributions. Statistics courses on these topics have been given at many universities, including Cornell, Georgia Tech, Minnesota, Ohio State, Purdue, Syracuse, and The University of California at Santa Barbara. Unfortunately, very few of these methodologies have found their way into textbooks and hence into usage. In part, this came about because very few of the articles were written with the statistically oriented practitioner in mind, and that may have inhibited their dissemination. Our intent with the present volume is to bridge this gap.

This volume is not a survey of the ranking and selection literature. Rather, we have focused on three types of procedures: selection procedures using the so-called indifference-zone approach, screening procedures using the subset approach, and some important multiple comparison procedures involving normal means. Concerning selection procedures, we show how they can be applied to three probability models that play a central role in many statistical studies of experimental data. These are the univariate normal distribution where interest lies in the mean, the Bernoulli distribution where we are concerned with the "success" probabilities, and the multinomial distribution where we study the probabilities associated with certain categories (events).

Among those cases that we do not consider are procedures for normal variances, means of Poisson distributions (or processes), or parameters of exponential or gamma distributions; selection procedures exist for such distributions, but treatment of these would have made the book unduly long. Our approach throughout is from the frequentist point of view; the presentation of related Bayesian procedures would require an additional book. However, we do make reference to some of these procedures in the Chapter Notes.

We would like to thank those students and colleagues whose comments and discussions have helped to improve the presentation in this book: C. W. Dunnett, M. Hartmann, A. J. Hayter, D. R. Hoover, C. Jennison, G.-H. Pan, A. C. Tamhane, and R. R. Wilcox. We would also like to thank two anonymous referees who provided feedback on the manuscript. We especially wish to thank Edward Paulson for making the manuscripts of his 1993 normal and Bernoulli sequential procedures with elimination available to us before publication; we have described these procedures and some of their properties with his permission. Lastly, we would like to thank Kate Roach and the editorial staff of John Wiley & Sons for their very professional assistance in the production phases of this book.

Research on these topics has been supported at Cornell University for many years by the Office of Naval Research and the Army Research Office. The work of D. Goldsman was supported by National Science Foundation Grant No. DDM-9012020. We are greatly indebted to those agencies for their strong support and their faith and encouragement.

We are also grateful to Cornell's School of Operations Research and Industrial Engineering, Georgia Tech's School of Industrial and Systems Engineering, and Ohio State's Department of Statistics. Their support and facilities furnished us with a pleasant environment to carry out our research and writing.

Finally, we would like to thank our families for their support and patience during the completion of this project—it was very much appreciated!

Cornell University R.E. BECHHOFER
The Ohio State University T.J. SANTNER
Georgia Institute of Technology D.M. GOLDSMAN

CHAPTER 1

The Rationale of Selection, Screening and Multiple Comparisons

1.1 INTRODUCTION

This book describes methods for designing and analyzing experiments involving several qualitative factors when the scientific objective is selection of the "best" treatments. For example, the best treatment might be the one having the largest population mean. More generally, an experimenter might wish to select several of the best treatments. Selection of some (all) of the treatments according to their ordered means is the problem of partial (complete) ranking of the treatments.

Some of the methods that we discuss are designed to select a single best treatment. Other methods are designed to screen a set of treatments by choosing a (random size) subset of the treatments containing the best one. Still other methods construct simultaneous confidence intervals for specific sets of mean treatment differences.

During the past 40 years a large number of new design methodologies for selection and screening problems have been introduced in the statistical literature. More recently a number of comparative studies have been made of the performance characteristics of competing procedures that can be used under the same statistical assumptions. This book is our attempt to present these results in a form useful for practitioners who are familiar with experimental design for more classical experimental objectives such as hypothesis testing or response surface analysis.

Selection problems pervade our everyday life. The statistician is called upon to help provide rational procedures for selecting the "best" of several alternatives. Thus, an agronomist studying a number of varieties of wheat may seek to select the variety that will produce the largest number of bushels per acre. An educator may be studying different teaching methods to determine which one aids a student in scoring higher on an SAT examination. A clinician may be interested in which type of drug is most effective in treating a certain disease. A polling agency may wish to determine which candidate is most preferred by the electorate. A manufacturing engineer may wish to find the least costly of a number of possible assembly line configurations.

To illustrate the types of problems that we are studying, consider the balanced one-way layout with unknown means given by the linear model $Y_{ij} = \mu + \tau_i + \epsilon_{ij}$ ($1 \leq i \leq t, 1 \leq j \leq n$), where $\sum_{i=1}^{t} \tau_i = 0$. Here μ is the overall mean, and $\tau_1, \ldots, \tau_t$ are the t so-called treatment "effects." The measurement errors ϵ_{ij} are independent normal random variables with $E\{\epsilon_{ij}\} = 0$ and constant $\text{Var}\{\epsilon_{ij}\} = \sigma^2$.

For this model, classical statistical analysis is most often concerned with testing homogeneity of the means (H: $\tau_1 = \cdots = \tau_t$) as in the Analysis of Variance. However, such homogeneity tests (whether or not they yield statistically significant results) usually do not supply the type of conclusion that the experimenter truly desires. For example, in an agricultural setting, the hypothesis that several *different* varieties of grain have the *same* population mean yield is unrealistic; and a sufficiently large sample will establish this fact at any preassigned level of significance. Moreover, should a significant result be obtained, the experimenter's problems have usually just begun. Having established that the varieties are different, the question of interest may then become that of identifying the variety which is "best." Here the best variety might be defined as the one having the *largest* (population) mean yield. In the one-way layout, this is the problem of identifying the treatment associated with $\max\{\tau_1, \ldots, \tau_t\}$.

The philosophy expounded in this book is that whenever the experimenter is ultimately faced with the task of choosing a *best* variety, the experiment should be designed with that goal in mind. What is needed is (1) a statistical procedure that will tell the experimenter which treatment or treatments to select and (2) an "operating characteristic function" for that procedure based on the probability of making a correct selection. The experiment should then be designed so as to control (in some sense) this probability at a specified level.

This book is intended for experimenters who are knowledgeable in classical experimental design. We assume that the reader has previously studied the fundamental notions of randomization, replication, and blocking in the context of designing experiments with qualitative factors for testing homogeneity of main effects and interactions, or estimating these quantities. The rational choice of sample size, which plays a major role in designing experiments from the viewpoint of power (say), also plays a critical role in selection experiments when controlling the probability of a correct selection.

This book restricts attention to experiments with one of three types of responses: normal, Bernoulli and multinomial. At the end of each chapter we provide a Chapter Notes section that gives references to the literature for other related selection procedures. The volume can serve as a handbook for practitioners, presenting them with choices of experimental objectives; and when several statistical procedures exist to accomplish a given experimental goal, we discuss these competing statistical procedures. In the latter case, the advantages and disadvantages of the procedures are listed, tables are provided of the constants necessary to implement the procedures, and the performance characteristics of the procedures are compared. We provide a summary of the statistical assumptions underlying the procedures and, where available, indicate the consequences when the assumptions are violated. For example, among other topics, Chapter 2 considers the problem of selecting the "level" of a normal response qualitative factor having the largest treatment mean when the competing treatments have known or unknown variance(s). We describe the circumstances under which single-stage, two-stage, and sequential procedures are appropriate, along with the advantages and disadvantages of each. Thus, while not encyclopedic in character, it is expected that the book will be useful in many experimental settings. Where possible, we either supply or describe sources of FORTRAN programs to supplement the tables provided in the text so that experimenters are not limited to those ranges of values covered by the tables.

In the case of multiple comparison formulations, our emphasis is on design and optimal allocation in the standard normal theory cases (the one-way layout and randomized blocks and related experimental designs). The reader should refer to the 1987 book *Multiple Comparison Procedures* by Hochberg and Tamhane for more complicated cases. There is relatively little available literature on exact small-sample multiple comparison procedures for the Bernoulli and multinomial models; almost all of the available methods are asymptotic.

There is no book on designing experiments for selection, screening and multiple comparisons that accomplishes what we seek to do. The most notable texts dealing with selection procedures are as follows:

1. Bechhofer, R. E., Kiefer, J. C., and Sobel, M. (1968). *Sequential Identification and Ranking Procedures*. Chicago: University of Chicago Press.
2. Büringer, H., Martin, H., and Schriever, K. H. (1980). *Nonparametric Sequential Selection Procedures*. Boston: Birkhauser.
3. Dourleijn, C. J. (1993). *On Statistical Selection in Plant Breeding*, Ph.D. Dissertation, Agricultural University, Wageningen, The Netherlands.
4. Driessen, S. G. (1992). *Statistical Selection: Multiple Comparison Approach*, Ph.D. Thesis, University of Technology, Eindhoven, The Netherlands.
5. Gibbons, J. D., Olkin, I., and Sobel, M. (1977). *Selecting and Ordering Populations: A New Statistical Methodology*. New York: John Wiley & Sons.
6. Gupta, S. S., and Huang, D.-Y. (1981). *Multiple Decision Theory: Recent Developments*, Lecture Notes in Statistics, Vol. 6. New York: Springer-Verlag.
7. Gupta, S. S., and Panchapakesan, S. (1979). *Multiple Decision Procedures*. New York: John Wiley & Sons.
8. Mukhopadhyay, N., and Solanky, T. K. S. (1994). *Multistage Selection and Ranking Procedures: Second-Order Asymptotics*. New York: Marcel Dekker.

References 1, 2, 4, 6, 7 and 8 are aimed at researchers who are interested in theoretical studies in the field of selection and ranking. Bechhofer, Kiefer and Sobel use a generalization of the sequential probability ratio test to derive a class of selection procedures for exponential families of distributions. Büringer, Martin and Schriever provide a comprehensive treatment of a large number of Bernoulli sequential selection procedures; their procedures afford a different probability guarantee than do those in the present text. Also included are subset-selection procedures based on linear rank-order statistics for continuous responses. Driessen derives equivalences between various selection and multiple comparison procedures for incomplete treatment designs. Gupta and Huang use loss functions and the principles of invariance and minimaxity to derive general ranking and selection rules. Gupta and Panchapakesan is an encyclopedia of work on indifference-zone and subset selection procedures as of 1979. Mukhopadhyay and Solanky is a recent addition to the literature that provides an asymptotic analysis of the average sample size performance of open multi-stage procedures.

Dourleijn's thesis describes many practical aspects of the process of genetic development in the sugar beet industry together with statistical selection procedures useful in selecting promising varieties. These sugar beet experiments are characterized by the large number of sugar beet varieties compared in each growing season, as well as by the experimenter's desire to compare them over many different types of growing environments.

The Gibbons, Olkin and Sobel (GOS) text is perhaps closest in spirit to this book. A special virtue of their book is that their descriptions of procedures are particularly clear and detailed. Furthermore, their text contains a large number of useful tables necessary to implement the procedures. Our book differs from GOS in a number of ways. We assume that the reader is knowledgeable about standard experimental design, whereas GOS assume that the reader has only a knowledge of elementary statistics. We discuss sequential procedures and subset selection formulations more extensively than GOS. Lastly, when several procedures for the same problem are available, we present the experimenter with options and make recommendations concerning their use based on extensive performance characteristic studies; GOS typically present a single procedure.

In addition to texts that explicitly deal with selection procedures, we point out that experimenters and researchers may find useful both the categorized index to selection procedures by Dudewicz and Koo (1982) and the *Current Index to Statistics*, particularly through its computerized search facility, for more recent references. The remainder of this chapter reviews several aspects of selection that are assumed in this book, gives a more detailed statement of the philosophy we expound in designing selection experiments, and provides an overview of the rest of the book.

1.2 BASIC CONCEPTS OF CLASSICAL EXPERIMENTAL DESIGN

1.2.1 Qualitative Versus Quantitative Factors

One classification of statistical problems is according to the types of explanatory variables available. We next discuss the traditional dichotomy between *qualitative* and *quantitative* explanatory variables.

Qualitative variables are measured on either a *nominal* or *ordinal* scale. A nominal scale categorizes the data into distinct groups. For example, the variables *plant variety*, *patient sex* and *type of fertilizer* (as opposed to amount of fertilizer) are measured on nominal scales. An ordinal scale both categorizes the values into groups *and* linearly orders the groups. Examples of variables measured using an ordinal scale are *pain* (with values *none, moderate, severe*) and *socioeconomic status* (with values *low, middle, high*).

BASIC CONCEPTS OF CLASSICAL EXPERIMENTAL DESIGN   5

Quantitative variables are measured on either an *interval* or *ratio* scale. Interval scales add another characteristic to ordinal scales; they categorize, order, and *quantify* comparisons between *pairs* of measurements. An example of a quantitative variable is temperature measured in degrees Fahrenheit.

This book is primarily, though not exclusively, concerned with *qualitative* factors measured on *nominal* scales. Designing experiments for selecting the best treatment (combination) can be viewed as the qualitative factor analog of the quantitative factor "optimal design" formulations pursued by George Box and his colleagues, and formalized by Jack Kiefer, Jacob Wolfowitz, and others.

1.2.2 Principles of Experimental Design

The planning of any experiment to investigate a research hypothesis requires several different types of decisions. First, the experimenter must determine what characteristics are to be measured and then what *treatment design* is to be used. The treatment design includes the specification of the factors to be studied in the experiment, the levels of each factor, and the determination of which treatment combinations are to be used (if there are several factors).

The second aspect of the experiment that must be addressed is the determination of the number of *times* each treatment (combination) is to be observed—that is, how much *replication* is to be performed. Adequate replication ensures that the experimenter will be able to achieve a desired design requirement for whatever goal and statistical procedure is to be employed.

Blocking is the third aspect of experimental design; it means that treatment comparisons must be made across homogeneous experimental units. Blocking is an explicit recognition by the experimenter of the presence of one or more nuisance factors affecting the response. The model that the experimenter uses to represent the data that come from an experiment with blocking will explicitly take into account potential block differences, and the analysis compares treatments only on uniform experimental material. For instance, the simplest such experiment might involve two treatments in blocks of size 2 for which a paired t-statistic is used to perform the analysis. More generally, the statistical procedures appropriate for experiments with blocking come to conclusions by accumulating evidence of treatment differences from the results of smaller experiments among uniform material that make up the entire experiment.

The final aspect of planning an experiment is the determination of the manner in which the treatment combinations are *randomized* to the experimental units. Randomization prevents unrecognized factors from systematically confounding the results. For example, in an experiment comparing several varieties, an agronomist would randomly assign varieties to test plots to prevent unrecognized fertility differences or gradients from masking possible variety effects.

In summary, the completed experiment is specified by its treatment design, amount of replication (to achieve the experiment's goals), use of blocking (to "balance" the effects of recognized nuisance factors), and manner of randomization of treatments to experimental units (to balance the effects of unrecognized nuisance factors). The

6   THE RATIONALE OF SELECTION, SCREENING AND MULTIPLE COMPARISONS

choice of these elements also helps determine the appropriate probabilistic model for describing the experimental response (and analysis of the experiment).

We illustrate these ideas with three examples. The first example shows the determination of sample size using a classical power requirement in a hypothesis testing setup, the second discusses blocking, and the last example illustrates randomization.

Example 1.2.1 (Replication)   This example employs a hypothesis testing formulation and determines the amount of replication (sample size) by using a power requirement. Suppose that we take a random sample $Y_1, \ldots, Y_n$ from a normal population having unknown mean μ and known variance σ^2. We wish to test the one-sided hypothesis H_O: $\mu = \mu_0$ versus the alternative H_A: $\mu > \mu_0$. Furthermore, suppose that we wish to guarantee that the probability of erroneously rejecting H_O is α, a Type I error, and that the probability of incorrectly accepting H_O when $\mu = \mu_1$, a Type II error, is β. In order to satisfy the power requirement, one must take $n = (z^{(\alpha)} + z^{(\beta)})^2 \sigma^2 / (\mu_1 - \mu_0)^2$ observations, where $z^{(\gamma)}$ is the $1 - \gamma$ quantile of the standard normal distribution.

Example 1.2.2 (Blocking)   McDonald (1979) and Gupta and Hsu (1980) applied subset selection procedures to a set of traffic fatality data in which the annual motor vehicle traffic fatality rates (VFRs) are recorded for the 48 contiguous states plus the District of Columbia over the period 1960–1976. Their goal was to select a set of the 49 regions with the largest (and smallest) mean VFRs. However, the VFRs have changed over time as a result of improvements in automobiles, road conditions, speed limit legislation, and public safety laws such as drunk driving statutes. Thus a regional comparison of VFRs can only be made within a short period of time—both papers take this to be a year-by-year comparison. Formally, they model the VFR for the ith region during the jth year as

$$Y_{ij} = \mu + R_i + A_j + \epsilon_{ij}$$

where R_i is the main effect of the ith region, A_j is the main effect of the jth year and ϵ_{ij} is the noise. The year effects are blocks that permit overall differential year levels, and there are constant region differences for each fixed year.

Example 1.2.3 (Randomization)   This example illustrates the importance of randomization in a simple case, as well as the potential for confounding that can occur when randomization is not used in designing experiments.

Five test varieties of sugar beets are to be compared in a field trial (call them varieties I–V). Each variety is to be grown on four test plots. The experiment is to take place in a rectangular array of 20 (= 4 × 5) test plots as illustrated in Figure 1.1.

In a completely randomized design the investigator randomly selects four test plots for each variety. Suppose that the experimenter decides that it is impractical to change the seed in the planter after each plot and decides on the following randomization scheme to decrease the number of times the seed must changed in the planter. An entire row of four plots is assigned at random to the seed of one of the varieties; then

FORMULATIONS CONSIDERED IN THIS BOOK

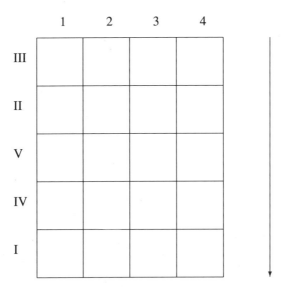

Figure 1.1. Rectangular field with 20 test plots and arrow showing direction of increasing fertility.

all of the plots in that row can be easily planted in a single set of passes of the planter. For example, Figure 1.1 shows a randomization in which all of the plots in the top row are planted with variety III.

There is a potential problem with such a restricted randomization. Suppose that the drainage in the field is such that there is a fertility gradient that increases in the direction indicated by the arrow in Figure 1.1. Then even in the absence of variety differences, the plots in the row labeled "I" will have greater yields, on average, than those in the row labeled "III."

The variety effect is *confounded* with the (unrecognized) fertility spatial pattern. If the latter factor were recognized, then the experimenter could block on the rows, as in Example 1.2.2, and use a suitable incomplete block design to perform a valid experiment and corresponding statistical model to analyze it.

But the point of this example is that randomization of treatments in space and time is used to prevent confounding of the treatments by unrecognized (nuisance) factors. This technique complements blocking which is used to model known (nuisance) factors.

1.3 FORMULATIONS CONSIDERED IN THIS BOOK

This book stresses three basic problem formulations for selection, screening and multiple comparison problems. It also provides descriptions of some variants of the basic formulations and provides references to the literature for additional ones. All of the formulations produce sharper inferences than do traditional testing approaches. Throughout we assume an experiment with one or more *qualitative* treatment factors.

In this chapter, we first illustrate and contrast three *formulations* for the case of a single normal treatment factor with t levels. Then additional examples are given to demonstrate the application to more complicated normal theory experiments. Lastly, we consider the indifference-zone formulation applied to treatments involving Bernoulli and multinomial responses.

Consider an experiment involving t treatments in which the jth response on the ith treatment is

$$Y_{ij} = \mu_i + \epsilon_{ij}$$

$(1 \le i \le t, 1 \le j \le n)$ where the measurement errors ϵ_{ij} are independent identically distributed *normal* random variables having mean zero and variance σ^2. Thus μ_i is the mean of the ith treatment. Let

$$\mu_{[1]} \le \cdots \le \mu_{[t]}$$

denote the ordered μ_i-values. Neither the values of the $\mu_{[s]}$ nor the *pairing* of the Π_i with the $\mu_{[s]}$ $(1 \le i, s \le t)$ is assumed to be known.

Indifference-Zone Formulation

The first formulation considered in this chapter is called the *indifference-zone* approach. The goal of the indifference-zone approach is to select the treatment associated with the largest mean $\mu_{[t]}$. This goal can be viewed as the qualitative analog of quantitative factor optimal design theory. It emphasizes that the experiment must be planned on an adequate scale so that specific design requirements can be achieved. The simplest design requirement for this problem is the following:

Probability Requirement: For given constants (δ^*, P^*) with $0 < \delta^* < \infty$ and $1/t < P^* < 1$, specified prior to the start of experimentation, we require

$$P\{CS\} \ge P^* \quad \text{whenever} \quad \mu_{[t]} - \mu_{[t-1]} \ge \delta^* \tag{1.3.1}$$

where CS denotes the event that the treatment associated with $\mu_{[t]}$ has been correctly identified (a "correct selection").

This intuitive design requirement implicitly specifies the choice of the common single-stage sample size n for each treatment. The reasoning that motivates the choice of (δ^*, P^*) is as follows. If the value of the true but unknown difference $\mu_{[t]} - \mu_{[t-1]}$ is *very small*, the experimenter may be *indifferent* as to which of the two associated treatments is selected. Thus, the experimenter is asked to specify a threshold value δ^* that can be regarded as an indifference point. True mean vectors $\boldsymbol{\mu}$ for which $0 \le \mu_{[t]} - \mu_{[t-1]} \le \delta^*$ can be thought of as belonging to an *indifference-zone*, in which the experimenter is indifferent as to which of the treatments associated with $\mu_{[t]}$ and $\mu_{[t-1]}$ is selected. True mean vectors $\boldsymbol{\mu}$ for which $\delta^* < \mu_{[t]} - \mu_{[t-1]} < \infty$ can be thought of as belonging to a *preference-zone* for *correct selection*, in which the experimenter strongly prefers selection of the treatment associated with $\mu_{[t]}$. Care must be taken in specifying the value δ^* because the design requirement (1.3.1) implicitly determines the common *single-stage* sample size required from the t competing treatments. If δ^* is (set too) small, then the number of observations required to guarantee (1.3.1) may be prohibitively expensive. Therefore, δ^* can be thought of as the smallest difference $\mu_{[t]} - \mu_{[t-1]}$ that is considered "worth detecting." Of course, large P^* may also require a large number of observations to guarantee (1.3.1). In Chapter 2, we propose single-stage, two-stage and multi-stage procedures that guarantee (1.3.1); we discuss the merits of each.

Subset Selection Formulation

The second formulation considered in this chapter is the *random subset size selection* approach. The goal of the subset selection approach is to select a (small) subset of the treatments that contains the treatment associated with the largest mean $\mu_{[t]}$; this can be thought of as a *screening* objective. A typical design requirement in subset selection requires that this goal be achieved *no matter what the values of the underlying treatment means*. Formally this requirement is stated as follows:

Probability Requirement: For a given constant P^* with $1/t < P^* < 1$, specified prior to the analysis, we require that

$$P\{CS\} \geq P^* \quad \text{for all } (\boldsymbol{\mu}, \sigma^2) \tag{1.3.2}$$

where CS denotes the event that the treatment associated with $\mu_{[t]}$ has been included in the selected subset (a "correct selection").

The rationale underlying this formulation is the following: If $\mu_{[t]} - \mu_{[t-s]}$ ($s \geq 1$) is very small, it may be prohibitively expensive to select the single treatment associated with $\mu_{[t]}$. Also, as noted above, the experimenter may be indifferent as to which treatment to select if one or more treatments have means that are very close to $\mu_{[t]}$. One can think of procedures employing the random subset selection approach as screening devices, dividing the t treatments into two sets, namely, a group of treatments that are established as being "inferior" and its complement. What can be claimed is that the best treatment is contained in the complementary set of treatments (but, in general, this collection of treatments will also contain inferior treatments).

Of course, the requirement (1.3.2) can be satisfied if the selected subset consists of all t treatments, and the usual procedures emphasize a second requirement, namely, that the number of treatments in the selected subset be "small" if $\mu_{[t]}$ is sufficiently separated from the other $\mu_{[i]}$. Thus, typical procedures employ rules that select a random number of treatments in such a way that the probability requirement is guaranteed.

Multiple Comparison Formulation

The *multiple comparison* goals we consider are to assess the magnitudes of two or more differences between treatment means. A single confidence statement is made about the *simultaneous* coverage of these intervals. Simultaneous confidence intervals have been proposed for a number of meaningful sets of treatment comparisons. Three specific goals involving treatment differences are as follows.

- To form simultaneous one-sided or two-sided $100 \times P^*\%$ confidence intervals for $\{\mu_i - \mu_j\}_{i \neq j}$. The purpose of this goal is to form joint confidence intervals for each pairwise difference between the treatment means.
- To form simultaneous one-sided or two-sided $100 \times P^*\%$ confidence intervals for $\{\mu_i - \mu_1\}_{i=2}^t$. In this case the treatment associated with μ_1 is viewed as a "control" treatment and $\mu_2, \ldots, \mu_t$ are "test" treatments; the purpose of this goal is to form joint confidence intervals for each of the $t - 1$ differences between the test treatment means and the control treatment mean.
- To form simultaneous one-sided or two-sided $100 \times P^*\%$ confidence intervals for $\{\mu_i - \max_{j \neq i} \mu_j\}_{i=1}^t$. The purpose of this goal is to form joint confidence intervals for the differences between each treatment mean and the best of the rest.

The design requirement depends on the specific set of quantities for which one desires to determine simultaneous confidence intervals.

Probability Requirement: For a given constant P^* with $0 < P^* < 1$, specified prior to the analysis, we require

$$P\{\text{Confidence region covers all differences specified}\} \geq P^* \quad \text{for all } \boldsymbol{\mu}. \quad (1.3.3)$$

We conclude our introduction with three additional examples that can be regarded as extensions of the above material. The first of these describes some selection, screening and multiple comparison goals for normal means in a *factorial* setting.

Example 1.3.1 (Factorial Normal Responses) An agronomist wishes to study the effects of two types of fertilizers (A and B, each applied at two levels) on the mean yield of a certain variety of grain. Thus there are $2^2 = 4$ fertilizer-level combinations. Suppose that n blocks of land are available for the experiment, each block being of homogeneous fertility and consisting of four plots.

There are several natural goals that the experimenter might consider. Using the usual notation (1), a, b and ab to denote the four treatment combinations and $\mu(1)$, $\mu(a)$, $\mu(b)$ and $\mu(ab)$ to denote the corresponding expected mean yields, these goals are as follows:

(i) Select the treatment combination associated with

$$\max\{\mu(1), \mu(a), \mu(b), \mu(ab)\}.$$

(ii) Estimate simultaneously the three (orthogonal) contrasts:

The main effect of A, that is, $\mu(A) = \frac{1}{2}[\mu(ab) - \mu(b) + \mu(a) - \mu(1)]$.
The main effect of B, that is, $\mu(B) = \frac{1}{2}[\mu(ab) + \mu(b) - \mu(a) - \mu(1)]$.
The AB interaction, that is, $\mu(AB) = \frac{1}{2}[\mu(ab) - \mu(b) - \mu(a) + \mu(1)]$.

How should the experiment be carried out? For each goal the four treatment combinations should be allocated at random to the four plots within a block. The particular goal that the experimenter wishes to achieve will, along with the associated probability requirement, determine the amount of replication required—that is, the number of observations required for each treatment combination.

Goal (i) can be formulated using either the indifference-zone approach or the random subset size approach. In Chapters 2 and 3, we also discuss generalizations of this basic goal—for example, selecting the treatment (or treatment combinations) with the s $(1 \leq s \leq t-1)$ largest treatment means, both with and without regard to order. Goal (ii) and related joint confidence interval estimation goals, discussed in Chapter 4, quantify the information available about relationships between the mean effects of the treatments. Procedures appropriate for normal responses are discussed in Chapters 2–6.

Example 1.3.2 (Bernoulli Responses) This example describes an indifference-zone selection goal for Bernoulli "success" probabilities in a single-factor setting. A clinician wishes to study the effectiveness of t different analgesic drugs to provide relief from headache pain. If the drug provides at least h hours of relief, it will be deemed effective; otherwise, it will be considered ineffective. Let p_i $(1 \leq i \leq t)$ denote the probability that the ith drug is effective and $p_{[1]} \leq \cdots \leq p_{[t]}$ the ordered probabilities. An important clinical goal is to

Select the drug associated with the greatest probability, $p_{[t]}$.

One possible design requirement for this problem is the following:

Probability Requirement: For given constants (θ^*, P^*) with $1 < \theta^* < \infty$ and $1/t < P^* < 1$, specified prior to the start of experimentation, we require

$$P\{CS\} \geq P^* \quad \text{whenever} \quad \frac{p_{[t]}(1 - p_{[t-1]})}{(1 - p_{[t]})p_{[t-1]}} \geq \theta^* \qquad (1.3.4)$$

where CS denotes the event that a correct choice of treatment has been made.

The fraction given in (1.3.4) is the ratio of the odds of at least h hours of relief for the best treatment to the odds of at least h hours of relief for the second best treatment. Unlike the normal means experiment, design requirements for Bernoulli experiments can be stated in several natural ways, depending on the area of application, that differ fundamentally in their mathematical analyses. Certain of these probability requirements can be guaranteed using single-stage procedures, whereas others can be guaranteed only by using open sequential procedures. Procedures appropriate for Bernoulli responses are discussed in Chapter 7.

Example 1.3.3 (Multinomial Responses)   This final example describes an indifference-zone goal for multinomial "event" probabilities in a single-factor setting. One month before a general election, a polling organization wishes to determine which of t candidates is most preferred by the electorate. To this end, a random sample of individuals is polled, each individual being asked to indicate the individual's *unique* preference. If $p_i > 0$ is the probability that the ith candidate is most preferred by an individual, and $p_{[1]} \leq \cdots \leq p_{[t]}$ are the ordered probabilities, then $\sum_{i=1}^{t} p_i = 1 = \sum_{i=1}^{t} p_{[i]}$. The goal of the polling organization is to

Identify the most preferred candidate, that is, the candidate associated with $p_{[t]}$.

One possible design requirement for this problem is the following:

Probability Requirement: For given constants (θ^*, P^*) with $1 < \theta^* < \infty$ and $1/t < P^* < 1$, specified prior to the start of experimentation, we require

$$P\{CS\} \geq P^* \quad \text{whenever} \quad p_{[t]}/p_{[t-1]} \geq \theta^*$$

where CS denotes the event that the multinomial category associated with $p_{[t]}$ has been selected. Procedures appropriate for multinomial responses are discussed in Chapter 8.

1.4   ORGANIZATION OF THE BOOK

This book discusses selection, screening and multiple comparison procedures for three classes of responses: Chapters 2–6 study normal responses, Chapter 7 considers Bernoulli responses and Chapter 8 surveys multinomial responses.

Chapters 2–5 consider normal theory *single* treatment factor experiments, whereas Chapter 6 discusses normal theory *factorial* experiments. Chapter 2 studies indifference-zone formulations, as introduced in Section 1.3. Sections 2.2–2.4 give procedures for selecting the best treatment when there is a common known variance and for various types of blocking. Section 2.5 considers procedures for the goal of selecting the s best treatments for the same variance assumption. Sections 2.6–2.8 return to the goal of selecting the single best treatment for three additional sets of assumptions concerning the variances of the responses: (i) the variances are known but possibly unequal, (ii) the variances are common but unknown, and (iii) the variances are completely unknown.

Chapter 3 considers the screening goal of selecting a subset containing the best treatment in a single treatment factor experiment. Procedures for the cases of common known and common unknown variance are given in Sections 3.2 and 3.3, respectively. The modifications of the procedures to accommodate various types of blocking are also given in these sections. Several alternate goals are considered in Section 3.4, including selection of a subset containing the s best treatments, selection of subsets that are bounded *a priori* by the experimenter, and selection of near-best treatments.

Chapter 4 discusses the construction of simultaneous confidence intervals for three sets of parameters: (i) orthogonal contrasts among the treatment means, (ii) all pairwise differences of treatment means and (iii) pairwise differences of each treatment mean with the largest of the remaining treatment means.

Chapter 5 provides procedures for problems in which it is desired to make comparisons of each treatment mean with either a standard or a control. Comparisons are made with respect to a *standard* when the value of the quantity with which each mean is being compared is *known*; the comparison is with respect to a *control* when the value of the quantity with which each mean is being compared is *unknown* and must be simultaneously estimated from the data. Sections 5.2–5.4 discuss procedures for comparison with respect to a standard and control using the indifference-zone formulation, the subset selection formulation and simultaneous confidence intervals, respectively.

Chapter 6 places special emphasis on two-factor experiments in which the mean effects are additive, that is, the two factors do not interact. Sections 6.2 and 6.3 consider the indifference-zone formulation for simultaneously selecting the best level of each of the two additive factors for completely randomized, blocked and split-plot experiments when the measurement error has a common known variance. Sections 6.4 and 6.5 consider the same problems for the case of common unknown variance. Lastly, Sections 6.6 and 6.7 consider the screening goal of selecting a subset of treatment combinations containing the best levels of the two factors for completely randomized, blocked and split-plot experiments.

Chapter 7 studies Bernoulli responses. Sections 7.2–7.4 discuss selection of the best treatment using the indifference-zone approach. Procedures are given for two methods of defining the indifference-zone: The first is defined in terms of the difference between the two largest event probabilities (a single-stage procedure is given for this case in Section 7.2), and the second is stated in terms of the odds ratio of the two largest event probabilities (a sequential procedure is given for this case in Section 7.4). Section 7.3 describes an adaptive method of curtailing sampling so that the resulting rule achieves the *same* probability of selecting the best treatment as does the single-stage procedure defined in Section 7.2 that takes exactly n observations per treatment, and it accomplishes this with a smaller expected total number of observations. Section 7.5 discusses single-stage subset selection procedures for screening in single-factor experiments.

Finally, Chapter 8 is concerned with selection procedures for multinomial and cross-classified multinomial data. The chapter considers both indifference-zone and subset selection formulations. In Section 8.2, single-stage and sequential indifference-zone procedures are given for selecting the cell with the largest or smallest probability in a multinomial distribution. Section 8.3 describes a single-stage procedure for selecting a subset containing the cell with the largest probability, discusses curtailment of the single-stage rule and concludes with a sequential rule. In Section 8.4 we propose a procedure for simultaneously identifying the row level and column level with the largest marginal probabilities in a two-way crossed classification satisfying independence of the row and column variables.

We have added Notes sections at the end of the chapters that describe developments beyond the basic ones given in this book. We recognize that the limitation in implementing many of the procedures we describe in this text is the availability of certain tabled constants. Thus, in addition to providing tables to implement the procedures, we also provide a number of FORTRAN programs to supplement our tables. In addition, we describe other public domain programs valuable for implementing certain selection, screening and simultaneous confidence interval procedures and state how to obtain them.

CHAPTER 2

Selecting the Best Treatment in a Single-Factor Normal Response Experiment Using the Indifference-Zone Approach

2.1 INTRODUCTION

In this chapter, we consider problems of designing and analyzing experiments for selecting the best (or several best) treatments in completely randomized or blocked experiments. We assume that there is a single treatment factor and normal, though not necessarily homoscedastic, errors.

If the experimental units are homogeneous, a completely randomized experiment can be conducted, while if the experimental units are nonhomogeneous, blocking must be introduced. The more knowledge that the experimenter has about the variability in the data, the easier it is to plan efficient procedures for selection. This chapter considers five different cases, detailed below in the "Statistical Assumptions," concerning the variance of the responses; one important case is the classical assumption of unknown homogeneous variances that is made in the Analysis of Variance.

A variety of methods, differing in their sampling mechanisms, are introduced to solve this selection problem. Depending on the circumstances, one or more of these methods may be useful in a given application. The methods are contrasted with respect to their performance characteristics—primarily their achieved probability of correctly selecting the best treatment, and the amount of sampling they require.

The basic assumptions used throughout the chapter, except for experiments with blocking, are as follows.

Statistical Assumptions: Independent random samples of observations $Y_{i1}, Y_{i2}, \ldots$ ($1 \leq i \leq t$) are taken from $t \geq 2$ normal treatments $\Pi_1, \ldots, \Pi_t$. The number of observations to be taken from each treatment depends on the goal of the experiment, the probability requirement to be guaranteed and the particular procedure employed, as explained in the subsequent sections. Here Π_i has *unknown* treatment mean μ_i and *known* or *unknown* variance σ_i^2. Procedures are given for five assumptions concerning the variances:

1. $\sigma_1^2 = \cdots = \sigma_t^2$ known
2. $\sigma_1^2, \ldots, \sigma_t^2$ known but not necessarily equal
3. $\sigma_1^2 = \cdots = \sigma_t^2$ unknown (the usual Analysis of Variance assumption)
4. $\sigma_1^2, \ldots, \sigma_t^2$ completely unknown but $\max\{\sigma_1^2, \ldots, \sigma_t^2\} \leq \sigma_U^2$, where σ_U^2 is known
5. $\sigma_1^2, \ldots, \sigma_t^2$ completely unknown and arbitrary

We denote the vector of treatment means by $\boldsymbol{\mu} = (\mu_1, \ldots, \mu_t)$ and the vector of treatment variances by $\boldsymbol{\sigma}^2 = (\sigma_1^2, \ldots, \sigma_t^2)$. The ordered μ_i-values are denoted by

$$\mu_{[1]} \leq \cdots \leq \mu_{[t]}.$$

Neither the values of the $\mu_{[s]}$ nor the *pairing* of the Π_i with the $\mu_{[s]}$ ($1 \leq i, s \leq t$) is assumed to be known. The treatment having mean $\mu_{[t]}$ is referred to as the "best" treatment. In cases where $\sigma_1^2 = \cdots = \sigma_t^2$, we denote the common variance by σ^2. Henceforth, let y_{ij} ($1 \leq i \leq t$, $j \geq 1$) denote the observed values of the Y_{ij}.

For most of Chapter 2, the experimental goal and the associated probability (design) requirement are stated in Goal 2.1.1 and Equation (2.1.1), respectively.

Goal 2.1.1 To select the treatment associated with mean $\mu_{[t]}$.

A *correct selection* (CS) is said to be made if Goal 2.1.1 is achieved.

Probability Requirement: For specified constants (δ^*, P^*) with $0 < \delta^* < \infty$ and $1/t < P^* < 1$, we require

$$P\{CS\} \geq P^* \quad \text{whenever } \mu_{[t]} - \mu_{[t-1]} \geq \delta^*. \tag{2.1.1}$$

The probability in Equation (2.1.1) depends on the differences $\mu_i - \mu_j$ ($i \neq j$, $1 \leq i, j \leq t$), the sample size n and σ^2. The constant δ^* can be thought of as the "smallest difference worth detecting." If $\mu_{[t-1]}$ and $\mu_{[t]}$ are very "close" in standardized units, that is, if $(\mu_{[t]} - \mu_{[t-1]})/\sigma$ is small, then the sampling cost required to distinguish between the associated treatments can be prohibitive. Furthermore, if $\mu_{[t-1]}$ is very close to $\mu_{[t]}$, then it may matter little which of the associated treatments is selected. Thus, δ^* is the smallest difference which can be detected at a reasonable sampling cost or which is of practical importance.

Clearly, it makes no sense to choose $P^* \leq 1/t$ because $P^* = 1/t$ can be achieved without taking *any* observations by rolling a fair t-sided die and selecting the treatment so identified as the best one. Also, we must have $P^* < 1$ because we cannot guarantee (2.1.1) with probability unity.

Parameter configurations $\boldsymbol{\mu}$ satisfying $\mu_{[t]} - \mu_{[t-1]} \geq \delta^*$ are said to be in the *preference-zone* for a correct selection; configurations satisfying $\mu_{[t]} - \mu_{[t-1]} < \delta^*$ are said to be in the *indifference-zone*. Any procedure that guarantees (2.1.1) is said to be employing the so-called *indifference-zone* approach.

If (possibly unequal) sample sizes $n_1, \ldots, n_t$ are available, then most texts on experimental design use the parameterization $\mu_i = \mu + \tau_i$ and adopt the identifiability constraint $\sum_{i=1}^{t} n_i \tau_i = 0$; this reduces to $\sum_{i=1}^{t} \tau_i = 0$ when $n_1 = \cdots = n_t$. With this parameterization, $\mu = \sum_{i=1}^{t} n_i \mu_i / \sum_{i=1}^{t} n_i$ is the weighted average of the t means μ_i, and $\tau_i = \mu_i - \mu$ is the difference between the mean response for the ith level of the treatment and the overall weighted average mean of the t treatments.

In either case, τ_i is referred to as the *effect* of the ith level of the treatment. If $\tau_{[1]} \leq \cdots \leq \tau_{[t]}$ are the ordered τ_i-values, then the level of the factor associated with $\tau_{[i]}$ is the same as the level associated with $\mu_{[i]}$ and, in particular, Goal (2.1.1) can be stated equivalently as that of selecting the factor-level associated with $\tau_{[t]}$. Furthermore, the probability requirement (2.1.1) can be rephrased as

$$P\{CS\} \geq P^* \quad \text{whenever} \quad \tau_{[t]} - \tau_{[t-1]} \geq \delta^*$$

because $\mu_{[t]} - \mu_{[t-1]} = \tau_{[t]} - \tau_{[t-1]}$.

The invariance of the ordering of the treatments based on their τ_i-values is an example of the more general *location invariance* of this selection problem. Location invariance means that the ordering of the treatments is unchanged if their means are all shifted by the same amount. In particular, the identity of the best treatment is unchanged by a common location shift. All of the differences $\mu_i - \mu_j$ between the treatment means are unchanged, and thus the membership of any configuration of means in the indifference-zone or preference-zone is unaffected by a common location shift. The selection rules defined in the following sections are also location invariant because the treatment selected does not change if all of the data are shifted by a common amount.

The variance cases and sampling situations studied in this chapter are summarized as follows: Sections 2.2–2.5 consider the case of common known variance. A number of procedures corresponding to different sampling schemes (single-stage, two-stage, and sequential) will be introduced in this chapter. In some applications, only the single-stage procedures discussed in Sections 2.2 and 2.5 can reasonably be used. In other applications, the multi-stage or sequential plans of Section 2.3 are feasible and can be more efficient. Comparisons among the performance characteristics of the procedures, primarily their achieved probabilities of correct selection, expected numbers of stages to terminate sampling, and expected total sample sizes, are given in Section 2.4.

The final three sections consider four different ways in which the common known variance assumption can be weakened. Of course, the weaker the assumption, the larger the expected sample size necessary to guarantee the probability requirement. Section 2.6 considers the case of arbitrary known variances, Section 2.7 the case of common unknown variance and Section 2.8 the case of arbitrary unknown variances (with and without known upper bounds).

2.2 A SINGLE-STAGE PROCEDURE FOR COMMON KNOWN VARIANCE

This section describes single-stage procedures for selecting the best treatment in the case of normally distributed observations having a common known variance. Completely randomized, balanced and unbalanced situations will be considered in Section 2.2.1; blocked experiments are discussed in Section 2.2.2. A qualitative assessment of the robustness of the procedures is made in Section 2.2.3.

2.2.1 The Completely Randomized Design

Bechhofer (1954) proposed the following *single-stage* procedure:

Procedure $\mathcal{N}_B$ For the given t and specified $(\delta^\star/\sigma, P^\star)$, determine n from Table 2.1 or by using Equation (2.2.1).

Sampling rule: Take a random sample of n observations Y_{ij} ($1 \leq j \leq n$) in a single stage from Π_i ($1 \leq i \leq t$).

Terminal decision rule: Calculate the t sample means $\bar{y}_i = \sum_{j=1}^{n} y_{ij}/n$ ($1 \leq i \leq t$). Select the treatment that yielded the largest sample mean, $\bar{y}_{[t]} = \max\{\bar{y}_1, \ldots, \bar{y}_t\}$, as the one associated with $\mu_{[t]}$.

For values of $(t; \delta^\star/\sigma, P^\star)$ not covered by Table 2.1, choose $Z_{t-1,1/2}^{(1-P^\star)}$ from Table B.1 corresponding to the $(t, P^\star)$ of interest and set

$$n = \left\lceil 2(\sigma Z_{t-1,1/2}^{(1-P^\star)}/\delta^\star)^2 \right\rceil \tag{2.2.1}$$

where $\lceil b \rceil$ is the smallest integer greater than or equal to b. For $(t; \delta^\star/\sigma, P^\star)$-combinations not available in Tables 2.1 or B.1, the FORTRAN program USENB in Appendix C calculates the smallest common single-stage sample size per treatment required to guarantee (2.1.1) using procedure $\mathcal{N}_B$.

The $Z_{t-1,1/2}^{(1-P^\star)}$ constants used to implement procedure $\mathcal{N}_B$ are a special case of the upper equicoordinate point of the multivariate normal distribution; if $(W_1, \ldots, W_p)$ has the p-variate multivariate normal distribution with mean vector zero, unit variances, and common correlation ρ, then

$$P\left\{\max_{1 \leq i \leq p} W_i \leq Z_{p,\rho}^{(\alpha)}\right\} = 1 - \alpha \tag{2.2.2}$$

Table 2.1. Values of the Smallest Common Single-Stage Sample Size n per Treatment Required for Procedure $\mathcal{N}_B$ to Guarantee $P\{CS \mid LF\} \geq P^*$ When Selecting the Treatment with the Largest Mean from t Normal Treatments

t	P^*	\multicolumn{10}{c}{δ^*/σ}
		0.1	0.2	0.3	0.4	0.5	0.6	0.7	0.8	0.9	1.0
2	0.75	91	23	11	6	4	3	2	2	2	1
	0.90	329	83	37	21	14	10	7	6	5	4
	0.95	542	136	61	34	22	16	12	9	7	6
	0.99	1083	271	121	68	44	31	23	17	14	11
3	0.75	206	52	23	13	9	6	5	4	3	3
	0.90	498	125	56	32	20	14	11	8	7	5
	0.95	735	184	82	46	30	21	15	12	10	8
	0.99	1309	328	146	82	53	37	27	21	17	14
4	0.75	283	71	32	18	12	8	6	5	4	3
	0.90	602	151	67	38	25	17	13	10	8	7
	0.95	851	213	95	54	35	24	18	14	11	9
	0.99	1442	361	161	91	58	41	30	23	18	15
5	0.75	341	86	38	22	14	10	7	6	5	4
	0.90	676	169	76	43	28	19	14	11	9	7
	0.95	934	234	104	59	38	26	20	15	12	10
	0.99	1537	385	171	97	62	43	32	25	19	16
6	0.75	388	97	44	25	16	11	8	7	5	4
	0.90	735	184	82	46	30	21	15	12	10	8
	0.95	998	250	111	63	40	28	21	16	13	10
	0.99	1610	403	179	101	65	45	33	26	20	17
7	0.75	426	107	48	27	18	12	9	7	6	5
	0.90	783	196	87	49	32	22	16	13	10	8
	0.95	1051	263	117	66	43	30	22	17	13	11
	0.99	1670	418	186	105	67	47	35	27	21	17
8	0.75	459	115	51	29	19	13	10	8	6	5
	0.90	824	206	92	52	33	23	17	13	11	9
	0.95	1096	274	122	69	44	31	23	18	14	11
	0.99	1721	431	192	108	69	48	36	27	22	18
9	0.75	487	122	55	31	20	14	10	8	7	5
	0.90	859	215	96	54	35	24	18	14	11	9
	0.95	1135	284	127	71	46	32	24	18	15	12
	0.99	1764	441	196	111	71	49	36	28	22	18
10	0.75	513	129	57	33	21	15	11	9	7	6
	0.90	890	223	99	56	36	25	19	14	11	9
	0.95	1169	293	130	74	47	33	24	19	15	12
	0.99	1803	451	201	113	73	51	37	29	23	19

defines the upper-α equicoordinate critical point $Z_{p,\rho}^{(\alpha)}$ of this distribution. The phrase *equicoordinate* is used to describe this value because

$$\left\{\max_{1 \le i \le p} W_i \le c\right\} = \{W_1 \le c, W_2 \le c, \ldots, W_p \le c\}.$$

The constant $Z_{p,\rho}^{(\alpha)}$ is determined to satisfy the probability requirement (2.1.1) for any true configuration of means satisfying

$$\mu_{[1]} = \mu_{[t-1]} = \mu_{[t]} - \delta^*. \tag{2.2.3}$$

Equation (2.2.3) specifies a *set* of μ-configurations in which the "worst" $t-1$ treatment means are fixed at some common value, say γ, and the mean of the best treatment is δ^* larger than that common value, that is, $\mu_{[1]} = \gamma = \mu_{[t-1]}$ and $\mu_{[t]} = \gamma + \delta^*$. It can be shown that for configurations μ satisfying Equation (2.2.3), the $P\{CS\}$ depends only on δ^* and *not* on the common value of $\mu_{[1]} = \mu_{[t-1]}$.

Configurations (2.2.3) are termed *least-favorable* (LF) because, for fixed n, they minimize the $P\{CS\}$ among all configurations satisfying the preference-zone requirement $\mu_{[t]} - \mu_{[t-1]} \ge \delta^*$. If the sample size is adequate to guarantee (2.1.1) when μ satisfies (2.2.3), then it will also guarantee (2.1.1) when the means are in configurations more favorable to the experimenter. Formally, if $\mu_{[1]} \le \cdots \le \mu_{[t-1]} \le \mu_{[t]} - \delta^*$ with at least one of the inequalities *strict*, then the n-value calculated in the sampling rule of procedure $\mathcal{N}_B$ will guarantee a $P\{CS\}$ strictly greater than P^*. Panels (a) and (b) of Figure 2.1 illustrate a least-favorable configuration and a configuration more favorable to the experimenter, respectively. Alternatively, the number of observations required to guarantee the probability requirement (2.1.1) is maximized for *any fixed* configuration μ satisfying (2.2.3).

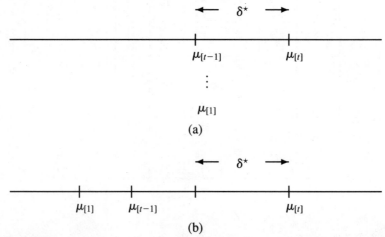

Figure 2.1. Illustration of an LF-configuration (a) and a configuration of means more favorable than the LF-configuration (b).

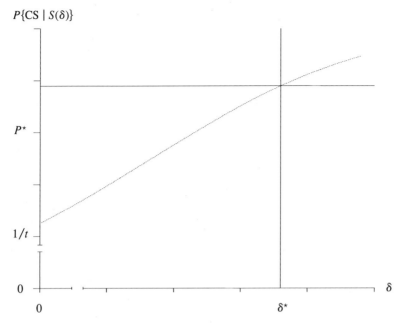

Figure 2.2. $P\{CS \mid S(\delta)\}$ versus δ, where $S(\delta) = (0, \ldots, 0, \delta)$ is the slippage configuration.

Another illustration of how the $P\{CS\}$ changes as a function of the true vector of treatment means is displayed in Figure 2.2. The vertical axis of Figure 2.2, $P\{CS \mid S(\delta)\}$, is the achieved $P\{CS\}$ when $\boldsymbol{\mu}$ satisfies $\mu_{[1]} = \cdots = \mu_{[t-1]} = \mu_{[t]} - \delta$, and δ is an arbitrary non-negative constant. Some authors refer to configurations of the treatment means of this form as *slippage* configurations; we denote such configurations by $S(\delta)$. As $\delta \to 0$, the configuration approaches the equal-means (EM) configuration and $P\{CS \mid S(\delta)\}$ approaches $1/t$. If $\delta = \delta^*$, that is, if the treatment means are in the LF-configuration, then $P\{CS \mid S(\delta)\}$ is specified to be at least P^*. As $\delta \to \infty$, the mean of the best treatment becomes arbitrarily larger than the remaining ones and $P\{CS \mid S(\delta)\}$ approaches unity.

In general, selection experiments can be costly when the number of competing treatments is large. The following examples illustrate this phenomenon.

Example 2.2.1 Suppose that $t = 4$ and that our goal is to detect a difference in means as small as 0.2 standard deviations with probability 0.99 (a very demanding requirement). Table 2.1 shows that procedure $\mathcal{N}_B$ calls for $n = 361$ observations per treatment. In general, increasing δ^* and/or decreasing P^* requires a smaller n. For example, when $\delta^*/\sigma = 0.6$ and $P^* = 0.95$, procedure $\mathcal{N}_B$ requires only $n = 24$ observations per treatment.

Example 2.2.2 Consider the effect of increasing t from 4 to 10 in the second part of Example 2.2.1, that is, when $\delta^*/\sigma = 0.6$ and $P^* = 0.95$. Then a common sample size of $n = 33$ observations must be collected. In general, the critical quantity is δ^*

since n is directly proportional to $1/(\delta^*)^2$ (Ramberg 1972, pp. 1977–1980). For fixed $(\delta^*/\sigma, P^*)$ the common sample size n increases rather slowly with t—approximately as $\ln(t-1)$ for P^* near unity where $\ln(\cdot)$ denotes the natural logarithm.

Relation to Hypothesis Testing

Experimenters who are used to thinking in terms of significance levels of $\alpha = 0.01$ or 0.05 for hypothesis testing problems may have to reorient their thinking for selection problems such as those considered here. A hypothesis testing problem is a two-decision problem in which the experimenter either accepts or rejects a null hypothesis. The experimenter can be assured of correctly rejecting *false* null hypotheses 50% of the time (that is, having power 0.5) by merely tossing a fair coin and rejecting the null hypothesis every time that heads (say) comes up. In such a situation, the experimenter is concerned with increasing the power of the test above 0.5. This is done by taking a sufficient number of observations and using an appropriate decision rule.

In contrast, the problem of selecting that one of t (≥ 2) treatments that has the largest treatment mean is a t-decision problem. The probability of selecting the best treatment is $1/t$ if the decision is based on the roll of a fair t-sided die. The experimenter collects data and uses an appropriate decision rule to increase the probability of a correct selection above $1/t$. Since $0.5 > 1/t$ for $t > 2$, it is clear that, in general, larger sample sizes will be required for the selection problem than for the hypothesis testing problem. Thus, the experimenter may be willing to settle for $P^* = 0.75$ when $t = 10$ for a 10-decision selection problem, but not for a power of 0.75 in a two-decision hypothesis testing problem. Alternatively, as noted above, the sample size can be lowered for fixed t and P^* by increasing δ^* to a value that might be larger than acceptable in the power computation of a hypothesis test.

Confidence Statement Formulation

There are a number of different ways of restating the event of CS that provide insight into what the experiment guarantees. If the common sample size is chosen to satisfy (2.1.1), then the reader may be concerned with the performance of procedure $\mathcal{N}_B$ when $\mu_{[t]} - \mu_{[t-1]} < \delta^*$. If μ_S denotes the mean of the treatment selected by $\mathcal{N}_B$, then it can be shown that

$$P\{\mu_{[t]} - \delta^* \leq \mu_S \leq \mu_{[t]}\} \geq P^* \tag{2.2.4}$$

for *all* $\boldsymbol{\mu}$. In other words, the experimenter can assert with confidence coefficient at least P^* that the mean of the *selected* treatment, μ_S, is within δ^* of the largest treatment mean.

Unbalanced Experiments

If the experiment is performed with (or terminates so that there are) unequal numbers of observations $n_1, \ldots, n_t$ from the treatments $\Pi_1, \ldots, \Pi_t$, respectively, then the achieved $P\{CS\}$ can be calculated as follows. Let $n_{(i)}$ denote the number of observations from the treatment having mean $\mu_{[i]}$ ($1 \leq i \leq t$); then the $P\{CS\}$ for procedure

A SINGLE-STAGE PROCEDURE FOR COMMON KNOWN VARIANCE

$\mathcal{N}_B$ is

$$P\{CS\} = \int_{-\infty}^{+\infty} \left[\prod_{i=1}^{t-1} \Phi \left(\frac{\sqrt{n_{(i)}}(\mu_{[t]} - \mu_{[i]})}{\sigma} + y\sqrt{n_{(i)}/n_{(t)}} \right) \right] d\Phi(y)$$

where $\Phi(\cdot)$ is the standard normal c.d.f. Unfortunately, for a fixed vector of treatment means $\boldsymbol{\mu}$, each of the $t!$ different assignments of the n_i to the treatments can yield a different $P\{CS\}$.

In particular, the minimum of the $P\{CS\}$ over the indifference-zone is

$$P\{CS \mid LF\} = \int_{-\infty}^{+\infty} \left[\prod_{i=1}^{t-1} \Phi \left(\frac{\sqrt{n_{(i)}}\delta^*}{\sigma} + y\sqrt{n_{(i)}/n_{(t)}} \right) \right] d\Phi(y). \qquad (2.2.5)$$

The LF-configuration described in (2.2.5) consists of $t - 1$ treatments with equal means, and a single (best) treatment whose mean is exactly δ^* larger than the common value. Notice that once the sample size is assigned to the treatment with the largest mean, $\mu_{[t]}$, then the probability (2.2.5) is the same for any permutation of the remaining sample sizes with the $t - 1$ worst treatments. If, as would usually be the case, the association between the n_i and $\mu_{[j]}$ ($1 \leq i, j \leq t$) is not known, then one must calculate the minimum of (2.2.5) over all t pairings of $n_1, \ldots, n_t$ with $\mu_{[t]}$ to determine the minimum of the $P\{CS\}$ over the indifference-zone. The FORTRAN program UNEQNB in Appendix C can be used to calculate the probability (2.2.5). This program can also be used to determine the degradation in the achieved $P\{CS \mid LF\}$ of procedure $\mathcal{N}_B$ when the experiment is designed to achieve a certain P^*-value and data are missing.

Example 2.2.3 Table 2.2 lists values of (2.2.5) for $t = 3$, $n_i = 5, 10, 15$, $\sigma = 1$, and $\boldsymbol{\mu} \in \{(0, 0, 0.8), (0, 0, 0.2)\}$. Observe that for $\delta^* = 0.8$, the minimum $P\{CS \mid LF\}$ occurs when $n_{(3)} = 5$, the *smallest* of the n_i. This might suggest the conjecture that the minimum $P\{CS \mid LF\}$ occurs when the least amount of information is collected on the best treatment. However, the calculations for $\delta^* = 0.2$ show that this does not always have to be the case, for here the minimum $P\{CS \mid LF\}$ occurs when $n_{(3)} = 15$, the *largest* n_i.

Table 2.2. Achieved $P\{CS \mid LF\}$ Using Equation (2.2.5) for $\delta^* = 0.2, 0.8$

			$P\{CS \mid LF\}$	
$n_{(1)}$	$n_{(2)}$	$n_{(3)}$	$\delta^* = 0.2$	$\delta^* = 0.8$
5	10	15	0.4908	0.9193
5	15	10	0.5051	0.9100
10	15	5	0.5306	0.8954
10	10	10	0.5234	0.9343

Example 2.2.4 (Determining Minimal Detectable Differences) Suppose that an experimenter wishes to select the best of $t = 5$ treatments with probability at least 0.85. The amount of funding available for the experiment permits $n = 15$ observations to be taken. If the standard deviation of each response is $\sigma = 1.0$, then FORTRAN program USENB in Appendix C calculates $\delta^* = 0.592$ as the minimal detectable difference for which (2.1.1) is achieved with probability $P^* = 0.85$. If $\delta^* = 0.592$ is too large to be of practical value, then the scientist might well be advised not to perform the experiment. Alternatively, one might compute δ^* for other sample sizes, as in Table 2.3, and then have a basis for obtaining additional funding.

Optimality Property

Hall (1959) and Eaton (1967) proved that if the experimenter is restricted to *single-stage* location invariant procedures that guarantee (2.1.1), then there is no procedure requiring fewer observations per treatment than procedure $\mathcal{N}_B$.

2.2.2 Designs with Blocking

Suppose that there are insufficient quantities of uniform experimental material available to perform an experiment using a completely randomized design. If the experimental units can be grouped into blocks of uniform material, then one of the traditional blocking designs can be employed, thereby minimizing possible bias (and reducing the residual variance).

Randomized Complete Block Designs

Suppose that there are sufficient experimental units so that each treatment can be used *at least* once in each block. Assume that

$$Y_{ijk} = \mu + \tau_i + \beta_j + \epsilon_{ijk} \qquad (2.2.6)$$

$(1 \leq i \leq t, 1 \leq j \leq b, 1 \leq k \leq r)$ where the ϵ_{ijk} are independent and identically distributed normal errors with zero mean and common known variance σ^2, $\sum_{i=1}^{t} \tau_i = \sum_{j=1}^{b} \beta_j = 0$, and r is the common number of times that a treatment is repeated in each block. This is the model for the *randomized complete block design* with *fixed* treatment effects in which the treatments have the same relative magnitude in all blocks (that is, no block × treatment interaction). The τ_i and the β_j are the treatment and block "effects," respectively, for this single-factor experiment.

Table 2.3. Minimum Detectable Differences δ^* Using Procedure $\mathcal{N}_B$ for Various Sample Sizes n When $(t, \sigma; P^*) = (5, 1.0; 0.85)$

n	Minimum Detectable δ^*
10	0.726
15	0.592
20	0.513
25	0.459

Procedure $\mathcal{N}_B$ (Randomized Complete Block Designs)
For the given t and specified $(\delta^*/\sigma, P^*)$, determine $Z_{t-1,1/2}^{(1-P^*)}$ from Table B.1 (or otherwise) and set $b = [(2/r)(\sigma Z_{t-1,1/2}^{(1-P^*)}/\delta^*)^2]$.

Sampling rule: Take r independent observations Y_{ijk} $(1 \leq k \leq r)$ on treatment i $(1 \leq i \leq t)$ in each of b blocks $(1 \leq j \leq b)$.

Terminal decision rule: Calculate the best linear unbiased estimates (BLUEs) of the treatment effects, namely, $\hat{\tau}_i = \bar{y}_{i..} - \bar{y}_{...}$ $(1 \leq i \leq t)$, where the "dot" notation has the usual interpretation, that is, $\bar{y}_{i..} = \sum_{j=1}^{b} \sum_{k=1}^{r} y_{ijk}/(br)$ and $\bar{y}_{...} = \sum_{i=1}^{t} \sum_{j=1}^{b} \sum_{k=1}^{r} y_{ijk}/(tbr)$. Select the treatment corresponding to $\hat{\tau}_{[t]} = \max\{\hat{\tau}_1, \ldots, \hat{\tau}_t\}$.

Recall that $Z_{t-1,1/2}^{(1-P^*)}$ can also be determined by using the FORTRAN programs MULTZ, in Appendix C, or MVNPRD, in Dunnett (1989). When procedure $\mathcal{N}_B$ is used, then the $P\{CS\}$ will be at least P^* for all τ_i and β_j whenever

$$\tau_{[t]} - \tau_{[t-1]} \geq \delta^*.$$

The probability requirement is guaranteed because the BLUEs of the τ_i $(1 \leq i \leq t)$ contain no block effects.

Balanced Incomplete Block Designs

Selection based on incomplete block designs is more complicated than it is for complete blocks. There is a cost, in terms of sample size, for not using a completely randomized design when there is adequate experimental material to make a complete block design feasible. However, that cost must be weighed against the potential benefit of increased precision of estimated treatment differences when there are, indeed, location differences among the experimental units in different blocks as postulated by the incomplete block design (see Cox 1958, pp. 229–230).

Consider the case of *balanced incomplete block designs* (BIBDs). Such designs have the property that every pair of treatments appears together in a block the same number of times, say λ, over the design. Throughout this discussion, also let k denote the block size, b the number of blocks in the design and r the common number of observations from each treatment in the design. For example, the panel below illustrates the layout of an experiment with $t = 3$ treatments in $b = 3$ blocks of size $k = 2$; each treatment pair appears together in the same block $\lambda = 1$ time.

BIBD with $t = 3 = b$, $k = 2$

Block		
1	2	3
Treatment 1	Treatment 3	Treatment 2
Treatment 2	Treatment 1	Treatment 3

A short table of selected BIBDs for 10 or fewer treatments in blocks of size 6 or less can be found in Appendix 8C of Box, Hunter and Hunter (1978). A more complete list of BIBDs is given in Cochran and Cox (1957). As usual, randomization must be used to assign treatments to the treatment labels, and the treatments must be assigned randomly to the experimental units within a block. Also, if there is more than one complete replication, the associated randomizations must be carried out independently.

Let y_{ij} be the observed response of treatment i in block j for $(i, j) \in D$, say, where D is the set of treatment by block combinations used in the design. The associated model is again assumed to be (2.2.6) for $(i, j) \in D$.

Procedure $\mathcal{N}_B$ (Balanced Incomplete Block Designs) For the given t and specified $(\delta^*/\sigma, P^*)$, determine $Z_{t-1,1/2}^{(1-P^*)}$ from Table B.1 (or otherwise) and set

$$\lambda = \left\lceil (\sigma Z_{t-1,1/2}^{(1-P^*)}/\delta^*)^2 (2k/t) \right\rceil$$

where k is the common block size.

Sampling rule: Take independent observations Y_{ij} in a BIBD with (at least) λ occurrences of each pair of treatments over the design.

Terminal decision rule: Calculate the BLUEs of the treatment effects τ_i ($1 \leq i \leq t$) as follows: Let T_i denote the *total* of the responses for the ith treatment, and let B_i denote the sum of the responses in all blocks that *contain* the ith treatment. Then the BLUEs of the τ_i (subject to the identifiability constraint $\sum_{i=1}^{t} \tau_i = 0$) are

$$\hat{\tau}_i = \frac{kT_i - B_i}{t\lambda}.$$

Select the treatment corresponding to $\hat{\tau}_{[t]} = \max\{\hat{\tau}_1, \ldots, \hat{\tau}_t\}$.

Example 2.2.5 Suppose that it is desired to detect the standardized difference $\delta^*/\sigma = 1/3$ in an experiment with $t = 3$ treatments in blocks of size $k = 2$ when $P^* = 0.75$. Then we require $\lambda = \lceil (\sigma Z_{2,1/2}^{(0.25)}/\delta^*)^2 (2k/t) \rceil = \lceil (3 \times 1.014)^2 \times \frac{4}{3} \rceil = \lceil 12.34 \rceil = 13$. The basic three-block design described earlier with $t = 3$ and $k = 2$ could be replicated 13 times to guarantee this requirement, yielding a design with a total of 39 blocks and a total of 78 observations with $26 = 13(2)$ observations per treatment. This is to be contrasted with the corresponding common sample size of $n = \lceil 2(\sigma Z_{2,1/2}^{(0.25)}/\delta^*)^2 \rceil = \lceil 2(3 \times 1.014)^2 \rceil = 19$ observations per treatment required by a completely randomized design.

For certain (t, k) combinations, several BIBDs exist that guarantee the probability requirement. For example, this is the case for $(t, k) = (6, 3)$, where we have Design I with $r = 5, b = 10, \lambda = 2$ and Design II with $r = 10, b = 20, \lambda = 4$ (plans 11.4 and 11.5 in Cochran and Cox 1957). In such situations one should ordinarily choose the design with the smaller λ-value, thereby obtaining greater flexibility. If one wanted to achieve $\lambda = 6$, this could be accomplished by taking three replications of Design I (or equivalently one replication of each of Designs I and II).

Other Block Designs

In a similar way, Latin squares or other balanced designs can be employed if the treatment effects are estimated by the BLUEs of the τ_i ($1 \leq i \leq t$), and the "sample size" n is calculated appropriately as the following examples illustrate.

Example 2.2.6 (Latin Squares and Generalizations) Suppose that $E\{Y_{ijs}\} = \mu + \tau_i + \beta_j + \gamma_s$ subject to the identifiability constraints $0 = \sum_i \tau_i = \sum_j \beta_j = \sum_s \gamma_s$. Here τ_i is the ith treatment effect, β_j is the jth column effect and γ_s is the sth row effect. Case (a) corresponds to a pair of (randomly selected) Latin squares each of which eliminates heterogeneity in the row and column directions and for which $0 = \sum_{j=1}^{3} \beta_j = \sum_{j=4}^{6} \beta_j$. The principal assumption underlying the use of this design is that there is no treatment $\times$ block interaction with either row or column blocks.

(a) Latin Square Design

	β_1	β_2	β_3
γ_1	τ_1	τ_3	τ_2
γ_2	τ_2	τ_1	τ_3
γ_3	τ_3	τ_2	τ_1

	β_4	β_5	β_6
γ_4	τ_1	τ_2	τ_3
γ_5	τ_3	τ_1	τ_2
γ_6	τ_2	τ_3	τ_1

Case (b) is a "crossover design" with 6 "levels" of the column factor and 3 of the row factor. A crossover design is characterized by the fact that each subject (here indicated by a different β) receives all t treatments in some predetermined order. In addition to treatment effects, this design allows a subject (experimental unit) effect and an order (in which the subject receives the treatment) effect. Thus, the design in Case (b) consists of 6 subjects, 3 orders and 3 treatments.

(b) Crossover Design

	β_1	β_2	β_3	β_4	β_5	β_6
γ_1	τ_1	τ_3	τ_2	τ_2	τ_1	τ_3
γ_2	τ_2	τ_1	τ_3	τ_3	τ_2	τ_1
γ_3	τ_3	τ_2	τ_1	τ_1	τ_3	τ_2

In practice, a crossover design is conducted with a "washout" period between the application of treatments to each subject, and the experimenter assumes that the effect of each treatment *does not depend on the number or identity of prior* treatments that the subject received. Suppose that Y_{ijs} is the response for the sth subject when administered the ith treatment in the jth order. The model for the mean of the crossover design assumes additive subject (one block effect), order (another block effect) and treatment effects, that is, $E\{Y_{ijs}\} = \mu + \tau_i + \beta_j + \gamma_s$ and the usual normality assumptions hold. From a practical viewpoint this model states that the comparison of any two treatments is the same within every subject and ordering combination.

Both the Latin square design and the crossover design assume additivity in the treatment, column and row effects (and thus can eliminate heterogeneity in two directions). The sample size is 6 for both Cases (a) and (b). For Case (a) the sample size can be increased by using a larger number of 3×3 Latin squares to p ($p > 2$). For Case (b) the sample size can be increased by supplementing the number of columns (subjects) to $3p$ ($p \geq 2$).

Selection is also possible for more general types of block designs called "connected" designs. The theses of Driessen (1992) and Dourleijn (1993) provide methods of indifference-zone selection, subset selection and multiple comparisons for these more-complicated designs.

2.2.3 Robustness of Normal Theory Procedure

We discuss the robustness of procedure $\mathcal{N}_B$ under three types of violations of the underlying assumptions on which the procedure is based. We consider the following:

- Lack of normality
- Lack of homogeneity of variances or incorrect knowledge about variances
- Lack of independence of the data

We address the effects of lack of normality first. For nonnormal location families, procedure $\mathcal{N}_B$ will have *nearly nominal performance characteristics* when n is not too small. Dudewicz and Mishra (1984) demonstrated this in their study of the location model; they suppose that the density of Y_{ij} is $F(y - \mu_i)$ ($1 \leq i \leq t$), where the distribution $F(\cdot)$ is arbitrary—it may be highly skewed, short-tailed or long-tailed. The quantities $Y_{i_1,j} - Y_{i_2,j}$ ($i_1 \neq i_2, 1 \leq i_1, i_2 \leq t, 1 \leq j \leq n$) on which procedure $\mathcal{N}_B$ is based have symmetrical distributions. It is known that jointly the $\sum_{j=1}^{n}(Y_{i_1,j} - Y_{i_2,j})$ approach the multivariate normal distribution as n increases. Thus, one would expect that the achieved $P\{CS\}$ would be close to the nominal value.

Table 2.4, extracted from Dudewicz and Mishra (1984), demonstrates $P\{CS \mid LF\}$-robustness for procedure $\mathcal{N}_B$ for three choices of $F(\cdot)$: the uniform distribution on the unit interval, $U(0, 1)$ (as an example of a short-tailed distribution), Student's t-distribution with 3 d.f., T_3 (as an example of a long-tailed distribution), and the standard normal distribution, $N(0, 1)$ (as the reference distribution). The uniform and Student's t-distributions are scaled to have unit variances (the uniform by $\sqrt{1/12}$ and the T_3 by $\sqrt{3}$). The location vector of means is $\boldsymbol{\mu} = (0, 0, \delta^\star)$. Clearly, the approach to normality is rapid with the $P\{CS \mid LF\}$ not greatly affected even for the small sample sizes used in this study. [The interested reader can see Domröse and Rasch

Table 2.4. $P\{CS \mid LF\}$ Using Procedure $\mathcal{N}_B$ for $t = 3$ Treatments Under the Assumption That the Distributions Are $N(0, 1)$ When the True $F(\cdot)$ Is $N(0, 1)$, $\sqrt{12}\, U(0, 1)$ or $T_3/\sqrt{3}$

n	$F(\cdot)$	$\delta^\star$		
		0.1	1.0	10.0
5	$N(0, 1)$	0.40	0.90	1.00
	$\sqrt{12}\, U(0, 1)$	0.40	0.90	1.00
	$T_3/\sqrt{3}$	0.42	0.93	1.00
10	$N(0, 1)$	0.43	0.98	1.00
	$\sqrt{12}\, U(0, 1)$	0.43	0.98	1.00
	$T_3/\sqrt{3}$	0.44	0.98	1.00

Reprinted from Dudewicz and Mishra (1984), p. 42, by courtesy of Marcel Dekker, Inc.

(1987) for a simulation study involving distributions with various choices of skewness and kurtosis. Hustý (1981) investigates the robustness of a general procedure that is related to procedure $\mathcal{N}_B$.]

Selection procedures are *very sensitive* to violations of the assumption of a *common known variance*, as are hypothesis tests and ANOVA techniques. Driessen, van der Laan and van Putten (1990) made a detailed analytic study of the effect of the violation of the homogeneity assumption on $P\{CS \mid LF\}$. They found a serious lack of robustness in the sense that, under heterogeneity, the achieved $P\{CS \mid LF\}$ is much less than $P^\star$ for $0.75 \leq P^\star \leq 0.99$. This finding was determined by varying the σ_i ($1 \leq i \leq t$) in the interval $[\sigma/\gamma, \gamma\sigma]$ ($\gamma \geq 1$), where σ is the assumed common known standard deviation. They found that the effect is substantial for large values of t and for large values of γ ($\gamma \sim 4$), but even for small values of t and γ the effect is appreciable. For example, for nominal $P^\star = 0.95$, $t = 3$ and $\gamma = 2$, they showed that the achieved $P\{CS \mid LF\} \sim 0.73$.

To the maximum extent possible, the deleterious effects on the achieved $P\{CS \mid LF\}$ of heteroscedastic variances can be minimized by choosing equal sample sizes per treatment (because then, very large or very small sample sizes cannot be associated with large or small variances). However, the absolute effect on the achieved $P\{CS \mid LF\}$ can still be large using this equi-sample-size design. [See Scheffé (1959, p. 364), for a similar admonition regarding the Analysis of Variance.]

The selection procedure $\mathcal{N}_B$ is also *very sensitive* to lack of independence. We consider two forms of dependence. First, suppose that successive observations from the same treatment are correlated but that the observations from different treatments are independent. Intuitively, in this case the data from each treatment contain less information than if the observations were independent, and one would suspect that the achieved $P\{CS\}$ would be lower than its nominal value.

One specific case where a within-treatment correlation model has been studied is the stationary first-order autoregressive process,

$$Y_{ij} = \rho Y_{i,j-1} + Z_{ij} \qquad (2.2.7)$$

($-1 < \rho < 1$) considered by Dudewicz and Zaino (1977). Here Y_{ij} is the jth observation from a normal treatment Π_i ($1 \leq i \leq t$) and the Z_{ij} are independent random normal variables with $E\{Z_{ij}\} = (1 - \rho)\mu_i$ and $\text{Var}\{Z_{ij}\} = \sigma^2$. Thus, observations from the ith treatment satisfy $E\{Y_{ij}\} = \mu_i$ for all j; all treatments have the same variance. The case $\rho = 0$ corresponds to independent observations from each treatment; however, the observations are correlated when $\rho \neq 0$.

The model (2.2.7) arises in the study of time series involving dependent observations, as well as in certain simulation studies of queueing systems having highly correlated customer waiting times. Dudewicz and Zaino (1977) showed that the smallest common sample size required to guarantee (2.1.1) for model (2.2.7) is the smallest integer $n(\rho)$ satisfying

$$\frac{1}{n(\rho)}\left\{\frac{1+\rho}{1-\rho} - \frac{2\rho(1-\rho^{n(\rho)})}{n(\rho)(1-\rho^2)}\right\} \leq \frac{1}{n(0)}$$

where $n(0)$ is the sample size required by procedure $\mathcal{N}_B$ when $\rho = 0$. A conservative approximation to $n(\rho)$ for moderate or large $n(0)$ is given by

$$n(\rho) = n(0)\frac{1+\rho}{1-\rho}.$$

For "high" P^*, their studies indicate that in order to guarantee (2.1.1) for $0 < \rho < 0.23$ one would need *twice* the sample size required for $\rho = 0$, while for $\rho > 0.9$ one would need up to *19* times the sample size required for $\rho = 0$. Thus, as intuition suggests, choosing n as if $\rho = 0$ will lead to lower-than-desired $P\{CS \mid LF\}$ when $\rho > 0$. [Dudewicz and Zaino (1977) also gave a heuristic selection procedure that could be used for the model (2.2.7) when ρ is unknown.] Although the results that Dudewicz and Zaino cite hold only for the first-order autoregressive model, they clearly demonstrate that if successive observations from the same treatment are correlated, the achieved $P\{CS \mid LF\}$ can be greatly affected for a given sample size.

A second example of the lack of independence occurs if the observations on the t treatments are taken in independent blocks for which there is within-block treatment correlation. Such correlation occurs in the crossover design because all of the t treatments are given to each subject after a suitable "washout" period. This situation also arises in simulation studies when the experimenter intentionally induces positive correlation between competing processes in order to more precisely estimate any differences in process means (see, e.g., Law and Kelton 1991).

As a theoretical framework in which to study within-treatment dependence, suppose that the set of ith observations from the t treatments, $Y_i = (Y_{i1}, \ldots, Y_{it})$, has a multivariate normal distribution with constant correlation ρ but that the n t-vectors of observations, $Y_1, \ldots, Y_n$, are independent. If the correlation $\rho = 0$, then the assumptions of procedure $\mathcal{N}_B$ are satisfied. In the constant correlation case, the $P\{CS \mid LF\}$ is strictly increasing in the true common correlation ρ, and it approaches 1 as $\rho \to 1$. Thus the procedure is conservative ($P\{CS \mid LF\} > P^*$) if $\rho > 0$ and liberal ($P\{CS \mid LF\} < P^*$) if $-1/(t-1) < \rho < 0$.

2.3 MULTI-STAGE PROCEDURES FOR COMMON KNOWN VARIANCE

Sometimes an experimenter may find that the single-stage sample sizes required by procedure $\mathcal{N}_B$ are unaffordable for values of $(\delta^\star, P^\star)$ considered to be important for the number of treatments under consideration. As noted in Section 2.2, the results of Hall (1959) and Eaton (1967) show that there is no alternative *single-stage* procedure that will require smaller sample sizes. In such situations a selection experiment guaranteeing the desired probability requirement (2.1.1) might still be affordable if the experimenter is willing and able to use *multi-stage* or *sequential* procedures. We will use these terms interchangeably; however, we distinguish the attributes of the procedures as follows. We say that a procedure is *open* if, prior to experimentation, no fixed upper bound can be placed on the number of observations to be taken from each treatment; otherwise, it is *closed*. A procedure is *eliminating* if the data taken in prior stages can be used to exclude treatments from further sampling and consideration, while the procedure is *non-eliminating* if data must be collected from all t treatments at each stage, in a vector-at-a-time fashion. Historically, the two earliest sequential procedures were the *open, non-eliminating* procedure $\mathcal{N}_{BKS}$ of Bechhofer, Kiefer and Sobel (1968) and a *closed, eliminating* procedure due to Paulson (1964).

One reason that it might be better to use a procedure with a multi-stage sampling scheme is that the single-stage procedure $\mathcal{N}_B$ can be very conservative. It chooses a sample size n so that $P\{CS\} = P^\star$ when the true means are in a least-favorable configuration, that is, $\mu_{[1]} = \mu_{[t-1]}$ and $\mu_{[t]} - \mu_{[t-1]} = \delta^\star$. If the true configuration $\boldsymbol{\mu}$ of treatment means is as in Figure 2.1(b) with $\mu_{[t]} - \mu_{[t-1]} > \delta^\star$, then the total sample size tn calculated under the LF assumption will result in an achieved $P\{CS\} > P^\star$. The overprotection, $P\{CS\} - P^\star$, can be quite large if $\boldsymbol{\mu}$ is very favorable to the experimenter. Of course, the experimenter does not know $\boldsymbol{\mu}$, although after *part* of the experiment has been conducted the data may indicate that something like Figure 2.1(b) is indeed the true state of affairs. This suggests the possibility of taking the observations in two or more stages and using the data taken in the early stages to suggest the true configuration of the treatment means. If the data taken in these early stages strongly suggest a favorable configuration of the treatment means, then the experimenter may be able to capitalize on this "peek" at Nature and guarantee $P^\star$ with an expected total number of observations smaller than tn. The earlier stages may also indicate that certain treatments are not "in contention" for having the largest treatment mean, and therefore that it is a waste of resources to take additional observations from these treatments.

Motivated by the above considerations, Section 2.3.1 describes a closed two-stage procedure with elimination, procedure $\mathcal{N}_{TB}$, due to Tamhane and Bechhofer (1977, 1979). Section 2.3.2 concerns a closed version of the open procedure $\mathcal{N}_{BKS}$ obtained by truncating the sampling from each treatment. This truncated procedure, denoted by $\mathcal{N}_{BG}$, is due to Bechhofer and Goldsman (1987, 1989a). Procedure $\mathcal{N}_{BG}$ is designed to have improved performance characteristics relative to procedure $\mathcal{N}_{BKS}$. Section 2.3.3 describes a closed multi-stage procedure with elimination, procedure $\mathcal{N}_P$, due to Paulson (1993); this new procedure is related to the original Paulson (1964) procedure and exhibits greatly superior performance characteristics. The operating

2.3.1 A Closed Two-Stage Procedure with Elimination

The simplest procedure having the ability to adapt to the data is a two-stage procedure. Cohen (1959), Alam (1970) and Tamhane and Bechhofer (1977, 1979) introduced two-stage procedures that guarantee the probability (design) requirement (2.1.1), eliminate treatments after the first stage that are indicated as being inferior, and select from among the remaining ones after a second stage of sampling. The first stage of their procedures is related to the subset selection procedure of Gupta (1956, 1965) that will be discussed in Section 3.2. We describe the Tamhane-Bechhofer procedure.

Procedure $\mathcal{N}_{TB}$ For the given t and specified $(\delta^*/\sigma, P^*)$, select constants (c_1, c_2, h) from Table 2.5. Calculate

$$n_1 = \lceil (c_1\sigma/\delta^*)^2 \rceil \quad \text{and} \quad n_2 = \lceil (c_2\sigma/\delta^*)^2 \rceil. \tag{2.3.1}$$

Stage 1: Take a random sample of n_1 observations Y_{ij} ($1 \leq j \leq n_1$) from each Π_i ($1 \leq i \leq t$). Calculate the t first-stage sample means $\overline{y}_i^{(1)} = \sum_{j=1}^{n_1} y_{ij}/n_1$. Let $\overline{y}_{[1]}^{(1)} < \cdots < \overline{y}_{[t]}^{(1)}$ denote their ordered values. If $\overline{y}_{[i]}^{(1)} < \overline{y}_{[t]}^{(1)} - h\sigma/\sqrt{n_1}$, eliminate treatment Π_i from further consideration. If only one treatment remains, then stop sampling at the end of the first stage and select the treatment that produced $\overline{y}_{[t]}^{(1)}$ as the one associated with $\mu_{[t]}$. Otherwise, let $s > 1$ denote the number of remaining treatments and then proceed to the second stage.

Stage 2: Take a random sample of n_2 observations from each of the $s > 1$ remaining treatments, that is, those that produced $\overline{y}_{[t-s+1]}^{(1)}, \ldots, \overline{y}_{[t]}^{(1)}$. Calculate the *cumulative* sample means $\overline{y}_i = \sum_{j=1}^{n_1+n_2} y_{ij}/(n_1 + n_2)$ from these s treatments. Let $\overline{y}_{[1]} < \cdots < \overline{y}_{[s]}$ denote the ordered *cumulative* sample means. Select the treatment that yielded $\overline{y}_{[s]}$ as the one associated with $\mu_{[t]}$.

Clearly, if $\mu_{[t]} \gg \mu_{[t-1]}$, then with high probability, sampling will stop after the first stage with a total of n_1 vector-observations having been taken.

Example 2.3.1 This (artificial) example shows how to carry out procedure $\mathcal{N}_{TB}$. Suppose that $t = 5$, $\sigma = 1.5$, and we specify $(\delta^*, P^*) = (0.5, 0.90)$. Then Table 2.5 yields the constants $c_1 = 1.845$, $c_2 = 1.963$ and $h = 1.537$. Using (2.3.1), we calculate

$$n_1 = \lceil (1.845 \times 1.5/0.5)^2 \rceil = \lceil 30.6 \rceil = 31.$$

Table 2.5. Constants (c_1, c_2, h) Required to Implement Procedure $\mathcal{N}_{TB}$ for Selected t and P^*

t	P^*	c_1	c_2	h	t	P^*	c_1	c_2	h
2	0.75	0.728	0.742	0.820	8	0.75	1.428	1.709	2.049
	0.90	1.427	1.313	0.853		0.90	2.011	2.340	1.342
	0.95	1.862	1.616	0.881		0.95	2.394	2.595	1.321
	0.99	2.719	2.091	0.919		0.99	3.163	2.989	1.294
3	0.75	0.999	0.949	3.989	9	0.75	1.463	1.824	1.861
	0.90	1.578	1.525	2.100		0.90	2.045	2.426	1.304
	0.95	1.999	1.846	1.552		0.95	2.425	2.673	1.315
	0.99	2.791	2.406	1.205		0.99	3.189	3.060	1.309
4	0.75	1.161	1.128	3.545	10	0.75	1.500	1.889	1.570
	0.90	1.760	1.777	1.700		0.90	2.067	2.507	1.342
	0.95	2.139	2.090	1.452		0.95	2.452	2.744	1.322
	0.99	2.965	2.508	1.222		0.99	3.194	3.142	1.322
5	0.75	1.260	1.277	2.821	12	0.75	1.541	2.100	1.468
	0.90	1.845	1.963	1.537		0.90	2.120	2.630	1.256
	0.95	2.252	2.257	1.362		0.95	2.492	2.858	1.318
	0.99	3.043	2.693	1.241		0.99	3.243	3.231	1.349
6	0.75	1.332	1.440	2.556	15	0.75	1.588	2.350	1.364
	0.90	1.916	2.121	1.478		0.90	2.174	2.791	1.330
	0.95	2.307	2.397	1.352		0.95	2.532	3.001	1.346
	0.99	3.087	2.815	1.261		0.99	3.272	3.376	1.384
7	0.75	1.386	1.582	2.276	25	0.75	1.704	2.809	1.256
	0.90	1.969	2.240	1.398		0.90	2.271	3.121	1.358
	0.95	2.355	2.505	1.332		0.95	2.621	3.302	1.411
	0.99	3.130	2.909	1.278		0.99	3.340	3.649	1.463

Reprinted from Tamhane and Bechhofer (1977), pp. 1024–1025, by courtesy of Marcel Dekker, Inc.

Thus, we take $n_1 = 31$ first-stage observations from all $t = 5$ treatments. Suppose that we obtain the following first-stage sample means:

$\overline{y}_1^{(1)}$	$\overline{y}_2^{(1)}$	$\overline{y}_3^{(1)}$	$\overline{y}_4^{(1)}$	$\overline{y}_5^{(1)}$
3.96	6.22	4.16	5.96	5.73

The corresponding ordered first-stage sample means are:

$\overline{y}_{[1]}^{(1)}$	$\overline{y}_{[2]}^{(1)}$	$\overline{y}_{[3]}^{(1)}$	$\overline{y}_{[4]}^{(1)}$	$\overline{y}_{[5]}^{(1)}$
3.96	4.16	5.73	5.96	6.22

The Stage 1 criterion says to eliminate treatment Π_i from further consideration if

$$\overline{y}_{[i]}^{(1)} < \overline{y}_{[t]}^{(1)} - h\sigma/\sqrt{n_1} = 6.22 - (1.537)(1.5)/\sqrt{31} = 5.81.$$

Thus, treatments Π_1, Π_3 and Π_5 are eliminated, whereas the $s = 2$ treatments Π_2 and Π_4 remain in contention. Using (2.3.1) and the rule for Stage 2, we find that

$$n_2 = \lceil (1.963 \times 1.5/0.5)^2 \rceil = \lceil 34.7 \rceil = 35.$$

So we take 35 Stage 2 observations from each of the two remaining treatments. Suppose that we obtain the following *cumulative* sample means after Stage 2 sampling. (Note that these sample means are based on $n_1 + n_2 = 66$ observations.)

$\bar{y}_2$	$\bar{y}_4$
6.70	5.51

In the notation of Stage 2, $\bar{y}_4 = \bar{y}_{[1]} = 5.51$ and $\bar{y}_2 = \bar{y}_{[2]} = 6.70$. Therefore, we select treatment Π_2 as best because it corresponds to the largest cumulative sample mean $\bar{y}_{[s]}$.

Notice that we collected $225 = 3 \times 31 + 2 \times 66$ observations in this example. However, in the (worst) case in which none of the treatments is eliminated in the first stage, it is possible that 330 (i.e., 5×66) observations would be required by procedure $\mathcal{N}_{TB}$. In contrast, the single-stage procedure $\mathcal{N}_B$ would have always required $n = \lceil 2 \times (1.5 Z_{4,1/2}^{(0.10)}/0.5)^2 \rceil = \lceil 2 \times (1.5(1.838)/0.5)^2 \rceil = 61$ observations per treatment, that is, a total of $305 = 5 \times 61$ observations for this same experiment. Thus the two-stage procedure saves observations for this experiment but, in principle, might be *more* costly to run in other circumstances, particularly when the μ_i are "close."

Choice of Constants

There are many choices of constants (n_1, n_2, h) that guarantee the probability requirement (2.1.1). For example, one can choose $(n_1, n_2, h) = (n, 0, 0)$ where n is the sample size used by procedure $\mathcal{N}_B$; this choice of constants yields the single-stage procedure since no treatment can enter the second stage when $h = 0$. This same single-stage procedure is obtained for $(n_1, n_2, h) = (n_1, n_2, \infty)$, where $n_1 + n_2$ is again the sample size used by procedure $\mathcal{N}_B$; then *all* treatments enter the second stage. The constants (n_1, n_2, h) computed from Table 2.5 are chosen in such a way that the probability requirement (2.1.1) is guaranteed *and* the secondary requirement

$$\max_{\boldsymbol{\mu}} E\{T \mid \boldsymbol{\mu}\} = \min_{n_1, n_2, h} \max_{\boldsymbol{\mu}} E\{T \mid \boldsymbol{\mu}\} \qquad (2.3.2)$$

is also guaranteed; here T is the random *total number of observations* used by procedure $\mathcal{N}_{TB}$. The minimum in the right-hand side of (2.3.2) is taken over all (n_1, n_2, h) that satisfy (2.1.1). Equation (2.3.2) shows that procedure $\mathcal{N}_{TB}$ has a *minimax* property. Thus by construction procedure $\mathcal{N}_{TB}$ is a *uniform improvement* over procedure $\mathcal{N}_B$ in the sense that *for all* $\boldsymbol{\mu}$, the expected total number of observations, $E\{T \mid \boldsymbol{\mu}\}$, used by the two-stage procedure, is *strictly less* than the total number, tn, required by the single-stage procedure $\mathcal{N}_B$ to guarantee the *same* probability requirement (2.1.1) for a given $(t; \delta^*/\sigma, P^*)$. Of course, in any given experiment the actual number of observations used by procedure $\mathcal{N}_{TB}$ can be greater than or less than tn.

Miscellaneous Comments

In the same way as for procedure $\mathcal{N}_B$, if the constants are chosen as in (2.3.1), then the confidence statement (2.2.4) holds with confidence coefficient at least P^*. A procedure that allows the experimenter to specify a maximum number of treatments to enter the second stage is provided by Santner and Behaxeteguy (1992).

2.3.2 A Closed Multi-Stage Procedure without Elimination

This subsection discusses a truncated sequential procedure $\mathcal{N}_{BG}$ due to Bechhofer and Goldsman (1987, 1989a). The procedure guarantees the probability requirement (2.1.1) and is designed to have improved performance relative to procedure $\mathcal{N}_{BKS}$ by decreasing $E\{T\}$ uniformly in $(t; \delta^*/\sigma, P^*)$ for all $\boldsymbol{\mu}$. The truncated procedure $\mathcal{N}_{BG}$ sets an upper bound n_0 on the number of vector-observations (or stages) that can be taken; thus the experimenter need not be concerned that procedure $\mathcal{N}_{BG}$ will require an arbitrarily large number of vector-observations.

Procedure $\mathcal{N}_{BG}$ For the given t, σ and specified (δ^*, P^*), find the truncation number n_0 in Table 2.6.

Sampling rule: At the mth stage of experimentation ($m \geq 1$), observe the random vector $(Y_{1m}, \ldots, Y_{tm})$ where Y_{im} is the mth observation from Π_i ($1 \leq i \leq t$).

Stopping rule: After the mth vector ($m \geq 1$) has been taken, calculate $x_{im} = \sum_{j=1}^{m} y_{ij}$ ($1 \leq i \leq t$). Denote the ordered values of the x_{im} by $x_{[1]m} < \cdots < x_{[t]m}$. Stop sampling when, for the first time, *either*

$$z_m \equiv \sum_{i=1}^{t-1} \exp\{-\delta^*(x_{[t]m} - x_{[i]m})/\sigma^2\} \leq (1 - P^*)/P^*$$

or

$$m = n_0$$

whichever occurs first.

Terminal decision rule: Let N denote the value of m at the termination of sampling. Select the treatment that yielded $x_{[t]N}$ as the one associated with $\mu_{[t]}$.

The truncation numbers in Table 2.6 are given for $t = 2(1)6$, $P^* = 0.75, 0.90, 0.95, 0.99$ and $\delta^*/\sigma = 0.2(0.1)0.8$. If it is desired to obtain a truncation number n_0 for a $(t, \sigma; \delta^*, P^*)$ not listed in Table 2.6, a good approximation to n_0 can be obtained (for P^* close to unity) by quadratic interpolation of $\ln(n_0)$ in $1/\delta^*$. This technique is illustrated in Example 2.3.3.

Table 2.6. Truncation Numbers n_0 Required to Implement Procedure $\mathcal{N}_{BG}$ for Selected t, $\delta^\star/\sigma$ and $P^\star$

t	$P^\star$	$\delta^\star/\sigma$						
		0.2	0.3	0.4	0.5	0.6	0.7	0.8
2	0.75	28	12	7	4	3	2	2
	0.90	114	47	25	16	11	8	6
	0.95	190	78	42	26	18	13	10
	0.99	370	154	84	52	35	26	19
3	0.75	68	28	15	10	7	5	4
	0.90	170	70	38	24	16	12	9
	0.95	255	106	56	35	24	17	13
	0.99	440	188	100	62	42	31	23
4	0.75	94	39	21	13	9	7	5
	0.90	205	86	46	29	19	14	11
	0.95	290	122	65	41	28	20	15
	0.99	480	200	110	68	46	34	25
5	0.75	114	47	25	16	11	8	6
	0.90	230	96	51	32	22	16	12
	0.95	320	132	71	44	30	22	17
	0.99	520	216	116	73	49	36	27
6	0.75	130	54	29	18	12	9	7
	0.90	250	104	56	35	24	17	13
	0.95	335	142	76	47	32	24	18
	0.99	530	224	122	76	52	37	29

Reprinted from Bechhofer and Goldsman (1989a), p. 77, by courtesy of Marcel Dekker, Inc.

Example 2.3.2 Suppose that $t = 3$, $\sigma = 2.0$ and that we specify $(\delta^\star, P^\star) = (1.0, 0.75)$. From Table 2.6, the truncation number for procedure $\mathcal{N}_{BG}$ is $n_0 = 10$. Using the following (artificial) data, we show how to apply the procedure.

m	y_{1m}	y_{2m}	y_{3m}	x_{1m}	x_{2m}	x_{3m}	$x_{[1]m}$	$x_{[2]m}$	$x_{[3]m}$	z_m
1	3.06	4.86	1.32	3.06	4.86	1.32	1.32	3.06	4.86	1.050
2	1.49	2.96	2.98	4.55	7.82	4.30	4.30	4.55	7.82	0.856
3	2.01	5.36	2.89	6.56	13.18	7.19	6.56	7.19	13.18	0.415
4	1.42	5.08	3.11	7.98	18.26	10.30	7.98	10.30	18.26	0.213

Since $z_4 \leq (1 - P^\star)/P^\star = 1/3$, we stop sampling after Stage 4 (but not before) and select treatment Π_2 as the best. In doing so, we guarantee the probability requirement (2.1.1).

If the data in this example had been such that termination had not occurred by stage $m = 9$, that is, $z_m > 1/3$ ($1 \leq m \leq 9$), then the truncation number would have forced termination at stage $m = n_0 = 10$.

Example 2.3.3 Suppose that we wish to estimate n_0 for $t = 4$, $\delta^\star = 0.35$, $P^\star = 0.95$. From Table 2.6 we find the estimated n_0-values associated with the three $\delta^\star$-values that are closest to $\delta^\star = 0.35$.

$P^\star$	$\delta^\star/\sigma$	Tabled n_0
0.95	0.4	65
0.95	0.3	122
0.95	0.2	290

Fitting a quadratic equation of the form

$$\ln(n_0) = \beta_0 + \beta_1/\delta^\star + \beta_2/(\delta^\star)^2$$

we obtain

$$\ln(n_0) = 1.4992 + 1.3060/\delta^\star - 0.09438/(\delta^\star)^2.$$

Thus for $\delta^\star = 0.35$, the estimate of $\ln(n_0)$ is 4.4602, or $n_0 = \lceil 86.5 \rceil = 87$. As a check, we also used Monte Carlo simulation to estimate the value of n_0; the estimate proved to be $n_0 = 87$, the same value as estimated for this example by quadratic interpolation.

Clearly, procedure $\mathcal{N}_{BG}$ can react to favorable $\boldsymbol{\mu}$-configurations (that is, widely spaced treatment means) leading to early termination of sampling. In fact, if $\mu_{[t]} - \mu_{[t-1]} \gg \sigma$ then, with high probability, sampling will terminate after the first stage (requiring therefore a total of only t observations). The expected number of stages (or vector-observations) to terminate sampling is a maximum, as might be conjectured, in the equal-means (EM) configuration, that is, $\mu_{[1]} = \mu_{[t]}$; however, even in the EM-configuration the original *untruncated* procedure $\mathcal{N}_{BKS}$ terminates sampling in a finite number of stages with probability one. Estimates of the expected number of stages to terminate sampling, $E\{N_S\}$, as well as the expected total number of observations, $E\{T\}$, are given in Table 2.7 (see p.43) for selected $(t, \sigma; \delta^\star, P^\star)$-combinations and $\boldsymbol{\mu}$-configurations.

Group Sequential Sampling

In some applications it is convenient to collect several observations from each treatment at each sampling stage; this is called *group sequential sampling*. Procedure $\mathcal{N}_{BG}$ can be applied to data in which a common number of observations, say k, are collected at each stage of sampling before application of the stopping rule. In this case the sample treatment means of the observations are computed at each sampling stage and play the role of the y_{ij} in the stopping rule; σ^2/k plays the role of the variance. We emphasize that there is one restriction when applying procedure $\mathcal{N}_{BG}$ in this manner: *The same number of observations must be collected from each treatment at all stages*. The reason is that the procedure assumes a common known variance, and this will be the case only if a common number of observations are collected at each stage. Group sequential sampling can increase the expected number of vector-observations to termination relative to one-vector-at-a-time sampling. This increase occurs when termination using the stopping rule could have taken place for one of the vectors within the final group, and it is an especially important issue if the group size k is large.

2.3.3 A Closed Multi-Stage Procedure with Elimination

Recall that a closed procedure has a known upper bound on the number of stages beyond which the procedure will not sample. This section describes a closed procedure that has the additional virtue that it can *eliminate* treatments that sampling indicates are not in contention for best. We describe a recent closed multi-stage procedure with elimination, denoted as procedure $\mathcal{N}_P$, due to Paulson (1993). Procedure $\mathcal{N}_P$ yields impressive savings relative to the earliest eliminating multi-stage procedure, Paulson (1964), in terms of decreased expected number of stages, $E\{N_S\}$, and expected total observations taken, $E\{T\}$, uniformly in $(t, \sigma; \delta^\star, P^\star)$ for all $\boldsymbol{\mu}$.

We introduce notation in terms of which the procedure is described. At the end of the mth stage of Paulson's procedure, we let R_m ($m \geq 1$) denote the set of indices of treatments still in contention (not yet eliminated); thus $R_1 = \{1, 2, \ldots, t\}$, initially. Also let N_j ($1 \leq j \leq t$) denote the stage at which treatment Π_j is eliminated; if Π_j has not yet been eliminated, we set $N_j = \infty$.

Procedure $\mathcal{N}_P$ For the given t, σ and specified $(\delta^\star, P^\star)$, choose an operating constant $0 < c < 1$.

Sampling rule: At the mth stage of experimentation ($m \geq 1$), observe the random vector $(Y_{im}: i \in R_m)$; here Y_{im} is the mth observation from Π_i ($i \in R_m$).

Stopping rule: After the mth vector-observation ($m \geq 1$) has been taken, calculate $x_{im} = \sum_{j=1}^{m} y_{ij}$ ($i \in R_m$). For each non-eliminated treatment $i \in R_m$ ($m \geq 2$), define

$$L(i, m) = \operatorname{argmax}_{1 \leq \ell \leq t} \{x_{\ell, m-1} \mid \ell \neq i \text{ and } \ell \in R_m\}$$

as the best (in the sense of maximum sample total) of the rest of the treatments remaining at stage m. Further define $x'_{i1} = 0$ and

$$x'_{im} = \sum_{j=2}^{m} y_{L(i,j),j} \quad (m \geq 2).$$

Eliminate from further sampling and consideration (and set $N_i = m$) any treatment Π_i ($i \in R_m$) for which

$$z_{im} \equiv \sum_{j=1}^{t} \exp\left[\frac{c\delta^\star}{\sigma^2}\left(x_{j,\min(m,N_j)} - x_{im} + x'_{im} - x'_{i,\min(m,N_j)}\right)\right]$$

$$> z_m \equiv \left(\frac{t-1}{1-P^\star}\right) \exp\left\{\frac{-mc(1-c)(\delta^\star)^2}{\sigma^2}\right\}.$$

Stop sampling when, for the first time, only one treatment remains. Otherwise, proceed to the $(m + 1)$st stage.

Terminal decision rule: Select the remaining treatment as the one associated with $\mu_{[t]}$.

While any choice of c ($0 < c < 1$) guarantees the probability requirement, Paulson recommends using $c = 0.85$. Our simulations of the performance of the procedure $\mathcal{N}_P$ show that this value is a good choice. The procedure $\mathcal{N}_P$ is closed because the number of stages required to terminate cannot exceed

$$1 + \frac{\sigma^2 \ln(\frac{t-1}{1-P^\star})}{c(1-c)(\delta^\star)^2}.$$

Roughly, procedure $\mathcal{N}_P$ compares each remaining treatment $i \in R_m$ with *all* other treatments through their sample totals after "adjusting" for the fact that some of the treatments may have terminated sampling.

Example 2.3.4 We show how to implement procedure $\mathcal{N}_P$ with an explicit numerical example. Suppose $t = 3$, $(\delta^\star, P^\star) = (1.0, 0.75)$ and $\sigma = 1$. Using Paulson's recommended choice for the operating constant, $c = 0.85$, we eliminate treatment Π_i ($1 \le i \le 3$) after stage m if

$$z_{im} = \sum_{j=1}^{t} \exp\left[0.85\left(x_{j,\min(m,N_j)} - x_{im} + x'_{im} - x'_{i,\min(m,N_j)}\right)\right] > z_m = 8.0(0.8803)^m.$$

With this elimination criterion in mind, suppose that we observe the following data sequence.

	Stage			
m	1	2	3	4
R_m	$\{1,2,3\}$	$\{1,2\}$	$\{1,2\}$	$\{1,2\}$
y_{1m}	2.1182	-0.8468	-0.2392	-0.5487
y_{2m}	1.0223	-0.9743	1.8442	1.8766
y_{3m}	0.3425	—	—	—
x_{1m}	2.1182	1.2714	1.0322	0.4835
x_{2m}	1.0223	0.0480	1.8922	3.7689
x_{3m}	0.3425	—	—	—
$L(1,m)$	—	2	2	2
$L(2,m)$	—	1	1	1
$L(3,m)$	—	—	—	—
x'_{1m}	0	-0.9743	0.8699	2.7465
x'_{2m}	0	-0.8468	-1.0860	-1.6347
x'_{3m}	0	—	—	—
z_m	7.0423	6.1993	5.4572	4.8040
z_{1m}	1.6150	1.5518	4.2429	26.4825
z_{2m}	4.0995	4.4543	1.5878	1.0748
z_{3m}	7.3059	—	—	—

At the end of Stage 3, we find that $N_1 = \infty = N_2$ and $N_3 = 1$. Then we can calculate, for example,

$$z_{1,4} = \sum_{j=1}^{3} \exp\left[0.85\left(x_{j,\min(4,N_j)} - x_{1,4} + x'_{1,4} - x'_{1,\min(4,N_j)}\right)\right]$$

$$= \exp[0.85(x_{1,4} - x_{1,4} + x'_{1,4} - x'_{1,4})] + \exp[0.85(x_{2,4} - x_{1,4} + x'_{1,4} - x'_{1,4})]$$

$$+ \exp[0.85(x_{3,1} - x_{1,4} + x'_{1,4} - x'_{1,1})]$$

$$= 1 + \exp[0.85(3.7689 - 0.4835)]$$

$$+ \exp[0.85(0.3425 - 0.4835 + 2.7465 - 0)]$$

$$= 26.48 > 4.80 = z_4.$$

For this reason, we eliminate treatment Π_1 at stage $m = 4$ and set $N_1 = 4$. Because only treatment Π_2 remains, we declare that it has the largest mean.

Blocking for Procedures $\mathcal{N}_{TB}$, $\mathcal{N}_{BG}$ and $\mathcal{N}_P$

Procedures $\mathcal{N}_{TB}$, $\mathcal{N}_{BG}$ and $\mathcal{N}_P$ can also be carried out using certain blocking designs. The reason for using these designs is the same as that for procedure $\mathcal{N}_B$, as explained in Section 2.2.2. We illustrate these ideas for the *randomized complete blocks* design.

For procedure $\mathcal{N}_{TB}$ the n_1 observations taken in Stage 1 from each of the t treatments can be taken in n_1 blocks, each of size t, that is, $(\overline{y}_{1j}^{(1)}, \ldots, \overline{y}_{tj}^{(1)})$ $(1 \leq j \leq n_1)$. Similarly, the n_2 observations taken in Stage 2 from each of the $s \geq 2$ treatments which enter the second stage should also be taken in vectors, corresponding to blocks of size s. We might denote the second-stage data from the jth block as $(\overline{y}_{1j}^{(2)}, \ldots, \overline{y}_{sj}^{(2)})$ $(1 \leq j \leq n_2)$ where, without loss of generality, we have relabeled the subscripts of the s treatments that entered the second stage as $1, \ldots, s$. Then the analysis can be carried out as in Section 2.3.1 for the s treatments that entered the second stage, the terminal decision being based on the cumulative sample means, with $n_1 + n_2$ observations contributing to each.

For procedures $\mathcal{N}_{BG}$ and $\mathcal{N}_P$, all of the observations in any *stage* should be taken from the same *block*, with the block allowed to change from stage to stage. Then the analysis can be carried out as in Section 2.2.2.

2.4 COMPARISON OF COMMON KNOWN VARIANCE PROCEDURES

Criteria for Comparisons

Sections 2.2 and 2.3 described four procedures ($\mathcal{N}_B$, $\mathcal{N}_{TB}$, $\mathcal{N}_{BG}$ and $\mathcal{N}_P$) all of which require the same statistical assumptions (in particular, a *common known* variance σ^2), achieve Goal 2.1.1, and guarantee the same probability requirement (2.1.1). However, the procedures are fundamentally different, and each plays a different role: Procedures $\mathcal{N}_B$ and $\mathcal{N}_{TB}$ are single-stage and two-stage, respectively, whereas procedures $\mathcal{N}_{BG}$

and $\mathcal{N}_P$ are closed multi-stage procedures (the number of stages usually being fairly large). The particular real-life situation in which a procedure is to be used will often dictate the one (or ones) that might be appropriate. For example, if the procedure is to be applied in an agricultural setting, say to determine the variety of grain that will produce the largest yield, the experimenter is usually limited to single-stage experiments where a stage is a growing season. If the experiment can extend over two growing seasons, the experimenter might well consider a two-stage procedure with a possible resultant reduction in the number of varieties to be tested in the second year, and an associated reduction in the total number of plots required over the two-year period. Multi-stage procedures are usually more feasible for scientific and engineering experiments as well as for certain classes of medical experiments.

Procedures can differ with respect to the number of *stages* required, the number of *vector-observations* required, and the *total* number of observations required. For a given procedure, let N_S denote the number of stages, N_V the number of vector-observations and T the total number of observations it requires. The number of stages a procedure uses may differ from the number of vector-observations it takes. For example, procedure $\mathcal{N}_B$ always requires $N_S = 1$ stage, $N_V = n$ vector-observations (from Table 2.1), and $T = t \times n$ total observations. By way of comparison, procedure $\mathcal{N}_{TB}$ always requires $N_S \leq 2$ stages, at most $N_V = n_1 + n_2$ vector-observations, and $t \times n_1 \leq T \leq t \times (n_1 + n_2)$ total observations (because some treatments may have been eliminated after the first stage). Procedure $\mathcal{N}_{BG}$ always requires the same number of stages, N_S, as vector-observations, N_V, and a total of $T = t \times N_S$ observations. Lastly, procedure $\mathcal{N}_P$ also always requires the same number of stages, N_S, as vector-observations, N_V, but with a total of $T \leq t \times N_S$ observations (because some treatments may have been eliminated during the course of sampling).

This section compares the four procedures in terms of their achieved $P\{CS\}$, $E\{N_S\}$ and $E\{T\}$ for various configurations of the treatment means. The choice of the most relevant performance measures will depend on the particular application under study and associated costs. The quantity $E\{N_S\}$ can often be equated to the expected *duration* of the study since one stage of an experiment can be performed in a given unit of time; $E\{T\}$ corresponds to the expected total amount of experimental resources used. For example, in an agricultural experiment, a growing season will constitute the natural time unit and multiple observations can be collected simultaneously from the set of test plots available for each variety. In this case it is most important to minimize the total number of stages. In a clinical trial, ethical considerations suggest that it is more important to minimize the total number of patients.

The configurations considered in this section are the equally-spaced δ^*/σ-apart configuration (ES(δ^*/σ) = $(0, \delta^*/\sigma, 2\delta^*/\sigma, \ldots, t\delta^*/\sigma)$), the least-favorable configuration with the largest mean δ^*/σ greater than the other means (LF = $(0, 0, \ldots, \delta^*/\sigma)$) and the equal-means configuration (EM = $(0, \ldots, 0)$).

Summary of Comparisons

In general, the performance measure $E\{T\}$ improves as one moves from single-stage to two-stage to multi-stage procedures. Thus, if a two-stage procedure is feasible for the problem at hand, procedure $\mathcal{N}_{TB}$ would appear to be preferable over procedure $\mathcal{N}_B$ for minimizing $E\{T\}$. If multi-stage procedures are feasible, then the experimenter should consider procedures $\mathcal{N}_{BG}$ or $\mathcal{N}_P$. Neither of these two sequential procedures dominates the other one for *all* configurations of $\boldsymbol{\mu}$ in terms of minimizing both $E\{N_S\}$ ($= E\{N_V\}$) and $E\{T\}$. However, we can make the following *general statement* concerning the relative performances of these two multi-stage procedures:

- To minimize $E\{T\}$, use procedure $\mathcal{N}_P$.
- To minimize $E\{N_S\}$, use procedure $\mathcal{N}_{BG}$.

Details

Table 2.7 can serve as a partial guide to the relative performances of the procedures $\mathcal{N}_B$, $\mathcal{N}_{TB}$, $\mathcal{N}_{BG}$ and $\mathcal{N}_P$. This table is abstracted from a large simulation study in Bechhofer and Goldsman (1989b). The values listed in Table 2.7 are Monte Carlo (MC) estimates of the quantities of interest (except for procedure $\mathcal{N}_B$ for which the calculations are exact); below each estimate (in parentheses) is the estimated standard error of the value above it. The table includes results for nominal $P^* = 0.75, 0.90, 0.95, 0.99$. Calculations are given for $t = 4$ and $\delta^*/\sigma = 0.2$ (which can be thought of as representative values of these quantities). Note that the $P\{CS\}$ is estimated only for the two configurations $ES(\delta^*/\sigma)$ and LF since these are both in the preference-zone defined by the probability requirement (2.1.1); all of the procedures guarantee the nominal P^* although some *overprotect* by a sizable amount. Table 2.7 uses the notation $\hat{p}$, $\hat{N}_S$ and $\hat{T}$ to denote our MC estimates of the $P\{CS\}$, $E\{N_S\}$ and $E\{T\}$, respectively.

We see that procedure $\mathcal{N}_{TB}$ dominates procedure $\mathcal{N}_B$ in terms of smaller estimated $E\{T\}$ uniformly in t, δ^*/σ, P^* and $\boldsymbol{\mu}$. Of course, $\mathcal{N}_{TB}$ was constructed with this attribute as the objective. Although Table 2.7 was constructed based on data for $t = 4$ and $\delta^*/\sigma = 0.2$, similar comparisons among the four procedures indicate that the *regions of dominance* remain essentially the same for different t and/or δ^*/σ. Clearly, any rough insights that the experimenter may have concerning $\boldsymbol{\mu}$ can provide a guide to the expected cost of experimentation and can assist in making an informed choice concerning the procedure to use.

2.5 A SINGLE-STAGE PROCEDURE FOR SELECTING THE s BEST OF t TREATMENTS HAVING A COMMON KNOWN VARIANCE

In certain applications, alternatives to Goal 2.1.1 may be appropriate. In this section, we consider three alternative goals and associated probability requirements to those of Section 2.2.1. The first of these goals, one that we have already considered, is that of selecting any treatment whose mean μ_i is within δ^* of $\mu_{[t]}$ subject to the design requirement that a correct selection must occur with probability at least P^* for *any* $\boldsymbol{\mu}$; for example, Equation (2.2.4) shows that procedure $\mathcal{N}_B$ achieves this goal. Thus, this goal is relevant when the experimenter is willing to accept any treatment

A SINGLE-STAGE PROCEDURE FOR THE s BEST

Table 2.7. Estimated Achieved $P\{CS\}$, $E\{N_S\}$ and $E\{T\}$ Under the ES($\delta^\star/\sigma$)-, LF- and EM-Configurations for Procedures $\mathcal{N}_B$, $\mathcal{N}_{TB}$, $\mathcal{N}_{BG}$ and $\mathcal{N}_P$ When $t = 4$ and $\delta^\star/\sigma = 0.2$

$P^\star$	Procedure	ES($\delta^\star/\sigma$)			LF			EM	
		$\hat{p}$	$\hat{N}_S$	$\hat{T}$	$\hat{p}$	$\hat{N}_S$	$\hat{T}$	$\hat{N}_S$	$\hat{T}$
0.75	$\mathcal{N}_B, n = 71$	0.8796	1	284	0.7508	1	284	1	284
	$\mathcal{N}_{TB}, h = 0.611$	0.8729	≤2	237.9	0.7361	≤2	258.9	≤2	261.8
	$n_1 = 34, n_2 = 32$	(0.0030)		(0.3)	(0.0028)		(≤0.1)		(≤0.1)
	$\mathcal{N}_{BG}$	0.8439	33.6	134.5	0.7507	50.2	200.8	60.5	242.2
	$n_0 = 94$	(0.0033)	(0.2)	(0.7)	(0.0004)	(0.1)	(0.4)	(0.2)	(1.0)
	$\mathcal{N}_P$	0.8984	50.5	138.7	0.7781	63.5	180.8	76.8	214.3
	$c = 0.85$	(0.0020)	(0.2)	(0.4)	(0.0027)	(0.2)	(0.5)	(0.2)	(0.6)
0.90	$\mathcal{N}_B, n = 151$	0.9588	1	604	0.9008	1	604	1	604
	$\mathcal{N}_{TB}, h = 1.71$	0.9583	≤2	394.7	0.9035	≤2	474.7	≤2	545.0
	$n_1 = 78, n_2 = 79$	(0.0018)		(0.8)	(0.0019)		(0.7)		(0.8)
	$\mathcal{N}_{BG}$	0.9350	61.8	247.0	0.9006	93.1	372.6	136.7	546.6
	$n_0 = 205$	(0.0023)	(0.3)	(1.3)	(0.0001)	(0.1)	(0.4)	(0.5)	(2.2)
	$\mathcal{N}_P$	0.9682	83.7	233.0	0.9134	108.8	326.4	157.7	455.2
	$c = 0.85$	(0.0011)	(0.3)	(0.6)	(0.0018)	(0.3)	(0.9)	(0.4)	(1.1)
0.95	$\mathcal{N}_B, n = 213$	0.9805	1	852	0.9502	1	852	1	852
	$\mathcal{N}_{TB}, h = 1.46$	0.9794	≤2	530.3	0.9495	≤2	614.2	≤2	748.8
	$n_1 = 115, n_2 = 110$	(0.0013)		(1.0)	(0.0014)		(1.0)		(1.2)
	$\mathcal{N}_{BG}$	0.9628	81.0	324.2	0.9504	120.6	482.6	202.9	811.7
	$n_0 = 290$	(0.0017)	(0.4)	(1.8)	(0.0002)	(0.3)	(1.2)	(0.8)	(3.0)
	$\mathcal{N}_P$	0.9848	105.8	296.2	0.9558	137.1	423.8	224.6	659.2
	$c = 0.85$	(0.0008)	(0.4)	(0.8)	(0.0013)	(0.4)	(1.1)	(0.6)	(1.6)
0.99	$\mathcal{N}_B, n = 361$	0.9964	1	1444	0.9901	1	1444	1	1444
	$\mathcal{N}_{TB}, h = 1.22$	0.9963	≤2	912.5	0.9894	≤2	964.8	≤2	1242.5
	$n_1 = 220, n_2 = 158$	(0.0006)		(0.9)	(0.0007)		(1.0)		(1.9)
	$\mathcal{N}_{BG}$	0.9921	122.4	489.6	0.9902	175.8	703.0	370.8	1483.3
	$n_0 = 480$	(0.0008)	(0.6)	(2.5)	(<0.0001)	(0.4)	(1.6)	(1.2)	(4.7)
	$\mathcal{N}_P$	0.9963	149.7	425.1	0.9914	192.1	622.6	389.3	1177.9
	$c = 0.85$	(0.0004)	(0.5)	(1.0)	(0.0006)	(0.5)	(1.5)	(1.0)	(2.6)

Reprinted from Bechhofer and Goldsman (1989b), pp. 310–313, by courtesy of Springer-Verlag, Inc.

that is essentially equivalent to that corresponding to $\mu_{[t]}$. Selection of treatments whose means are near-best in the sense that their means are within $\delta^\star$ of $\mu_{[t]}$ will be discussed for other formulations in Section 3.4.

The purpose of this section is to provide procedures for the other two alternative goals, both of which involve selection of the s ($s \geq 2$) best treatments. Specifically, Section 2.5.1 concerns selection of the s best treatments without regard to order, and Section 2.5.2 deals with selection of the s best with regard to order. Both of the latter two goals are generalizations of Goal 2.1.1.

2.5.1 Selection without Regard to Order

Bechhofer (1954) considered the following goal and procedure.

Goal 2.5.1 To select the s ($1 \leq s \leq t-1$) treatments associated with $\mu_{[t-s+1]}, \ldots, \mu_{[t]}$ *without regard to order.*

A *correct selection* (CS) is said to be made if Goal 2.5.1 is achieved. Notice that there is no guarantee that all of the best s treatments need be particularly "good." In fact, it might be the case that none of the $s - 1$ next best treatments have means close to the largest mean, that is, $\mu_{[t]} - \mu_{[t-1]}$ may be large. Thus, the experimenter should consider carefully whether Goal 2.5.1 or even Goal 2.5.2 (below) is appropriate for the application at hand. With this warning we proceed to describe the probability requirement we use for Goal 2.5.1.

Probability Requirement: For specified integer s ($1 \leq s \leq t - 1$) and constants (δ^*, P^*) with $0 < \delta^* < \infty$, $1/\binom{t}{s} < P^* < 1$, we require

$$P\{CS\} \geq P^* \quad \text{whenever } \mu_{[t-s+1]} - \mu_{[t-s]} \geq \delta^*.$$

Procedure $\mathcal{N}_B$ (Selection of s Best without Regard to Order)
For the given t, s, σ and specified (δ^*, P^*), set $n = \lceil (\sigma c_{t,s}^{(P^*)}/\delta^*)^2 \rceil$, where $c_{t,s}^{(P^*)}$ is from Table 2.8 or is the solution to Equation (2.5.2).

Sampling rule: Take a random sample of n observations Y_{ij} ($1 \leq j \leq n$) in a *single* stage from each Π_i ($1 \leq i \leq t$).

Terminal decision rule: Calculate the t sample means $\bar{y}_i = \sum_{j=1}^{n} y_{ij}/n$. Let $\bar{y}_{[1]} < \cdots < \bar{y}_{[t]}$ denote the ordered values of the $\bar{y}_i$. Select the treatments associated with $\{\bar{y}_{[t-s+1]}, \ldots, \bar{y}_{[t]}\}$ as the ones associated with $\{\mu_{[t-s+1]}, \ldots, \mu_{[t]}\}$.

The entries in Table 2.8 are calculated under the assumption that the treatment means are in an LF-configuration which, for Goal 2.5.1, is

$$\mu_{[1]} = \mu_{[t-s]} = \mu_{[t-s+1]} + \delta^* = \mu_{[t]} + \delta^*$$

The value $c = c_{t,s}^{(P^*)}$ is the solution of the equation

$$s \int_{-\infty}^{\infty} \Phi^{t-s}(x+c)[1 - \Phi(x)]^{s-1} \, d\Phi(x) = P^* \qquad (2.5.1)$$

which can be expressed equivalently in probabilistic terms as

$$s \times P\{U_j \leq c/\sqrt{2} \ (1 \leq j \leq t-s);\ U_j \geq 0 \ (t-s+1 \leq j < t)\} = P^* \qquad (2.5.2)$$

where $(U_1, \ldots, U_{t-1})$ has the multivariate normal distribution with mean vector zero, unit variances and common correlation $1/2$. Equation (2.5.2) can be solved for c using the FORTRAN program MVNPRD in Dunnett (1989).

Table 2.8. Constants $c_{t,s}^{(P^*)}$ Required to Calculate the Smallest Single-Stage Sample Size per Treatment to Guarantee $P\{CS \mid LF\} \geq P^*$ When Selecting the s Treatments with the Largest Means from t Normal Treatments *without* Regard to Order

	P^*					
(t, s)	0.60	0.75	0.85	0.90	0.95	0.99
(4, 2)	1.406	1.904	2.338	2.635	3.081	3.932
(5, 2) or (5, 3)	1.671	2.147	2.564	2.850	3.280	4.106
(6, 2) or (6, 4)	1.844	2.309	2.715	2.995	3.415	4.224
(7, 2) or (7, 5)	1.972	2.428	2.827	3.102	3.516	4.314
(8, 2) or (8, 6)	2.072	2.522	2.916	3.188	3.597	4.386
(9, 2) or (9, 7)	2.154	2.598	2.989	3.258	3.663	4.445
(10, 2) or (10, 8)	2.222	2.663	3.051	3.318	3.720	4.496
(6, 3)	1.934	2.389	2.788	3.063	3.477	4.276
(7, 3) or (7, 4)	2.107	2.549	2.937	3.205	3.610	4.393
(8, 3) or (8, 5)	2.234	2.667	3.048	3.311	3.709	4.481
(9, 3) or (9, 6)	2.334	2.760	3.136	3.395	3.788	4.551
(10, 3) or (10, 7)	2.415	2.836	3.208	3.465	3.854	4.610
(8, 4)	2.279	2.707	3.085	3.346	3.741	4.508
(9, 4) or (9, 5)	2.406	2.825	3.195	3.452	3.840	4.595
(10, 4) or (10, 6)	2.505	2.917	3.282	3.535	3.918	4.665
(10, 5)	2.532	2.942	3.305	3.556	3.938	4.681

Reprinted from Bechhofer (1954), pp. 30–34, by courtesy of the Institute of Mathematical Statistics.

Example 2.5.1 Suppose that it is desired to select the two out of six (or four out of six) normal treatments with the largest treatment means using procedure $\mathcal{N}_B$ for Goal 2.5.1 when $\mu_{[5]} - \mu_{[4]} = \delta^* = 3$ and σ is known to be 4.5. What common sample size must be taken from each of the six treatments if the probability requirement for Goal 2.5.1 is to be guaranteed with $P^* = 0.95$? Referring to Table 2.8 in the column headed 0.95 and the row headed "(6,2) or (6,4)" we find that $c_{6,2}^{(0.95)} = 3.415$. Thus $n = \lceil (4.5 \times 3.415/3)^2 \rceil = \lceil 26.2 \rceil = 27$ observations must be taken from each treatment.

Remark 2.5.1 The sample size necessary to guarantee Goal 2.5.1 for specified (δ^*, P^*) when the common variance σ^2 is known may be prohibitive. Mahamunulu (1967) proposed a weaker goal that generalized Goal 2.5.1 and required smaller sample sizes. The new goal is to select a *fixed-size* subset of r treatments that contains at least q of the s best treatments. Here, q, r, s are integers such that $\max\{1, r + s + 1 - t\} \leq q \leq \min\{r, s\}$. (Goal 2.5.1 corresponds to $q = r = s$.) Two special cases of the general goal are of particular interest. The first is to select a fixed-size subset of r treatments that contains the s best treatments ($q = r$ with $r \geq s$); the second goal is to select a fixed-size subset of s treatments that includes any r of the s best treatments ($q = s$ with $r \leq s$). The sample sizes necessary to implement these goals can be determined using Table D in Milton (1970).

2.5.2 Selection with Regard to Order

Next we consider a stronger version of Goal 2.5.1.

Goal 2.5.2 To select the s $(1 \leq s \leq t - 1)$ treatments associated with $\mu_{[t-s+1]}, \ldots, \mu_{[t]}$ *with regard to order*.

A CS is said to have been made if Goal 2.5.2 is achieved. Notice that if $s = 1$, then Goals 2.5.1 and 2.5.2 coincide (as well as Goal 2.1.1); of special interest is the case $s = t - 1$ which calls for a *complete ordering* of the μ_i $(1 \leq i \leq t)$.

Probability Requirement: For specified integer s $(1 \leq s \leq t - 1)$ and constants $(\delta_1^\star, \ldots, \delta_s^\star, P^\star)$ with $0 < \delta_i^\star < \infty$ $(1 \leq i \leq s)$ and $(t - s)!/t! < P^\star < 1$, we require

$$P\{CS\} \geq P^\star \quad \text{whenever} \quad \mu_{[t-s+i]} - \mu_{[t-s+i-1]} \geq \delta_i^\star \quad (1 \leq i \leq s). \quad (2.5.3)$$

The most important special case of the probability requirement (2.5.3) is when $\delta_i^\star = \delta^\star$ $(1 \leq i \leq s)$. For example, below we present a procedure for determining a complete ordering $(s = t - 1)$ that has $P\{CS\} \geq P^\star$ when all successive treatment differences are at least $\delta^\star$.

Procedure $\mathcal{N}_B$ (Selection of $s = t - 1$ Best with Regard to Order—Complete Ordering)

For the given t, σ and specified $(\delta^\star, P^\star)$, set $n = \lceil (\sigma d_t^{(P^\star)}/\delta^\star)^2 \rceil$ where the constant $d_t^{(P^\star)}$ is determined as described below.

Sampling rule: Take a random sample of n observations Y_{ij} $(1 \leq j \leq n)$ in a *single* stage from each Π_i $(1 \leq i \leq t)$.

Terminal decision rule: Calculate $\bar{y}_i = \sum_{j=1}^n y_{ij}/n$ $(1 \leq i \leq t)$. Let $\bar{y}_{[1]} < \cdots < \bar{y}_{[t]}$ denote the ordered values of the $\bar{y}_i$. Associate the treatment yielding $\bar{y}_{[i]}$ with that having mean $\mu_{[i]}$ $(1 \leq i \leq t)$.

Table 2.9 gives values of $d_t^{(P^\star)}$ required to implement procedure $\mathcal{N}_B$; the table is abstracted from Table P.1 of Gibbons, Olkin and Sobel (1977). Each entry in the body of Table 2.9 is the probability, P, of a correct complete ordering of t normal treatments as a function of t and $d_t^{(P^\star)} = \sqrt{n}\delta^\star/\sigma$ when the treatment means are equally-spaced $\delta^\star$ apart. To use Table 2.9 to determine sample sizes corresponding to a given $(t, \sigma; \delta^\star, P^\star)$, the experimenter first enters the table in the column under the t of interest and finds the largest P_1 and P_2 such that $P_1 < P^\star < P_2$. Let $d_t^{(P_1)}$ and $d_t^{(P_2)}$ be the associated row constants; these values must satisfy $d_t^{(P_1)} < d_t^{(P_2)}$. Use linear interpolation to determine the approximate value of the constant $d_t^{(P^\star)}$ associated with the specified $P^\star$. Then $n = \lceil (d_t^{(P^\star)}\sigma/\delta^\star)^2 \rceil$ is the required common sample size.

Table 2.9. Probability (P) of a Correct Complete Ordering of t Normal Treatments with Respect to Their Treatment Means When Procedure $\mathcal{N}_B$ Is Used for Goal 2.5.2, the Treatment Means Are Equally-Spaced δ^* Apart, the Treatments Have a Common Known Variance σ^2, and n Independent Observations Are Taken from Each Treatment

$d_t^{(P)}$	t								
	2	3	4	5	6	7	8	9	10
0.0	0.500	0.167	0.041	0.008	0.001	0.000	0.000	0.000	0.000
0.1	0.528	0.196	0.056	0.014	0.003	0.001	0.000	0.000	0.000
0.2	0.556	0.228	0.077	0.023	0.006	0.002	0.000	0.000	0.000
0.3	0.584	0.263	0.101	0.036	0.012	0.004	0.001	0.000	0.000
0.4	0.611	0.299	0.130	0.052	0.020	0.008	0.003	0.001	0.000
0.5	0.638	0.337	0.162	0.074	0.033	0.014	0.006	0.003	0.001
0.6	0.664	0.376	0.192	0.100	0.050	0.025	0.012	0.006	0.003
0.7	0.690	0.416	0.237	0.132	0.073	0.040	0.022	0.012	0.006
0.8	0.714	0.456	0.279	0.168	0.101	0.060	0.036	0.021	0.013
0.9	0.738	0.496	0.324	0.208	0.134	0.086	0.055	0.035	0.022
1.0	0.760	0.536	0.369	0.252	0.172	0.117	0.080	0.054	0.037
1.1	0.782	0.574	0.415	0.298	0.214	0.154	0.110	0.079	0.057
1.2	0.802	0.612	0.461	0.346	0.260	0.195	0.146	0.110	0.082
1.3	0.821	0.647	0.506	0.395	0.308	0.240	0.187	0.146	0.114
1.4	0.839	0.681	0.550	0.444	0.358	0.288	0.232	0.187	0.151
1.5	0.855	0.714	0.593	0.492	0.408	0.338	0.281	0.233	0.193
1.6	0.871	0.744	0.633	0.539	0.458	0.390	0.332	0.282	0.240
1.7	0.885	0.772	0.671	0.584	0.507	0.441	0.384	0.334	0.290
1.8	0.898	0.797	0.707	0.626	0.555	0.492	0.436	0.386	0.342
1.9	0.910	0.821	0.740	0.667	0.601	0.541	0.488	0.439	0.396
2.0	0.921	0.843	0.770	0.704	0.644	0.589	0.538	0.492	0.450
2.1	0.931	0.862	0.798	0.739	0.684	0.633	0.586	0.543	0.503
2.2	0.940	0.880	0.824	0.771	0.722	0.676	0.632	0.592	0.554
2.3	0.948	0.896	0.846	0.800	0.756	0.715	0.675	0.638	0.603
2.4	0.955	0.910	0.867	0.826	0.788	0.750	0.715	0.681	0.649
2.5	0.961	0.923	0.886	0.850	0.816	0.783	0.752	0.721	0.692

Reprinted from Gibbons, Olkin and Sobel (1977), p. 489, by courtesy of John Wiley & Sons.

Example 2.5.2 Suppose that it is desired to achieve a correct complete ordering of the μ_i when these means are equally-spaced δ^* units apart. How large must n be to guarantee $P^* = 0.75$ for $t = 6$ when $\delta^*/\sigma = 0.2$? From Table 2.9, we obtain

$d_6^{(P)}$	P
2.2	0.722
$d_6^{(0.75)} = ?$	$0.75 =$ target P^*
2.3	0.756

Using linear interpolation we calculate that $d_6^{(0.75)} = 2.282$ which we set equal to $\sqrt{n}\delta^*/\sigma$. Then for $\delta^*/\sigma = 0.2$, we find that $n = \lceil (2.282/0.2)^2 \rceil = \lceil 130.2 \rceil = 131$.

Table 2.9 can also be used to determine the probability of a correct complete ordering for a given equally-spaced configuration when the sample size has been chosen using other considerations. The final example of this section illustrates this use.

Example 2.5.3 Suppose that it is desired to completely order four normal treatments with respect to their treatment means when the treatment means are equally-spaced 2 units apart and procedure $\mathcal{N}_B$ for Goal 2.5.2 is used. If σ is known to be 10 units, and $n = 60$ independent observations are taken from each treatment, what probability of a correct complete ordering can be achieved? Since $\sqrt{n}\delta^*/\sigma = \sqrt{60}(2)/10 = 1.549$ we see from Table 2.9 for $t = 4$ that the achieved probability of a correct complete ordering, P, satisfies $0.593 < P < 0.633$. For these same treatment parameters, if $n = 100$ we obtain $\sqrt{n}\delta^*/\sigma = 2$ and $P = 0.770$ is achieved, while $n = 225$ yields $P = 0.950$. Thus by taking n sufficiently large it is clear that any desired probability of a correct complete ordering can be guaranteed. However, inspection of Table 2.9 shows that very large sample sizes are required to guarantee a moderately high probability if δ^*/σ is small and t is large.

2.6 A SINGLE-STAGE PROCEDURE FOR UNEQUAL KNOWN VARIANCES

This section returns to the consideration of Goal 2.1.1 and the associated probability (design) requirement in Equation (2.1.1). If treatment Π_i has *known* variance σ_i^2 ($1 \leq i \leq t$) but the values of the σ_i^2 are not equal, then the sample size n_i from Π_i should reflect the value of σ_i^2, with larger numbers of observations being associated with the larger variances. For reasons described below we treat the cases $t = 2$ and $t \geq 3$ differently.

Procedure $\mathcal{N}_B$ (Unequal Known Variances)
If $t = 2$, set

$$n_i = \left\lceil \sigma_i \sqrt{2(\sigma_1^2 + \sigma_2^2)}\,(Z_{1,1/2}^{(1-P^*)}/\delta^*)^2 \right\rceil \quad (2.6.1)$$

($i = 1, 2$) and if $t \geq 3$, set

$$n_i = \left\lceil 2(\sigma_i Z_{t-1,1/2}^{(1-P^*)}/\delta^*)^2 \right\rceil \quad (2.6.2)$$

($1 \leq i \leq t$), where $Z_{t-1,1/2}^{(1-P^*)}$ is obtained from Table B.1. Take n_i independent observations from treatment Π_i ($1 \leq i \leq t$), and denote the sample means by $\bar{y}_i = \sum_{j=1}^{n_i} y_{ij}/n_i$.

Terminal decision rule: Select the treatment associated with the largest sample mean $\bar{y}_{[t]} = \max\{\bar{y}_1, \ldots, \bar{y}_t\}$ as the one associated with $\mu_{[t]}$.

A SINGLE-STAGE PROCEDURE FOR UNEQUAL KNOWN VARIANCES

For $t = 2$, it can be shown that (n_1, n_2) determined from (2.6.1) minimizes $n_1 + n_2$ and guarantees (2.1.1) while approximately satisfying the allocation $\sigma_1/n_1 = \sigma_2/n_2$. When $t \geq 3$, one can show that $(n_1, \ldots, n_t)$ determined from (2.6.2) minimizes $\sum_{i=1}^{t} n_i$ and guarantees (2.1.1) while approximately satisfying the allocation

$$\frac{\sigma_1^2}{n_1} = \frac{\sigma_2^2}{n_2} = \cdots = \frac{\sigma_t^2}{n_t} \tag{2.6.3}$$

that yields equal variances for all t sample means (the "equal variance" allocation). Note that both (2.6.1) and (2.6.2) reduce to the "usual" choice of sample size given in (2.2.1) when $\sigma_1^2 = \cdots = \sigma_t^2$.

Example 2.6.1 Suppose that $t = 2$, $\sigma_1^2 = 2.0$ and $\sigma_2^2 = 3.0$ and that it is desired to satisfy probability requirement (2.1.1) with $(\delta^*, P^*) = (0.5, 0.95)$. How large should n_1 and n_2 be? From Table B.1, we find that $Z_{1,1/2}^{(1-0.95)} = 1.645$. Applying Equation (2.6.1) for $t = 2$, we obtain

$$n_1 = \left\lceil \sqrt{2\sigma_1^2(\sigma_1^2 + \sigma_2^2)} (Z_{1,1/2}^{(1-P^*)}/\delta^*)^2 \right\rceil = \left\lceil \sqrt{2(2)(2+3)}(1.645/0.5)^2 \right\rceil = 49$$

and

$$n_2 = \left\lceil \sqrt{2(3)(2+3)}(1.645/0.5)^2 \right\rceil = 60.$$

Example 2.6.2 Suppose that $t = 4$, $\sigma_1^2 = 2.0$, $\sigma_2^2 = 2.4$, $\sigma_3^2 = 1.7$ and $\sigma_4^2 = 2.3$ and that $(\delta^*, P^*) = (0.8, 0.90)$ are specified. How large should the n_i $(1 \leq i \leq 4)$ be? From Table B.1, we find that $Z_{3,1/2}^{(1-0.90)} = 1.734$. Applying Equation (2.6.2) for $t \geq 3$, we obtain

$$n_1 = \left\lceil 2\sigma_1^2 (Z_{3,1/2}^{(1-0.90)}/\delta^*)^2 \right\rceil = \left\lceil 4.0(1.734/0.8)^2 \right\rceil = 19.$$

Similarly, we find that $n_2 = 23$, $n_3 = 16$ and $n_4 = 22$.

Optimality

The choice of (n_1, n_2) in (2.6.1) is optimal for $t = 2$ in the sense that it maximizes the minimum of $P\{CS \mid (\boldsymbol{\mu}, \boldsymbol{\sigma}^2)\}$ over $\boldsymbol{\mu}$ satisfying $\mu_{[2]} - \mu_{[1]} \geq \delta^*$ for fixed $n_1 + n_2$. Equivalently, the allocation (2.6.1) is optimal in that for any given (δ^*, P^*), it requires the smallest total sample size $n_1 + n_2$ to achieve (2.1.1).

For $t \geq 3$, the equal variance allocations, Equations (2.6.2)–(2.6.3), effectively reduce the sample size problem to the common known variance case. The calculation of $(n_1, \ldots, n_t)$ achieving (2.6.2) and (2.6.3) is easy to implement because it only requires the entries in Table B.1, but it is *not optimal* in the same sense as (2.6.1) is for $t = 2$. The optimal allocation for $t \geq 3$ is difficult to characterize analytically; in principle, it can be calculated numerically by trial and error for any $(t; \delta^*, P^*)$ and $\sum_{i=1}^{t} n_i$.

Bechhofer, Hayter and Tamhane (1991) have partially characterized the optimal rule when $t \geq 3$. To describe one of their main conclusions, first find the largest P_U-value from Table 2.10 such that $P^* \geq P_U$; for that choice of (t, P_U), obtain β_U from the table. Compute $\beta = t \min_{1 \leq i \leq t} \sigma_i^2 / \sum_{j=1}^{t} \sigma_j^2$. If $\beta \geq \beta_U$, then the allocation satisfying (2.6.3) is locally optimal. Otherwise, if $\beta < \beta_U$, then the allocation for which

$$\frac{\sigma_1}{n_1} = \frac{\sigma_2}{n_2} = \cdots = \frac{\sigma_t}{n_t} \qquad (2.6.4)$$

is generally superior to (2.6.3) and is recommended. One can intuitively state their finding as follows. Use the equal variance allocation (2.6.3) when the vector of variances $(\sigma_1^2, \ldots, \sigma_t^2)$ is less "spread out" and use allocation (2.6.4) when the vector of variances is more "spread out." To implement (2.6.4) for $t \geq 3$, set $n_1 = \lceil 2(\sigma_1 Z_{t-1,1/2}^{(1-P^*)} / \delta^*)^2 \rceil$ and take

$$n_i = \left\lceil \frac{\sigma_i}{\sigma_1} n_1 \right\rceil \qquad (2 \leq i \leq t)$$

to make $n_1/\sigma_1 \approx n_i/\sigma_i$ $(2 \leq i \leq t)$.

Example 2.6.3 Suppose that $t = 3$ and $\sigma^2 = (5, 6, 7)$ and that the experimenter specifies $(\delta^*, P^*) = (2, 0.95)$. We take $P_U = 0.95$ since it is the largest P_U-value from Table 2.10 such that $P^* \geq P_U$. For $(t, P_U) = (3, 0.95)$, we obtain $\beta_U = 0.806$ from the table. Now $\beta = 3 \times 5/(5 + 6 + 7) = 0.833 \geq \beta_U$ so that allocation (2.6.2), satisfying (2.6.3), yields the locally optimal allocation (for all δ^*). In particular for $\delta^* = 2$, we have $Z_{2,1/2}^{(1-0.95)} = 1.916$ so that $n_1 = \lceil 2(5)(1.916/2)^2 \rceil = 10$, $n_2 = \lceil 2(6)(1.916/2)^2 \rceil = 12$ and $n_3 = \lceil 2(7)(1.916/2)^2 \rceil = 13$, for a total of 35 observations.

Example 2.6.4 (Example 2.6.3 Continued) Now suppose that $t = 3$ and $\sigma^2 = (4, 6, 8)$ and that the experimenter specifies $(\delta^*, P^*) = (2, 0.95)$. For this more "spread out" variance case, we have $\beta = 3 \times 4/(4 + 6 + 8) = 0.667$. From Table 2.10, we see that $\beta < \beta_U$ for $P_U = 0.95$; thus the superior allocation (2.6.4) is recommended. In this case, $n_1 = \lceil 2(4)(1.916/2)^2 \rceil = \lceil 7.34 \rceil = 8$, $n_2 = \lceil \sqrt{6/4}\,(7.34) \rceil = 9$ and $n_3 = \lceil \sqrt{8/4}\,(7.34) \rceil = 11$, for a total of 28 observations.

Table 2.10. Value of β_U such that if $P^* \geq P_U$ and $\beta \geq \beta_U$, then (2.6.3) is the Locally Optimal Allocation

	t							
P_U	3	4	5	6	7	8	9	10
0.80	0.907	0.854	0.826	0.809	0.798	0.791	0.785	0.781
0.90	0.838	0.778	0.748	0.731	0.720	0.713	0.707	0.704
0.95	0.806	0.740	0.708	0.690	0.678	0.670	0.664	0.660
0.99	0.774	0.700	0.664	0.643	0.629	0.620	0.613	0.608

What would it have cost had we used the suboptimal allocation (2.6.2) in this example? Equation (2.6.2) would have called for $n_1 = \lceil 2(4)(1.916/2)^2 \rceil = 8$, $n_2 = \lceil 2(6)(1.916/2)^2 \rceil = 12$ and $n_3 = \lceil 2(8)(1.916/2)^2 \rceil = 15$, for a total of 35 observations, that is, 7 more than the 28 required by (2.6.4). Notice that the average variance is 6 in both the $\sigma^2 = (5, 6, 7)$ case (from Example 2.6.3) and the more spread out $\sigma^2 = (4, 6, 8)$ case (from the current example); however, the total sample size for the case $\sigma^2 = (4, 6, 8)$ is 28 as compared to 35 for $\sigma^2 = (5, 6, 7)$.

2.7 MULTI-STAGE PROCEDURES FOR COMMON UNKNOWN VARIANCE

The classical problem studied for the one-way ANOVA model is that of testing homogeneity of treatment means when $\sigma_1^2 = \cdots = \sigma_t^2 = \sigma^2$ is unknown. This section studies selection problems when there is a common unknown variance.

There are at least three formulations that can be adopted when considering Goal 2.1.1, that is, the selection of the treatment with the largest mean. One approach is to replace the probability requirement (2.1.1) by

$$P\{CS\} \geq P^* \quad \text{whenever } \mu_{[t]} - \mu_{[t-1]} \geq \delta^* \sigma \tag{2.7.1}$$

where $1/t < P^* < 1$ and $\delta^* > 0$ are specified values. However, the role of δ^* in (2.1.1) is fundamentally different from that in (2.7.1). In the former, δ^* is the smallest difference deemed to be worth detecting (in the units of Y_{ij}) whereas in the latter, δ^* is the (unitless) number of *standard deviations* worth detecting. Formally, the single-stage procedure $\mathcal{N}_B$ with $n = \lceil 2(Z^{(1-P^*)}_{t-1,1/2}/\delta^*)^2 \rceil$ will guarantee (2.7.1). Nevertheless, the adoption of the requirement (2.7.1) is discouraged on the grounds that it postulates an indifference-zone of *unknown* width. The procedure need not guarantee the nominal $P\{CS\}$ for true treatment differences of practical importance.

An approach that avoids this difficulty is to require that a given probability of correct selection be guaranteed for σ^2 no greater than a known amount whenever the difference between the two largest means exceeds a given value. Formally, given P^*

we seek to guarantee

$$P\{CS\} \geq P^\star \quad \text{whenever } \mu_{[t]} - \mu_{[t-1]} \geq \delta^\star \text{ and } \sigma^2 \leq \sigma_U^2 \qquad (2.7.2)$$

where $\sigma_U^2 > 0$ and $\delta^\star > 0$ are specified values. As described in Section 2.8.1, if $n = \lceil 2(\sigma_U Z_{t-1,1/2}^{(1-P^\star)}/\delta^\star)^2 \rceil$, then procedure $\mathcal{N}_B$ guarantees (2.7.2) provided that $P^\star > 0.5$. This procedure is conservative if the true σ^2 is smaller than σ_U^2 or if the true $\boldsymbol{\mu}$-configuration is more favorable than the LF-configuration (2.2.3). Moreover, σ_U^2 is ordinarily not known to the experimenter.

The third formulation, and the one that we advocate, is to require that (2.1.1) be guaranteed for all $\sigma^2 > 0$, that is,

$$P\{CS\} \geq P^\star \quad \text{whenever } \mu_{[t]} - \mu_{[t-1]} \geq \delta^\star \quad \text{for } all\ \sigma^2. \qquad (2.7.3)$$

As background, consider the problem of testing homogeneity of means with a power requirement that is to be guaranteed whenever $\mu_{[t]} - \mu_{[1]} > c$ for *all* $\sigma^2 > 0$ (where $c > 0$ is a specified constant). For this problem, Dantzig (1940) proved that it is *not* possible to devise a *single-stage* test, the power of which is independent of the common unknown σ^2. Similarly, it is not possible to devise a single-stage selection procedure that guarantees probability requirement (2.1.1) for all $\sigma^2 > 0$. (See Dudewicz 1971.)

Stein (1945) devised an *open two-stage* hypothesis test of homogeneity of treatment means that guarantees a prespecified power at a given alternative $\boldsymbol{\mu}$ when σ^2 is *unknown*. Recall that an *open* procedure is one for which, prior to experimentation, no fixed upper bound can be placed on the number of observations to be taken from each treatment.

Bechhofer, Dunnett and Sobel (1954), in the spirit of Stein, proposed an *open two-stage* selection procedure (described in Section 2.7.1) for Goal 2.1.1 that guarantees (2.7.3) when σ^2 is *unknown*. It is important to contrast the role of this two-stage procedure with the two-stage procedure $\mathcal{N}_{TB}$ introduced in Section 2.3.1. The procedure $\mathcal{N}_{TB}$ was proposed as an *improvement* over the single-stage procedure $\mathcal{N}_B$ for the σ^2 known case because it guarantees (2.1.1) with a smaller expected total number of observations than does procedure $\mathcal{N}_B$. In contrast, procedure $\mathcal{N}_{BDS}$ was proposed because single-stage procedures cannot guarantee (2.7.3) when σ^2 is *unknown*. Section 2.7.2 considers an open multi-stage procedure $\mathcal{N}_H$, due to Hartmann (1991), that permits elimination at every stage and guarantees (2.7.3); this procedure is based on Paulson's (1964) procedure. In Section 2.7.3, we compare the performance of the procedures and make recommendations concerning the circumstances under which each procedure should be used.

2.7.1 An Open Two-Stage Procedure without Elimination

We now describe the Bechhofer, Dunnett and Sobel (1954) *open two-stage* selection procedure for the common unknown variance case.

Procedure $\mathcal{N}_{BDS}$ For the given t, specify (δ^*, P^*). Fix a number of observations $n_1 \geq 2$ to be taken in Stage 1. Choose the constant $g = T_{t-1,\nu,1/2}^{(1-P^*)}$ from Table B.3 where $\nu = t(n_1 - 1)$.

Stage 1: Take a random sample of n_1 observations Y_{ij} $(1 \leq j \leq n_1)$ from each Π_i $(1 \leq i \leq t)$.

Stage 2: Calculate the observed first-stage sample means $\bar{y}_i^{(1)} = \sum_{j=1}^{n_1} y_{ij}/n_1$ $(1 \leq i \leq t)$ and $s_\nu^2 = \sum_{i=1}^{t} \sum_{j=1}^{n_1} (y_{ij} - \bar{y}_i^{(1)})^2/\nu$, the unbiased pooled estimate of σ^2 based on $\nu = t(n_1 - 1)$ d.f. Take a random sample of $N - n_1 \geq 0$ *additional* independent observations from each of the Π_i $(1 \leq i \leq t)$ where

$$N - n_1 = \begin{cases} 0 & \text{if } 2(gs_\nu/\delta^*)^2 < n_1 \\ \lceil 2(gs_\nu/\delta^*)^2 \rceil - n_1 & \text{if } 2(gs_\nu/\delta^*)^2 \geq n_1. \end{cases}$$

Calculate the t cumulative sample means $\bar{y}_i = \sum_{j=1}^{N} y_{ij}/N$ $(1 \leq i \leq t)$. Select the treatment that yielded $\bar{y}_{[t]} = \max\{\bar{y}_1, \ldots, \bar{y}_t\}$ as the treatment associated with $\mu_{[t]}$.

The constant used to implement procedure $\mathcal{N}_{BDS}$ is a special case of the upper-α equicoordinate point of an equicorrelated multivariate central t-distribution. Recall that if $(T_1, \ldots, T_p)$ has the p-variate central t-distribution with unit variances, common correlation ρ and ν degrees of freedom, then

$$P\left\{\max_{1 \leq i \leq p} T_i \leq T_{p,\nu,\rho}^{(\alpha)}\right\} = 1 - \alpha \quad (2.7.4)$$

defines the upper-α equicoordinate point $T_{p,\nu,\rho}^{(\alpha)}$ of this distribution.

The procedure $\mathcal{N}_{BDS}$ is open. The total number of vector-observations sampled per treatment is an unbounded random variable because s_ν^2, computed in Stage 1, cannot be bounded a priori. For given t, specified (δ^*, P^*), fixed unknown σ^2 and first-stage common sample size $n_1 \geq 2$, the distribution of the random second-stage sample size is skewed to the right (following the distribution of s_ν^2). If the first-stage sample size n_1 is chosen to be "too small," then large values of the second-stage sample size will occur with sizable probability; this is undesirable in most practical settings. If the first-stage sample size n_1 is chosen to be "too large," then the procedure will behave as it would in the common *known* variance case. And if n_1 is *larger* than would be required by the single-stage procedure $\mathcal{N}_B$ to guarantee (2.1.1) for the common *known* variance case, then the procedure will tend to terminate after the first stage. One way of achieving a compromise in the choice of n_1 is to conjecture (if possible) the *single-stage* sample size that would be necessary to guarantee (2.1.1) *if σ^2 were known*, and then to choose n_1 as (say) two-thirds of that sample size. Moshman (1958) proposed an analytic formulation of a method for choosing the first-stage sample size.

Remark 2.7.1 Other two-stage procedures have been proposed in the literature for the common unknown variance case. For example, Gupta and Kim (1984) adopted the group sampling strategy of Tamhane and Bechhofer (1977, 1979); Gupta and Kim proposed a two-stage procedure that selects a subset of the treatments in the first stage for further sampling, retaining only those treatments indicated as being in contention for final selection. The intent of this strategy is to reduce the *expected total number of observations* required in the experiment. Their choice of constants to implement the procedure was based on a lower bound for the $P\{CS\}$; however, simulation studies from Bechhofer, Dunnett, Goldsman and Hartmann (1990) showed that the Gupta and Kim procedure is sometimes dominated by the procedures described in the current section in terms of the expected total number of observations, $E\{T\}$.

Procedure $\mathcal{N}_{BDS}$ cannot capitalize on favorable configurations of the treatment means. That is, the distribution of the number of vector-observations $(Y_{1j}, \ldots, Y_{tj})$ ($j \geq 1$) to terminate sampling is the same for all configurations of the treatment means even if, e.g., $\mu_{[t]} - \mu_{[t-1]} \gg \sigma$. Also, the same number of observations is taken from *all* of the treatments even if, after the first stage, one or more treatments have sample means that differ by many standard errors from the largest sample mean (and thus are indicated as being non-contending for selection). These drawbacks are addressed in the next subsection, where we discuss Hartmann's open multi-stage procedure with elimination.

2.7.2 An Open Multi-Stage Procedure with Elimination

This subsection describes a variant of a selection procedure due to Paulson (1964). Hartmann (1991) generalized Paulson's (1964) *closed* sequential procedure with elimination for the case of common *known* variance to an *open* sequential procedure with elimination for the case of common *unknown* variance.

The first stage of Hartmann's procedure $\mathcal{N}_H$ is used mainly to provide an estimate of σ^2; it takes observations one vector-at-a-time, each vector consisting of one observation from all treatments not previously eliminated. Treatments can be eliminated at the second and *every* subsequent stage. This procedure has the highly desirable feature of reacting to favorable configurations of the treatment means by eliminating treatments indicated as not being in contention for selection, and concentrating sampling on treatments still remaining in contention.

Procedure $\mathcal{N}_H$ For the given t, specify $(\delta^\star, P^\star)$. Fix a number of observations $n_1 \geq 2$ to be taken in Stage 1.

Stage 1: Take a random sample of n_1 observations Y_{ij} ($1 \leq j \leq n_1$) from each Π_i ($1 \leq i \leq t$). Calculate $\bar{y}_i^{(1)} = \sum_{j=1}^{n_1} y_{ij}/n_1$ ($1 \leq i \leq t$), the observed first-stage sample means, and $s_\nu^2 = \sum_{i=1}^{t} \sum_{j=1}^{n_1} (y_{ij} - \bar{y}_i^{(1)})^2/\nu$, the unbiased pooled estimate of σ^2 based on $\nu = t(n_1 - 1)$ d.f. Next let

$$\gamma = (2[1 - (P^\star)^{1/(t-1)}])^{-2/\nu} - 1$$

MULTI-STAGE PROCEDURES FOR COMMON UNKNOWN VARIANCE 55

and $a^\star = \gamma v s_\nu^2/\delta^\star$, and set $W^\star$ equal to the largest integer less than $2\gamma v s_\nu^2/(\delta^\star)^2$. If $n_1 > W^\star$, then the experiment is terminated and the treatment associated with $\max\{\bar{y}_1^{(1)},\ldots,\bar{y}_t^{(1)}\}$ is selected as the treatment associated with $\mu_{[t]}$. If $n_1 \leq W^\star$, then all treatments Π_j for which

$$\sum_{s=1}^{n_1} y_{js} < \max_{1 \leq i \leq t} \sum_{s=1}^{n_1} y_{is} - a^\star + n_1 \delta^\star/2$$

are permanently eliminated. If only one treatment remains, then stop and declare that treatment to be the one associated with $\mu_{[t]}$. Otherwise, let $r = n_1 + 1$ and proceed to the next stage.

Stage r: Take observations $\{Y_{ir}\}$ a vector-at-a-time from treatments Π_i that have not already been eliminated. At any stage r, eliminate any remaining treatment Π_j for which

$$\sum_{s=1}^{r} y_{js} < \max_{i} \sum_{s=1}^{r} y_{is} - a^\star + r\delta^\star/2$$

where the maximum is taken over the remaining treatments. If, at some stage $r \leq W^\star$, only one treatment remains, then that treatment is selected as the one associated with $\mu_{[t]}$. If $r = W^\star$ and two or more treatments remain, then one more observation is taken from each of these treatments, and the one with the largest cumulative sum is selected as the treatment associated with $\mu_{[t]}$. Otherwise, if $r < W^\star$, increment r by one and repeat.

The choice of the first-stage sample size is important in determining the performance of procedure $\mathcal{N}_H$ as it is for procedure $\mathcal{N}_{BDS}$. See the discussion on this point following the definition of procedure $\mathcal{N}_{BDS}$.

Example 2.7.1 We show how to implement procedure $\mathcal{N}_H$ with a numerical example. Suppose that $t = 4$ and $(\delta^\star, P^\star) = (0.8, 0.75)$, and that we take the first-stage sample size $n_1 = 5$. Thus, $\nu = t(n_1 - 1) = 16$ and $\gamma = (2[1 - (P^\star)^{1/(t-1)}])^{-2/\nu} - 1 = (2[1 - (0.75)^{1/3}])^{-2/16} - 1 = 0.2366$. Let $x_{ir} = \sum_{j=1}^{r} y_{ij}$ $(1 \leq i \leq t, r \geq 1)$ denote the treatment sample sums after r vector-observations have been taken. Suppose that after the first stage of sampling, we obtain $x_{1,5} = -1.560$, $x_{2,5} = -1.089$, $x_{3,5} = 5.361$, $x_{4,5} = 3.628$ and $s_\nu^2 = 1.540$. This yields $a^\star = \gamma v s_\nu^2/\delta^\star = (0.2366)(16)(1.540)/0.8 = 7.287$.

After having taken vector-observation r, we eliminate any treatment Π_j for which

$$\max_{i \text{ alive}} x_{ir} - x_{jr} > a^\star - r\delta^\star/2 = 7.287 - 0.4r.$$

Suppose that we observe the following data:

	Stage					
r	5	6	7	8	9	10
y_{1r}		—	—	—	—	—
y_{2r}		−1.326	−0.385	—	—	—
y_{3r}		−1.598	1.161	0.247	−1.209	−0.607
y_{4r}		−0.341	0.723	0.211	2.408	2.319
x_{1r}	−1.560	—	—	—	—	—
x_{2r}	1.089	−0.237	−0.622	—	—	—
x_{3r}	5.361	3.763	4.924	5.171	3.962	3.355
x_{4r}	3.628	3.287	4.010	4.221	6.628	8.948
$\max_{i\,\text{alive}} x_{ir} - x_{1r}$	6.921	—	—	—	—	—
$\max_{i\,\text{alive}} x_{ir} - x_{2r}$	4.450	4.000	5.546	—	—	—
$\max_{i\,\text{alive}} x_{ir} - x_{3r}$	0	0	0	0	2.666	5.593
$\max_{i\,\text{alive}} x_{ir} - x_{4r}$	1.733	0.476	0.914	0.950	0	0
$a^* - r\delta^*/2$	5.287	4.887	4.487	4.087	3.687	3.287

We immediately eliminate treatment Π_1 after taking $r = 5$ vector-observations. Treatment Π_2 is eliminated after $r = 7$ vector-observations; and treatment Π_3 is eliminated when $r = 10$. Thus, treatment Π_4 is declared to have the largest mean. Note that we took a total of 32 observations.

Remark 2.7.2 (A Conditional Optimality Property) Procedure $\mathcal{N}_H$ is actually a member of a *class* of procedures described by Hartmann. Conditional on the value of s_ν^2 calculated from the first-stage observations, procedure $\mathcal{N}_H$ is a minimax procedure in the sense that, among all of the procedures in that class, it *minimizes the maximum possible number of stages to terminate sampling* (cf. Hartmann 1991 and Bechhofer, Dunnett, Goldsman and Hartmann 1990).

Remark 2.7.3 One could also consider a *heuristic* generalization of Paulson's (1993) common known variance procedure $\mathcal{N}_P$ (described in Section 2.3.3) to the common unknown variance case. For this heuristic procedure, take an initial sample from the t competing treatments, assume that the sample variance from this first stage, s_ν^2, is the true σ^2, and then proceed as in Section 2.3.3. For sufficiently large ν (say, $\nu > 50$), Monte Carlo simulation indicates that this heuristic procedure exhibits impressive performance characteristics that are quite similar to those of procedure $\mathcal{N}_P$ (which *assumes* a common known variance).

2.7.3 Comparison of Procedures

In this section, we study performance characteristics of the procedures and make recommendations concerning the use of procedures $\mathcal{N}_{BDS}$ and $\mathcal{N}_H$.

Criteria for Comparisons

In Sections 2.7.1 and 2.7.2, we discussed procedures $\mathcal{N}_{BDS}$ and $\mathcal{N}_H$. These procedures require the same statistical assumptions (in particular, a *common unknown* σ^2), achieve Goal 2.1.1 and guarantee the same probability requirement (2.7.3). As in Section 2.4, the key question that must be answered is which procedure to use in a given environment. In some cases, certain procedures may not be feasible to implement. For example, procedure $\mathcal{N}_H$ should not be used when it is not practical to take observations a vector-at-a-time.

In cases where it is feasible to use either procedure, the choice of procedure will depend in large part on which performance measure the experimenter wishes to minimize, either $E\{N_V\} = E\{$Number of *vector-observations* to terminate sampling$\}$ or $E\{T\} = E\{Total$ number of *observations* to terminate sampling$\}$. (Notice that for procedures $\mathcal{N}_{BDS}$ and $\mathcal{N}_H$, the number of stages is given by $N_S = N_V - n_1 + 1$, where n_1 is the first-stage sample size.) The choice of measure will in turn depend on whether the cost of observation-taking is more closely related to the number of vector-observations (which might be the case if only one vector-observation can be taken per unit time, e.g., per day) or the *total* number of observations (which could be the case if individual observations are very expensive).

However, here the choice of procedure is somewhat more complicated than in Section 2.4 because the performance characteristics of procedures $\mathcal{N}_{BDS}$ and $\mathcal{N}_H$ depend on many factors. Some of these factors are under the control of the experimenter: the number of treatments t, the specified value of $P^\star$ and the first-stage sample size n_1. Finally, the choice of procedure will depend on any (even partial) information that may be available concerning the value of $\sigma/\delta^\star$ (where σ is unknown and $\delta^\star$ is specified) and the unknown spacing of the treatment means; two configurations of $\boldsymbol{\mu}$ are of particular interest, namely, the least-favorable (LF, that is, $\mu_{[1]} = \mu_{[t-1]} = \mu_{[t]} - \delta^\star$), and the equal-means (EM, that is, $\mu_{[1]} = \mu_{[t]}$).

Summary of Comparisons

Neither procedure $\mathcal{N}_{BDS}$ nor $\mathcal{N}_H$ dominates the other in all environments. But we can make the following *general statements* concerning the relative performances of these two procedures (see Tables 2.11 and 2.12).

- To minimize $E\{T\}$ for "high" $P^\star$, use procedure $\mathcal{N}_H$.
- To minimize $E\{N_V\}$ in the LF-configuration for "low" $P^\star$ ["high" $P^\star$], use procedure $\mathcal{N}_{BDS}$ [procedure $\mathcal{N}_H$].
- To minimize $E\{N_V\}$ in the EM-configuration, use procedure $\mathcal{N}_{BDS}$.

Details

Tables 2.11 and 2.12 can serve as partial guides to the relative performances of procedures $\mathcal{N}_{BDS}$ and $\mathcal{N}_H$. These tables are abstracted from a large comparative simulation experiment reported in Bechhofer, Dunnett, Goldsman and Hartmann (1990). In that article, the levels of the several factors of the problem were varied over reasonable ranges in order to study their effects on the performances of the

Table 2.11. Estimated Achieved $P\{CS\}$, $E\{N_V\}$ and $E\{T\}$ Under the LF- and EM-Configurations for Procedures $\mathcal{N}_{BDS}$ and $\mathcal{N}_H$ When $t = 5$, $P^\star = 0.90$, $\sigma/\delta^\star = 2, 3$ and $n_1 = 5(5)20$

		LF						EM			
$\sigma/\delta^\star$		$\hat{p}$		$\hat{N}_V$		$\hat{T}$		$\hat{N}_V$		$\hat{T}$	
(n_B)	n_1	$\mathcal{N}_{BDS}$	$\mathcal{N}_H$	$\mathcal{N}_{BDS}$	$\mathcal{N}_H$	$\mathcal{N}_{BDS}$	$\mathcal{N}_H$	$\mathcal{N}_{BDS}$	$\mathcal{N}_H$	$\mathcal{N}_{BDS}$	$\mathcal{N}_H$
	5	0.9059	0.9355	30.3	25.0	151.7	100.4	30.3	35.6	151.7	140.4
		(0.0029)	(0.0012)	(0.1)	(0.1)	(0.5)	(0.2)	(0.1)	(0.2)	(0.5)	(0.6)
	10	0.9025	0.9326	28.7	23.3	143.6	94.1	28.7	32.1	143.6	126.4
2		(0.0030)	(0.0025)	(0.1)	(0.1)	(0.3)	(0.3)	(0.1)	(0.1)	(0.3)	(0.4)
(27)	15	0.9024	0.9387	28.3	23.3	141.6	98.9	28.3	31.5	141.6	127.0
		(0.0030)	(0.0024)	(<0.1)	(0.1)	(0.2)	(0.2)	(<0.1)	(0.1)	(0.2)	(0.3)
	20	0.8954	0.9396	28.1	24.8	140.6	112.3	28.1	31.7	140.6	133.6
		(0.0031)	(0.0024)	(<0.1)	(0.1)	(0.2)	(0.2)	(<0.1)	(0.1)	(0.2)	(0.3)
	5	0.8974	0.9329	67.3	54.9	336.4	220.0	67.3	78.7	336.4	309.2
		(0.0030)	(0.0013)	(0.2)	(0.1)	(1.1)	(0.5)	(0.2)	(0.2)	(1.1)	(0.7)
	10	0.9092	0.9315	64.0	50.8	320.2	201.5	64.0	70.1	320.2	272.7
3		(0.0029)	(0.0013)	(0.1)	(0.1)	(0.7)	(0.4)	(0.1)	(0.1)	(0.7)	(0.5)
(61)	15	0.8958	0.9311	63.0	49.9	315.0	197.6	63.0	68.0	315.0	264.2
		(0.0031)	(0.0013)	(0.1)	(0.1)	(0.5)	(0.3)	(0.1)	(0.1)	(0.5)	(0.4)
	20	0.9022	0.9291	62.7	49.6	313.3	198.0	62.7	66.9	313.3	260.7
		(0.0030)	(0.0013)	(0.1)	(0.1)	(0.4)	(0.3)	(0.1)	(0.1)	(0.4)	(0.4)

Reprinted from Bechhofer, Dunnett, Goldsman and Hartman (1990), pp. 989–990, by courtesy of Marcel Dekker, Inc.

procedures. Table 2.11 provides a typical numerical summary for the case $t = 5$, $P^\star = 0.90$. Table 2.12 summarizes the relative performances of the procedures based on the entire Bechhofer et al. paper.

Table 2.11 gives estimates $\hat{p}$, $\hat{N}_V$ and $\hat{T}$ of the achieved $P\{CS\}$, $E\{N_V\}$ and $E\{T\}$, respectively, under the LF- and EM-configurations when $t = 5$, $P^\star = 0.90$, $\sigma/\delta^\star = 2, 3$ and $n_1 = 5(5)20$ for procedures $\mathcal{N}_{BDS}$ and $\mathcal{N}_H$. The upper number in the first column is the value of $\sigma/\delta^\star$, while the lower one, in parentheses (n_B), is the smallest common number of integer observations per population necessary to guarantee the $(\delta^\star/\sigma, P^\star)$-requirement using the single-stage procedure $\mathcal{N}_B$; the n_B are taken from Table 2.1. (We use procedure $\mathcal{N}_B$ as a reference point for procedures $\mathcal{N}_{BDS}$ and $\mathcal{N}_H$.) Each of the numbers in parentheses in the body of Table 2.11 is the estimated standard error of the value above it.

The n_1 observations per treatment taken in the first stage are used primarily to estimate the value of σ^2; these same observations also contribute to the estimation of the treatment means. If the first-stage d.f. available for estimating σ^2 is sufficiently large ($t(n_1 - 1)$ greater than 50, say), then σ^2 can be estimated with high precision. If, in addition, $n_1 \approx n_B$, the performances of procedures $\mathcal{N}_{BDS}$ and $\mathcal{N}_H$ mimic that of procedure $\mathcal{N}_B$ in that the achieved $P\{CS \mid LF\} \approx P^\star$ and $E\{N_V \mid LF\} \approx n_B$. On the other hand, if $n_1 > n_B$, then the achieved $P\{CS \mid LF\} > P^\star$, and the higher-than-specified achieved $P\{CS \mid LF\}$ is purchased at the expense of increased $E\{N_V \mid LF\}$. For both procedures $\mathcal{N}_{BDS}$ and $\mathcal{N}_H$, $E\{N_V\}$ and $E\{T\}$ decrease monotonically as

Table 2.12. For the Cases $\sigma/\delta^* = 2, 3$, $P^* = 0.75, 0.90$, $t = 3, 5, 10$ and $n_1 = 5(5)20$, Procedure Among $\mathcal{N}_{BDS}$ and $\mathcal{N}_H$ Having Minimum Estimated $E\{N_V\}$ or $E\{T\}$ for the LF- and EM-Configurations

			$\sigma/\delta^* = 2$				$\sigma/\delta^* = 3$			
			LF		EM		LF		EM	
P^*	t	n_1	$E\{N_V\}$	$E\{T\}$	$E\{N_V\}$	$E\{T\}$	$E\{N_V\}$	$E\{T\}$	$E\{N_V\}$	$E\{T\}$
0.75	3	5	$\mathcal{N}_{BDS}$	$\mathcal{N}_H$	$\mathcal{N}_{BDS}$	$\mathcal{N}_{BDS}$	$\mathcal{N}_{BDS}$	$\mathcal{N}_H$	$\mathcal{N}_{BDS}$	$\mathcal{N}_{BDS}$
		10	$\mathcal{N}_{BDS}$	$\mathcal{N}_{BDS}$	$\mathcal{N}_{BDS}$	$\mathcal{N}_{BDS}$	$\mathcal{N}_H$	$\mathcal{N}_H$	$\mathcal{N}_{BDS}$	$\mathcal{N}_{BDS}$
		15	$\mathcal{N}_{BDS}$	$\mathcal{N}_{BDS}$	$\mathcal{N}_{BDS}$	$\mathcal{N}_{BDS}$	$\mathcal{N}_{BDS}$	$\mathcal{N}_H$	$\mathcal{N}_{BDS}$	$\mathcal{N}_{BDS}$
		20	$\mathcal{N}_{BDS}$	$\mathcal{N}_{BDS}$	$\mathcal{N}_{BDS}$	$\mathcal{N}_{BDS}$	$\mathcal{N}_{BDS}$	$\mathcal{N}_{BDS}$	$\mathcal{N}_{BDS}$	$\mathcal{N}_{BDS}$
	5	5	$\mathcal{N}_{BDS}$	$\mathcal{N}_H$	$\mathcal{N}_{BDS}$	$\mathcal{N}_{BDS}$	$\mathcal{N}_{BDS}$	$\mathcal{N}_H$	$\mathcal{N}_{BDS}$	$\mathcal{N}_{BDS}$
		10	$\mathcal{N}_{BDS}$	$\mathcal{N}_H$	$\mathcal{N}_{BDS}$	$\mathcal{N}_{BDS}$	$\mathcal{N}_{BDS}$	$\mathcal{N}_H$	$\mathcal{N}_{BDS}$	$\mathcal{N}_H$
		15	$\mathcal{N}_{BDS}$	$\mathcal{N}_{BDS}$	$\mathcal{N}_{BDS}$	$\mathcal{N}_{BDS}$	$\mathcal{N}_{BDS}$	$\mathcal{N}_H$	$\mathcal{N}_{BDS}$	$\mathcal{N}_H$
		20	$\mathcal{N}_{BDS}$	$\mathcal{N}_{BDS}$	$\mathcal{N}_{BDS}$	$\mathcal{N}_{BDS}$	$\mathcal{N}_{BDS}$	$\mathcal{N}_H$	$\mathcal{N}_{BDS}$	$\mathcal{N}_{BDS}$
	10	5	$\mathcal{N}_{BDS}$	$\mathcal{N}_H$	$\mathcal{N}_{BDS}$	$\mathcal{N}_H$	$\mathcal{N}_{BDS}$	$\mathcal{N}_H$	$\mathcal{N}_{BDS}$	$\mathcal{N}_H$
		10	$\mathcal{N}_{BDS}$	$\mathcal{N}_H$	$\mathcal{N}_{BDS}$	$\mathcal{N}_H$	$\mathcal{N}_{BDS}$	$\mathcal{N}_H$	$\mathcal{N}_{BDS}$	$\mathcal{N}_H$
		15	$\mathcal{N}_{BDS}$	$\mathcal{N}_H$	$\mathcal{N}_{BDS}$	$\mathcal{N}_H$	$\mathcal{N}_{BDS}$	$\mathcal{N}_H$	$\mathcal{N}_{BDS}$	$\mathcal{N}_H$
		20	$\mathcal{N}_{BDS}$	$\mathcal{N}_{BDS}$	$\mathcal{N}_{BDS}$	$\mathcal{N}_{BDS}$	$\mathcal{N}_{BDS}$	$\mathcal{N}_H$	$\mathcal{N}_{BDS}$	$\mathcal{N}_H$
0.90	3	5	$\mathcal{N}_H$	$\mathcal{N}_H$	$\mathcal{N}_{BDS}$	$\mathcal{N}_H$	$\mathcal{N}_H$	$\mathcal{N}_H$	$\mathcal{N}_{BDS}$	$\mathcal{N}_H$
		10	$\mathcal{N}_H$	$\mathcal{N}_H$	$\mathcal{N}_{BDS}$	$\mathcal{N}_H$	$\mathcal{N}_H$	$\mathcal{N}_H$	$\mathcal{N}_{BDS}$	$\mathcal{N}_H$
		15	$\mathcal{N}_H$	$\mathcal{N}_H$	$\mathcal{N}_{BDS}$	$\mathcal{N}_H$	$\mathcal{N}_H$	$\mathcal{N}_H$	$\mathcal{N}_H$	$\mathcal{N}_H$
		20	$\mathcal{N}_H$	$\mathcal{N}_H$	$\mathcal{N}_{BDS}$	$\mathcal{N}_{BDS}$	$\mathcal{N}_H$	$\mathcal{N}_H$	$\mathcal{N}_H$	$\mathcal{N}_H$
	5	5	$\mathcal{N}_H$	$\mathcal{N}_H$	$\mathcal{N}_{BDS}$	$\mathcal{N}_H$	$\mathcal{N}_H$	$\mathcal{N}_H$	$\mathcal{N}_{BDS}$	$\mathcal{N}_H$
		10	$\mathcal{N}_H$	$\mathcal{N}_H$	$\mathcal{N}_{BDS}$	$\mathcal{N}_H$	$\mathcal{N}_H$	$\mathcal{N}_H$	$\mathcal{N}_{BDS}$	$\mathcal{N}_H$
		15	$\mathcal{N}_H$	$\mathcal{N}_H$	$\mathcal{N}_{BDS}$	$\mathcal{N}_H$	$\mathcal{N}_H$	$\mathcal{N}_H$	$\mathcal{N}_{BDS}$	$\mathcal{N}_H$
		20	$\mathcal{N}_H$	$\mathcal{N}_H$	$\mathcal{N}_{BDS}$	$\mathcal{N}_H$	$\mathcal{N}_H$	$\mathcal{N}_H$	$\mathcal{N}_{BDS}$	$\mathcal{N}_H$
	10	5	$\mathcal{N}_H$	$\mathcal{N}_H$	$\mathcal{N}_{BDS}$	$\mathcal{N}_H$	$\mathcal{N}_H$	$\mathcal{N}_H$	$\mathcal{N}_{BDS}$	$\mathcal{N}_H$
		10	$\mathcal{N}_H$	$\mathcal{N}_H$	$\mathcal{N}_{BDS}$	$\mathcal{N}_H$	$\mathcal{N}_H$	$\mathcal{N}_H$	$\mathcal{N}_{BDS}$	$\mathcal{N}_H$
		15	$\mathcal{N}_H$	$\mathcal{N}_H$	$\mathcal{N}_{BDS}$	$\mathcal{N}_H$	$\mathcal{N}_H$	$\mathcal{N}_H$	$\mathcal{N}_{BDS}$	$\mathcal{N}_H$
		20	$\mathcal{N}_H$	$\mathcal{N}_H$	$\mathcal{N}_{BDS}$	$\mathcal{N}_H$	$\mathcal{N}_H$	$\mathcal{N}_H$	$\mathcal{N}_{BDS}$	$\mathcal{N}_H$

the first-stage sample size n_1 increases from 5 to 15, but they change little for $n_1 \geq 20$; however, the sample standard deviations of N_V and T continue to decrease monotonically, even for $n_1 \geq 20$. Thus, values of n_1 larger than 15 play only a modest role in stabilizing $E\{N_V\}$ and $E\{T\}$, but a major role in decreasing the variances of N_V and T.

If n_1 is chosen "too large," then, on the average, many more observations and vector-observations will be taken to guarantee (2.7.3) than if the value of σ^2 were known. However, it is dangerous to start with "very small" n_1 if $\nu = t(n_1 - 1)$ is not sufficiently large (say > 50), for then $S_\nu^2 \gg \sigma^2$ with high probability, resulting in large values of $E\{N_V\}$ and $E\{T\}$ and occasional excessively large values of N_V and T. If t is large, then a moderate n_1 will protect the experimenter against unnecessarily large values of S_ν^2. If the experimenter has an order-of-magnitude approximate value of σ^2, then n_1 could be chosen as (say) $2n_B/3$ subject, of course, to the recommendation that $t(n_1 - 1) > 50$.

2.8 PROCEDURES FOR UNEQUAL UNKNOWN VARIANCES

This section considers the problem of selecting the treatment having the largest mean when the variances of each treatment are unknown and not necessarily the same. This, of course, is the *most common* situation encountered in practice; indeed, the weak assumptions on the variances allow for wide applicability of the procedures to be discussed in this section. In particular, in Section 2.8.1, we present a procedure that can be used when the experimenter is prepared to set an upper bound σ_U^2 on the unknown treatment variances $\sigma_1^2, \ldots, \sigma_t^2$; Section 2.8.2 studies a procedure that requires no assumptions concerning the unknown variances.

We caution the reader that if the treatment variances are too different, then the goal of selecting the treatment based on its mean value alone may not be the most appropriate one for many applications. For example, consider the problem of purchasing one of t measuring instruments. Suppose that each instrument is used to measure a series of test signals of known intensity and frequency. It may be more appropriate to adopt the device that has the smallest mean square error in measuring the test signals (a quantity that accounts for both the variability and mean of the instrument) than to examine only the bias of the instrument in measuring these signals.

2.8.1 Variances Bounded

Hayter (1989) proposed a single-stage procedure for the case when $\sigma_1^2, \ldots, \sigma_t^2$ are unknown but $\max\{\sigma_1^2, \ldots, \sigma_t^2\} < \sigma_U^2$ where σ_U^2 is *known*.

Procedure $\mathcal{N}_{Ha}$ For the given t and specified $(\delta^\star, P^\star)$, choose $Z_{t-1,1/2}^{(1-P^\star)}$ from Table B.1 corresponding to the $(t; P^\star)$ of interest. Calculate $n = \lceil 2(\sigma_U Z_{t-1,1/2}^{(1-P^\star)}/\delta^\star)^2 \rceil$ and proceed as in the single-stage procedure $\mathcal{N}_B$ of Section 2.2.

The choice of n in procedure $\mathcal{N}_{Ha}$ will guarantee (2.1.1) provided that $P^\star > 0.5$. Recall that the single-stage procedure $\mathcal{N}_B$ is conservative when the true vector of treatment means $\boldsymbol{\mu}$ is more favorable to the experimenter than the LF (slippage) configuration, that is, $P\{CS \mid \boldsymbol{\mu}\} > P^\star$. In a similar way, procedure $\mathcal{N}_{Ha}$ has *two* sets of circumstances for which the choice of n can result in a conservative procedure: if either (1) the true configuration of the means $\boldsymbol{\mu}$ is favorable to the experimenter, or (2) σ_U^2 is quite a bit larger than the largest unknown σ_i^2. In either case, the experimenter might prefer to use the two-stage procedure of Rinott (1978), discussed in the next subsection, if that is feasible.

2.8.2 Variances Unbounded

Dudewicz and Dalal (1975) and Rinott (1978) proposed open two-stage procedures without elimination for the case in which the t variances $(\sigma_1^2, \ldots, \sigma_t^2)$ are completely unknown. We describe the Rinott procedure; it is easier to apply than the Dudewicz-Dalal procedure.

PROCEDURES FOR UNEQUAL UNKNOWN VARIANCES

Procedure $\mathcal{N}_R$ Fix a number of observations $n_1 \geq 2$ to be taken in Stage 1. For the given t and specified $(\delta^\star, P^\star)$, choose the constant $g = g(t, P^\star, \nu)$ from Table 2.13, where $\nu = n_1 - 1$.

Stage 1: Take a random sample of n_1 observations Y_{ij} $(1 \leq j \leq n_1)$ from each Π_i $(1 \leq i \leq t)$.

Stage 2: Calculate $\bar{y}_i^{(1)} = \sum_{j=1}^{n_1} y_{ij}/n_1$ and $s_i^2 = \sum_{j=1}^{n_1}(y_{ij} - \bar{y}_i^{(1)})^2/\nu$, an unbiased estimate of σ_i^2 $(1 \leq i \leq t)$ based on $\nu = n_1 - 1$ d.f. Take $N_i - n_1$ additional observations from Π_i $(1 \leq i \leq t)$ where

$$N_i - n_1 = \begin{cases} 0 & \text{if } (gs_i/\delta^\star)^2 < n_1 \\ \lceil (gs_i/\delta^\star)^2 \rceil - n_1 & \text{if } (gs_i/\delta^\star)^2 \geq n_1 \end{cases}.$$

Calculate $\bar{y}_i = \sum_{j=1}^{N_i} y_{ij}/N_i$ $(1 \leq i \leq t)$ based on the combined results of the Stage 1 and Stage 2 samples. Select the treatment that yielded $\bar{y}_{[t]} = \max\{\bar{y}_1, \ldots, \bar{y}_t\}$ as the one associated with $\mu_{[t]}$.

The constant $g = g(t, P^\star, \nu)$ is the solution to

$$\int_0^\infty \int_0^\infty \left[\Phi\left(\frac{g}{\sqrt{\nu(1/x + 1/y)}}\right) f_\nu(x) \right]^{t-1} f_\nu(y)\, dy\, dx = P^\star$$

where $\Phi(\cdot)$ is the standard normal c.d.f. and $f_\nu(\cdot)$ is the p.d.f. of the χ^2-distribution with ν d.f. The FORTRAN program RINOTT in Appendix C calculates values of $g(t, P^\star, \nu)$ for additional cases not covered by Table 2.13. See Rinott (1978) and Wilcox (1984) for more computational details.

We emphasize two characteristics of this procedure. First, it does not eliminate treatments at the end of Stage 1, but instead uses each treatment's Stage 1 data to estimate that treatment's variance. Second, similar to the Bechhofer, Dunnett and Sobel (1954) two-stage procedure $\mathcal{N}_{BDS}$, one cannot bound *a priori* the number of observations that procedure $\mathcal{N}_R$ will require. Rather, the total number of observations procedure $\mathcal{N}_R$ samples from each treatment is a random variable, possibly different for each treatment.

Example 2.8.1 (A Simulation Study of Airline Reservation Systems) Goldsman, Nelson and Schmeiser (1991) studied $t = 4$ different airline reservation systems with the objective of determining the system with the *largest* expected time to failure (E[TTF]). Let μ_i denote the E[TTF] for system i $(1 \leq i \leq 4)$. From past experience it was known that the E[TTF]'s were roughly 100,000 minutes (about 70 days) for all four systems. It was desired to select the best system with probability at least $P^\star = 0.90$ if the difference in the expected failure times for the best and second best systems was greater than or equal to $\delta^\star = 3000$ minutes (about two days).

Table 2.13. Values of $g(t, P^\star, \nu)$ Required by Procedure $\mathcal{N}_R$ to Determine the Second-Stage Sample Sizes

$P^\star$	$n_1 = $ $\nu + 1$	\multicolumn{9}{c}{t}								
		2	3	4	5	6	7	8	9	10
0.75	5	1.138	1.927	2.367	2.675	2.912	3.105	3.269	3.411	3.537
	6	1.098	1.844	2.253	2.535	2.750	2.924	3.071	3.197	3.307
	7	1.072	1.792	2.182	2.449	2.652	2.814	2.950	3.067	3.170
	8	1.054	1.757	2.134	2.391	2.585	2.741	2.870	2.981	3.078
	9	1.040	1.731	2.099	2.349	2.537	2.688	2.813	2.920	3.013
	10	1.030	1.711	2.073	2.318	2.501	2.648	2.770	2.874	2.964
	11	1.022	1.695	2.052	2.293	2.473	2.617	2.736	2.838	2.927
	12	1.016	1.683	2.036	2.273	2.451	2.592	2.710	2.809	2.896
	13	1.010	1.673	2.022	2.257	2.432	2.572	2.688	2.786	2.872
	14	1.006	1.664	2.011	2.243	2.417	2.555	2.669	2.767	2.851
	15	1.002	1.657	2.001	2.232	2.404	2.541	2.654	2.750	2.834
	16	0.999	1.650	1.993	2.222	2.393	2.529	2.641	2.736	2.819
	17	0.996	1.645	1.986	2.213	2.383	2.518	2.630	2.724	2.807
	18	0.993	1.640	1.979	2.206	2.375	2.509	2.620	2.714	2.795
	19	0.991	1.636	1.974	2.199	2.367	2.501	2.611	2.704	2.786
	20	0.989	1.632	1.969	2.193	2.361	2.493	2.603	2.696	2.777
	30	0.977	1.609	1.938	2.158	2.321	2.450	2.556	2.646	2.724
	40	0.971	1.598	1.924	2.141	2.302	2.429	2.533	2.622	2.699
	50	0.967	1.591	1.916	2.131	2.291	2.417	2.521	2.608	2.685
0.90	5	2.291	3.058	3.511	3.837	4.093	4.305	4.486	4.644	4.786
	6	2.177	2.871	3.270	3.552	3.771	3.951	4.103	4.235	4.352
	7	2.107	2.758	3.126	3.384	3.582	3.744	3.881	3.999	4.103
	8	2.059	2.682	3.031	3.273	3.459	3.609	3.736	3.845	3.941
	9	2.025	2.628	2.963	3.195	3.372	3.515	3.635	3.738	3.829
	10	1.999	2.587	2.913	3.137	3.307	3.445	3.560	3.659	3.746
	11	1.978	2.556	2.874	3.092	3.258	3.391	3.503	3.598	3.682
	12	1.962	2.531	2.843	3.056	3.218	3.349	3.457	3.551	3.632
	13	1.948	2.510	2.817	3.027	3.186	3.314	3.420	3.512	3.592
	14	1.937	2.493	2.796	3.003	3.160	3.285	3.390	3.480	3.558
	15	1.928	2.479	2.779	2.983	3.138	3.261	3.364	3.453	3.530
	16	1.919	2.467	2.764	2.966	3.119	3.241	3.343	3.430	3.506
	17	1.912	2.456	2.751	2.951	3.102	3.223	3.324	3.410	3.485
	18	1.906	2.447	2.739	2.938	3.088	3.208	3.308	3.393	3.467
	19	1.901	2.438	2.729	2.926	3.075	3.194	3.293	3.378	3.451
	20	1.896	2.431	2.720	2.916	3.064	3.182	3.280	3.364	3.437
	30	1.866	2.387	2.666	2.855	2.997	3.110	3.204	3.284	3.354
	40	1.852	2.366	2.641	2.827	2.966	3.077	3.169	3.247	3.315
	50	1.844	2.354	2.627	2.810	2.948	3.057	3.148	3.225	3.292

PROCEDURES FOR UNEQUAL UNKNOWN VARIANCES

Table 2.13. *Continued*

P^*	$n_1 =$ $\nu + 1$	2	3	4	t 5	6	7	8	9	10
0.95	5	3.107	3.905	4.390	4.744	5.025	5.259	5.461	5.638	5.797
	6	2.910	3.602	4.010	4.303	4.533	4.722	4.884	5.025	5.150
	7	2.791	3.424	3.791	4.051	4.253	4.419	4.559	4.681	4.789
	8	2.712	3.308	3.649	3.889	4.074	4.225	4.353	4.463	4.561
	9	2.656	3.226	3.550	3.776	3.950	4.091	4.210	4.313	4.404
	10	2.614	3.166	3.476	3.693	3.859	3.993	4.106	4.204	4.290
	11	2.582	3.119	3.420	3.629	3.789	3.918	4.027	4.121	4.203
	12	2.556	3.082	3.376	3.579	3.734	3.860	3.965	4.055	4.135
	13	2.534	3.052	3.340	3.539	3.690	3.812	3.915	4.003	4.080
	14	2.517	3.027	3.310	3.505	3.654	3.773	3.874	3.960	4.035
	15	2.502	3.006	3.285	3.477	3.623	3.741	3.839	3.924	3.998
	16	2.489	2.988	3.264	3.453	3.597	3.713	3.810	3.893	3.966
	17	2.478	2.973	3.246	3.433	3.575	3.689	3.785	3.867	3.938
	18	2.468	2.959	3.230	3.415	3.556	3.669	3.763	3.844	3.914
	19	2.460	2.948	3.216	3.399	3.539	3.650	3.744	3.824	3.894
	20	2.452	2.937	3.203	3.385	3.523	3.634	3.727	3.806	3.875
	30	2.407	2.874	3.129	3.303	3.434	3.539	3.626	3.701	3.766
	40	2.386	2.845	3.094	3.264	3.392	3.495	3.580	3.652	3.716
	50	2.373	2.828	3.074	3.242	3.368	3.469	3.553	3.624	3.687
0.99	5	5.136	6.124	6.751	7.220	7.599	7.918	8.195	8.440	8.661
	6	4.603	5.370	5.842	6.188	6.463	6.693	6.891	7.065	7.221
	7	4.303	4.959	5.353	5.638	5.863	6.050	6.209	6.348	6.472
	8	4.114	4.703	5.052	5.302	5.498	5.659	5.796	5.916	6.022
	9	3.983	4.529	4.848	5.076	5.253	5.399	5.522	5.629	5.724
	10	3.887	4.403	4.702	4.914	5.079	5.213	5.327	5.426	5.514
	11	3.815	4.308	4.593	4.793	4.949	5.075	5.182	5.275	5.357
	12	3.758	4.234	4.507	4.699	4.848	4.969	5.071	5.159	5.237
	13	3.712	4.175	4.439	4.625	4.768	4.884	4.982	5.066	5.141
	14	3.675	4.126	4.383	4.564	4.702	4.815	4.910	4.991	5.063
	15	3.643	4.086	4.337	4.513	4.648	4.758	4.850	4.929	4.999
	16	3.616	4.051	4.298	4.470	4.602	4.709	4.799	4.877	4.945
	17	3.593	4.022	4.265	4.434	4.563	4.668	4.756	4.832	4.899
	18	3.573	3.996	4.235	4.402	4.529	4.633	4.719	4.794	4.859
	19	3.556	3.974	4.210	4.374	4.500	4.602	4.687	4.760	4.825
	20	3.540	3.954	4.188	4.350	4.474	4.574	4.658	4.731	4.794
	30	3.448	3.838	4.056	4.207	4.322	4.414	4.492	4.558	4.616
	40	3.405	3.785	3.996	4.142	4.253	4.342	4.416	4.480	4.536
	50	3.381	3.754	3.962	4.104	4.213	4.300	4.373	4.436	4.490

The competing systems were sufficiently complicated that computer simulation was required to analyze their behavior. As will be seen below, the large E[TTF]'s, the highly variable nature of rare failures, the similarity of the systems, and the relatively small δ^* yielded a problem with reasonably large computational costs.

Let T_{ij} ($1 \leq i \leq 4$, $j \geq 1$) denote the observed time to failure from the jth independent simulation replication of system i. Application of the Rinott procedure $\mathcal{N}_R$ requires independent and identically distributed (i.i.d.) normal observations from each system. If each simulation replication is initialized from a particular system under the same operating conditions, but with independent random number seeds, the resulting $T_{i1}, T_{i2}, \ldots$ will be i.i.d. for each system. However, the T_{ij} cannot be justified as being normally distributed and, in fact, are somewhat skewed to the right.

Thus, instead of using the raw T_{ij} in procedure $\mathcal{N}_R$, Goldsman et al. (1991) applied the procedure to the so-called *macroreplication* estimators of the μ_i. These estimators group the $\{T_{ij}: j \geq 1\}$ into disjoint *batches* and use the batch averages as the "data" to which procedure $\mathcal{N}_R$ is applied. More formally, they fix an integer number, m, of simulation replications that comprise each macroreplication (that is, m is the batch size) and let

$$Y_{ij} \equiv \frac{1}{m} \sum_{k=1}^{m} T_{i,(j-1)m+k}$$

($1 \leq i \leq 4$, $1 \leq j \leq b_i$), where b_i is the number of macroreplications to be taken from system i. The macroreplication estimators from the ith system, $Y_{i1}, Y_{i2}, \ldots, Y_{ib_i}$, are i.i.d. with expectation μ_i. If m is sufficiently large, say at least 20, then the Central Limit Theorem yields approximate normality for each Y_{ij}. No assumptions are made concerning the variances of the macroreplications.

To apply procedure $\mathcal{N}_R$, the authors conducted a pilot study to serve as the first stage of the procedure; each system was run for $n_1 = 20$ macroreplications with each macroreplication consisting of the averages of $m = 20$ simulations of the system. The results are summarized in the first two rows of Table 2.14. (Shapiro-Wilks tests on the 20 macroreplications from each system showed no evidence of non-normality of the Y_{ij} ($1 \leq j \leq 20$) from each system.) From Table 2.13, the critical constant for $t = 4$ and $P^* = 0.90$ is $g = 2.720$. The total sample sizes N_i were computed for each system and are displayed in row 3 of Table 2.14. For example, System 2

Table 2.14. Summary of First- and Second-Stage Results for the Airline Reservation Experiment Based on $n_1 = 20$, Where $s_{\bar{y}_i}$ is the Estimated Standard Error of the Sample Mean $\bar{y}_i$

i	1	2	3	4
$\bar{y}_i^{(1)}$	108286.	107686.	96167.7	89747.9
s_i	29157.3	24289.9	25319.5	20810.8
N_i	699	485	527	356
$\bar{y}_i$	110816.5	106411.8	99093.1	86568.9
$s_{\bar{y}_i}$	872.0	1046.5	894.2	985.8

required an *additional* $N_2 - 20 = 465$ macroreplications in the second stage (each macroreplication again being the average of $m = 20$ system simulations). In all, a total of about 40,000 simulations of the four systems were required to implement procedure $\mathcal{N}_R$. The combined sample means for each system are listed in row 4 of Table 2.14, and the standard errors of each mean are listed in row 5. They clearly establish System 1 as having the largest E[TTF].

2.9 CHAPTER NOTES

Section 2.2.1 expresses the sample size to achieve Goal 2.1.1 for the procedure $\mathcal{N}_B$ in terms of the equicoordinate critical point of a certain multivariate normal distribution. The following presents an alternative (implicit) expression for n in terms of the standard normal c.d.f. This expression is a formula for the $P\{CS\}$ of procedure $\mathcal{N}_B$ at any least favorable configuration, that is, any $\boldsymbol{\mu}$ satisfying $\mu_{[1]} = \mu_{[t-1]} = \mu_{[t]} - \delta^*$. We present this expression since readers who consult the selection literature are much more likely to find this representation than the multivariate normal distribution representation that is the basis for the sample size formula presented in Section 2.2. In the subsequent derivation, $\Phi(\cdot)$ denotes the standard normal c.d.f. Furthermore, we assume without loss of generality that treatment t is associated with the largest mean $\mu_{[t]}$.

$$P\{CS \mid LF\} = P\{\overline{Y}_i \leq \overline{Y}_t, \quad 1 \leq i \leq t-1 \mid \mu_{[1]} = \mu_{[t-1]} = \mu_{[t]} - \delta^*\}$$

$$= P\left\{\frac{\overline{Y}_i - \mu_t}{\sqrt{\text{Var}(\overline{Y}_t)}} \leq \frac{\overline{Y}_t - \mu_t}{\sqrt{\text{Var}(\overline{Y}_t)}}, \quad 1 \leq i \leq t-1 \mid \mu_{[1]} = \mu_{[t-1]} = \mu_{[t]} - \delta^*\right\}$$

$$= \int_{-\infty}^{\infty} P\left\{\frac{\overline{Y}_i - \mu_t}{\sqrt{\text{Var}(\overline{Y}_t)}} \leq z, \quad 1 \leq i \leq t-1 \mid \mu_{[1]} = \mu_{[t-1]} = \mu_{[t]} - \delta^*\right\} d\Phi(z)$$

$$= \int_{-\infty}^{\infty} P\left\{\frac{\overline{Y}_i - \mu_i}{\sqrt{\text{Var}(\overline{Y}_i)}} \leq z + \frac{\delta^*}{\sqrt{\text{Var}(\overline{Y}_i)}}, \quad 1 \leq i \leq t-1\right\} d\Phi(z)$$

$$= \int_{-\infty}^{\infty} \Phi^{t-1}\left(z + \frac{\sqrt{n}\delta^*}{\sigma}\right) d\Phi(z). \tag{2.9.1}$$

To satisfy the probability requirement (2.1.1), we determine the smallest integer value n for which the expression (2.9.1) is greater than or equal to P^*. The integral (2.9.1) is related to the equicoordinate critical point $c = Z_{t-1,1/2}^{(1-P^*)}$ of the multivariate normal distribution (see (2.2.2)) as follows:

$$P\{W_1 \leq c, W_2 \leq c, \ldots, W_{t-1} \leq c\} = \int_{-\infty}^{\infty} \Phi^{t-1}\left(z + c\sqrt{2}\right) d\Phi(z).$$

The vector $(W_1, \ldots, W_{t-1})$ has the $(t-1)$-variate multivariate normal distribution with mean vector zero, unit variances, and common correlation $1/2$.

Given that a sample of common size n has been taken from each treatment, one method of assessing the adequacy of the sample size when using procedure $\mathcal{N}_B$ is to estimate the achieved $P\{CS\}$ using this procedure. This is essentially an issue of *analysis* rather than *design* of the experiment. The reader can consult the papers of Faltin (1980), Faltin and McCulloch (1983), Olkin, Sobel and Tong (1982), Kim (1986), Gupta, Leu and Liang (1990) and Gupta and Liang (1991a) and the references therein for estimators of this quantity.

A number of authors have formulated the problem of choosing the sample size in selection problems by using a loss function approach that balances the benefit of a greater probability of a correct selection associated with larger sample size against the cost of collecting additional data. Somerville (1954) and Fairweather (1968) adopted a minimax approach based on a loss function that exhibits these characteristics, while Dunnett (1960), Raiffa and Schlaiffer (1961) and Tiao and Afonja (1976) considered a Bayesian formulation. Somerville (1954) considered a two-stage sampling approach with the loss increasing in the sample size and the probability of making a wrong decision. He chose the sample size to minimize the maximum expected loss over the parameter space. Fairweather (1968) generalized Somerville's formulation to an arbitrary finite number of stages of sampling.

Dunnett (1960) considered a single-stage procedure for the same problem but assumed that the treatment means $\mu_1, \ldots, \mu_t$ have a multivariate normal prior distribution with means $\beta_1, \ldots, \beta_t$, respectively, that are known up to an additive constant and have equal variances and equal covariances, the values of which are known. He selected n to minimize the risk (the integrated expected loss) for a loss function that is linear in the difference between the selected means and $\mu_{[t]}$, and he chose a sampling cost proportional to the common sample size n. Raiffa and Schlaiffer (1961) and Guttman and Tiao (1964) used a similar approach, while Tiao and Afonja (1976) investigated sensitivity of the optimal design to the specification of the experimenter's loss function.

A related Bayesian formulation of the selection problem focuses on the calculation of posterior distributions of important quantities for selection. For example, Berger and Deely (1988) emphasized hierarchical Bayesian modeling for a setup in which there is positive probability for the two hypotheses $H_0: \mu_1 = \cdots = \mu_t$ and H_A: not H_0; for each treatment they calculated the (posterior) probability that it has the greatest mean given the data and that H_0 is false. Fong and Chow (1991) gave a computer program to implement the approach of Berger and Deely. Fong (1992a, 1992b) presented related results.

Sehr (1988) proved that the Tamhane-Bechhofer two-stage procedure has the same least favorable configuration (2.1.1) as the single-stage procedure $\mathcal{N}_B$ while Santner and Hayter (1993) proved the same result for the Santner and Behaxeteguy (1992) two-stage procedure mentioned in Section 2.3. Both of these two-stage procedures use a minimax approach to select the first- and second-stage sample sizes, that is, they minimize the maximum expected total sample size among procedures that achieve the basic P^* probability requirement. Gupta and Liang (1989c) adopted a Bayesian, loss function approach for this problem that is similar in spirit to the Dunnett (1960) formulation. See also Gupta and Miescke (1984) for an earlier development of Bayes rules for certain loss functions when the first- and second-stage sample sizes are *fixed*.

For the case of selecting the normal treatment with unknown and possibly unequal variances, Mukhopadhyay (1979) described a modification of the Rinott's (1978) two-stage procedure that is "asymptotically efficient" when $t = 2$. Two-stage procedures that employ screening in the first stage have been derived for other location and scale models. For example, Gupta and Han (1991) and Gupta, Miao and Sun (1993) developed procedures for the logistic distribution.

In Section 2.2 we noted that procedure $\mathcal{N}_B$ also achieves the goal of what might be called the "δ^*-near-best" formulation of the selection problem. For the procedure that selects a single treatment, the idea of this formulation is to select *any* treatment whose mean is sufficiently close to the best treatment. In particular, if the means are all close, then the selection of any treatment achieves the goal. As we saw in the Confidence Statement of Section 2.2, a specific example of such a goal is satisfied by procedure $\mathcal{N}_B$; this procedure selects a treatment whose mean satisfies $\mu_i \geq \mu_{[t]} - \delta^*$. In general, whether the selection of a particular treatment has achieved this or some other "δ^*-near-best" goal depends on the underlying $\boldsymbol{\mu}$. Further, the assertion of the equivalence of a given IZ goal for some procedure and a related δ^*-near-best goal requires a problem specific analysis and need not always hold. Parnes and Srinivasan (1986) presented single-stage procedures for which the P^* condition holds for the IZ approach but for which the mean of the selected treatment is not δ^*-near-best with probability at least P^*. However, there are many papers that provide positive results giving δ^*-near-best type goals that are satisfied by IZ procedures. Perhaps the earliest such result is that of Fabian (1962) who proposed a δ^*-near-best goal for procedure $\mathcal{N}_B$. Chiu (1977) identified a δ^*-near-best type goal that the untruncated Bechhofer, Kiefer and Sobel (1968) procedure satisfies when sampling from arbitrary exponential families. Hsu and Edwards (1983) introduced a sequential procedure that can be used to assert that the selected normal treatment mean satisfies $\mu_i \geq \mu_{[t]} - \delta^*$. Edwards (1987) provided an adaptation of the Paulson (1964) normal means selection procedure that satisfies the same goal. Chiu (1974) gave a δ^*-near-best goal for the selection procedure $\mathcal{N}_B$ that selects the s best treatments. Feigin and Weissman (1981) gave an extension of such a goal for the procedure that selects the s treatments corresponding to the s largest summary statistics when the distribution of these statistics follows an arbitrary stochastically increasing family.

Turnbull, Kaspi and Smith (1978) studied *adaptive* sequential procedures for selecting the normal population associated with $\mu_{[t]}$ when there is a common known variance, and the procedure is required to guarantee (2.1.1). The objective of their study was to find a procedure with smaller $E\{N_V\}$ than that required by the vector-at-a-time procedure $\mathcal{N}_{BKS}$. Such procedures exist for the so-called identification problem wherein $\mu_{[1]}, \ldots, \mu_{[t]}$ are known and satisfy the LF-configuration $\mu_{[1]} = \mu_{[t-1]} = \mu_{[t]} - \delta^*$. But the same slippage configuration of means need not be LF for *all* sampling rules when $t \geq 3$ if the $\mu_{[i]}$ are not known, as in the ranking problem. Hoel, Sobel and Weiss (1975) provide a well-written introduction to research on adaptive sampling in the normal theory case.

When the data are not normally distributed, there are several alternative nonparametric selection procedures that can be used. Assume that $Y_{i1}, \ldots, Y_{in}$ is a random sample from the distribution $F(x - \theta_i)$ ($1 \leq i \leq t$) where $F(\cdot)$ is an unknown but symmetric c.d.f. and $\theta_{[1]} \leq \cdots \leq \theta_{[t]}$ are the unknown, ordered (centers) of the t treatment distributions. The literature contains a number of procedures for selecting the treatment with location parameter $\theta_{[t]}$ based on this model. Historically, Lehmann (1963) was the first to propose a single-stage procedure for the location model based on the (joint) ranks of the pooled observations; the configuration $\theta_{[1]} = \cdots = \theta_{[t-1]} = \theta_{[t]} - \delta^*$ is not least favorable (see Rizvi and Woodworth 1970 and the references therein), and other procedures have superseded it (see also Bartlett and Govindarajulu 1968). The alternative procedures differ according to their choice of θ_i estimator. Among these, we mention (single-stage) procedures based on the two-sample Hodges-Lehmann estimators of $\theta_i - \theta_j$ (Randles 1970), procedures based on U-statistics (Bhapkar and Gore 1971), procedures based on the one-sample Hodges-Lehmann estimator (Ghosh 1973) and procedures based on adaptive M-estimators that first assess the tail weight of $F(\cdot)$ and use the corresponding M-estimator of θ_i (Randles, Ramberg and Hogg 1973; Moberg, Ramberg and Randles 1978).

CHAPTER 3

Selecting a Subset Containing the Best Treatment in a Normal Response Experiment

3.1 INTRODUCTION

This chapter discusses techniques for selecting a (random-size) subset of t treatments that contains the treatment with the largest mean. To contrast the approach of this chapter with that of Chapter 2, recall that the indifference-zone approach is ordinarily employed when *designing* experiments; the choice of the sample size n is critical to guaranteeing the probability requirement. Subset selection is concerned with screening a set of treatments to identify a subset containing the best one, and thus it can be employed when *analyzing* results from an experiment with arbitrary sample sizes. However, the consequence of using too small an n may be a lack of screening power in that a large subset is selected; thus we also describe a rational method of choosing n when sample size determination is possible.

Initially we assume a completely randomized experiment with normal errors, a common known variance, and equal numbers of observations from all treatments. Then extensions of the basic method to unbalanced data, blocked experiments, and unknown variances will be considered. We also consider related goals with emphasis on the selection of treatments, the means of which are *near* the largest treatment mean.

Statistical Assumptions: Independent random samples of n observations $Y_{i1}, \ldots, Y_{in}$ ($1 \leq i \leq t$) are taken from $t \geq 2$ normal treatments, $\Pi_1, \ldots, \Pi_t$. Treatment Π_i has unknown mean μ_i and common known or common unknown variance σ^2. It is assumed that $n \geq 2$ if σ^2 is unknown so that an estimate of σ^2 can be calculated; otherwise, $n \geq 1$ if σ^2 is known.

The ordered values of the elements of $\boldsymbol{\mu} = (\mu_1, \ldots, \mu_t)$ are denoted by $\mu_{[1]} \leq \cdots \leq \mu_{[t]}$. The values of the $\mu_{[i]}$ are assumed to be unknown as is the *pairing* of the Π_i with the $\mu_{[s]}$ ($1 \leq i, s \leq t$). The treatment having mean $\mu_{[t]}$ is referred to as the "best" treatment. Let y_{ij} denote the observed value of Y_{ij} ($1 \le i \le t$, $1 \le j \le n$). The purpose of the experiment and the associated probability (design) requirement are stated in Goal 3.1.1 and Equation (3.1.1), respectively.

Goal 3.1.1 To select a (random-size) subset that contains the treatment associated with $\mu_{[t]}$.

A *correct selection* (CS) is said to have been made if Goal 3.1.1 is achieved. If σ^2 is unknown the $P\{CS\}$ depends on (μ, σ^2) while if σ^2 is known the $P\{CS\}$ depends only on μ. As written, the probability requirement (3.1.1) assumes that σ^2 is unknown.

Probability Requirement: For specified constant P^* with $1/t < P^* < 1$, we require that

$$P\{CS \mid (\mu, \sigma^2)\} \ge P^* \qquad (3.1.1)$$

for all (μ, σ^2).

If σ^2 is known, then the probability of correct selection must be at least P^* for all μ. Any procedure that guarantees (3.1.1) is said to employ the *(random-size) subset selection* formulation for selection problems.

One use of subset selection procedures is for *screening*. The objective is to eliminate ("screen out") inferior treatments (those with the smallest μ_i-values) and to retain superior treatments. After the inferior treatments have been eliminated, then the selected treatments can be subjected to further studies, for example, to identify the best one or ones among them using the indifference-zone approach (say). For common known σ^2, this can be accomplished by using, for example, the two-stage procedure $\mathcal{N}_{TB}$ of Tamhane and Bechhofer (1977, 1979) or the two-stage procedure $\mathcal{N}_{SB}$ of Santner and Behaxeteguy (1992), each of which is discussed in Section 2.3.1.

The probability requirement (3.1.1) does not mention the size of the selected subset. Smaller subsets are clearly more informative than larger ones. In a sense, a subset of size unity is maximally informative in that it contains no extraneous treatments. In fact, the bounds on P^* in (3.1.1) can be viewed as arising from two (no-data) procedures that are extreme in their approach to subset size. One procedure chooses a "subset" of size unity consisting of a single randomly selected treatment; it achieves a $P\{CS\}$ equal to $1/t$. The second procedure selects all t treatments; this procedure achieves a $P\{CS\}$ equal to unity but with a subset of size t. Clearly there is a vast difference between the two procedures. The first chooses a subset that is maximally informative, while the second procedure chooses a subset that is minimally informative.

In general, it is desirable that the sample size be such that, on the average, the number of treatments in the selected subset will be (very) small if the $t - 1$ smallest treatment means differ (substantially) from $\mu_{[t]}$. Thus the true but unknown configuration of the treatment means should be reflected in the distribution of the random number of treatments in the selected subset. This fact provides the basis for one

SINGLE-STAGE PROCEDURES

method of choosing n, namely, by setting an upper bound on the expected number of treatments selected or the expected average rank of the selected treatments when the true treatment means are sufficiently "spread out." The implementation of this idea is discussed in Section 3.2.

Lastly, we note that unlike the indifference-zone probability requirement (2.1.1), it will be seen that the subset selection probability requirement (3.1.1) can be guaranteed using a *single-stage* procedure even if the common variance is *unknown*. By way of contrast, when the common variance is unknown, the indifference-zone approach must employ at least a two-stage procedure to guarantee its probability requirement (2.1.1) when selecting the treatment associated with $\mu_{[t]}$.

The order of presentation in the remainder of Chapter 3 is as follows. Section 3.2 discusses completely randomized experiments with common known or common unknown variance. Section 3.3 considers experiments that involve blocking. Section 3.4 concludes with a description of alternative goals.

3.2 SINGLE-STAGE PROCEDURES

This section describes statistical procedures that have as their goal the selection of a random-size subset that contains the treatment associated with $\mu_{[t]}$.

3.2.1 Balanced Experiments

Gupta (1956, 1965) proposed a single-stage procedure for this problem that is applicable to data obtained either from designed experiments or from observational studies. For situations in which the sample size n can be specified by the experimenter, a criterion for choosing n is discussed.

Procedure $\mathcal{N}_G$ (Completely Randomized Design) Calculate the t sample means $\bar{y}_i = \sum_{j=1}^{n} y_{ij}/n$ $(1 \leq i \leq t)$. Let $\bar{y}_{[1]} \leq \cdots \leq \bar{y}_{[t]}$ denote their ordered values. If σ^2 is unknown also calculate

$$s_\nu^2 = \sum_{i=1}^{t}\sum_{j=1}^{n}(y_{ij} - \bar{y}_i)^2/\nu,$$

the unbiased pooled estimate of σ^2 based on $\nu = t(n-1)$ d.f.

Case 1 (σ^2 Known): Include Π_i in the selected subset if and only if

$$\bar{y}_i \geq \bar{y}_{[t]} - h\sigma\sqrt{2/n} \qquad (3.2.1)$$

where $h = Z_{t-1,1/2}^{(1-P^*)}$ is from Table B.1.

Case 2 (σ^2 Unknown): Include Π_i in the selected subset if and only if

$$\bar{y}_i \geq \bar{y}_{[t]} - h s_\nu \sqrt{2/n} \qquad (3.2.2)$$

where $h = T_{t-1,\nu,1/2}^{(1-P^*)}$ is from Table B.3.

If σ^2 is known, the constant h used in (3.2.1) is the upper-$(1 - P^*)$ equicoordinate critical point of the equicorrelated multivariate standard normal distribution that was introduced in Section 2.2. If σ^2 is unknown, the constant h used in (3.2.2) is the equicoordinate critical point of the equicorrelated multivariate central t-distribution that was introduced in Section 2.7.1. We remind the reader that both critical points can be computed using the FORTRAN programs MVNPRD and MVTPRD of Dunnett (1989). It can be shown that as $\nu \to \infty$, both s_ν^2 converges stochastically to σ^2 and $T_{p,\nu,\rho}^{(\alpha)}$ decreases to $Z_{p,\rho}^{(\alpha)}$ for any α and ρ. Thus the procedures for σ^2 known and for σ^2 unknown become identical as the treatment variability is estimated more precisely, that is, with more degrees of freedom.

Visualizing Procedure $\mathcal{N}_G$

The decision rule used by procedure $\mathcal{N}_G$ can be visualized as follows: The treatment Π_i is selected if its sample mean, $\bar{y}_i$, falls "sufficiently close" to that of the largest sample mean, $\bar{y}_{[t]}$. The quantity $h\sigma\sqrt{2/n}$ [or $h s_\nu \sqrt{2/n}$] is the "yardstick" used to measure this distance, where h is obtained from Table B.1 [Table B.3]. The length of the yardstick is a constant times the true (or estimated) standard error of the difference between (any two) treatment means $\bar{Y}_i - \bar{Y}_j$, namely, $[\text{Var}(\bar{Y}_i - \bar{Y}_j)]^{1/2} = \sigma\sqrt{2/n}$. For example, for the data illustrated in Figure 3.1 with σ^2 unknown, the procedure $\mathcal{N}_G$ would select the treatments associated with $\bar{y}_{[t-2]}, \bar{y}_{[t-1]},$ and $\bar{y}_{[t]}$.

Procedure $\mathcal{N}_G$ achieves the goal of selecting a subset containing the treatment associated with $\mu_{[t]}$ with probability at least P^* no matter what the underlying $\boldsymbol{\mu}$. An important reason why procedure $\mathcal{N}_G$ can guarantee this probability requirement is that it selects a *random* number of treatments between 1 and t. As the standard deviation σ increases or as the sample size n decreases, the length of the yardstick increases, and hence the number of treatments selected tends to increase. In the worst-case situation for the experimenter, the procedure $\mathcal{N}_G$ can select *all* t treatments which becomes more likely for the combination of a "large" σ or a "small" n or a configuration of "close" means μ_i. In such a case, the only real interpretation is that the standard error

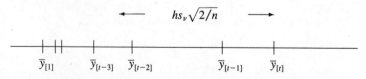

Figure 3.1. Yardstick used by procedure $\mathcal{N}_G$ to select treatments.

SINGLE-STAGE PROCEDURES

of the difference between treatment means, $\sigma\sqrt{2/n}$, is "large" relative to the true treatment differences $\mu_{[t]} - \mu_{[i]}$ ($1 \leq i < t$).

In one sense, procedure $\mathcal{N}_G$ operates similarly to that of a normal means confidence interval procedure. To illustrate this similarity, suppose that σ^2 is unknown. Then the *number* of treatments that procedure $\mathcal{N}_G$ selects *increases* as the inference conditions become more difficult (that is, σ increases, n decreases, or $\mu_{[t]} - \mu_{[i]}$ decreases for any $1 \leq i < t$). This behavior of the size of the selected subset is similar to that of the *length* of the normal theory t-interval for a mean; the length of the confidence interval increases as inferences become more difficult (that is, σ increases or n decreases).

However, the analogy with confidence intervals is not perfect because normal theory intervals achieve their nominal (targeted) confidence level *exactly* no matter what the true underlying mean and variance. By way of contrast, the value of the probability of correct selection for procedure $\mathcal{N}_G$ depends on the true treatment differences; procedure $\mathcal{N}_G$ achieves at least its nominal confidence level for all $\boldsymbol{\mu}$, but, as indicated below, it is *conservative* whenever $\mu_{[t]} > \mu_{[t-1]}$.

Performance of Procedure $\mathcal{N}_G$

Let S denote the (random) number of treatments in the selected subset; clearly S can take values $1, 2, \ldots, t$. For fixed σ^2 and $\boldsymbol{\mu}$ with $\mu_{[t-1]} < \mu_{[t]}$, we have

$$E\{S \mid (\boldsymbol{\mu}, \sigma^2)\} \to 1 \text{ and } P\{CS \mid (\boldsymbol{\mu}, \sigma^2)\} \to 1 \text{ as } n \to \infty.$$

This limiting behavior has several applications. First, the fact that the $P\{CS \mid (\boldsymbol{\mu}, \sigma^2)\} \to 1$ illustrates the important difference between the confidence guarantee for subset selection procedures and that for confidence interval procedures. The *achieved* confidence level of procedure $\mathcal{N}_G$ depends on the true $\boldsymbol{\mu}$ and is at least equal to its nominal level, but it can be arbitrarily close to unity depending on the difference between $\mu_{[t]}$ and $\mu_{[t-1]}$.

Second, the fact that $E\{S \mid (\boldsymbol{\mu}, \sigma^2)\} \to 1$ as $n \to \infty$ suggests the use of n to control the number of treatments in the selected subset (when the experimenter has that option). For example, for $P^* = 0.90$ and 0.95, Table 3.1 lists the expected proportion of treatments, $E\{S \mid (\boldsymbol{\mu}, \sigma^2)\}/t$, that procedure $\mathcal{N}_G$ selects when σ^2 is known and the means are equally spaced $\delta\sigma$ units apart, that is, $\boldsymbol{\mu}$ is of the form $(\mu, \mu + \delta\sigma, \ldots, \mu + (t-1)\delta\sigma)$.

More generally the FORTRAN program EVALNG in Appendix C uses simulation to calculate $E\{S \mid (\boldsymbol{\mu}, \sigma^2)\}$ at arbitrary slippage or equally spaced configurations.

Example 3.2.1 (Designing a Screening Procedure) To illustrate the use of Table 3.1, consider a study in which there are $t = 10$ treatments and suppose that one is interested in the behavior of procedure $\mathcal{N}_G$ for the configuration in which the means are equally spaced one-quarter of a standard deviation apart, that is, $\delta = 1/4$. Table 3.2 lists the expected proportion of the 10 treatments selected by procedure $\mathcal{N}_G$ when $P^* = 0.90$ and $\boldsymbol{\mu} = (\mu, \mu + \sigma/4, \ldots, \mu + 9\sigma/4)$ for four sample sizes n. For example, the row of Table 3.2 corresponding to $n = 4$ is computed by noting that $\sqrt{n}\delta = 2/4 = 0.50$ and thus from Table 3.1 we obtain the expected proportion

Table 3.1. Expected Proportion of Treatments Selected by Procedure $\mathcal{N}_G$ with Known σ in the Equally-Spaced Configuration with Means $\delta\sigma$ Apart

$P^\star$	t	$\delta\sqrt{n}$						
		0.5	1.0	1.5	2.0	3.0	4.0	5.0
0.90	2	0.886	0.847	0.789	0.722	0.600	0.530	0.506
	3	0.864	0.768	0.651	0.552	0.432	0.368	0.342
	4	0.833	0.674	0.529	0.438	0.339	0.284	0.259
	5	0.793	0.582	0.441	0.363	0.279	0.232	0.209
	10	0.550	0.329	0.244	0.199	0.151	0.124	0.108
	15	0.393	0.231	0.171	0.139	0.105	0.085	0.073
	20	0.307	0.180	0.132	0.107	0.080	0.066	0.056
0.95	2	0.939	0.908	0.859	0.795	0.658	0.559	0.515
	3	0.923	0.844	0.730	0.619	0.476	0.394	0.351
	4	0.900	0.757	0.598	0.490	0.373	0.306	0.268
	5	0.868	0.661	0.497	0.404	0.307	0.250	0.217
	10	0.630	0.371	0.272	0.219	0.165	0.134	0.113
	15	0.449	0.259	0.189	0.152	0.114	0.093	0.077
	20	0.348	0.200	0.146	0.117	0.087	0.071	0.059

From Gupta (1965), p. 242. Reprinted with permission from *Technometrics*. Copyright 1965 by the American Statistical Association and the American Society for Quality Control. All rights reserved.

of treatments selected as 0.550. The rows corresponding to $n = 9$ and $n = 36$ in Table 3.2 are obtained by interpolation in Table 3.1. As n increases from 4 to 36 observations, the expected proportion of treatments selected decreases from 0.550 to 0.244. Of course, if the experimenter had the option of planning a two-stage experiment at the outset, then the Tamhane and Bechhofer (1977, 1979) procedure $\mathcal{N}_{TB}$ or the Santner and Behaxeteguy (1992) procedure $\mathcal{N}_{SB}$ could be used to determine simultaneously the amount of sampling in both stages of the experiment.

Example 3.2.2 We illustrate the use of Appendix C's FORTRAN program EVALNG to evaluate the performance characteristics of procedure $\mathcal{N}_G$. Suppose that an experimenter wishes to use the procedure to evaluate screening in an experiment with $t = 6$ treatments when the measurement error standard deviation is $\sigma = 4.0$ and the desired $P\{CS\}$ is at least $P^\star = 0.85$. First suppose that the resources available allow $n = 20$

Table 3.2. Expected Proportion of Treatments Selected in the Equally-Spaced Configuration with Means $\sigma/4$ Apart When $t = 10$

n	$E\{S\}/t$
4	0.550
9	0.439
16	0.329
36	0.244
64	0.199

SINGLE-STAGE PROCEDURES

Table 3.3. $P\{S \geq k\}$ for $k = 2(1)6$ When $\sigma = 4.0$ and $\mu = (0,\ldots,0,\delta)$ for $\delta = 0.75(0.25)2.25$, $P^* = 0.85$ and $n = 20$

	k				
δ	2	3	4	5	6
0.75	0.99	0.95	0.87	0.70	0.41
1.00	0.98	0.93	0.83	0.65	0.36
1.25	0.97	0.90	0.78	0.59	0.31
1.50	0.94	0.85	0.72	0.52	0.26
1.75	0.91	0.78	0.63	0.43	0.20
2.00	0.87	0.72	0.54	0.35	0.16
2.25	0.81	0.62	0.44	0.27	0.11

observations to be taken from each treatment. Table 3.3 lists the probabilities that S is *at least* $2,\ldots,6$ treatments when the treatment means are in the slippage configuration $(0,\ldots,0,\delta)$ for $\delta = 0.75(0.25)2.25$. For example, when $\delta = 2.0$, the table shows that there is still over a 50% chance that the selected subset will contain four or more treatments. Suppose that the experimenter decides that this is inadequate screening power and elects to increase the sample size to 30 per treatment. Then $P\{S \geq 4\}$ decreases to 0.37 when $\delta = 2.0$.

Example 3.2.3 (Selection of a Subset Containing the Smallest Mean) Table 3.4 reproduces the results of a completely randomized experiment described in Lowe (1935) and analyzed in Snedecor and Cochran (1967, pp. 258–262); these data are the amounts of fat absorbed by doughnuts during cooking using one of $t = 4$ brands of shortening (A, B, C, D). For each of the four fats, six batches of doughnuts were prepared, a batch consisting of 24 doughnuts. The sample variance pooled over the four brands is $s_\nu^2 = 100.9$ based on $\nu = 4(6 - 1) = 20$ d.f. Suppose that our objective is to determine, with confidence coefficient 0.95, a subset containing the brand that absorbs the *least* amount of fat. Notice that this is the symmetric version of Goal (3.1.1); we would like to select the treatment with the *smallest* mean $\mu_{[1]}$. A minor

Table 3.4. Grams of Fat Absorbed per Batch for Four Brands of Shortening

	Brand			
Observation	A	B	C	D
1	164	178	175	155
2	172	191	193	166
3	168	197	178	149
4	177	182	171	164
5	156	185	163	170
6	195	177	176	168
Means	172	185	176	162

Reprinted from Snedecor and Cochran (1967), p. 259, by courtesy of Iowa State University Press.

modification of procedure $\mathcal{N}_G$ selects those brands for which

$$\bar{y}_i \leq \bar{y}_{[1]} + hs_{20}\sqrt{2/6}$$

where $h = T^{(0.05)}_{4-1,20,1/2} = 2.19$ is from Table B.3. Therefore, procedure $\mathcal{N}_G$ selects brands A and D because their means satisfy

$$\bar{y}_i \leq 162 + 2.19\sqrt{100.9}\sqrt{2/6} = 174.7.$$

Remark 3.2.1 As was the case for the indifference-zone probability requirement (2.1.1) and procedure $\mathcal{N}_B$, the subset selection probability requirement (3.1.1) and selection procedure $\mathcal{N}_G$ are location invariant. Thus, procedure $\mathcal{N}_G$ yields the *same* subset under a common shift of all the data and has the same $P\{CS \mid (\boldsymbol{\mu}, \sigma^2)\}$ under a common shift of all the t treatment means.

3.2.2 Unbalanced Experiments

Statistical procedures for selecting a subset of treatments from an unbalanced one-way completely randomized experiment have been studied by a number of authors. We will describe the procedure proposed by Gupta and Huang (1976).

Assume that independent random samples are taken from the t treatments. There are n_i normally distributed observations, Y_{ij} ($1 \leq j \leq n_i$), taken from the ith treatment Π_i ($1 \leq i \leq t$); observation Y_{ij} has mean μ_i and variance σ^2 ($1 \leq i \leq t, 1 \leq j \leq n_i$). In the same way as for procedure $\mathcal{N}_G$, we consider two cases: σ^2 known and σ^2 unknown. Throughout we let $n_{[1]} \leq \cdots \leq n_{[t]}$ denote the ordered sample sizes. The pairing of the $n_{[i]}$ and the ordered means $\mu_{[i]}$ is assumed to be unknown.

Procedure $\mathcal{N}_{GH}$ (Completely Randomized Design) Calculate $\bar{y}_i = \sum_{j=1}^{n_i} y_{ij}/n_i$ ($1 \leq i \leq t$); and if σ^2 is unknown, also calculate

$$s_\nu^2 = \sum_{i=1}^{t} \sum_{j=1}^{n_i} (y_{ij} - \bar{y}_i)^2 / \nu,$$

the unbiased pooled estimate of σ^2 based on $\nu = \sum_{i=1}^{t}(n_i - 1)$ d.f.

Case 1 (σ^2 Known): Include Π_i in the selected subset if and only if

$$\bar{y}_i \geq \max_{1 \leq j \leq t} \left\{ \bar{y}_j - h\sigma \sqrt{\frac{1}{n_i} + \frac{1}{n_j}} \right\}$$

where h solves Equation (3.2.3), below.

SINGLE-STAGE PROCEDURES

Case 2 (σ^2 Unknown): Include Π_i in the selected subset if and only if

$$\bar{y}_i \geq \max_{1 \leq j \leq t} \left\{ \bar{y}_j - hs_\nu \sqrt{\frac{1}{n_i} + \frac{1}{n_j}} \right\}$$

where h solves Equation (3.2.4), below.

As usual, the form of the yardstick is a constant times the true (or estimated) standard error of the estimated treatment differences, $\bar{Y}_i - \bar{Y}_j$, namely, $\sigma \sqrt{1/n_i + 1/n_j}$. Procedure $\mathcal{N}_{GH}$ reduces to procedure $\mathcal{N}_G$ in the special case $n_1 = \cdots = n_t$.

The constants required to implement procedure $\mathcal{N}_{GH}$ are more difficult to determine than those for the balanced procedure $\mathcal{N}_G$. Let

$$\lambda_i = \sqrt{n_{[i]}/(n_{[i]} + n_{[t]})} \qquad (1 \leq i < t).$$

Then $\lambda_1 \leq \cdots \leq \lambda_{t-1}$ because the $n_{[i]}$ are ordered. If σ^2 is known (Case 1), then the constant h is defined as the upper-$(1 - P^*)$ equicoordinate critical point

$$P \left\{ \max_{1 \leq i \leq t-1} W_i \leq h \right\} = P^* \qquad (3.2.3)$$

where $(W_1, \ldots, W_{t-1})$ has the $(t-1)$-variate normal distribution with mean vector zero, unit variances, and correlation matrix having ijth element $\lambda_i \times \lambda_j$ for $i \neq j$.

If σ^2 is unknown (Case 2), then the constant h is defined as the upper-$(1 - P^*)$ equicoordinate critical point

$$P \left\{ \max_{1 \leq i \leq t-1} T_i \leq h \right\} = P^* \qquad (3.2.4)$$

where $(T_1, \ldots, T_{t-1})$ has the $(t-1)$-variate central t-distribution with the covariance matrix given in the previous paragraph and $\nu = \sum_{i=1}^{t}(n_i - 1)$ d.f.

The exact critical points required to implement procedure $\mathcal{N}_{GH}$ can be computed using the FORTRAN programs MVNPRD and MVTPRD of Dunnett (1989). Lacking access to this program, the following conservative choice of critical point (derived by an application of Slepian's inequality) can be used. Recall that $Z_{p,\rho}^{(\alpha)}$ is the upper-α equicoordinate critical point of the multivariate normal distribution with common correlation ρ defined by Equation (2.2.2) and that $T_{p,\nu,\rho}^{(\alpha)}$ is the corresponding value for the multivariate t-distribution with common correlation ρ defined by Equation (2.7.4). Then

$$\rho^* = \min_{r \neq s} \lambda_r \lambda_s = \lambda_1 \lambda_2$$

is the minimum correlation for the multivariate normal or multivariate t-distribution defined in Equations (3.2.3) and (3.2.4), respectively. The critical points

$$h = Z_{t-1,\rho^*}^{(1-P^*)} \quad \text{or} \quad h = T_{t-1,\nu,\rho^*}^{(1-P^*)} \qquad (3.2.5)$$

Table 3.5. Coagulation Times (seconds) for Blood Drawn from 24 Animals Randomly Allocated to Four Different Diets

Observation	Diet			
	A	B	C	D
1	62	63	68	56
2	60	67	66	62
3	63	71	71	60
4	59	64	67	61
5		65	68	63
6		66	68	64
7				63
8				59
Means	61	66	68	61

Reprinted from Box, Hunter and Hunter (1978), p. 166, by courtesy of John Wiley & Sons.

are conservative constants for Cases 1 and 2, respectively. Tables B.2 and B.4 list the equicoordinate upper percentiles required in Equation (3.2.5) for a selection of α, p and $\rho^\star$; Table A of Bechhofer and Dunnett (1988) lists them for a much denser set of α, p and $\rho^\star$.

Example 3.2.4 (Subset Selection in the Unbalanced Case) Box, Hunter and Hunter (1978, p. 166) list the coagulation times of blood from 24 animals, each fed one of four diets A, B, C, and D. For convenience, the data are reproduced in Table 3.5. Suppose that it is desired to identify a subset of the diets that contains the diet with the longest mean coagulation time. Unequal numbers of observations had been allocated to the four diets. The estimated variance is $s_\nu^2 = 5.6$ based on $\nu = 20$ d.f. The ordered sample sizes are $n_{[1]} = 4$, $n_{[2]} = 6$, $n_{[3]} = 6$, and $n_{[4]} = 8$. The quantities used to define the correlation matrix that determines the exact critical point for procedure $\mathcal{N}_{GH}$ are $\lambda_1 = \sqrt{4/(4+8)} = 0.577$ and $\lambda_2 = \sqrt{6/(6+8)} = 0.655 = \lambda_3$. Thus the correlation matrix is

$$\begin{bmatrix} 1 & \lambda_1\lambda_2 & \lambda_1\lambda_3 \\ \lambda_1\lambda_2 & 1 & \lambda_2\lambda_3 \\ \lambda_1\lambda_3 & \lambda_2\lambda_3 & 1 \end{bmatrix} = \begin{bmatrix} 1 & 0.378 & 0.378 \\ 0.378 & 1 & 0.429 \\ 0.378 & 0.429 & 1 \end{bmatrix}.$$

First, we illustrate procedure $\mathcal{N}_{GH}$ using the *conservative* constant h defined by the central multivariate t-distribution with correlation $\rho^\star = \min\{0.378, 0.429\} = 0.378$ and 20 d.f. From Equation (3.2.5) we obtain $h = T_{3,20,0.378}^{(0.05)} = 2.222$. For each treatment, Table 3.6 lists the sample mean and corresponding value of the maximum that defines the yardstick for procedure $\mathcal{N}_{GH}$. We conclude that, with confidence coefficient at least 0.95, only diets B and C need be considered in further research to determine the diet with the longest coagulation time. Lastly, we note (using Dunnett's program) that the exact critical point based on $(\lambda_1, \lambda_2, \lambda_3) = (0.577, 0.655, 0.655)$ is 2.218 in contrast to the 2.222 value used above. Thus the *approximate* critical point is very close to the exact critical point for this example.

Table 3.6. Summary of Calculations to Implement Procedure $\mathcal{N}_{GH}$ for the Unbalanced Experiment of Box, Hunter and Hunter (1978, p. 166)

Treatment	i	n_i	$\bar{y}_i$	$\max_{1 \leq j \leq t} \left\{ \bar{y}_j - hs_\nu \sqrt{\frac{1}{n_i} + \frac{1}{n_j}} \right\}$
A	1	4	61	64.59
B	2	6	66	64.95
C	3	6	68	64.95
D	4	8	61	65.15

3.2.3 Robustness of Procedure $\mathcal{N}_G$

The performance characteristics of subset selection procedure $\mathcal{N}_G$ and the indifference-zone procedure $\mathcal{N}_B$ are closely related; indeed both use the same tables to determine the constants required for their implementation. Van der Laan, his co-workers, and several other authors have performed simulation studies of the robustness of $\mathcal{N}_G$ (and $\mathcal{N}_B$) to violations of the normality of the true location model and to heterogeneity of the treatment distribution variances. They considered simultaneously the robustness of both procedures in a series of articles (Driessen, van der Laan and van Putten 1990; van der Laan 1992a; van der Laan and van Putten 1988; Listing and Rasch 1993; Mengersen 1992). Their main conclusions are that the results of Section 2.2 concerning the performance of $\mathcal{N}_B$ under non-normality and heterogeneity of variances also hold for procedure $\mathcal{N}_G$.

Specifically, for non-normal location families, procedure $\mathcal{N}_G$ is *not sensitive* to lack of normality; for such families, procedure $\mathcal{N}_G$ will have nearly nominal performance characteristics when n is not too small. However, the achieved P{CS} of procedure $\mathcal{N}_G$ is *very sensitive* to violations of the assumption of a common known variance. For even moderate values of the ratio of the maximum treatment variance to the minimum variance and small t, the achieved P{CS} can be much less than the nominal level. The experimenter will be afforded the greatest protection against deleterious effects of heteroscedastic variances on the achieved P{CS} under the LF-configuration of means $\mu_{[1]} = \mu_{[t-1]} = \mu_{[t]} - \delta^*$ if the sample sizes per treatment are equal.

Goldsman and Petit (1994) used analytical and Monte Carlo techniques to study the performance of procedure $\mathcal{N}_G$ under violations of independence. They found that the conclusions concerning $\mathcal{N}_B$ also generally apply to procedure $\mathcal{N}_G$. In particular, the selection procedure $\mathcal{N}_G$ is *very sensitive* to lack of independence: If successive observations on the t treatments are correlated, then the achieved P{CS} can be either less than or greater than the nominal value (that is, "liberal" or "conservative," respectively). For instance, suppose that the observations within a treatment arise from a first-order autoregressive process [AR(1)], as described in Section 2.2.3. If the AR(1) coefficient $\rho > 0$ [$\rho < 0$], then the authors found that P{CS} < P^* [P{CS} > P^*]. As another example, suppose that the observations consist of n independent and identically distributed t-vectors, each having a multivariate normal distribution with constant correlation ρ. Then the procedure $\mathcal{N}_G$ is liberal if $\rho > 0$ and conservative if $-1/(t-1) < \rho < 0$.

3.3 EXPERIMENTS WITH BLOCKING

As with the indifference-zone approach, subset selection can be implemented in experiments with one or more blocking factors. This section illustrates the appropriate modifications of procedure $\mathcal{N}_G$ to be used with randomized complete blocks, balanced incomplete blocks, and Latin square designs.

In all cases the modified procedure selects those treatments that have best linear unbiased estimates (BLUEs) of treatment effects within a specified yardstick of the maximum estimated treatment effect. Following the pattern set in Section 3.2, the yardstick is chosen to be a constant times the true (or estimated) standard error of the estimated difference between two treatment effects. Any unknown σ^2 is estimated by the residual mean square in the ANOVA table corresponding to the design employed and has the associated d.f.

3.3.1 Randomized Complete Block Designs

Suppose that the experimental units can be grouped into sufficiently large blocks of uniform material so that all t treatments can be used exactly once in each block. This is the randomized complete block design (RCBD) with fixed treatment effects.

Statistical Assumptions (RCBD): The observation Y_{ij} on the ith treatment in the jth block satisfies

$$Y_{ij} = \mu + \tau_i + \beta_j + \epsilon_{ij} \qquad (3.3.1)$$

where the ϵ_{ij} are independent and identically distributed normal random variables with mean zero and common known or common unknown variance σ^2 ($1 \le i \le t$, $1 \le j \le b$).

The τ_i and β_j in the model statement (3.3.1) are the *treatment* and *block* effects, respectively. The RCBD model assumes that every fixed pair of treatments has the *same* difference between treatment effects in *all* blocks; that is, there is no block × treatment interaction. The identifiability conditions $\sum_{i=1}^{t} \tau_i = \sum_{j=1}^{b} \beta_j = 0$ uniquely define the treatment effects τ_i and block effects β_j. The ordered treatment effects are denoted by $\tau_{[1]} \le \cdots \le \tau_{[t]}$. As usual, y_{ij} denotes the observed value of Y_{ij}. Henceforth, a dot replacing a subscript denotes that an average has been computed with respect to that subscript; for example, $\bar{y}_{\cdot j} = \sum_{i=1}^{t} y_{ij}/t$.

The treatment having effect $\tau_{[t]} = \max\{\tau_1, \ldots, \tau_t\}$ is referred to as the "best" treatment. The purpose of the experiment and the associated probability (design) requirement are stated in Goal 3.3.1 and Equation (3.3.2), respectively.

Goal 3.3.1 To select a (random-size) subset that contains the treatment effect associated with $\tau_{[t]}$.

A *correct selection* (CS) is said to have been made if Goal 3.3.1 is achieved.

EXPERIMENTS WITH BLOCKING

Probability Requirement: For specified constant P^* with $1/t < P^* < 1$, we require that

$$P\{CS\} \geq P^* \tag{3.3.2}$$

for all $(\mu, \tau_i, \beta_j, \sigma^2)$.

Procedure $\mathcal{N}_G$ (Randomized Complete Block Designs) Calculate the BLUEs of the treatment effects, that is, $\hat{\tau}_i = \bar{y}_{i\cdot} - \bar{y}_{\cdot\cdot}$ ($1 \leq i \leq t$). Let $\hat{\tau}_{[1]} \leq \cdots \leq \hat{\tau}_{[t]}$ denote the ordered values of the $\hat{\tau}_i$. If σ^2 is unknown, also calculate

$$s_\nu^2 = \sum_{i=1}^{t}\sum_{j=1}^{b}(y_{ij} - \bar{y}_{i\cdot} - \bar{y}_{\cdot j} + \bar{y}_{\cdot\cdot})^2/\nu,$$

the unbiased estimate of σ^2 based on $\nu = (t-1)(b-1)$ d.f.

Case 1 (σ^2 Known): Include Π_i in the selected subset if and only if

$$\hat{\tau}_i \geq \hat{\tau}_{[t]} - h\sigma\sqrt{2/b}$$

where $h = Z_{t-1,1/2}^{(1-P^*)}$ is from Table B.1.

Case 2 (σ^2 Unknown): Include Π_i in the selected subset if and only if

$$\hat{\tau}_i \geq \hat{\tau}_{[t]} - hs_\nu\sqrt{2/b}$$

where $h = T_{t-1,\nu,1/2}^{(1-P^*)}$ is from Table B.3.

Note that the variance of the difference between the estimated treatment effects $\hat{\tau}_i - \hat{\tau}_j$ is $\mathrm{Var}(\bar{Y}_{i\cdot} - \bar{Y}_{j\cdot}) = 2\sigma^2/b$ for every pair of treatments. In both the known and unknown σ^2 cases, the $P\{CS\}$ does *not* depend on μ or on the block effects β_j and is at least P^* for all τ_i. The number of blocks in the experiment is the analog of the number of replications for each treatment in the balanced randomized complete block designs discussed in Section 3.2. The number of treatments selected depends on the true treatment effects τ_i, σ^2 (or s_ν^2), the number of blocks b, and the specified P^*. The values in Table 3.1 can be used as a guide to the number of treatments selected by experiments with various numbers of blocks.

3.3.2 Balanced Incomplete Block Designs

The simplest incomplete block design for which subset selection can be implemented is the balanced incomplete block design (BIBD). The modification of procedure $\mathcal{N}_G$ required for subset selection in a BIBD requires knowledge of both the BLUEs of the treatment effects and the variance of the differences between the estimated treatment effects.

Let k denote the block size, b the number of blocks in the design, r the (common) number of observations taken on each treatment in the design, and λ the common number of times each pair of treatments appears together in the same block over the whole design. Lastly, let D denote the set of treatment × block pairs (i, j) that appear in the design. For example, if the rows represent treatments and columns represent blocks, then Table 3.7 contains a design with $D = \{(1, B1), (2, B1), (3, B2), (1, B2), (2, B3), (3, B3)\}$.

Statistical Assumptions (BIBD): For $(i, j) \in D$ the response Y_{ij} on the ith treatment in the jth block satisfies

$$Y_{ij} = \mu + \tau_i + \beta_j + \epsilon_{ij}$$

where the ϵ_{ij} are independent and identically distributed normal random variables with mean zero and common known or common unknown variance σ^2. The treatment effects τ_i and block effects β_j are assumed to satisfy the identifiability conditions $\sum_{i=1}^{t} \tau_i = 0 = \sum_{j=1}^{b} \beta_j$.

The purpose of the experiment and the associated probability (design) requirement are the same as those given by Goal 3.3.1 and Equation (3.3.2), respectively.

Procedure $\mathcal{N}_G$ (Balanced Incomplete Block Designs) Calculate the BLUEs of the treatment effects (subject to the identifiability constraint $\sum_{i=1}^{t} \hat{\tau}_i = 0$) as

$$\hat{\tau}_i = \frac{kT_i - B_i}{t\lambda}$$

where T_i is the sum of the responses for the ith treatment and B_i is the sum of the block totals for those blocks that *contain* the ith treatment ($1 \leq i \leq t$). If σ^2 is unknown, also calculate the usual unbiased estimate s_ν^2 of σ^2 for this design, given by Equation (3.3.5) below, and which has $\nu = kb - t - b + 1$ d.f. Let $\hat{\tau}_{[1]} \leq \cdots \leq \hat{\tau}_{[t]}$ denote the ordered values of the estimated treatment effects $\hat{\tau}_i$.

Case 1 (σ^2 Known): Include Π_i in the selected subset if and only if

$$\hat{\tau}_i \geq \hat{\tau}_{[t]} - h\sigma\sqrt{\frac{2k}{t\lambda}} \tag{3.3.3}$$

where $h = Z_{t-1,1/2}^{(1-P^*)}$ is from Table B.1.

Case 2 (σ^2 Unknown): Include Π_i in the selected subset if and only if

$$\hat{\tau}_i \geq \hat{\tau}_{[t]} - hs_\nu\sqrt{\frac{2k}{t\lambda}} \tag{3.3.4}$$

where $h = T_{t-1,\nu,1/2}^{(1-P^*)}$ is from Table B.3.

EXPERIMENTS WITH BLOCKING

Table 3.7. Assignments of $t = 3$ Treatments to $b = 3$ Blocks of Size $k = 2$

	Block	
B1	B2	B3
1	3	2
2	1	3

Note that the variance of the difference between pairs of estimated treatment effects is $\text{Var}(\hat{\tau}_i - \hat{\tau}_j) = 2k\sigma^2/t\lambda$ for $i \neq j$, which determines the form of the yardstick. When σ^2 is unknown, the numerator of its unbiased estimator is a residual sum of squares that simplifies to

$$\nu \times s_\nu^2 = \sum_{i=1}^{t}\sum_{j=1}^{b} y_{ij}^2 - \sum_{j=1}^{b} Q_j^2/k - \sum_{i=1}^{t}(kT_i - B_i)^2/\lambda kt \qquad (3.3.5)$$

where Q_j is total for the jth block.

In either the known or unknown σ^2 cases, the probability of a correct selection does not depend on μ or on the block effects β_j and is at least P^* for all τ_i. However, the size of the subset selected by procedure $\mathcal{N}_G$ depends on the variance of each estimated treatment difference. Qualitatively, choosing a design with a smaller value of $2k/t\lambda$ is more advantageous. Quantification of this fact, by using Table 3.1 to assess the expected number of treatments selected by procedure $\mathcal{N}_G$ under a given true configuration of τ_i for a particular design, is more difficult.

Example 3.3.1 (Selecting a Subset Containing the Smallest Mean) To illustrate the calculations for the BIBD case, we consider the $t = 7$ treatment example of Box, Hunter and Hunter (1978, p. 277). They describe an experiment that had been conducted to determine the weight loss, under a mechanical test, for seven types of cloth (A–G). The experimental procedure consisted of measuring the weight loss, in tenths of a milligram, for each cloth following 1000 revolutions on a Martindale wear tester. Only four test pieces could be compared simultaneously in the machine. Table 3.8 lists the results of $b = 7$ blocks (runs) on the machine, each with four pieces of cloth. Based on the data collected, suppose that it is desired to determine a subset of the treatments that contains the cloth having the *smallest* mean weight loss.

This experiment was conducted using a BIBD with block size $k = 4$, $r = 4$ replications of each treatment, and $\lambda = 2$ times in which each treatment pair appears together in the blocks of the experiment. The estimated treatment effects $\hat{\tau}_i$ are listed in Table 3.9.

The estimate of σ^2 is $s_\nu^2 = 1471.42 = (38.35)^2$ with $\nu = 15$ d.f. The standard error of each treatment difference is $\sigma\sqrt{2k/t\lambda}$, which is estimated by $s_\nu\sqrt{(2 \times 4)/(7 \times 2)}$ = $38.35 \times 0.76 = 29.15$. At the 95% confidence level, procedure $\mathcal{N}_G$, modified to select a subset containing the *smallest* treatment mean, chooses all treatments with $\hat{\tau}_i$ below $\hat{\tau}_{[1]}$ *plus* a yardstick of the form (3.3.3) or (3.3.4). Because $h = T_{6,15,1/2}^{(0.05)} = 2.52$,

Table 3.8. Weight Loss (in 0.1 mg) Following 1000 Revolutions on a Martindale Wear Tester for Seven Types of Cloth

Cloth	Block						
	1	2	3	4	5	6	7
A		344		337		369	396
B	627			537	520		602
C		233	251		278		240
D	248		211			196	273
E			160	195	199	185	
F	563	442			595	606	
G	252	226	297	300			
Total Q_j	1690	1245	919	1369	1592	1356	1511

Reprinted from Box, Hunter and Hunter (1978), p. 260, by courtesy of John Wiley & Sons.

the procedure $\mathcal{N}_G$ selects Π_i if

$$\hat{\tau}_i \leq \hat{\tau}_{[1]} + h \times (29.15) = -162.86 + (2.52)(29.15) = -89.40. \qquad (3.3.6)$$

Thus C, D and E are selected as the subset of cloth types containing the type with the smallest mean weight loss.

As in the analysis of problems employing indifference-zone selection, if the experimental units are sufficiently homogeneous so that an RCBD could be used, then adopting a BIBD is less efficient than conducting an RCBD experiment with the same number of experimental units. Here efficiency is measured by the achieved $P\{CS\}$ when using a BIBD compared to a completely randomized experiment having the same number of experimental units. Conversely, if the experimental material is not sufficiently uniform for an RCBD to be appropriate, then the (incorrect) use of an RCBD will cause an increase in the variance of the estimated differences between the treatment effects, thus lowering the achieved $P\{CS\}$.

3.3.3 Latin Square Designs

Latin square designs (LSDs) are used for single treatment-factor experiments in which two additive blocking factors are to be eliminated. Selection can be performed

Table 3.9. Estimated Treatment Effects $\hat{\tau}_i$ (in 0.1 mg) for Seven Types of Cloth

Cloth	$\hat{\tau}_i$
A	21.64
B	213.00
C	−89.93
D	−126.00
E	−162.86
F	210.07
G	−65.93

Reprinted from Box, Hunter and Hunter (1978), p. 277, by courtesy of John Wiley & Sons.

EXPERIMENTS WITH BLOCKING

for such an experiment in the following way: For a $t \times t$ LSD involving t treatments and two blocking factors each at t levels, let D denote the set of treatment–row block–column block combinations (i, j, k) present in the experiment.

Statistical Assumptions (LSD): For $(i, j, k) \in D$ the response Y_{ijk} on the ith treatment in the jth row block and kth column block satisfies

$$Y_{ijk} = \mu + \tau_i + \beta_j + \delta_k + \epsilon_{ijk}$$

where the ϵ_{ijk} are independent normal random variables with mean zero and common known or common unknown variance σ^2. The treatment effects, τ_i, and row and column effects, β_j and δ_k, are defined by the identifiability conditions $\sum_{i=1}^{t} \tau_i = 0 = \sum_{j=1}^{t} \beta_j = \sum_{k=1}^{t} \delta_k$.

The treatment having effect $\tau_{[t]} = \max\{\tau_1, \ldots, \tau_t\}$ is referred to as the "best" treatment. The purpose of the experiment and the associated probability (design) requirement are stated in Goal 3.3.1 and Equation (3.3.7), respectively. A *correct selection* (CS) is said to have been made if the goal is achieved.

Probability Requirement: For specified constant P^* with $1/t < P^* < 1$, we require that

$$P\{CS\} \geq P^* \tag{3.3.7}$$

for all $(\mu, \tau_i, \beta_j, \delta_k, \sigma^2)$.

Procedure $\mathcal{N}_G$ (Latin Square Designs) Calculate $\hat{\tau}_i = \bar{y}_{i..} - \bar{y}_{...}$ ($1 \leq i \leq t$), the BLUEs of the treatment effects. Let $\hat{\tau}_{[1]} \leq \cdots \leq \hat{\tau}_{[t]}$ denote the ordered values of the $\hat{\tau}_i$. If σ^2 is unknown, also calculate

$$s_\nu^2 = \sum_{(i,j,k) \in D} (y_{ijk} - \bar{y}_{i..} - \bar{y}_{.j.} - \bar{y}_{..k} + \bar{y}_{...})^2 / \nu,$$

the unbiased estimate of σ^2 based on $\nu = t^2 - 3t + 2$ d.f.

Case 1 (σ^2 Known): Include Π_i in the selected subset if and only if

$$\hat{\tau}_i \geq \hat{\tau}_{[t]} - h\sigma\sqrt{2/t}$$

where $h = Z_{t-1, 1/2}^{(1-P^*)}$ is from Table B.1.

Case 2 (σ^2 Unknown): Include Π_i in the selected subset if and only if

$$\hat{\tau}_i \geq \hat{\tau}_{[t]} - hs_\nu\sqrt{2/t}$$

where $h = T_{t-1, \nu, 1/2}^{(1-P^*)}$ is from Table B.3.

In a similar way, procedure $\mathcal{N}_G$ can be modified to accommodate other designs—such as Youden squares, Graeco-Latin squares, or Latin cubes—that eliminate heterogeneity in two or more directions.

3.4 ALTERNATIVE GOALS

We consider two goals in this section. The first is that of selecting a subset containing the s (given) best treatments; the parallel problem of selecting the s best treatments for IZ selection is discussed in Section 2.5. The second goal concerns selecting a subset containing "near-best" treatments. After defining this concept, we first consider a version of procedure $\mathcal{N}_G$ that selects near-best treatments using a modification of the constant in Equation (3.2.1). Then we consider an *alternative* procedure for this same goal; the new procedure allows the experimenter to control the maximum size of the selected subset. The use of the IZ procedure $\mathcal{N}_B$ to select near-best treatments is presented in the Confidence Statement discussion of Section 2.2.

3.4.1 Selection of the s Best Treatments

The statistical assumptions and notation of this subsection are the same as those given in Section 3.1. The purpose of the experiment and the associated probability (design) requirement are stated in Goal 3.4.1 and Equation (3.4.1), respectively.

Goal 3.4.1 To select a (random-size) subset that contains the s ($1 \leq s \leq t - 1$) treatments associated with $\mu_{[t-s+1]}, \ldots, \mu_{[t]}$.

A *correct selection* (CS) is said to have been made if Goal 3.4.1 is achieved. If σ^2 is unknown, the $P\{CS\}$ depends on $(\boldsymbol{\mu}, \sigma^2)$, whereas if σ^2 is known, the $P\{CS\}$ depends only on $\boldsymbol{\mu}$. As written, the probability requirement (3.4.1) assumes that σ^2 is unknown.

Probability Requirement: For specified constant P^* with $1/t < P^* < 1$, we require that

$$P\{CS \mid (\boldsymbol{\mu}, \sigma^2)\} \geq P^* \qquad (3.4.1)$$

for all $(\boldsymbol{\mu}, \sigma^2)$.

Carroll, Gupta and Huang (1975) proposed a procedure of the following form with a critical point h that made the procedure *conservative*, that is, it guarantees Equation (3.4.1) with strict inequality. Later, Bofinger and Mengersen (1986) determined a less conservative choice of critical point. We describe the Carroll, Gupta and Huang procedure employing the Bofinger and Mengersen critical value.

ALTERNATIVE GOALS

Procedure $\mathcal{N}_{CGHBM}$ Calculate the t sample means $\bar{y}_i = \sum_{j=1}^n y_{ij}/n \, (1 \leq i \leq t)$. Let $\bar{y}_{[1]} \leq \cdots \leq \bar{y}_{[t]}$ denote the ordered sample means. If σ^2 is unknown, also calculate $s_\nu^2 = \sum_{i=1}^t \sum_{j=1}^n (y_{ij} - \bar{y}_i)^2/\nu$, the unbiased pooled estimate of σ^2 based on $\nu = t(n-1)$ d.f.

Case 1 (σ^2 Known): Include Π_i in the selected subset if and only if

$$\bar{y}_i \geq \bar{y}_{[t-s+1]} - h\sigma\sqrt{2/n}$$

where h is given in Table 3.10 with $\nu = \infty$ or, equivalently, solve Equation (3.4.2), below.

Case 2 (σ^2 Unknown): Include Π_i in the selected subset if and only if

$$\bar{y}_i \geq \bar{y}_{[t-s+1]} - hs_\nu\sqrt{2/n}$$

where h is given in Table 3.10 or, equivalently, solve Equation (3.4.3), below.

When σ^2 is known, the constant h defining the yardstick is the solution of

$$(t-s+1)P_{t-1,s-1}(h) - (t-s)P_{t,s-1}(h) + P_{t-s+1,1}(h) - 1 = P^\star \quad (3.4.2)$$

where

$$P_{k,\ell}(h) = \ell \times \int_{-\infty}^{\infty} \Phi^{k-\ell}(x + h\sqrt{2})[1 - \Phi(x)]^{\ell-1} d\Phi(x)$$

and $\Phi(\cdot)$ denotes the cumulative distribution function of a standard normal random variable. The integral $P_{k,\ell}(h)$ can be expressed as the probability that a certain multivariate normal vector lies in a rectangle. Specifically,

$$P_{k,\ell}(h) = \ell \times P\{U_j \leq h \, (1 \leq j \leq k - \ell); \quad 0 < U_j < \infty \, (k - \ell + 1 \leq j < k)\}$$

where $(U_1, \ldots, U_{k-1})$ has the multivariate normal distribution with mean vector zero, unit variances, and common correlation $1/2$. Table 3.10 provides a list of solutions to Equation (3.4.2), for selected values of t, s and $P^\star$. Alternatively, the probability $P_{k,\ell}(h)$, and hence the lower bound in Equation (3.4.2), can be evaluated using the FORTRAN program MVNPRD in Dunnett (1989) (see Appendix C for a short description of this program).

When σ^2 is unknown, the constant h defining the yardstick is the solution of

$$(t-s+1)P_{t-1,s-1}(h,\nu) - (t-s)P_{t,s-1}(h,\nu) + P_{t-s+1,1}(h,\nu) - 1 = P^\star \quad (3.4.3)$$

Table 3.10. Values of the Critical Point h That Solves Equation (3.4.2) ($\nu = \infty$) and Equation (3.4.3) for $P^* = 0.75$

s	t	\multicolumn{9}{c}{ν}								
		5	10	15	20	24	30	40	60	120	∞
2	3	1.04	0.98	0.97	0.96	0.95	0.95	0.95	0.95	0.93	0.93
	4	1.43	1.33	1.30	1.29	1.28	1.27	1.27	1.27	1.26	1.24
	5	1.65	1.53	1.48	1.47	1.46	1.45	1.44	1.44	1.42	1.41
	6	1.80	1.65	1.61	1.59	1.58	1.57	1.56	1.55	1.53	1.53
	7	1.92	1.75	1.70	1.68	1.67	1.66	1.65	1.63	1.62	1.61
	8	2.01	1.83	1.78	1.75	1.74	1.73	1.72	1.70	1.69	1.68
	9	2.09	1.90	1.84	1.82	1.80	1.79	1.77	1.76	1.75	1.73
	10	2.15	1.95	1.90	1.87	1.85	1.84	1.82	1.81	1.80	1.78
3	4	1.19	1.12	1.10	1.08	1.07	1.07	1.07	1.06	1.05	1.05
	5	1.58	1.46	1.43	1.41	1.41	1.40	1.39	1.38	1.37	1.36
	6	1.82	1.66	1.61	1.59	1.58	1.57	1.56	1.56	1.54	1.53
	7	1.97	1.80	1.74	1.72	1.70	1.69	1.68	1.67	1.66	1.64
	8	2.09	1.90	1.84	1.81	1.80	1.78	1.77	1.75	1.75	1.73
	9	2.18	1.97	1.92	1.88	1.87	1.85	1.84	1.82	1.82	1.79
	10	2.27	2.04	1.97	1.94	1.93	1.91	1.90	1.88	1.87	1.85
4	5	1.29	1.20	1.17	1.17	1.16	1.15	1.15	1.14	1.13	1.12
	6	1.69	1.56	1.51	1.49	1.48	1.48	1.46	1.46	1.44	1.44
	7	1.92	1.75	1.70	1.68	1.66	1.65	1.64	1.63	1.61	1.61
	8	2.09	1.89	1.83	1.80	1.79	1.77	1.76	1.75	1.73	1.72
	9	2.21	1.99	1.92	1.90	1.88	1.86	1.85	1.83	1.82	1.80
	10	2.31	2.07	2.00	1.97	1.95	1.94	1.92	1.90	1.89	1.87
5	6	1.36	1.27	1.24	1.22	1.22	1.21	1.20	1.20	1.19	1.18
	7	1.77	1.62	1.58	1.56	1.55	1.53	1.53	1.51	1.50	1.49
	8	2.01	1.82	1.77	1.74	1.73	1.71	1.70	1.68	1.67	1.66
	9	2.17	1.96	1.90	1.87	1.85	1.83	1.82	1.80	1.79	1.77
	10	2.30	2.06	1.99	1.96	1.94	1.92	1.91	1.90	1.87	1.86
6	7	1.42	1.32	1.29	1.27	1.27	1.26	1.25	1.24	1.24	1.23
	8	1.84	1.68	1.63	1.61	1.60	1.58	1.57	1.56	1.55	1.53
	9	2.07	1.87	1.82	1.79	1.77	1.76	1.75	1.73	1.72	1.70
	10	2.24	2.02	1.94	1.92	1.90	1.89	1.87	1.85	1.84	1.82
7	8	1.47	1.36	1.33	1.32	1.31	1.30	1.29	1.29	1.27	1.27
	9	1.89	1.73	1.68	1.65	1.63	1.63	1.61	1.60	1.58	1.58
	10	2.13	1.92	1.86	1.83	1.82	1.80	1.79	1.77	1.76	1.75
8	9	1.51	1.40	1.36	1.35	1.34	1.34	1.33	1.32	1.31	1.30
	10	1.94	1.76	1.71	1.68	1.67	1.66	1.65	1.63	1.62	1.61
9	10	1.55	1.44	1.40	1.38	1.37	1.36	1.36	1.34	1.34	1.33

ALTERNATIVE GOALS

Table 3.10. *Continued*, $P^* = 0.90$

s	t	\$\nu\$ = 5	10	15	20	24	30	40	60	120	∞
2	3	1.71	1.55	1.51	1.48	1.47	1.46	1.45	1.44	1.43	1.42
	4	2.13	1.90	1.83	1.80	1.78	1.77	1.75	1.73	1.72	1.70
	5	2.38	2.09	2.01	1.97	1.95	1.93	1.92	1.90	1.87	1.86
	6	2.55	2.23	2.14	2.09	2.06	2.04	2.02	2.00	1.99	1.97
	7	2.68	2.33	2.23	2.18	2.16	2.13	2.11	2.09	2.06	2.04
	8	2.79	2.40	2.30	2.25	2.22	2.20	2.18	2.15	2.13	2.11
	9	2.87	2.47	2.36	2.31	2.28	2.26	2.23	2.21	2.18	2.16
	10	2.95	2.53	2.42	2.35	2.33	2.31	2.28	2.25	2.23	2.21
3	4	1.85	1.66	1.61	1.58	1.56	1.55	1.54	1.53	1.51	1.51
	5	2.28	2.02	1.93	1.90	1.87	1.85	1.84	1.82	1.80	1.79
	6	2.54	2.21	2.11	2.06	2.04	2.02	2.00	1.98	1.96	1.94
	7	2.72	2.35	2.24	2.18	2.16	2.14	2.11	2.09	2.06	2.05
	8	2.86	2.45	2.33	2.28	2.25	2.23	2.20	2.17	2.15	2.13
	9	2.97	2.53	2.41	2.35	2.32	2.29	2.27	2.24	2.21	2.19
	10	3.06	2.60	2.47	2.41	2.38	2.35	2.33	2.29	2.27	2.24
4	5	1.94	1.74	1.68	1.65	1.63	1.62	1.61	1.59	1.58	1.57
	6	2.40	2.09	2.00	1.97	1.94	1.92	1.90	1.88	1.86	1.85
	7	2.66	2.29	2.18	2.14	2.11	2.09	2.06	2.04	2.02	2.00
	8	2.84	2.43	2.31	2.26	2.23	2.21	2.18	2.16	2.13	2.11
	9	2.98	2.54	2.41	2.35	2.33	2.29	2.26	2.23	2.21	2.18
	10	3.10	2.62	2.49	2.43	2.40	2.36	2.33	2.30	2.27	2.25
5	6	2.02	1.80	1.73	1.70	1.69	1.68	1.66	1.65	1.63	1.61
	7	2.48	2.16	2.06	2.02	1.99	1.97	1.95	1.93	1.91	1.90
	8	2.75	2.36	2.25	2.19	2.17	2.14	2.12	2.09	2.06	2.05
	9	2.93	2.50	2.38	2.32	2.29	2.26	2.23	2.21	2.18	2.15
	10	3.08	2.61	2.47	2.41	2.38	2.35	2.32	2.29	2.26	2.23
6	7	2.09	1.85	1.78	1.75	1.73	1.72	1.70	1.69	1.68	1.65
	8	2.55	2.21	2.11	2.06	2.04	2.02	1.99	1.97	1.95	1.93
	9	2.82	2.41	2.30	2.24	2.21	2.18	2.16	2.14	2.11	2.09
	10	3.01	2.56	2.43	2.36	2.33	2.31	2.28	2.25	2.22	2.19
7	8	2.14	1.89	1.82	1.79	1.77	1.75	1.74	1.73	1.71	1.69
	9	2.61	2.25	2.15	2.10	2.08	2.06	2.03	2.01	1.99	1.97
	10	2.88	2.46	2.34	2.28	2.25	2.23	2.20	2.17	2.15	2.12
8	9	2.18	1.93	1.85	1.82	1.80	1.78	1.77	1.75	1.74	1.72
	10	2.66	2.29	2.18	2.14	2.11	2.09	2.06	2.04	2.02	1.99
9	10	2.23	1.96	1.88	1.85	1.83	1.81	1.80	1.78	1.77	1.74

Table 3.10. *Continued*, $P^* = 0.95$

s	t	ν									
		5	10	15	20	24	30	40	60	120	∞
2	3	2.23	1.95	1.87	1.84	1.82	1.80	1.78	1.77	1.75	1.74
	4	2.69	2.31	2.19	2.14	2.12	2.09	2.07	2.04	2.02	2.01
	5	2.96	2.50	2.38	2.31	2.28	2.26	2.23	2.20	2.17	2.15
	6	3.15	2.64	2.50	2.43	2.40	2.37	2.33	2.31	2.28	2.25
	7	3.30	2.74	2.59	2.52	2.48	2.45	2.42	2.38	2.35	2.33
	8	3.41	2.83	2.67	2.59	2.55	2.52	2.48	2.45	2.41	2.38
	9	3.51	2.90	2.73	2.65	2.61	2.57	2.54	2.50	2.47	2.43
	10	3.59	2.96	2.78	2.70	2.66	2.62	2.58	2.55	2.51	2.47
3	4	2.36	2.05	1.96	1.92	1.90	1.87	1.85	1.84	1.82	1.80
	5	2.84	2.40	2.28	2.22	2.19	2.16	2.14	2.11	2.09	2.07
	6	3.13	2.61	2.46	2.40	2.36	2.33	2.30	2.27	2.23	2.21
	7	3.33	2.75	2.59	2.52	2.47	2.44	2.41	2.38	2.34	2.31
	8	3.49	2.86	2.69	2.60	2.57	2.53	2.49	2.45	2.42	2.39
	9	3.61	2.95	2.76	2.68	2.64	2.60	2.56	2.52	2.48	2.45
	10	3.72	3.02	2.83	2.74	2.69	2.65	2.61	2.57	2.53	2.50
4	5	2.47	2.12	2.02	1.98	1.95	1.93	1.91	1.89	1.87	1.85
	6	2.96	2.48	2.35	2.28	2.26	2.22	2.19	2.16	2.14	2.11
	7	3.25	2.69	2.53	2.46	2.43	2.39	2.35	2.32	2.29	2.26
	8	3.46	2.84	2.66	2.58	2.54	2.50	2.46	2.43	2.39	2.36
	9	3.62	2.94	2.76	2.67	2.63	2.59	2.55	2.51	2.47	2.44
	10	3.75	3.03	2.84	2.74	2.70	2.66	2.62	2.57	2.53	2.50
5	6	2.55	2.18	2.07	2.02	2.00	1.98	1.95	1.93	1.91	1.90
	7	3.05	2.54	2.40	2.33	2.30	2.27	2.24	2.21	2.18	2.16
	8	3.35	2.75	2.59	2.51	2.47	2.43	2.40	2.36	2.33	2.30
	9	3.56	2.90	2.72	2.63	2.59	2.55	2.51	2.47	2.43	2.40
	10	3.73	3.01	2.81	2.72	2.68	2.64	2.60	2.55	2.51	2.47
6	7	2.62	2.23	2.12	2.06	2.04	2.02	1.99	1.97	1.94	1.93
	8	3.13	2.60	2.45	2.38	2.34	2.31	2.28	2.24	2.21	2.19
	9	3.43	2.81	2.63	2.55	2.51	2.47	2.44	2.40	2.36	2.33
	10	3.65	2.96	2.76	2.67	2.63	2.59	2.55	2.51	2.47	2.43
7	8	2.67	2.27	2.16	2.10	2.08	2.05	2.02	2.00	1.97	1.96
	9	3.19	2.64	2.48	2.41	2.38	2.34	2.31	2.28	2.24	2.22
	10	3.50	2.85	2.67	2.59	2.55	2.51	2.47	2.43	2.39	2.36
8	9	2.72	2.31	2.18	2.14	2.11	2.08	2.05	2.03	2.00	1.99
	10	3.25	2.67	2.52	2.45	2.40	2.37	2.34	2.31	2.27	2.24
9	10	2.77	2.34	2.21	2.16	2.14	2.11	2.08	2.05	2.02	2.01

where

$$P_{k,\ell}(h, \nu) = \ell \times \int_0^\infty \int_{-\infty}^\infty \Phi^{k-\ell}(x + h\sqrt{2}z)[1 - \Phi(x)]^{\ell-1} \, d\Phi(x) \, dG_\nu(z)$$

and $G_\nu(\cdot)$ denotes the cumulative distribution function of W, where W^2 is distributed as $\nu\chi_\nu^2$. The integral $P_{k,\ell}(h, \nu)$ can be expressed as the probability that a random variable from a certain multivariate t-distribution lies in a rectangle. Specifically,

$$P_{k,\ell}(h, \nu) = \ell \times P\{V_j \leq h \, (1 \leq j \leq k - \ell); \quad 0 < V_j < \infty \, (k - \ell + 1 \leq j < k)\}$$

where $(V_1, \ldots, V_{k-1})$ has the multivariate t-distribution with mean vector zero, unit variances, common correlation $1/2$ and ν d.f. Table 3.10 also provides a short list of solutions to Equation (3.4.3); in principle, it can be solved using the FORTRAN program MVTPRD in Dunnett (1989).

Remark 3.4.1 The least favorable configuration, the $\boldsymbol{\mu}$ where the global minimum of the $P\{CS\}$ is attained, is not known for either the case σ^2 known or σ^2 unknown. Procedure $\mathcal{N}_{CGHBM}$ is conservative because it employs a lower bound on the minimum of the $P\{CS\}$. Bofinger and Mengersen (1986) provide a small simulation study of procedure $\mathcal{N}_{CGHBM}$. They conclude that the procedure is less conservative for large P^* than small P^* and that it can perform fairly well. They also note that its behavior is a complicated function of the true underlying mean vector $\boldsymbol{\mu}$, t and s. For example, when a nominal 90% procedure is employed, the minimum of the $P\{CS\}$ appears to be about 0.92 when $(t, s) = (6, 2)$. Obviously, there is room for additional theoretical work on this problem.

Example 3.4.1 Suppose that an experiment has been performed in which $n = 10$ observations have been taken from each of six normal treatments with unknown means and common but unknown variance. It is desired to select a subset containing the two treatments with the largest treatment means using procedure $\mathcal{N}_{CGHBM}$. To achieve a 90% confidence level we must determine the yardstick h to solve the equation

$$0.90 = 5 \times P_{5,1}(h, 54) - 4 \times P_{6,1}(h, 54) + P_{5,1}(h, 54) - 1.$$

From Table 3.10 we obtain $h = 2.01$ by interpolation between the entries for $\nu = 40$ and $\nu = 60$ degrees of freedom. Thus we select all those treatments for which

$$\bar{y}_i \geq \bar{y}_{[5]} - 2.01 \times s_\nu \sqrt{2/10}.$$

3.4.2 Selection of δ^*-Near-Best Treatments

As noted in Section 2.5, there is no guarantee that the best s treatments all need to be good relative to the best treatment. This fact suggests that the experimental goal of selecting one or all "near-best" treatments may be more useful in some applications than that of selecting the best s treatments.

In particular, suppose that an experimenter is willing to consider a treatment Π_i as "equivalent" to the best treatment if its mean is sufficiently close to the mean of the best treatment. Formally, we say a treatment Π_i is δ^*-*near-best* if μ_i is within a specified amount $\delta^* > 0$ of the largest treatment mean, that is, $\mu_i \geq \mu_{[t]} - \delta^*$. This section and the next consider selection of near-best treatments. The first researcher to investigate the selection of near-best treatments was Fabian (1962).

The present section first considers the efficiency of the basic procedure $\mathcal{N}_G$ for selecting δ^*-near-best treatments. The statistical assumptions and notation are the same as those in Section 3.1 except that only the case σ^2 *known* is discussed. Section 3.4.3 considers a procedure that permits the experimenter to bound the maximum size of the selected subset and yet retains the flexibility inherent in the random subset procedure $\mathcal{N}_G$. The experimental goal and the associated probability (design) requirement are stated in Goal 3.4.2 and Equation (3.4.4), respectively.

Goal 3.4.2 To select a (random-size) subset that contains at least one treatment Π_i satisfying $\mu_i > \mu_{[t]} - \delta^*$.

If δ^*-CS denotes the event of correctly achieving Goal (3.4.2), then the probability requirement is to guarantee δ^*-CS *no matter what the underlying true vector of means*.

Probability Requirement: For specified constants (δ^*, P^*) with $\delta^* > 0$ and $1/t < P^* < 1$, we require that

$$P\{\delta^*\text{-CS}\} \geq P^* \tag{3.4.4}$$

for all $\boldsymbol{\mu}$.

With the modified constant defined in (3.4.5), below, the basic procedure of Section 3.2 will guarantee probability requirement (3.4.4) [see Roth (1978) and van der Laan (1992b)].

Procedure $\mathcal{N}_G$ (δ^*-Near-Best Selection) Calculate the t sample means $\bar{y}_i = \sum_{j=1}^{n} y_{ij}/n$ $(1 \leq i \leq t)$. Let $\bar{y}_{[1]} \leq \cdots \leq \bar{y}_{[t]}$ denote the ordered sample means. Include Π_i in the selected subset if and only if

$$\bar{y}_i \geq \bar{y}_{[t]} - h(\delta^*)\sigma\sqrt{2/n}$$

where

$$h(\delta^*) = Z_{t-1,1/2}^{(1-P^*)} - \frac{\delta^*}{\sigma}\sqrt{\frac{n}{2}} \tag{3.4.5}$$

and $Z_{t-1,1/2}^{(1-P^*)}$ is from Table B.1.

ALTERNATIVE GOALS

Intuitively, it is easier to select a δ^*-near-best treatment than to select the treatment associated with $\mu_{[t]}$. Thus it should be no surprise that the procedure $\mathcal{N}_G$ uses a *shorter* yardstick to achieve Goal 3.4.2 than that used for selecting the best treatment. Mathematically,

$$h(\delta^*) = Z^{(1-P^*)}_{t-1,1/2} - \frac{\delta^*}{\sigma}\sqrt{\frac{n}{2}} < Z^{(1-P^*)}_{t-1,1/2} = h$$

where, it will be recalled, h is the yardstick used by procedure $\mathcal{N}_G$ to select a subset of treatments containing the best treatment in accordance with Goal 3.1.1.

The minimum value of $P\{\delta^*\text{-CS}\}$ over all vectors $\boldsymbol{\mu}$ occurs when

$$\mu_{[1]} = \cdots = \mu_{[t-1]} = \mu_{[t]} - \delta^*. \tag{3.4.6}$$

When (3.4.6) holds, there is only one δ^*-near-best treatment, namely, the treatment associated with mean $\mu_{[t]}$; hence the procedure is designed in such a way that this treatment is selected with probability at least P^*.

We complete the description of procedure $\mathcal{N}_G$ with the constant $h(\delta^*)$ from (3.4.5) by contrasting its performance with that of procedure $\mathcal{N}_G$ with the constant $h = Z^{(1-P^*)}_{t-1,1/2}$ from (3.2.1) (chosen to satisfy Goal 3.1.1). Suppose that the constant $h(\delta^*)$ is chosen to guarantee $P\{\delta^*\text{-CS}\} \geq P^* = 0.90$. Table 3.11 (from van der Laan 1992b) lists the probability that procedure $\mathcal{N}_G$ selects the treatment with mean $\mu_{[t]}$ (the unique best treatment) when the means are in the slippage configuration (3.4.6). For example, when there are $t = 10$ treatments and $h(\delta^*)$ satisfies (3.4.5) for $(\delta^*, P^*) = (0.5\sigma, 0.90)$, then the achieved probability of selecting the treatment having the largest mean is 0.81. When $h(\delta^*)$ is modified so that (3.4.4) holds when $\delta^* = \sigma$, then the achieved probability of selecting the treatment having the largest mean is 0.67. Therefore, by relaxing the procedure to select a δ^*-near-best treatment rather than the best one, the procedure can substantially increase the probability of correct selection.

Remark 3.4.2 Santner (1976) and Panchapakesan and Santner (1977) presented procedures for the goals of selecting a subset of treatments containing *all* δ^*-near-best treatments and *at least* one δ^*-near-best treatment, respectively, when σ^2 is unknown.

Table 3.11. Probability that the Best Treatment Is Selected When $\boldsymbol{\mu} = (0,\ldots,0,\delta^*)$ and Procedure $\mathcal{N}_G$ Is Used with $h(\delta^*)$ Chosen to Guarantee δ^*-CS with Probability 0.90

t	$\delta^* = 0.5$	$\delta^* = 1.0$
2	0.83	0.71
4	0.82	0.69
10	0.81	0.67
50	0.80	0.65

3.4.3 A Bounded Procedure for δ^*-Near-Best Treatments

One potential drawback of procedure $\mathcal{N}_G$ is that it can select *all* t treatments; this latter outcome permits only the inference that the treatments means are "close" relative to the standard error of the differences between treatment means, $\sigma\sqrt{2/n}$.

Gupta and Santner (1973) introduced a procedure that selects a random number of treatments subject to an *a priori* specified upper bound, say q ($q < t$), on the number of treatments selected. Initially, they stressed a formulation with a goal and probability requirement that selects a random-size subset of size no greater than q that contains the best treatment with a probability at least P^* whenever $\boldsymbol{\mu}$ satisfies the condition $\mu_{[t]} - \mu_{[t-1]} \geq \delta^*$. However, their procedure achieves the *stronger* goal of selecting any δ^*-near-best treatment *no matter what the true configuration of $\boldsymbol{\mu}$* (see Hooper and Santner 1979). We stress the latter interpretation in this section.

The statistical assumptions are the same as those of Section 3.1 except that σ^2 is assumed to be *known*. The experimental goal and the associated probability (design) requirement are stated in Goal 3.4.2 and Equation (3.4.4), respectively. Gupta and Santner (1973) proposed the following procedure to achieve Goal 3.4.2 subject to guaranteeing (3.4.4). (See also Santner 1975.)

Procedure $\mathcal{N}_{GSa}$ Calculate the t sample means $\bar{y}_i = \sum_{j=1}^{n} y_{ij}/n$ ($1 \leq i \leq t$). Let $\bar{y}_{[1]} \leq \cdots \leq \bar{y}_{[t]}$ denote the ordered sample means. Include Π_i in the selected subset if and only if

$$\bar{y}_i \geq \max\left\{\bar{y}_{[t-q+1]}, \bar{y}_{[t]} - h\sigma\sqrt{2/n}\right\}$$

where h solves Equation (3.4.7), below.

As usual, the width of the yardstick for procedure $\mathcal{N}_{GSa}$, is a multiple of the standard error of the difference between pairs of sample means, namely, $\sigma\sqrt{2/n}$. The constant h in the definition of the yardstick is the solution of

$$\int_{-\infty}^{+\infty} \left[\Phi(x + h\sqrt{2} + \delta^*\sqrt{n}/\sigma)\right]^{t-1} I\left(\frac{\Phi(x + \delta^*\sqrt{n}/\sigma)}{\Phi(x + h\sqrt{2} + \delta^*\sqrt{n}/\sigma)}; t-q, q\right) d\Phi(x) = P^* \quad (3.4.7)$$

where

$$I(y; a, b) = \frac{\Gamma(a+b)}{\Gamma(a)\Gamma(b)} \int_0^y x^{a-1}(1-x)^{b-1} dx \quad (0 \leq y \leq 1)$$

is the incomplete beta function and

$$\Gamma(w) = \int_0^\infty x^{w-1} e^{-x} dx \quad (w > 0)$$

is the gamma function.

ALTERNATIVE GOALS

In practice, for given (t, q, σ) and specified $(\delta^\star, P^\star)$, either the constant h can be chosen to solve (3.4.7) for given n *or* the sample size n can be chosen to solve (3.4.7) for given h. In the first case, tables of h would be prohibitive to construct because the constant h depends on the sample size n. Thus we provide the FORTRAN program USEGSA, listed in Appendix C, to solve (3.4.7) for h given $(t, q, n, \sigma; \delta^\star, P^\star)$. In the second case, we provide Table 3.12 which can be used to determine the sample size

Table 3.12. Values of $\sqrt{n}\delta/\sigma$ that Guarantee Probability Levels $P^\star = 0.75, 0.90, 0.975$ for selected t, q and h for Procedure $\mathcal{N}_{GSa}$

| | | $h = 0.4/\sqrt{2}$ | | | $h = 0.8/\sqrt{2}$ | | | $h = 1.2/\sqrt{2}$ | | | $h = 1.6/\sqrt{2}$ | | |
| | | $P^\star$ | | | $P^\star$ | | | $P^\star$ | | | $P^\star$ | | |
t	q	0.75	0.90	0.975	0.75	0.90	0.975	0.75	0.90	0.975	0.75	0.90	0.975
3	2	1.070	1.859	2.750	0.863	1.645	2.520	0.590	1.351	2.211	0.462	1.223	2.075
4	2	1.335	2.093	2.952	1.156	1.906	2.781	0.954	1.684	2.504	0.875	1.602	2.414
	3	1.287	2.057	2.932	1.017	1.767	2.642	0.599	1.345	2.189	0.359	1.093	1.921
5	2	1.510	2.252	3.096	1.348	2.098	2.973	1.181	1.892	2.696	1.123	1.830	2.627
	3	1.454	2.204	3.063	1.192	1.942	2.817	0.841	1.560	2.380	0.669	1.376	2.173
	4	1.447	2.201	3.060	1.151	1.932	2.807	0.686	1.429	2.272	0.375	1.101	1.929
6	2	1.639	2.370	3.198	1.490	2.209	3.084	1.340	2.041	2.830	1.295	1.994	2.775
	3	1.577	2.319	3.163	1.335	2.054	2.929	1.016	1.719	2.516	0.881	1.572	2.353
	4	1.569	2.312	3.155	1.283	2.033	2.908	0.849	1.572	2.400	0.597	1.300	2.097
	5	1.569	2.311	3.155	1.281	2.030	2.907	0.783	1.521	2.365	0.428	1.155	1.983
7	2	1.739	2.461	3.282	1.601	2.319	3.194	1.464	2.157	2.939	1.426	2.115	2.897
	3	1.676	2.406	3.242	1.438	2.157	3.032	1.153	1.844	2.633	1.039	1.722	2.488
	4	1.664	2.398	3.234	1.389	2.139	3.014	0.980	1.687	2.499	0.768	1.456	2.237
	5	1.663	2.397	3.233	1.370	2.120	2.995	0.896	1.623	2.451	0.588	1.295	2.100
8	2	1.822	2.537	3.349	1.684	2.403	3.278	1.565	2.251	3.024	1.530	2.213	2.987
	3	1.756	2.480	3.308	1.529	2.248	3.123	1.261	1.946	2.728	1.164	1.840	2.597
	4	1.742	2.473	3.301	1.468	2.212	3.087	1.087	1.786	2.583	0.904	1.584	2.349
	5	1.741	2.472	3.300	1.462	2.181	3.056	0.995	1.708	2.520	0.722	1.413	2.202
9	2	1.892	2.603	3.408	1.766	2.453	3.328	1.648	2.329	3.095	1.617	2.297	3.062
	3	1.824	2.543	3.363	1.609	2.296	3.171	1.353	2.033	2.799	1.266	1.934	2.692
	4	1.810	2.533	3.361	1.541	2.260	3.135	1.178	1.868	2.649	1.016	1.688	2.446
	5	1.808	2.531	3.359	1.526	2.245	3.120	1.080	1.783	2.588	0.836	1.518	2.291
10	2	1.953	2.656	3.460	1.837	2.525	3.337	1.719	2.397	3.163	1.691	2.367	3.132
	3	1.882	2.597	3.410	1.666	2.385	3.198	1.431	2.105	2.870	1.352	2.018	2.768
	4	1.868	2.587	3.407	1.601	2.319	3.194	1.256	1.940	2.721	1.110	1.778	2.528
	5	1.864	2.583	3.403	1.585	2.304	3.179	1.153	1.851	2.647	0.933	1.607	2.372
15	2	2.169	2.860	3.641	2.056	2.744	3.494	1.969	2.637	3.387	1.949	2.613	3.363
	3	2.094	2.793	3.590	1.894	2.582	3.394	1.704	2.365	3.107	1.650	2.303	3.037
	4	2.074	2.778	3.582	1.825	2.542	3.391	1.537	2.201	2.951	1.435	2.083	2.818
	5	2.069	2.776	3.581	1.794	2.513	3.388	1.428	2.100	2.866	1.273	1.923	2.666
20	2	2.309	2.992	3.766	2.207	2.895	3.645	2.130	2.788	3.530	2.113	2.769	3.511
	3	2.232	2.924	3.705	2.045	2.732	3.482	1.879	2.529	3.263	1.834	2.478	3.205
	4	2.209	2.905	3.702	1.968	2.656	3.468	1.716	2.368	3.110	1.635	2.276	3.003
	5	2.203	2.903	3.699	1.935	2.622	3.449	1.607	2.265	3.015	1.484	2.122	2.849

n to guarantee probability levels $P^* = 0.75, 0.90, 0.975$ of δ^*-CS for several choices of (h, t, q) and arbitrary (σ, δ^*). The program USEGSA can also be used to compute n given $(t, q, \sigma; \delta^*, P^*, h)$ or to compute δ^* given $(t, q, n, \sigma; P^*, h)$.

Remark 3.4.3 Santner (1976) studied a two-stage procedure for the case of σ^2 *unknown*. Sullivan and Wilson (1984, 1989) devised a bounded subset selection procedure for selecting a δ^*-near-best normal treatment when the treatments have different means and unknown and not necessarily equal variances. They also proposed a procedure to allow selection of a δ^*-near-best normal treatment when the data from the t treatments come from t stationary normal processes with unknown means and unknown covariance structures based on correlated sampling within each process.

Example 3.4.2 In a study of $t = 5$ fertilizing methods, an experimenter measures the yields on a set of 10 test plots planted with each fertilizer. Let Y_{ij} denote the yield on the jth test plot that employs the ith fertilizing method, and let μ_i denote the mean of Y_{ij} ($1 \leq i \leq 5, 1 \leq j \leq 10$). We assume that the yields are independent and normally distributed with a common known standard deviation of $\sigma = 3$ bushels per test plot.

Suppose that the experimenter wishes to select a subset containing any fertilizing method whose mean yield is within 2 bushels of the fertilizing method having the largest mean yield, that is, associated with any treatment having $\mu_i \geq \mu_{[5]} - \delta^* = \mu_{[5]} - 2$. Further assume that the experimenter would like the selected subset to contain no more than $q = 3$ treatments. If the experimenter wishes to guarantee this goal with probability at least 0.90 using procedure $\mathcal{N}_{GSa}$, then program USEGSA gives the value $h = 0.354$ that is required to achieve this probability requirement. The rule selects all those treatments satisfying

$$\bar{y}_i \geq \max\left\{\bar{y}_{[3]}, \bar{y}_{[5]} - h\sigma\sqrt{2/n}\right\} = \max\left\{\bar{y}_{[3]}, \bar{y}_{[5]} - (0.354)(3)\sqrt{2/10}\right\}$$
$$= \max\left\{\bar{y}_{[3]}, \bar{y}_{[5]} - 0.475\right\}.$$

3.5 CHAPTER NOTES

A number of authors have proposed competitors to the Gupta (1956, 1965) subset selection procedure $\mathcal{N}_G$. One class of alternatives are frequentist competitors to procedure $\mathcal{N}_G$ that seek to achieve the (same) probability requirement (3.1.1); another class adopts a loss function and uses the minimax principle, while a third group of competitors is based on the Bayesian paradigm.

Among the earliest frequentist competitors to procedure $\mathcal{N}_G$ was the class of procedures proposed by Seal (1955); this class included the Gupta procedure $\mathcal{N}_G$ and an alternative procedure that Seal advocated using. Deely and Gupta (1988) compared the expected size of the selected subsets of these two procedures and showed that in certain slippage configurations the Gupta procedure dominates (see also Gupta and Miescke 1981). More recently, authors have focused attention on constructing subset selection procedures that utilize multiple stages to compare the remaining sample means with $\bar{y}_{[t]}$ using several critical points depending on the number of means in the particular comparison. Examples of such procedures are Somerville (1984), Driessen (1992) and Finner and Giani (1994). For example, Driessen (1992) utilized a duality principle between subset selection procedures and multiple comparison procedures to construct subset selection procedures for the class of *autocongruent* designs. At the expense of computational complexity, it is possible to decrease $E\{S\}$ uniformly in μ for some values of t and $P^\star$, although the gain may not be large [see du Preez et al. (1985), who compared procedure $\mathcal{N}_G$ with the procedure of Somerville (1984)]. The Gupta procedure appears to be a good overall rule that is easily implemented and does nearly as well as more complicated competitors. As will be seen below, the same conclusion holds in comparisons with various Bayesian rules.

Among competitors to procedure $\mathcal{N}_G$ that adopt a minimax-loss function approach are Berger (1979), who measured loss by subset size, and Berger and Gupta (1980), who measured expected loss by the probability of including a non-best treatment (see also Gupta and Huang 1980). Examples of papers that assume both a loss function and a prior distribution, in a fully Bayesian formulation, are Deely and Gupta (1988), Goel and Rubin (1977), Miescke (1979) and Bickel and Yahav (1977, 1982). For example, Bickel and Yahav (1977, 1982) showed that the Gupta procedure $\mathcal{N}_G$ is asymptotically optimal (that is, it is asymptotically Bayes as $t \to \infty$) for a wide class of loss functions and prior distributions. For supplementary calculations of Bayes risk in these situations, see Chernoff and Yahav (1977) and Gupta and Hsu (1978); these studies showed that the Gupta procedure does nearly as well as the exact Bayes procedure for a wide variety of t and loss settings.

We have presented the selection procedure due to Gupta and Huang (1976) for selecting the best treatment when the experiment is unbalanced. There are several competing subset procedures for unbalanced experiments that have been proposed in the literature. These are given in the papers by Gupta and Huang (1974), Chen, Dudewicz and Lee (1976) and Gupta and Wong (1982). No comprehensive simulation comparison has been made of these rules, but it appears that no one rule dominates all the others. Berger and Gupta (1980) showed that the rule proposed in Gupta and Wong (1982) is minimax with respect to a certain intuitive loss function.

A class of problems related to the unequal sample size problem is that of selection from normal treatments when both the means and standard deviations of the treatments differ [Π_i has mean μ_i and variance σ_i^2 ($1 \leq i \leq t$)]. In this case, Dudewicz and Dalal (1975) and Dudewicz and Chen (1992) discussed the problem of selecting a subset of treatments to contain the treatment with mean $\mu_{[t]}$. Huang (1976) devised a procedure to select a subset of treatments having the greatest coefficient of variation, $\max_{1 \leq i \leq t}\{\mu_i/\sigma_i\}$.

We have used mean models that assume independent observations between and within each treatment. There have been several papers that have considered dependent responses. Gnanadesikan (1966) studied the case of dependent observations among the treatments; she assumed that $(Y_{1j}, \ldots, Y_{tj})$ are independent multivariate normal random variables with either a known arbitrary covariance matrix or an unknown arbitrary covariance matrix (see also Gupta and Panchapakesan 1993). A few special correlation structures have been considered in the literature for this setup. In particular, Gnanadesikan (1966) discussed the case of equicorrelated $(Y_{1j}, \ldots, Y_{tj})$. Santner and Pan (1994) proposed subset selection procedures for split-plot and strip-plot two-factor experiments (see Section 6.7).

We presented a subset selection procedure for selecting the s best treatments in Section 3.4.1. Mengersen and Bofinger (1989) studied procedures for selecting a subset containing the s treatments having the smallest scale parameters and for selecting a subset containing only treatments with scale parameters among the s smallest.

In some situations it may be of interest to select a *fixed-size* subset of the means of t normal treatments. Remark 2.5.1 describes one formulation of this problem due to Mahamunulu (1967). He determined the smallest common sample size required to achieve the goal of selecting a subset of size s that contains at least q of the r best treatments with a prespecified probability P^* (Goal 2.5.1) using the naive procedure that selects the s treatments with largest sample means. Here q, r and s are given and must satisfy $\max\{1, s + r + 1 - t\} \leq q \leq \min\{s, r\}$ for this formulation to be meaningful. Later Desu and Sobel (1968) considered the following inverse of this problem: Given a common sample size n from each normal treatment and a nonnegative integer r, they determined the smallest possible s so that the subset consisting of the s treatments with the largest sample means contains the r best of the t treatments ($r \leq s \leq t$) with a prespecified probability of at least P^*. Mengersen and Bofinger (1988) presented a lower confidence bound for the minimum selected parameter when s fixed treatments are selected from the t treatments.

We have focused on selection in terms of means. Some applications naturally lead to selection of the treatment with the smallest (or largest) variance. Gupta and Sobel (1962) studied procedures to select a subset of t normal treatments with the smallest variance, and Gupta (1963) considered the more general case of selecting the gamma distribution with the largest scale parameter.

We have restricted our discussion to normal theory related models. However, there have been a substantial number of papers that have considered nonparametric location models in which the data for the ith treatment has the form $F(x - \theta_i)$, where $\theta_1, \ldots, \theta_t$ are unknown as is $F(\cdot)$, and the goal is to select a subset containing the treatment with the largest θ_i. For example, McDonald (1969, 1985) and Gupta and McDonald (1970, 1980) studied subset selection procedures based on joint ranks of observations. Husty (1984) considered subset selection based on L-estimators of location. Listing and Rasch (1993) compared the achieved probabilities of correct selection for alternative subset selection rules and different families $F(\cdot)$ including rules based on R-estimators and other robust estimators of location.

One other important thrust in subset selection for nonparametric models has been in the study of reliability-motivated nonparametric families $F_{\theta_i}(\cdot)$ such as the increasing failure rate and increasing failure rate average families (Barlow and Gupta 1969, Gupta and Lu 1979 and Hooper and Santner 1979).

Desu (1970), Santner (1976), Naik (1977), Panchapakesan and Santner (1977), Lam (1986), Lin and Jen (1987), and Gupta and Liang (1991b) proposed formulations based on the notion of selecting near-best treatments (see Section 3.4.2). The goals studied in these papers differ. For instance, Panchapakesan and Santner (1977) considered the problems of (i) selecting a subset containing at least one δ^*-near-best treatment and (ii) selecting a subset of treatments *all of which* are δ^*-near-best. Naik (1977) also considered the latter goal. Lin and Jen (1987) studied selection of treatments that are both δ^*-near-best *and* sufficiently larger than the smallest mean. For a large class of goals including those above, Finner and Giani (1994) showed that, under mild conditions, the class of procedures achieving the goal with probability at least P^* is equivalent to a certain class of simultaneous tests for a related system of hypothesis testing problems. These results were used to suggest alternative selection procedures.

CHAPTER 4

Multiple Comparison Approaches for Normal Response Experiments

4.1 INTRODUCTION

This chapter considers multiple comparison approaches, the third alternative formulation to hypothesis tests among treatment means discussed in this book. The text by Hochberg and Tamhane (1987) and the forthcoming book by Hsu (1995) contain comprehensive treatments of multiple comparison methods. Hence this book will only introduce several basic multiple comparison goals and the corresponding procedures. We introduce methods to design experiments and form simultaneous confidence intervals for three families of comparisons among treatment means in the one-way layout. The reason to form *simultaneous* confidence intervals for comparisons of interest is that they provide more useful information to the experimenter than do the results of corresponding hypothesis tests. This is true despite the fact that there is a well-known one-to-one correspondence between the construction of confidence intervals and hypothesis tests.

Specifically, for a set of t treatment means, Sections 4.2, 4.3 and 4.4 determine simultaneous confidence interval estimates for a given set of orthogonal contrasts among the treatment means, the set of all pairwise differences between the treatment means, and the set of differences between each treatment mean and the largest of the remaining treatment means, respectively. The intervals are simultaneous in the sense that the probability that they jointly cover all the comparisons in the family of interest is set at a given, prespecified level. The first two problems differ from the third in that they use families of linear combinations of the means while the third family consists of nonlinear functions of the means. The primary design issue is to determine the scale of the experiment so that the widths of a specific interval or several of the intervals can be controlled at a given value.

4.2 SIMULTANEOUS CONFIDENCE INTERVALS FOR ORTHOGONAL CONTRASTS

In many experimental situations the most meaningful summary statistics are *contrasts*, often orthogonal, among treatment means. For example, this is the case when an experimenter examines the $2^n - 1$ orthogonal "effects" in a 2^n factorial experiment. This section discusses the formation of simultaneous confidence intervals when an experimenter is interested in a specified set of contrasts either orthogonal (Bechhofer and Dunnett 1982) or not. Five applications of the method are provided:

- Simultaneous confidence intervals for orthogonal contrasts (Example 4.2.1)
- A completely randomized balanced 2^3 experiment (Example 4.2.2)
- A completely randomized balanced 2^3 experiment in blocks (Example 4.2.3)
- Simultaneous confidence intervals for individual treatment means (Example 4.2.4)
- An experiment involving equally-spaced quantitative variables (Example 4.2.5)

Definition 4.2.1 The p linear combinations $c_{1j}x_1 + c_{2j}x_2 + \cdots + c_{tj}x_t$ ($1 \le j \le p$) of the vector $(x_1, \ldots, x_t)$ are said to be *orthogonal* provided that $\sum_{i=1}^{t} c_{ir} c_{iq} = 0$ ($1 \le r < q \le p$) (all pairs are orthogonal); they are *contrasts* provided that $\sum_{i=1}^{t} c_{ij} = 0$ ($1 \le j \le p$).

We require the following notation to describe the basic intervals for contrasts. If $W_1, \ldots, W_p$ are independent and identically distributed normal random variables with mean 0 and variance 1, and V has a χ_ν^2 distribution, independent of $(W_1, \ldots, W_p)$, then

$$P\left\{ \bigcap_{j=1}^{p} \left[\frac{|W_j|}{\sqrt{V/\nu}} \le |M|_{p,\nu}^{(\alpha)} \right] \right\} = 1 - \alpha \qquad (4.2.1)$$

defines $|M|_{p,\nu}^{(\alpha)}$, the upper-α critical point of the p-dimensional *studentized maximum modulus* distribution with ν degrees of freedom. Values of the studentized maximum modulus critical point can be obtained in several ways. Table B.7 lists values of $|M|_{p,\nu}^{(\alpha)}$ for selected α, p and ν. Bechhofer and Dunnett (1988) provide a more extensive table of $|M|_{p,\nu}^{(\alpha)}$-values. Lastly, the Dunnett (1989) program MVTPRD can be used to calculate these values because they are a special case of the two-sided studentized t-critical point, $|T|_{p,\nu,\rho}^{(\alpha)}$, which will be introduced in Section 5.4. The relationship between these two values is $|M|_{p,\nu}^{(\alpha)} = |T|_{p,\nu,0.0}^{(\alpha)}$.

We now describe the basic intervals in terms of a generic set of observations.

Statistical Assumptions: Assume that $Z_1, \ldots, Z_t$ are independent normally distributed observations with possibly different means $\mu_1, \ldots, \mu_t$, and that the Z_j ($1 \le j \le t$) have common variance $b^2 \sigma^2$ where σ^2 is *unknown* and $b > 0$ is *known*.

Suppose that it is desired to form a set of simultaneous confidence interval estimates for p *specified* orthogonal contrasts among the means $\theta_j = c_{1j}\mu_1 + \cdots + c_{tj}\mu_t$ ($1 \leq j \leq p$). We consider Goal 4.2.1 and the associated probability (design) requirement (4.2.2).

Goal 4.2.1 To obtain *joint* two-sided confidence interval estimates for the p contrasts $\theta_1, \ldots, \theta_p$.

Probability Requirement: For specified constant P^* with $0 < P^* < 1$, we require

$$P\{\text{Confidence region covers } \theta_j \ (1 \leq j \leq p)\} = P^* \quad (4.2.2)$$

for all $(\theta_1, \ldots, \theta_p)$.

Simultaneous Confidence Intervals for Contrasts Consider intervals based on the corresponding contrasts $\hat{\theta}_j = c_{1j}Z_1 + \cdots + c_{tj}Z_t$ ($1 \leq j \leq p$); $\hat{\theta}_j$ is the best linear unbiased estimator (BLUE) of θ_j. The random variables $\hat{\theta}_1, \ldots, \hat{\theta}_p$ are mutually independent and normally distributed with $\hat{\theta}_j$ having expected value θ_j and variance $\sigma^2 b^2 \sum_{i=1}^{t} c_{ij}^2$ ($1 \leq j \leq p$). Let S_ν^2 be an unbiased estimator of σ^2 having a $\sigma^2 \chi_\nu^2 / \nu$ distribution independent of $(\hat{\theta}_1, \ldots, \hat{\theta}_p)$. Then

$$P\left\{ \bigcap_{j=1}^{p} \left[\frac{|\hat{\theta}_j - \theta_j|}{S_\nu b \sqrt{\sum_{i=1}^{t} c_{ij}^2}} \leq |M|_{p,\nu}^{(1-P^*)} \right] \right\} = P^*. \quad (4.2.3)$$

Hence,

$$\left\{ \theta_j \in \hat{\theta}_j \pm |M|_{p,\nu}^{(1-P^*)} S_\nu b \sqrt{\sum_{i=1}^{t} c_{ij}^2} \ (1 \leq j \leq p) \right\} \quad (4.2.4)$$

are $100 \times P^*\%$ simultaneous two-sided confidence intervals for $\theta_1, \ldots, \theta_p$.

Example 4.2.1 (Simultaneous Confidence Intervals for Orthogonal Contrasts) Brownlee (1965, p. 315) discusses the units in the third decimal place of determinations by Heyl (1930) of the gravitational constant G. Balls of three different materials were used in the experimental determinations. Table 4.1 records a set of these data from a series of experiments; the value 83, for example, corresponds to an observation of 6.683.

We denote the random variables associated with gold (Au), platinum and glass by Y_{Aj}, Y_{Pj} and Y_{Gj} ($1 \leq j \leq 5$), respectively. We assume that the Y_{Aj}, Y_{Pj}, Y_{Gj} are random samples from normal treatments with unknown means μ_A, μ_P, μ_G and a common unknown variance σ^2. We are interested in obtaining 95% two-sided joint confidence interval estimates for the two orthogonal contrasts $\theta_1 = \mu_A + \mu_P - 2\mu_G$

Table 4.1. Third and Fourth Significant Digits of the Gravitational Constant Using Experimental Apparatus of Three Different Materials

Gold	Platinum	Glass
83	61	78
81	61	71
76	67	75
78	67	72
79	64	74

Reprinted from Brownlee (1965), p. 315, by courtesy of John Wiley & Sons.

(metal versus glass) and $\theta_2 = \mu_A - \mu_P$ (gold versus platinum). Intuitively, the first compares the measurements obtained with metal balls versus those obtained with glass balls, and the second compares the results with two different types of metal balls.

We take the sample means $\hat{\mu}_A$, $\hat{\mu}_P$ and $\hat{\mu}_G$ to be the Z_j in the general setup. Notice that the variances of the Z_j are of the form $b^2\sigma^2$ with $b^2 = 1/5$. From Table 4.1 we calculate the BLUEs of μ_A, μ_P, μ_G and the pooled unbiased estimate of σ^2, namely, $\hat{\mu}_A = 79.4$, $\hat{\mu}_P = 64.0$, $\hat{\mu}_G = 74.0$ and $s_\nu^2 = 7.933$ with $\nu = 12$ d.f. The BLUEs of θ_1 and θ_2 are thus $\hat{\theta}_1 = -4.6$ and $\hat{\theta}_2 = 15.4$ with variances $(\frac{1}{5} + \frac{1}{5} + \frac{4}{5})\sigma^2 = 1.2\sigma^2$ and $0.4\sigma^2$, respectively. For $P^* = 0.95$ we find from Table B.7 that $|M|_{2,12}^{(0.05)} = 2.54$. Thus, the required two-sided joint confidence intervals (with joint confidence coefficient 0.95) are:

$$\left\{ \begin{array}{l} \theta_1 \in -4.6 \pm 2.54\sqrt{(1.2)(7.933)} \\ \theta_2 \in 15.4 \pm 2.54\sqrt{(0.4)(7.933)} \end{array} \right\}.$$

or -4.6 ± 7.84 and 15.4 ± 4.52 for θ_1 and θ_2, respectively.

Example 4.2.2 (A Completely Randomized Balanced 2^3 Experiment) This example illustrates another standard application of the basic intervals (4.2.4). Consider an experiment involving three factors, A, B and C, each of which is to be studied at two levels with n observations to be taken at each of the eight treatment combinations. Let μ_{111} denote the treatment expected value when all three factors are at their "low" level, μ_{211} the treatment expected value when the first factor is at its "high" level and the second and third factors are at their low levels, with the remaining expected values denoted by similar notation. The *main effects* of A, B and C, the *two-factor interactions* of each pair of factors, and the *three-factor interaction* of A, B and C are defined by (4.2.5). The notation in which a dot replaces a subscript in μ_{ijk} and a "bar" is placed over the object means that an average has been computed over the values associated with that subscript; for example, $\bar{\mu}_{2..} = \sum_{j=1}^{2} \sum_{k=1}^{2} \mu_{2jk}/4$. We denote the three treatment main effects of A, B and C by $\mu_{(A)}$, $\mu_{(B)}$ and $\mu_{(C)}$, the three two-factor interactions by $\mu_{(AB)}$, $\mu_{(AC)}$ and $\mu_{(BC)}$ and the three-factor interaction by $\mu_{(ABC)}$. Their definitions are:

$$\mu_{(A)} = \bar{\mu}_{2\cdot\cdot} - \bar{\mu}_{1\cdot\cdot}$$
$$= \{\mu_{222} + \mu_{211} + \mu_{221} + \mu_{212} - \mu_{122} - \mu_{111} - \mu_{121} - \mu_{112}\}/4$$
$$\mu_{(B)} = \bar{\mu}_{\cdot 2\cdot} - \bar{\mu}_{\cdot 1\cdot}$$
$$= \{\mu_{222} - \mu_{211} + \mu_{221} - \mu_{212} + \mu_{122} - \mu_{111} + \mu_{121} - \mu_{112}\}/4$$
$$\mu_{(C)} = \bar{\mu}_{\cdot\cdot 2} - \bar{\mu}_{\cdot\cdot 1}$$
$$= \{\mu_{222} - \mu_{211} - \mu_{221} + \mu_{212} + \mu_{122} - \mu_{111} - \mu_{121} + \mu_{112}\}/4$$
$$\mu_{(AB)} = \{(\bar{\mu}_{22\cdot} - \bar{\mu}_{12\cdot}) - (\bar{\mu}_{21\cdot} - \bar{\mu}_{11\cdot})\}/2$$
$$= \{\mu_{222} - \mu_{211} + \mu_{221} - \mu_{212} - \mu_{122} + \mu_{111} - \mu_{121} + \mu_{112}\}/4$$
$$\mu_{(AC)} = \{(\bar{\mu}_{2\cdot 2} - \bar{\mu}_{1\cdot 2}) - (\bar{\mu}_{2\cdot 1} - \bar{\mu}_{1\cdot 1})\}/2$$
$$= \{\mu_{222} - \mu_{211} - \mu_{221} + \mu_{212} - \mu_{122} + \mu_{111} + \mu_{121} - \mu_{112}\}/4$$
$$\mu_{(BC)} = \{(\bar{\mu}_{\cdot 22} - \bar{\mu}_{\cdot 12}) - (\bar{\mu}_{\cdot 21} - \bar{\mu}_{\cdot 11})\}/2$$
$$= \{\mu_{222} + \mu_{211} - \mu_{221} - \mu_{212} + \mu_{122} + \mu_{111} - \mu_{121} - \mu_{112}\}/4$$
$$\mu_{(ABC)} = \{\mu_{222} + \mu_{211} - \mu_{221} - \mu_{212} - \mu_{122} - \mu_{111} + \mu_{121} + \mu_{112}\}/4$$

(4.2.5)

Recall that each of the main effects $\mu_{(A)}$, $\mu_{(B)}$ and $\mu_{(C)}$ is a difference between the average response over the high and low levels of that variable. The two-factor interactions $\mu_{(AB)}$, $\mu_{(AC)}$ and $\mu_{(BC)}$ are each the average of the effect (difference) of one variable at the high and low levels of the other variable. The three-factor interaction $\mu_{(ABC)}$ is a difference between (any) of the two-factor interactions at the high and low levels of the third variable.

Each effect is a linear contrast of the μ_{ijk} with coefficients $\pm 1/4$. Furthermore, it is simple to verify that each pair of effects is orthogonal. Thus the seven effects form a set of mutually orthogonal contrasts among the μ_{ijk}. Our goal is to obtain *joint* two-sided confidence interval estimates that guarantee the probability requirement (4.2.2) for the seven treatment effects $\mu_{(A)}$, $\mu_{(B)}$, $\mu_{(C)}$, $\mu_{(AB)}$, $\mu_{(AC)}$, $\mu_{(BC)}$, $\mu_{(ABC)}$.

The (raw) data consist of random samples of $n \geq 2$ normal observations Y_{ijks} ($i, j, k = 1, 2; 1 \leq s \leq n$) that are taken from the ijkth treatment combination which has unknown treatment mean μ_{ijk}. The Y_{ijks} have a common unknown variance σ^2. Let y_{ijks} denote the observed values of the Y_{ijks}, $\bar{y}_{ijk\cdot} = \sum_{s=1}^{n} y_{ijks}/n$, and let

$$s_\nu^2 = \frac{1}{\nu} \sum_{i=1}^{2} \sum_{j=1}^{2} \sum_{k=1}^{2} \sum_{s=1}^{n} (y_{ijks} - \bar{y}_{ijk\cdot})^2$$

denote the unbiased pooled estimate of σ^2 based on $\nu = 8(n-1)$ d.f.

The means $\bar{Y}_{ijk\cdot}$ play the role of the Z_j in the description of the general theory. The expected value of $\bar{Y}_{ijk\cdot}$ is μ_{ijk} and thus the contrasts (4.2.5) are defined in terms of the expected values of the basic random variables. The variance of $\bar{Y}_{ijk\cdot}$ is σ^2/n and thus $b^2 = 1/n$ in this application.

The $\hat{\theta}_j$, the BLUEs of the seven effects defined in Equation (4.2.5), are

$$\hat{\mu}_{(A)} = \{\bar{y}_{222\cdot} + \bar{y}_{211\cdot} + \bar{y}_{221\cdot} + \bar{y}_{212\cdot} - \bar{y}_{122\cdot} - \bar{y}_{111\cdot} - \bar{y}_{121\cdot} - \bar{y}_{112\cdot}\}/4$$

$$\hat{\mu}_{(B)} = \{\bar{y}_{222\cdot} - \bar{y}_{211\cdot} + \bar{y}_{221\cdot} - \bar{y}_{212\cdot} + \bar{y}_{122\cdot} - \bar{y}_{111\cdot} + \bar{y}_{121\cdot} - \bar{y}_{112\cdot}\}/4$$

$$\hat{\mu}_{(C)} = \{\bar{y}_{222\cdot} - \bar{y}_{211\cdot} - \bar{y}_{221\cdot} + \bar{y}_{212\cdot} + \bar{y}_{122\cdot} - \bar{y}_{111\cdot} - \bar{y}_{121\cdot} + \bar{y}_{112\cdot}\}/4$$

$$\hat{\mu}_{(AB)} = \{\bar{y}_{222\cdot} - \bar{y}_{211\cdot} + \bar{y}_{221\cdot} - \bar{y}_{212\cdot} - \bar{y}_{122\cdot} + \bar{y}_{111\cdot} - \bar{y}_{121\cdot} + \bar{y}_{112\cdot}\}/4$$

$$\hat{\mu}_{(AC)} = \{\bar{y}_{222\cdot} - \bar{y}_{211\cdot} - \bar{y}_{221\cdot} + \bar{y}_{212\cdot} - \bar{y}_{122\cdot} + \bar{y}_{111\cdot} + \bar{y}_{121\cdot} - \bar{y}_{112\cdot}\}/4$$

$$\hat{\mu}_{(BC)} = \{\bar{y}_{222\cdot} + \bar{y}_{211\cdot} - \bar{y}_{221\cdot} - \bar{y}_{212\cdot} + \bar{y}_{122\cdot} + \bar{y}_{111\cdot} - \bar{y}_{121\cdot} - \bar{y}_{112\cdot}\}/4$$

$$\hat{\mu}_{(ABC)} = \{\bar{y}_{222\cdot} + \bar{y}_{211\cdot} - \bar{y}_{221\cdot} - \bar{y}_{212\cdot} - \bar{y}_{122\cdot} - \bar{y}_{111\cdot} + \bar{y}_{121\cdot} + \bar{y}_{112\cdot}\}/4.$$

(4.2.6)

In the notation of the general description, we have $\sum_{i=1}^{t} c_{ij}^2 = 8/4^2 = 1/2$; recall that $b^2 = 1/n$ gives the variance of each estimated effect as $\sigma^2 b^2 \sum_{i=1}^{t} c_{ij}^2 = \sigma^2/2n$. Applying (4.2.4), we choose the constant $|M|_{7,\nu}^{(1-P^*)}$ from Table B.7. Then with joint confidence coefficient P^*, we can state

$$\left\{\begin{array}{l} \mu_{(A)} \in \hat{\mu}_{(A)} \pm |M|_{7,8(n-1)}^{(1-P^*)} s_\nu/\sqrt{2n} \\ \mu_{(B)} \in \hat{\mu}_{(B)} \pm |M|_{7,8(n-1)}^{(1-P^*)} s_\nu/\sqrt{2n} \\ \quad\quad\quad \vdots \\ \mu_{(ABC)} \in \hat{\mu}_{(ABC)} \pm |M|_{7,8(n-1)}^{(1-P^*)} s_\nu/\sqrt{2n} \end{array}\right\}. \quad (4.2.7)$$

In some situations the experimenter may desire to form joint interval estimates for *fewer* than the full set of seven contrasts defined in Goal 4.2.1. If p contrasts are to be studied, then the critical value used to form the intervals is $|M|_{p,\nu}^{(1-P^*)}$. For example, in the 2^3 experiment that we are considering, if it is desired to form joint confidence interval estimates with joint confidence coefficient P^* for the three main effects only, then

$$\left\{\begin{array}{l} \mu_{(A)} \in \hat{\mu}_{(A)} \pm |M|_{3,8(n-1)}^{(1-P^*)} s_\nu/\sqrt{2n} \\ \mu_{(B)} \in \hat{\mu}_{(B)} \pm |M|_{3,8(n-1)}^{(1-P^*)} s_\nu/\sqrt{2n} \\ \mu_{(C)} \in \hat{\mu}_{(C)} \pm |M|_{3,8(n-1)}^{(1-P^*)} s_\nu/\sqrt{2n} \end{array}\right\} \quad (4.2.8)$$

is the correct set of joint confidence intervals. The intervals in (4.2.8) are shorter than those in (4.2.7) because $|M|_{3,8(n-1)}^{(1-P^*)} < |M|_{7,8(n-1)}^{(1-P^*)}$.

As another example for which the experimenter would be interested in fewer than the full set of seven contrasts defined in (4.2.5), consider a 2^3 experiment in which there are two blocking ("stratifying") factors and a third treatment factor. In particular, suppose that a study on the effect of diet (Diet 1 versus Diet 2), gender (Male versus Female), and age (Old versus Young) on weight gain is conducted. Here diet is the treatment variable of research interest, and gender and age are stratifying variables. If the variables diet, gender and age are denoted by D, G and A, respectively, then the experimenter would be interested in the $p = 4$ contrasts corresponding to diet with the stratifying variables: $\mu_{(D)}$, $\mu_{(DG)}$, $\mu_{(DA)}$ and $\mu_{(DGA)}$.

Warning: The experimenter is cautioned that the decision to consider only particular sets of effects must be made before, rather than after, tests have been made for the presence of interactions. The nominal joint confidence coefficient is not valid in the latter case.

Two Extensions of the Basic Theory
Two extensions of the intervals (4.2.4) can be derived by checking that the fundamental probability equality (4.2.3) holds for an appropriate set of ratios.

In the first extension of the method, suppose that Z_j has expected value $\gamma + \mu_j$ instead of expected value μ_j, but that its variance is $b^2\sigma^2$ and the goal is as above. Then the intervals (4.2.4) again satisfy the probability requirement (4.2.2). The reason is that $\sum_{i=1}^{t} c_{ij}(\gamma + \mu_j) = \sum_{i=1}^{t} c_{ij}\mu_j$ and thus each $\hat{\theta}_j$ is unbiased for θ_j; the probability statement (4.2.3) also holds in this case. The use of this extension will be illustrated in Example 4.2.3.

In the second extension of the method, we return to the case in which Z_j has expected value μ_j and (common) variance $b^2\sigma^2$. However, now suppose that the θ_j, while orthogonal, need not be contrasts, that is, $\sum_{i=1}^{t} c_{ij} \neq 0$ for at least one j. The goal remains that of forming two-sided simultaneous confidence intervals for the set of θ_j. In this case it is straightforward to again verify that the $\hat{\theta}_j$ are independent and normally distributed, and thus the intervals (4.2.4) satisfy the probability requirement (4.2.2). Example 4.2.4 illustrates this extension of the basic method.

Example 4.2.3 (A Completely Randomized Balanced 2^3 Experiment in Blocks)
Snedecor and Cochran (1967, Example 12.9, pp. 359–360) analyzed a 2^3 experiment conducted at the Iowa Agricultural Experiment Station to study the effects of two supplements to a corn ration for feeding pigs. The experiment was conducted in eight randomized complete blocks on pigs of both sexes so that gender was a third factor. To summarize, the three factors were:

- Lysine (L): 0% and 0.6%
- Protein from soybean meal (P): 12% and 14%
- Sex (S): male and female

The data are listed in Table 4.2.

Let $Y_{ijk\ell}$ be the average daily weight gain for the pig in block i having the jth level of lysine, and the kth level of protein and of the ℓth sex. We assume the normal response mean model

$$E\{Y_{ijk\ell}\} = \mu + \beta_i + \mu_{jk\ell} \qquad (4.2.9)$$

Table 4.2. Average Daily Weight Gain of Pigs in a 2^3 Factorial Arrangement of Treatments in a Randomized Block Design (L = % Lysine and P = % Protein)

L	P	Sex	Block 1	2	3	4	5	6	7	8	Sum
0	12	M	1.11	0.97	1.09	0.99	0.85	1.21	1.29	0.96	8.47
		F	1.03	0.97	0.99	0.99	0.99	1.21	1.19	1.24	8.61
	14	M	1.52	1.45	1.27	1.22	1.67	1.24	1.34	1.32	11.03
		F	1.48	1.22	1.53	1.19	1.16	1.57	1.13	1.43	10.71
0.6	12	M	1.22	1.13	1.34	1.41	1.34	1.19	1.25	1.32	10.20
		F	0.87	1.00	1.16	1.29	1.00	1.14	1.36	1.32	9.14
	14	M	1.38	1.08	1.40	1.21	1.46	1.39	1.17	1.21	10.30
		F	1.09	1.09	1.47	1.43	1.24	1.17	1.01	1.13	9.63

Reprinted from Snedecor and Cochran (1967), p. 359, by courtesy of Iowa State University Press.

where $\sum_{i=1}^{8} \beta_i = 0$ with constant variance $\text{Var}(Y_{ijk\ell}) = \sigma^2$. Here β_i is the ith block effect which is assumed to not interact with the effects of the treatment combinations. We are interested in forming 95% simultaneous confidence intervals for the main effects and interactions among the means $\mu_{jk\ell}$.

In the notation of the general theory, let Z_j be the block average $Y._{jk\ell}$ which has expected value $\mu + \mu_{jk\ell}$ because $\sum_{i=1}^{8} \beta_i = 0$. The variance of $Y._{jk\ell}$ is $\sigma^2/8$, and thus $b^2 = 1/8$. The estimated treatment effects defined in terms of the $Y._{jk\ell}$ are given in Table 4.3, and the estimated variance is $s_\nu^2 = 0.0224$ based on 49 d.f. Each treatment effect $\hat{\mu}$ has variance $\sigma^2/(2 \times 8)$. Using the 95% critical point $|M|^{(0.05)}_{7,49} = 2.80$ interpolated from Table B.7, simultaneous confidence intervals for each effect are calculated to have half-width

$$|M|^{(0.05)}_{7,49} s_\nu / \sqrt{2n} = 2.80\sqrt{0.0224/16} = 0.1048.$$

Simultaneous confidence intervals for the main effects and interactions are listed in the two right-hand columns of Table 4.3.

Clearly there are large Protein and Sex main effects and a smaller Lysine main effect. To properly interpret this experiment, the researcher must also note that the two-factor interactions (LS and LP) and the three-factor interaction are statistically significant.

Table 4.3. Estimated Orthogonal Effects and 95% Simultaneous Confidence Intervals for $\mu_{(L)}, \mu_{(P)}, \mu_{(S)}, \mu_{(LP)}, \mu_{(LS)}, \mu_{(PS)}, \mu_{(LPS)}$

Effect	$\hat{\mu}$	Lower Limit	Upper Limit
L	0.11	0.01	0.21
P	1.31	1.21	1.41
S	−0.48	−0.58	−0.38
LP	−1.02	−1.12	−0.92
LS	−0.39	−0.49	−0.29
PS	−0.02	−0.12	0.08
LPS	0.21	0.11	0.31

Example 4.2.4 (Simultaneous Confidence Intervals for Individual Treatment Means) This example illustrates the application of the second extension of the basic theory. Continuing with the setup from Example 4.2.1, we consider the balanced data given in Table 4.1. We are now interested in obtaining 95% two-sided joint confidence interval estimates for the three individual means μ_A, μ_P and μ_G. We take the sample means $\hat{\mu}_A$, $\hat{\mu}_P$ and $\hat{\mu}_G$ to be the Z_j in the general setup. The variances of the Z_j are each $\sigma^2/5$ so that $b^2 = 1/5$. However, the linear combinations

$$\theta_A = 1 \times \mu_A + 0 \times \mu_P + 0 \times \mu_G$$
$$\theta_P = 0 \times \mu_A + 1 \times \mu_P + 0 \times \mu_G \quad \text{and}$$
$$\theta_G = 0 \times \mu_A + 0 \times \mu_P + 1 \times \mu_G$$

are *not contrasts*, although they are orthogonal. Thus the intervals (4.2.4) are valid $100P^*\%$ simlutaneous confidence intervals for μ_A, μ_P and μ_G.

From Table 4.1 we calculate the BLUEs of μ_A, μ_P, μ_G and the pooled unbiased estimate of σ^2 to be $\hat{\mu}_A = 79.4$, $\hat{\mu}_P = 64.0$, $\hat{\mu}_G = 74.0$, and $s_\nu^2 = 7.933$ with $\nu = 12$ d.f. For $P^* = 0.95$ we find from Table B.7 that $|M|_{3,12}^{(0.05)} = 2.75$. Thus, the required two-sided joint confidence intervals are:

$$\begin{cases} \mu_A \in 79.4 \pm 2.75\sqrt{7.933}\sqrt{(1^2+0^2+0^2)/5} \\ \mu_P \in 64.0 \pm 2.75\sqrt{7.933}\sqrt{(0^2+1^2+0^2)/5} \\ \mu_G \in 74.0 \pm 2.75\sqrt{7.933}\sqrt{(0^2+0^2+1^2)/5} \end{cases}$$

or 79.4 ± 3.46, 64.0 ± 3.46 and 74.0 ± 3.46 for μ_A, μ_P and μ_G, respectively.

We conclude this section with a standard example concerning quantitative variables.

Example 4.2.5 (An Experiment Involving Equally-Spaced Quantitative Variables) This example illustrates a situation in which orthogonal contrasts are formed among the treatment components of a quantitative variable. Snedecor and Cochran (1967, Section 12.6) discuss an experiment in which a (fixed) mixture of fertilizers was applied at four equally-spaced levels (0, 4, 8, 12 cwt per acre) to determine their effects on the mean yield of sugar beets. Eight replicates of each mixture were used in the experiment. For each level, Table 4.4 lists the sample mean yield of sugar beets, each mean being based on the eight observations. The estimated variance pooled (over the four mixtures) was reported as $s_\nu^2 = 11.9$ based on 28 d.f.

Table 4.4. Sample Mean Yields (Based on Eight Replicates) of Sugar Beets

Mixed Fertilizer (x) (cwt per acre)	0	4	8	12
Mean Yields ($\bar{y}_x$.)	34.8	41.1	42.6	41.8

Reprinted from Snedecor and Cochran (1967), p. 350, by courtesy of Iowa State University Press.

Let Y_{xj} ($x = 0, 4, 8, 12$, $1 \le j \le 8$) denote the 32 observed sugar beet yields for this experiment. We assume that the Y_{xj} are random samples from a normal treatment with mean μ_x and common unknown variance σ^2. The means $\overline{Y}_{x\cdot}$ play the role of the generic variables Z_j in the general theory; thus $b^2 = 1/8$.

A cubic in the level of application x, $E\{Y_{xj}\} = \alpha + \beta x + \gamma x^2 + \delta x^3$, can be fit to these data. This cubic can be transformed into another polynomial $A + BX_1 + CX_2 + DX_3$ in which the coefficients A, B, C, D are mutually orthogonal linear functions of the μ_x, and X_1, X_2 and X_3 are linear, quadratic and cubic functions of the variable x, respectively. (See, for example, Davies 1967, Appendix 8C.)

For a cubic polynomial with equally-spaced x values, the following quantities define the linear, quadratic and cubic components:

$$\begin{aligned}
\text{Linear component} &= -3\mu_0 - \mu_4 + \mu_8 + 3\mu_{12} = \mu(L) \\
\text{Quadratic component} &= \mu_0 - \mu_4 - \mu_8 + \mu_{12} = \mu(Q) \qquad (4.2.10)\\
\text{Cubic component} &= -\mu_0 + 3\mu_4 - 3\mu_8 + \mu_{12} = \mu(C)
\end{aligned}$$

Note that the three components form a mutually orthogonal set of contrasts among the μ_x ($x = 0, 4, 8, 12$). Our goal is to obtain two-sided joint confidence interval estimates of the components $\mu(L)$, $\mu(Q)$ and $\mu(C)$, the intervals satisfying the probability requirement (4.2.2).

From Table 4.4 we calculate the BLUEs of the linear, quadratic and cubic components in terms of the $\overline{y}_{x\cdot}$ as $\hat{\mu}(L) = 22.5$, $\hat{\mu}(Q) = -7.1$ and $\hat{\mu}(C) = 2.5$, respectively, with variances $20\sigma^2/8$, $4\sigma^2/8$ and $20\sigma^2/8$, respectively. The multiplier of σ^2 for the variance of the linear coefficient is $20 = (-3)^2 + (-1)^2 + (1)^2 + (3)^2$, and similar calculations give the multipliers for the variances of the two other coefficients.

If $P^* = 0.95$ is specified, then from Table B.7 we find, by interpolation, that $|M|_{3,28}^{(0.05)} = 2.533$. Thus the 95% two-sided joint confidence intervals are given by

$$\begin{cases} \mu(L) \in 22.5 \pm 2.533 \sqrt{\tfrac{20}{8}(11.9)} \\ \mu(Q) \in -7.1 \pm 2.533 \sqrt{\tfrac{4}{8}(11.9)} \\ \mu(C) \in 2.5 \pm 2.533 \sqrt{\tfrac{20}{8}(11.9)} \end{cases}. \qquad (4.2.11)$$

4.3 SIMULTANEOUS CONFIDENCE INTERVALS FOR ALL PAIRWISE DIFFERENCES

This section presents the classic simultaneous confidence intervals for all pairwise differences of treatment means in a single-factor normal theory experiment. The basic assumptions used throughout this section are stated first.

Statistical Assumptions: Independent random samples of observations $Y_{i1}, Y_{i2}, \ldots$ ($1 \le i \le t$) are taken from $t \ge 2$ normal treatments $\Pi_1, \ldots, \Pi_t$. Here Π_i has unknown treatment mean μ_i and all the Π_i have common unknown variance σ^2.

The number n_i of observations taken on the ith treatment ($1 \leq i \leq t$) is typically determined by economic or other reasons. Let y_{ij} ($1 \leq i \leq t$, $j \geq 1$) denote the observed value of Y_{ij}. The purpose of the experiment is given by Goal 4.3.1, and the associated probability (design) requirement is stated in (4.3.1).

Goal 4.3.1 To obtain joint two-sided confidence interval estimates of the $\binom{t}{2}$ differences $\mu_i - \mu_j$ ($1 \leq i \neq j \leq t$).

Probability Requirement: For specified constant P^* with $0 < P^* < 1$, we require

$$P\{\text{Confidence region covers } \mu_i - \mu_j \quad (1 \leq i \neq j \leq t)\} = P^* \tag{4.3.1}$$

for all $\boldsymbol{\mu}$ and σ^2.

Tukey (1953) proposed a single-stage procedure for achieving Goal 4.3.1 subject to guaranteeing (4.3.1).

Procedure $\mathcal{N}_T$ (Balanced Experiments) For the given t, specify the P^* and $n = n_1 = \cdots = n_t$ of interest. Let $\nu = t(n-1)$ if σ^2 is unknown, and $\nu = \infty$ if σ^2 is known.

Sampling rule: Take a random sample of $n \geq 2$ observations Y_{ij} ($1 \leq j \leq n$) in a *single* stage from each Π_i ($1 \leq i \leq t$). Calculate the t sample means $\bar{y}_i = \sum_{j=1}^n y_{ij}/n$ ($1 \leq i \leq t$). If σ^2 is unknown, calculate $s_\nu^2 = \sum_{i=1}^t \sum_{j=1}^n (y_{ij} - \bar{y}_i)^2/\nu$, the unbiased pooled estimate of σ^2 based on $\nu = t(n-1)$ d.f.

Confidence intervals: The required joint two-sided $100 \times P^*\%$ confidence intervals are

$$\left\{ \mu_i - \mu_j \in \bar{y}_i - \bar{y}_j \pm Q_{t,\nu}^{(1-P^*)} s_\nu / \sqrt{n} \quad (1 \leq i \neq j \leq t) \right\} \tag{4.3.2}$$

where the constant $Q_{t,\nu}^{(\alpha)}$ is the upper-α critical point of the studentized range distribution.

The constant used to implement Procedure $\mathcal{N}_T$ is the upper-α critical point of the studentized range distribution. It is defined implicitly by Equation (4.3.3). If $W_1, \ldots, W_p$ have independent normal distributions with mean vector zero and unit variances and V has a χ_ν^2 distribution independent of $(W_1, \ldots, W_p)$, then

$$P\left\{ \frac{W_{[p]} - W_{[1]}}{\sqrt{V/\nu}} \leq Q_{p,\nu}^{(\alpha)} \right\} = 1 - \alpha \tag{4.3.3}$$

defines $Q_{p,\nu}^{(\alpha)}$. Here $W_{[p]} - W_{[1]}$ is the range of the W_i. Values of $Q_{t,\nu}^{(\alpha)}$ can be found in Table B.6 and in many other standard reference books.

The reader should note that the form of the yardstick used by procedure $\mathcal{N}_T$ is different than that of the other procedures in this book. Instead of the yardstick being a constant times an estimate of the standard error of the treatment difference, $\bar{Y}_i - \bar{Y}_j$, that is, of $\sigma\sqrt{2/n}$, procedure $\mathcal{N}_T$ uses a constant times an estimate of $\sigma/\sqrt{n}$. This makes the constants differ by a factor of $\sqrt{2}$. Clearly this is a question of aesthetics, because the product of the appropriate constant and corresponding yardstick must be the same in either case for (4.3.1) to hold. We note this difference simply to emphasize that the reader cannot use the same mnemonic to remember the form of the procedure as in previous cases.

Optimality Property
Gabriel (1969) showed that in a balanced one-way layout, Tukey's procedure (4.3.2) gives the shortest intervals for all pairwise differences among all procedures that yield equal-width intervals with joint confidence coefficient P^*.

Tukey Procedure for Unbalanced Experiments
If the experiment is performed with unequal numbers of observations $n_1, \ldots, n_t$ from the treatments $\Pi_1, \ldots, \Pi_t$, respectively, then (4.3.2) cannot be used. However, Hayter (1984) proved that (4.3.2) can be replaced by

$$\left\{ \mu_i - \mu_j \in \bar{y}_i - \bar{y}_j \pm Q_{t,\nu}^{(1-P^*)} s_\nu \sqrt{\frac{1}{2}\left(\frac{1}{n_i} + \frac{1}{n_j}\right)} \quad (1 \leq i \neq j \leq t) \right\} \quad (4.3.4)$$

with an associated joint confidence coefficient *at least* P^*. This procedure, which is known as the Tukey–Kramer procedure, had earlier been conjectured by these authors to be conservative. It reduces to (4.3.2) if $n_1 = \cdots = n_t = n$.

Fixed-Width Confidence Intervals When σ^2 is Known
Suppose that $n_1 = \cdots = n_t = n$ is to be chosen and that it is desired to replace (4.3.2) by intervals of the form

$$\{\mu_i - \mu_j \in \bar{y}_i - \bar{y}_j \pm A \quad (1 \leq i \neq j \leq t)\} \quad (4.3.5)$$

where the half-width $A > 0$ is specified prior to the start of experimentation; that is, the intervals are to have a prespecified common width, and the two-sided confidence intervals are to have joint confidence coefficient P^*. From (4.3.2), this goal can be accomplished if the experiment is designed on such a scale that

$$n = \left\lceil \left(Q_{t,\infty}^{(1-P^*)} \sigma/A \right)^2 \right\rceil. \quad (4.3.6)$$

Fixed-Width Confidence Intervals When σ^2 is Unknown
Suppose that it is desired to construct intervals of the form (4.3.5) when σ^2 is unknown. To accomplish this, at least a two-stage procedure is required. Hochberg

and Lachenbruch (1976), paralleling Stein (1945), proposed the following two-stage procedure to guarantee (4.3.1).

> **Procedure** $\mathcal{N}_{HL}$ For the given t, specify the P^* of interest and first-stage sample size n_0.
>
> *Sampling rule:* Stage 1: Take a random sample of $n_0 \geq 2$ observations from each treatment. Calculate the usual pooled unbiased estimate s_ν^2 of σ^2 based on $\nu = t(n_0 - 1)$ d.f. Let
>
> $$N = \max\left\{\left\lceil\left(Q_{t,\nu}^{(1-P^*)} s_\nu/A\right)^2\right\rceil, n_0\right\}.$$
>
> *Stage 2:* In the second stage take $N - n_0$ additional observations from each of the t treatments.
>
> *Confidence intervals:* Compute the *overall* sample means $\bar{y}_i = \sum_{j=1}^N y_{ij}/N$ ($1 \leq i \leq t$). The required $100 \times P^*\%$ simultaneous confidence intervals are then given by (4.3.5).

In the same way as for the two-stage selection procedure of Section 2.7.1, the experimenter must strike a balance between "too small" a first-stage sample size n_0 (which can lead to excessively large second-stage sample sizes) and "too large" an n_0 (which is wasteful of the first-stage observations). One way of achieving a compromise in the choice of n_0 is to conjecture (if possible) the *single-stage* sample size that would be necessary to guarantee (4.3.5) with confidence P^* if σ^2 *were known*, and then to choose the first-stage sample size n_0 as (say) two-thirds of the conjectured single-stage sample size.

Blocking plays the same role here as it does in selection experiments, that is, it minimizes possible bias and reduces the residual variance. See the discussion of this topic in Section 2.2.

4.4 SIMULTANEOUS CONFIDENCE INTERVALS FOR COMPARING ALL TREATMENTS WITH THE BEST

In some situations the most useful comparisons may be those with respect to the unknown *best* treatment. We consider one method of making comparisons with respect to the best treatment in the present section and provide references to a related method in the Chapter Notes, Section 4.5.

The statistical assumptions are the same as in Section 4.3 except that we now require the data to be balanced with n observations, say, per treatment. As usual, the sample size n may be determined by economic or other reasons. Let y_{ij} ($1 \leq i \leq t$, $j \geq 1$) denote the observed values of Y_{ij}. We consider the goal of forming simultaneous confidence intervals for the t differences $\{\mu_i - \max_{j \neq i} \mu_j\}_{i=1}^t$. Informally, this might be thought of as comparisons of each treatment mean with the *best of the rest* of the treatment means. However, to properly interpret this family, first consider the $t - 1$ cases when $i \neq [t]$; in these cases $\mu_i - \max_{j \neq i} \mu_j = \mu_i - \mu_{[t]}$ so that we are making comparisons of each of the nonbest treatments with the best one. When $i = [t]$, then the best of the remaining treatments has mean equal to the second largest mean and so $\mu_i - \max_{j \neq i} \mu_j = \mu_{[t]} - \mu_{[t-1]}$.

When considering the formation of (one-sided) confidence bounds, the most meaningful type are simultaneous one-sided lower confidence bounds. The reason is that, assuming a unique best treatment, we know by their definition that the differences $\mu_i - \mu_{[t]}$ must be *negative*, and only the difference $\mu_{[t]} - \mu_{[t-1]}$ is *positive*.

Remark 4.4.1 The reader should notice that unlike Sections 4.2 and 4.3, both families of parameters considered in the present section are *nonlinear* in the treatment means μ_i. This implies, for example, that the function $h(\boldsymbol{\mu}) = \mu_{[t]} - \mu_i = \max_{1 \leq j \leq t} \mu_j - \mu_i$ does not satisfy $h(a\boldsymbol{\mu} + b) = ah(\boldsymbol{\mu}) + b$ for all a and b, the definition of linearity. In general, linear functions are easier to study theoretically than nonlinear ones.

The purpose of the experiment is given by Goal 4.4.1, and the associated probability (design) requirement is stated in (4.4.1).

Goal 4.4.1 To obtain joint one-sided or two-sided confidence interval estimates of the t differences $\mu_i - \max_{j \neq i} \mu_j$ $(1 \leq i \leq t)$.

Probability Requirement: For specified constant P^* with $0 < P^* < 1$, we require

$$P\{\text{Confidence region covers } \mu_i - \max_{j \neq i} \mu_j \ (1 \leq i \leq t)\} \geq P^* \quad (4.4.1)$$

for all $\boldsymbol{\mu}$ and σ^2.

Hsu (1984a) derived the following two-sided simultaneous confidence intervals for achieving Goal 4.4.1 subject to guaranteeing (4.4.1). In the following, we use the notation $(x)^- = \min\{x, 0\}$ and $(x)^+ = \max\{x, 0\}$.

Procedure $\mathcal{N}_{JH}$ For the given t, specify the P^* and n of interest.

Sampling rule: Take a random sample of $n \geq 2$ observations Y_{ij} $(1 \leq j \leq n)$ in a *single* stage from each Π_i $(1 \leq i \leq t)$. Calculate the t sample means $\bar{y}_i = \sum_{j=1}^{n} y_{ij}/n$. If σ^2 is unknown, calculate $s_\nu^2 = \sum_{i=1}^{t} \sum_{j=1}^{n} (y_{ij} - \bar{y}_i)^2 / \nu$, the unbiased pooled estimate of σ^2 based on $\nu = t(n-1)$ d.f.

Confidence intervals: If σ^2 is unknown, the intervals

$$\left\{ \left[\left(\bar{y}_i - \max_{j \neq i} \bar{y}_j - hs_\nu \sqrt{2/n} \right)^-, \left(\bar{y}_i - \max_{j \neq i} \bar{y}_j + hs_\nu \sqrt{2/n} \right)^+ \right] (1 \leq i \leq t) \right\}$$

(4.4.2)

cover the differences $\mu_i - \max_{j \neq i} \mu_j$ ($1 \leq i \leq t$) with joint confidence coefficient $\geq P^*$ where $h = T_{t-1,\nu,1/2}^{(1-P^*)}$ is from Table B.3. If σ^2 is known, then replace $T_{t-1,\nu,1/2}^{(1-P^*)}$ by $Z_{t-1,1/2}^{(1-P^*)}$ from Table B.1 and s_ν by σ.

Example 4.4.1 (Example 3.2.3 Continued) The absorption of fat during cooking in four different mediums was analyzed in Example 3.2.3. The data are presented in Table 3.4. Suppose that it is desired to obtain 95% simultaneous confidence intervals for the differences between each mean and the best of the remaining means. Table 4.5 summarizes the computations stated in Equation (4.4.2). Notice that the confidence interval difference $\mu_i - \max_{j \neq i} \mu_j$ for i = B (cooking medium B) consists only of positive values, in this case (0, 14.80), whereas the corresponding confidence intervals for the differences for the other three mediums consist only of negative values. This provides the unambiguous interpretation that medium B maximizes the amount of fat absorbed (a dubious distinction)!

Unbalanced Experiments

For *unbalanced* experiments Hochberg and Tamhane (1987, pp. 154–155) show how to obtain simultaneous P^*-level confidence intervals for $\mu_i - \max_{j \neq i} \mu_j$ based on work by Edwards and Hsu (1983) for the two-sided case and Hsu (1984b) for the one-sided case.

4.5 CHAPTER NOTES

The forthcoming book by Hsu (1995) will contain a comprehensive discussion of procedures for multiple comparisons with the best (MCB) whereas the currently available book by Hochberg and Tamhane (1987) devotes greater coverage to other multiple comparison methods. Hence we will restrict the majority of our comments to references concerning MCB methodology.

For the problem of forming simultaneous confidence intervals for all pairwise means, Dunnett (1982) reported the results of an extensive simulation comparison of the robustness of Procedure $\mathcal{N}_T$ and several competitors. One modification was based on the k-sample rank sum test, and the others used alternative estimators of location with their respective variance estimators. For a variety of non-normal location models, Dunnett compared the methods with respect to the accuracy of their nominal experimentwise error rates and the expected average lengths of the confidence intervals they produced.

Table 4.5. Summary of Calculations in Equation (4.4.2), where $L_i = \bar{y}_i - \max_{j \neq i} \bar{y}_j - hs_\nu \sqrt{2/n}$ and $U_i = \bar{y}_i - \max_{j \neq i} \bar{y}_j + hs_\nu \sqrt{2/n}$

i	$\bar{y}_i$	$\max_{j \neq i} \bar{y}_j$	L_i	$(L_i)^-$	U_i	$(U_i)^+$
A	72	85	−18.80	−18.80	−7.20	0
B	85	76	3.20	0	14.80	14.80
C	76	85	−14.80	−14.80	−3.20	0
D	62	85	−28.80	−28.80	−17.20	0

Hsu (1981) derived $100 \times P^*\%$ simultaneous lower confidence bounds for the family of parameters $\mu_i - \mu_{[t]}$ ($1 \leq i \leq t$). Edwards and Hsu (1983) generalized the intervals to simultaneous $100 \times P^*\%$ two-sided confidence limits; their two-sided limits reduce to the lower confidence bounds of Hsu (1981) when the upper limits are set equal to ∞.

Section 4.4 discusses MCB intervals for balanced or unbalanced one-way layouts. Extensions of MCB intervals have been made to various complete and incomplete block designs and multi-stage procedures. Hsu (1982) provided $100 \times P^*\%$ simultaneous upper confidence bounds for the family of parameters $\mu_{[t]} - \mu_i$ ($1 \leq i \leq t$) for randomized complete block designs, while Driessen (1991, 1992) considered other incomplete block designs. Extensions of MCB intervals have been constructed for the general linear model by Hsu (1988, 1989, 1992). Matejcik and Nelson (1993) derived the following MCB intervals that can be used after running the Rinott procedure $\mathcal{N}_R$ (see Section 2.8). With probability greater than or equal to P^*,

$$\mu_i - \max_{j \neq i} \mu_j \in \left[-\left(\bar{y}_i - \max_{j \neq i} \bar{y}_j - \delta^*\right)^-, \left(\bar{y}_i - \max_{j \neq i} \bar{y}_j + \delta^*\right)^+ \right] \quad (4.5.1)$$

($1 \leq i \leq t$). Finally, we note that software is available to implement basic single-sample MCB intervals in the form of a user-contributed SAS procedure (Gupta and Hsu 1984, 1985; Aubuchon, Gupta and Hsu 1986).

One important application of selection and multiple comparison methods is in stochastic simulation to compare several complicated systems or several operating methods of a single system. Matejcik and Nelson (1993) illustrated the use of procedure $\mathcal{N}_R$ and the formation of simultaneous confidence intervals in this setting. Nelson and Matejcik (1993) considered the use of *common random number streams* in simulation studies to reduce computational effort when selection and multiple comparisons is the goal. Their paper assumed normal observations with correlation between observations in the same stage. Nelson and Matejcik permitted simultaneous indifference-zone selection of the best treatment and construction of simultaneous confidence intervals with the best mean; they illustrated their method by designing a simulation study to determine which of five inventory policies had the smallest expected cost per period.

CHAPTER 5

Problems Involving a Standard or a Control Treatment in Normal Response Experiments

5.1 INTRODUCTION

This chapter presents statistical methods for designing experiments and analyzing data from studies involving a standard or a control treatment when the data are normal and have a common known or common unknown variance. Let $\mu_1, \ldots, \mu_t$ denote the means of t "experimental" treatments which are to be compared with a value μ_0. The quantity μ_0 may be a *given* standard or the *unknown* mean of a control treatment; in the latter case, μ_0 must be estimated from the data. The data for comparing the treatments with a control can arise either from a designed experiment or from an observational study.

The first selection goal that we consider is exemplified by the problem of designing an experiment to select a *single* method of heat treating steel from among t new technologies, the best method being deemed unsatisfactory unless its mean tensile strength exceeds μ_0, the mean tensile strength of the best available technology. Thus, the objective is to select the treatment associated with $\mu_{[t]}$, the largest mean among the t new methods, provided that $\mu_{[t]}$ is greater than μ_0. A second possible formulation for the same setup is to select a *subset* of the new treatments containing all those heat treatment methods that have mean tensile strengths that are at least as large as μ_0. A third formulation, which would be of interest in certain circumstances, is to construct *simultaneous confidence intervals* for the set of t treatment means minus μ_0, that is, $\{\mu_1 - \mu_0, \ldots, \mu_t - \mu_0\}$.

In all three formulations described in the previous paragraph, the cases of known and unknown μ_0 must be distinguished. A problem in which μ_0 is *known* is called a *comparison with a standard*. A problem in which μ_0 is the *unknown* mean of a control treatment is referred to as a *comparison with a control*. In both cases, data must be collected from the t experimental treatments to estimate $\mu_1, \ldots, \mu_t$; in the latter case, data must also be collected from the control treatment to estimate μ_0.

Experiments that make comparisons with a given standard or control treatment differ in several important respects from the experiments discussed in Chapter 2. Consider the tensile strength example introduced above (and discussed, in general, in Section 5.2). The experimenter must now make one of $t + 1$ decisions rather than one of t decisions as was the case in Section 2.2, that is, either Π_1 or Π_2 or ... or Π_t will be selected *or* none will be selected. Also, there are two fundamentally different ways of making a correct selection; these are (1) selecting the treatment associated with $\mu_{[t]}$ if $\mu_{[t]} > \mu_0$, or (2) selecting *no* treatment if $\mu_{[t]} \leq \mu_0$. As will be seen below, these considerations lead to a different probability (design) requirement replacing (2.1.1).

Lastly, problems of comparison with a standard *are not* location invariant, and hence blocking designs *cannot* be used; for example, if $E\{Y_{ij}\} = \mu + \beta_i + \tau_j$, where the β_i are block effects and the τ_j are treatment effects, then *only* the *differences* $\tau_r - \tau_s$ ($1 \leq r, s \leq t$) and functions of these differences can be estimated. Problems involving a control treatment *are* location invariant; the experimenter must sample from the control treatment as well as from the t experimental treatments. Blocking designs *can* be used for problems involving a control. The same two cases occur with respect to the possibility of using blocking designs for subset selection formulations (discussed in Section 5.3) and simultaneous confidence interval formulations (discussed in Section 5.4).

Section 5.2 discusses indifference-zone (IZ) procedures for selecting the single treatment associated with $\mu_{[t]}$ provided that $\mu_{[t]}$ is greater than a standard or control treatment mean μ_0 and for selecting no treatment if $\mu_{[t]} \leq \mu_0$. Section 5.3 discusses procedures for selecting a subset of treatments, all of which have means greater than the given standard or the control. In Section 5.4 we study problems of forming simultaneous confidence intervals for the differences between the t treatment means and (i) the unknown mean μ_0 of a single control treatment or (ii) the unknown means of two control treatments.

5.2 SELECTING THE BEST TREATMENT USING THE IZ APPROACH

This section studies the problem of selecting the best treatment provided that its associated mean is greater than a given standard or control treatment.

Statistical Assumptions: Independent observations $Y_{i1}, Y_{i2}, \ldots$ ($1 \leq i \leq t$) are available from $t \geq 1$ normal treatments $\Pi_1, \ldots, \Pi_t$. Treatment Π_i has unknown mean μ_i; the Π_i have common known or common unknown variance σ^2. In addition, the value of a standard, μ_0, is given *or* a sample $Y_{01}, Y_{02}, \ldots$ is available from a normal control treatment, Π_0, having unknown mean μ_0 and the same variance σ^2 as the experimental treatments.

Let y_{ij} ($0 \leq i \leq t, j \geq 1$) denote the observed value of Y_{ij}. The ordered μ_i-values are denoted by $\mu_{[0]} \leq \cdots \leq \mu_{[t]}$. The pairing of the μ_i with the $\mu_{[s]}$ ($0 \leq i, s \leq t$)

is assumed to be unknown. The experimental objective and probability (design) requirement are stated in Goal 5.2.1 and (5.2.1)–(5.2.2), respectively.

Goal 5.2.1 To select the treatment associated with $\mu_{[t]}$ if $\mu_{[t]} > \mu_0$, or to select the standard (or control treatment) if $\mu_{[t]} \leq \mu_0$.

If Goal 5.2.1 is achieved, we say that a *correct selection* (CS) has been made. We denote the event of selecting the treatment associated with $\mu_{[t]}$ by $\Pi_{[t]}$ and that of selecting the standard (or control treatment) by $\Pi_{[0]}$, that is, selecting none of the experimental treatments.

Probability Requirement: For specified constants $\{\delta^\star, P_0^\star, P_1^\star\}$ with $0 < \delta^\star < \infty$, $2^{-t} < P_0^\star < 1$ and $(1 - 2^{-t})/t < P_1^\star < 1$, we require

$$P\{\Pi_{[0]}\} \geq P_0^\star \quad \text{whenever } \mu_{[t]} \leq \mu_0 \tag{5.2.1}$$

and

$$P\{\Pi_{[t]}\} \geq P_1^\star \quad \text{whenever } \mu_{[t]} \geq \max\{\mu_0, \mu_{[t-1]}\} + \delta^\star \tag{5.2.2}$$

for all σ^2, if σ^2 is unknown.

Equation (5.2.1) requires that the standard (or control) be selected as best with probability at least $P_0^\star$ whenever μ_0 exceeds all of the treatment means; (5.2.2) requires that the treatment with the largest mean be selected with probability at least $P_1^\star$ whenever its mean, $\mu_{[t]}$, exceeds both the standard (or control) and the $t - 1$ other treatment means by at least $\delta^\star$.

The lower bounds 2^{-t} and $(1 - 2^{-t})/t$ for $P_0^\star$ and $P_1^\star$, respectively, arise from the fact that they can be achieved by the following no-data procedure that uses randomization. Consider the procedure in which t fair coins are flipped to decide whether each of the t treatments has a mean greater than μ_0. If none of the treatments is declared to have mean greater than μ_0, then the standard (or control treatment) is selected; if any of the t treatments is declared to have mean larger then μ_0, then one of the t treatments is chosen at random.

Section 5.2.1 describes a single-stage procedure for Goal 5.2.1 in the case of a standard when there is a common *known* variance. Section 5.2.2 presents a two-stage procedure in the case of a standard and a common *unknown* variance. Section 5.2.3 gives a single-stage procedure in the case of a control and a common *known* variance.

5.2.1 Selection Involving a Standard (Common Known σ^2)

We first present a procedure due to Bechhofer and Turnbull (1978) for selecting the best treatment relative to a given standard when the responses are normal with common known variance σ^2.

SELECTING THE BEST TREATMENT USING THE IZ APPROACH

Procedure $\mathcal{N}_{BT}$ (σ^2 Known) For the given (t, μ_0) and specified (δ^*, P_0^*, P_1^*), choose constants $g = g(t; P_0^*, P_1^*, \nu)$ and $h = h(t; P_0^*, P_1^*, \nu)$ from the $\nu = \infty$ row of Table 5.1.

Sampling rule: Take a random sample of $n = \lceil(g\sigma/\delta^*)^2\rceil$ observations Y_{ij} ($1 \leq j \leq n$) in a single stage from Π_i ($1 \leq i \leq t$).

Terminal decision rule: Calculate the t sample means $\bar{y}_i = \sum_{j=1}^{n} y_{ij}/n$ ($1 \leq i \leq t$) and let $\bar{y}_{[t]} = \max\{\bar{y}_1, \ldots, \bar{y}_t\}$ denote the largest sample mean. If $\bar{y}_{[t]} > \mu_0 + h\delta^*/g$, select the treatment that yielded $\bar{y}_{[t]}$ as the one associated with $\mu_{[t]}$; otherwise, select no treatment, that is, select the standard as best.

The constants g and h are given in Table 5.1 in the $\nu = \infty$ rows for $t = 2(1)5$ and selected (P_0^*, P_1^*). The values of g and h are chosen to solve simultaneously

$$\Phi^t(h) = P_0^*$$

and

$$\int_{h-g}^{\infty} \Phi^{t-1}(y + g)\, d\Phi(y) = P_1^*$$

where $\Phi(y)$ is the standard normal distribution function.

Example 5.2.1 To illustrate the calculations, suppose that four treatment means are to be compared with a standard $\mu_0 = 20$. Suppose further that the common standard deviation σ of the treatment responses is 0.8. The experimenter desires that no treatment be selected as superior to the standard with probability at least 0.95, whenever the standard is larger than all of the treatment means μ_i ($1 \leq i \leq 4$), that is,

$$P\{\Pi_{[0]}\} \geq 0.95 \quad \text{whenever } \mu_{[4]} \leq 20.$$

In addition, if any of the treatment means exceeds the standard by 0.2 or more and is also greater than the other treatment means by 0.2 or more, then the experimenter wishes to select the treatment with mean $\mu_{[4]}$ with probability at least 0.90, that is,

$$P\{\Pi_{[4]}\} \geq 0.90 \quad \text{whenever } \mu_{[4]} \geq 20 + 0.2 \text{ and } \mu_{[4]} - \mu_{[3]} \geq 0.2.$$

For this example, $t = 4$, $P_0^* = 0.95$, $P_1^* = 0.90$, $\sigma = 0.8$ and $\delta^* = 0.2$. From Table 5.1 we find $g(4, 0.95, 0.90, \infty) = 3.5341$ and $h(4, 0.95, 0.90, \infty) = 2.2340$. Thus a random sample of size $n = \lceil((3.5341)(0.8)/0.2)^2\rceil = 200$ must be taken from each treatment. Calculating $h\delta^*/g = (2.2340)(0.2)/3.5341 = 0.126$ we determine that the standard is deemed best if $\bar{y}_{[4]} \leq 20 + 0.126 = 20.126$; otherwise, the treatment

Table 5.1. Values of the Constants $g(t; P_0^\star, P_1^\star, \nu)$ (top entries) and $h = h(t; P_0^\star, P_1^\star, \nu)$ (bottom entries) Required to Implement Procedure $\mathcal{N}_{BT}$ for Comparing Normal Treatments with a Given Standard for $t = 2(1)5$ and Selected ν and $(P_0^\star, P_1^\star)$

		$(P_0^\star, P_1^\star)$					
t	ν	(0.75, 0.75)	(0.90, 0.75)	(0.90, 0.90)	(0.95, 0.75)	(0.95, 0.90)	(0.95, 0.95)
2	2	2.4037 1.4523	3.6087 2.7433	4.7751 2.7433	4.9150 4.0750	6.0326 4.0750	7.1781 4.0750
	4	2.1033 1.2601	2.8488 2.0722	3.6793 2.0722	3.4793 2.7215	4.2912 2.7215	4.9204 2.7215
	6	2.0166 1.2053	2.6543 1.9047	3.4013 1.9047	3.1496 2.4170	3.8846 2.4170	4.4058 2.4170
	8	1.9751 1.1795	2.5658 1.8294	3.2761 1.8294	3.0053 2.2849	3.7056 2.2849	4.1810 2.2849
	10	1.9512 1.1645	2.5153 1.7866	3.2041 1.7866	2.9242 2.2113	3.6050 2.2113	4.0552 2.2113
	14	1.9241 1.1477	2.4599 1.7399	3.1259 1.7399	2.8371 2.1321	3.4964 2.1321	3.9200 2.1321
	18	1.9094 1.1386	2.4301 1.7149	3.0844 1.7149	2.7907 2.0902	3.4384 2.0902	3.8479 2.0902
	22	1.9002 1.1329	2.4114 1.6993	3.0579 1.6993	2.7620 2.0643	3.4025 2.0643	3.8038 2.0643
	∞	1.8597 1.1078	2.3310 1.6322	2.9451 1.6322	2.6398 1.9545	3.2499 1.9545	3.6154 1.9545
3	3	2.5515 1.6363	3.4666 2.6481	4.4157 2.6481	4.3414 3.5508	5.2527 3.5508	6.0408 3.5508
	6	2.3089 1.4708	2.9310 2.1716	3.6876 2.1716	3.4327 2.6957	4.1723 2.6957	4.7013 2.6957
	9	2.2362 1.4218	2.7829 2.0421	3.4872 2.0421	3.1991 2.4787	3.8912 2.4787	4.3570 2.4787
	12	2.2012 1.3984	2.7136 1.9820	3.3938 1.9820	3.0928 2.3805	3.7630 2.3805	4.2006 2.3805
	15	2.1805 1.3846	2.6734 1.9473	3.3401 1.9473	3.0318 2.3245	3.6897 2.3245	4.1120 2.3245
	21	2.1576 1.3692	2.6290 1.9089	3.2805 1.9089	2.9650 2.2632	3.6084 2.2632	4.0135 2.2632
	27	2.1447 1.3608	2.6050 1.8881	3.2475 1.8881	2.9292 2.2303	3.5651 2.2303	3.9608 2.2303
	33	2.1368 1.3554	2.5896 1.8751	3.2274 1.8751	2.9064 2.2097	3.5378 2.2097	3.9277 2.2097
	∞	2.1011 1.3319	2.5237 1.8183	3.1386 1.8183	2.8095 2.1212	3.4199 2.1212	3.7860 2.1212

From Bechhofer and Turnbull (1978), pp. 388–389. Reprinted with permission from *JASA*. Copyright 1978 by the American Statistical Association. All rights reserved.

Table 5.1. *Continued*

t	ν	\(P_0^*, P_1^*\) (0.75, 0.75)	(0.90, 0.75)	(0.90, 0.90)	(0.95, 0.75)	(0.95, 0.90)	(0.95, 0.95)
4	4	2.6426 1.7507	3.4143 2.6201	4.2693 2.6201	4.1063 3.3401	4.9300 3.3401	5.5817 3.3401
	8	2.4378 1.6065	2.9927 2.2433	3.7141 2.2433	3.4300 2.7035	4.1357 2.7035	4.6190 2.7035
	12	2.3748 1.5626	2.8710 2.1357	3.5542 2.1357	3.2449 2.5310	3.9167 2.5310	4.3555 2.5310
	16	2.3442 1.5413	2.8133 2.0849	3.4783 2.0849	3.1586 2.4511	3.8143 2.4511	4.2385 2.4511
	20	2.3262 1.5288	2.7793 2.0553	3.4341 2.0553	3.1090 2.4051	3.7500 2.4051	4.1632 2.4051
	28	2.3058 1.5147	2.7146 2.0222	3.3844 2.0222	3.0537 2.3541	3.6898 2.3541	4.0849 2.3541
	36	2.2947 1.5069	2.7209 2.0042	3.3577 2.0042	3.0238 2.3265	3.6540 2.3265	4.0426 2.3265
	44	2.2875 1.5020	2.7801 1.9929	3.3410 1.9929	3.0052 2.3092	3.6314 2.3092	4.0160 2.3092
	∞	2.2559 1.4803	2.6510 1.9432	3.2664 1.9432	2.9233 2.2340	3.5341 2.2340	3.9003 2.2340
5	5	2.7098 1.8340	3.3951 2.6159	4.1969 2.6159	3.9861 3.2343	4.7612 3.2343	5.3424 3.2343
	10	2.5311 1.7057	3.0430 2.2999	3.7427 2.2999	3.4406 2.7205	4.1263 2.7205	4.5838 2.7205
	15	2.4752 1.6658	2.9381 2.2068	3.6082 2.2068	2.2855 2.5755	3.9447 2.5755	4.3690 2.5755
	20	2.4478 1.6464	2.8881 2.1623	3.5436 2.1623	3.2121 2.5074	3.8587 2.5074	4.2672 2.5074
	25	2.4318 1.6349	2.8586 2.1362	3.5060 2.1362	3.1695 2.4678	3.8089 2.4678	4.2084 2.4678
	35	2.4135 1.6219	2.8255 2.1070	3.4632 2.1070	3.1221 2.4237	3.7530 2.4237	4.1424 2.4237
	45	2.4033 1.6148	2.8071 2.0910	3.4403 2.0910	3.0964 2.3997	3.7228 2.3997	4.1070 2.3997
	55	2.3968 1.6102	2.7959 2.0810	3.4255 2.0810	3.0802 2.3847	3.7035 2.3847	4.0841 2.3847
	∞	2.3682 1.5900	2.7453 2.0365	3.3608 2.0365	3.0086 2.3187	3.6195 2.3187	3.9858 2.3187

associated with sample mean $\bar{y}_{[4]}$ is selected as being better than both the standard and the other treatments.

Example 5.2.2 (A Quality Control Problem) This example shows how procedure $\mathcal{N}_{BT}$ can be adapted to select the treatment that produces the greatest fraction of conforming product. Suppose that the tensile strengths of bolts produced by five manufacturing techniques are to be studied to determine whether any of the techniques should be adopted as the manufacturing method of choice. Individual bolts are required to achieve at least the lower specification limit of 60,000 psi. Suppose that the mean tensile strength of the ith manufacturing method is μ_i ($1 \leq i \leq 5$) and that all five methods have a standard deviation of 1000 psi. The manufacturer decides on the following design requirement. If none of the methods produces at least 75% conforming product, then with probability at least 0.95 none of the treatments should be adopted. If the manufacturing method with the greatest percent conforming product achieves at least 90% conforming product, and the second best method produces no more than 75% conforming product, then the best treatment should be selected with probability at least 0.90.

The design requirement can be translated into the formulation (5.2.1)–(5.2.2) by first noting that if the mean μ of a process is less than 60,675 psi, then no more than 75% of the product will meet the specification limit, while if the mean is 61,282 psi or greater, then at least 90% of the product will attain the lower limit. This follows because

$$P\{N(\mu, 1000^2) \geq 60,000\} = 1 - \Phi\left(\frac{60,000 - \mu}{1000}\right) \leq 0.75$$

whenever $\mu \leq 60,675$ psi, and

$$P\{N(\mu, 1000^2) \geq 60,000\} = 1 - \Phi\left(\frac{60,000 - \mu}{1000}\right) \geq 0.90$$

whenever $\mu \geq 61,282$ psi where $N(\mu, 1000^2)$ denotes a normal random variable with mean μ and variance 1000^2. Thus taking $\mu_0 = 60,675$ psi, $\delta^* = 61,282$ psi $- 60,675$ psi $= 607$ psi, $P_0^* = 0.95$ and $P_1^* = 0.90$, the original design requirement can be stated as

$$P\{\Pi_{[0]}\} \geq 0.95 \quad \text{whenever } \mu_{[5]} \leq 60,675$$

and

$$P\{\Pi_{[5]}\} \geq 0.90 \quad \text{whenever } \mu_{[5]} \geq 61,282 \text{ and } \mu_{[5]} \geq \mu_{[4]} + 607.$$

To guarantee this requirement we determine from Table 5.1 that $g(5, 0.95, 0.90, \infty) = 3.6195$ and $h(5, 0.95, 0.90, \infty) = 2.3187$ and take a random sample of size $n = \lceil ((3.6195)(1000)/607)^2 \rceil = 36$ from each treatment. The yardstick $h\delta^*/g =$

$(2.3187)(607)/3.6195 = 389$ psi is used to make the comparison with the standard. If $\bar{y}_{[5]} \leq 60{,}675 + 389 = 61{,}064$ psi, then none of the methods is adopted; otherwise, the treatment associated with sample mean $\bar{y}_{[5]}$ is selected.

Remark 5.2.1 The ability of procedure $\mathcal{N}_{BT}$ to guarantee (5.2.1) and (5.2.2) depends heavily on the validity of the normality assumption.

5.2.2 Selection Involving a Standard (Common Unknown σ^2)

We now present a procedure due to Bechhofer and Turnbull (1978) for selecting the best treatment relative to a given standard when the responses are normal with common unknown variance σ^2. The procedure below is an analogue of the Stein (1945) two-stage sampling procedure that was introduced in Section 2.7.1. It requires that an initial sample of $n_1 \geq 2$ observations be taken from each treatment in order to estimate σ^2; these data are supplemented by additional observations, the number of which depends on the estimated standard deviation of the observations calculated from the first sample.

Procedure $\mathcal{N}_{BT}$ (σ^2 Unknown) For the given (t, μ_0) and specified (δ^*, P_0^*, P_1^*), fix a number of observations $n_1 \geq 2$ to be taken in Stage 1. Choose constants (g, h) from Table 5.1 corresponding to the $(t; P_0^*, P_1^*, \nu)$ of interest, where $\nu = t(n_1 - 1)$.

Sampling rule:

Stage 1: Take a random sample of n_1 observations Y_{ij} $(1 \leq j \leq n_1)$ from Π_i $(1 \leq i \leq t)$. Calculate $\bar{y}_i^{(1)} = \sum_{j=1}^{n_1} y_{ij}/n_1$ $(1 \leq i \leq t)$, and $s_\nu^2 = \sum_{i=1}^{t}\sum_{j=1}^{n_1} \left(y_{ij} - \bar{y}_i^{(1)}\right)^2/\nu$, the unbiased pooled estimate of σ^2 based on $\nu = t(n_1 - 1)$ d.f.

Stage 2: Take a random sample of $N - n_1$ additional observations from each of the Π_i $(1 \leq i \leq t)$, where

$$N \stackrel{.}{=} \begin{cases} n_1 & \text{if } (gs_\nu/\delta^*)^2 \leq n_1 \\ \lceil (gs_\nu/\delta^*)^2 \rceil & \text{if } (gs_\nu/\delta^*)^2 > n_1. \end{cases}$$

Terminal decision rule: Calculate the t cumulative sample means $\bar{y}_i = \sum_{j=1}^{N} y_{ij}/N$. Let $\bar{y}_{[t]} = \max\{\bar{y}_1, \ldots, \bar{y}_t\}$. If $\bar{y}_{[t]} > \mu_0 + h\delta^*/g$, select the treatment that yielded $\bar{y}_{[t]}$ as the one associated with $\mu_{[t]}$; otherwise, select the standard as best.

When σ^2 is unknown, g and h are chosen to solve simultaneously

$$\int_0^\infty \Phi^t(hz) f_\nu(z)\, dz = P_0^*$$

and

$$\int_0^\infty \left[\int_{(h-g)z}^\infty \Phi^{t-1}(y + gz)\phi(y)\,dy \right] f_\nu(z)\,dz = P_1^*$$

where $f_\nu(\cdot)$ is the p.d.f. of a $(\chi_\nu^2/\nu)^{1/2}$ random variable.

Remark 5.2.2 Beginning with Stein (1945), several authors have made theoretical studies of how to choose the first-stage sample size n_1 (see Moshman 1958). This problem can, for example, be formulated mathematically as one of choosing n_1 to minimize the maximum expected total sample size $E\{N\}$ over a given range of σ^2 values. However, a simple heuristic for choosing n_1 is to require that n_1 be sufficiently large that a reasonably large number of degrees of freedom ν, say at least 50, are available for estimating σ^2.

Example 5.2.3 This example illustrates the calculations to implement procedure $\mathcal{N}_{BT}$ in a setting similar to that of Example 5.2.2. Suppose that a manufacturer is considering purchasing steel from one of three suppliers. The strength of the steel from the ith supplier ($1 \leq i \leq 3$) is assumed to be normally distributed with unknown mean μ_i, and unknown variance σ^2 that is common for all of the suppliers.

Suppose that the manufacturer seeks a sampling plan that guarantees the two requirements: (i) With probability at least 0.95, none of the suppliers will be chosen if all produce steel with mean tensile strength less than 60,000 psi, and (ii) with probability at least 0.90, the supplier with the largest mean $\mu_{[3]}$ will be selected whenever $\mu_{[3]} \geq 60{,}250$ psi and $\mu_{[3]} - \mu_{[2]} \geq 250$ psi. This probability requirement can be stated as

$$P\{\Pi_{[0]}\} \geq 0.95 \quad \text{whenever } \mu_{[3]} \leq 60{,}000 \text{ for all } \sigma^2$$

and

$$P\{\Pi_{[3]}\} \geq 0.90 \quad \text{whenever } \mu_{[3]} \geq 60{,}250 \text{ and } \mu_{[3]} - \mu_{[2]} \geq 250 \text{ for all } \sigma^2.$$

This is the probability requirement (5.2.1)–(5.2.2) where $\mu_0 = 60{,}000$ psi, $\delta^* = 250$ psi, $P_0^* = 0.95$ and $P_1^* = 0.90$.

Suppose further that $n_1 = 12$ test specimens are selected initially from each supplier, and $s_\nu = 753$ psi is calculated from the initial sample based on $\nu = 3(11) = 33$ d.f. From Table 5.1 we obtain $(g, h) = (3.5378, 2.2097)$ and calculate $(gs_\nu/\delta^*)^2 = (3.5378 \times 753/250)^2 = 113.5 > 12$. Thus an additional $N - 12 = \lceil 113.5 \rceil - 12 = 102$ observations must be taken from each supplier. Let $\bar{y}_{[i]}$ denote the ith ordered mean based on all $N = 114$ observations. If $\bar{y}_{[3]} > 60{,}000 + (2.2097 \times 250)/3.5378 = 60{,}156$, select the treatment that yielded $\bar{y}_{[3]}$ as the one associated with $\mu_{[3]}$; otherwise, select the standard as best.

5.2.3 Selection Involving a Control (Common Known σ^2)

The problem considered in the present section differs from that in the previous sections because here we consider selection with respect to a *control* treatment. The following single-stage procedure was proposed by Paulson (1952).

Procedure $\mathcal{N}_{P/C}$ (σ^2 Known) For the given t and specified (δ^*, P_0^*, P_1^*), let $h = Z_{t,1/2}^{(1-P_0^*)}$ be determined from Table B.1 and let n be the solution of (5.2.3) below.

Sampling rule: Take a random sample of n observations $Y_{i1}, \ldots, Y_{in}$ in a single stage from Π_i ($0 \leq i \leq t$).

Terminal decision rule: Calculate the $t+1$ sample means $\bar{y}_i = \sum_{j=1}^{n} y_{ij}/n$ ($0 \leq i \leq t$) and let $\bar{y}_{[t]} = \max\{\bar{y}_1, \ldots, \bar{y}_t\}$ denote the largest (noncontrol) sample mean. If $\bar{y}_{[t]} > \bar{y}_0 + h\sigma\sqrt{2/n}$, select the treatment $\Pi_{[t]}$; otherwise, select the control Π_0.

The sample size n is the smallest integer for which

$$\int_{-\infty}^{\infty} \Phi\left(x + \frac{\delta^*\sqrt{n}}{\sigma} - h\sqrt{2}\right) \Phi^{t-1}\left(x + \frac{\delta^*\sqrt{n}}{\sigma}\right) d\Phi(x) \geq P_1^*. \quad (5.2.3)$$

Equivalently, (5.2.3) has the representation

$$P\left\{h - \frac{\delta^*\sqrt{n}}{\sigma\sqrt{2}} \leq U_0; \quad -\frac{\delta^*\sqrt{n}}{\sigma\sqrt{2}} \leq U_i \ (1 \leq i \leq t-1)\right\} \geq P_1^* \quad (5.2.4)$$

where $(U_0, U_1, \ldots, U_{t-1})$ has a multivariate normal distribution with mean vector zero, unit variances, and common correlation $1/2$. The quadrant probability on the left-hand side of Equation (5.2.4) can be computed using the FORTRAN program MVNPRD in Dunnett (1989), and hence n can be determined easily by trial and error.

Example 5.2.4 Suppose $t = 5$ normal treatments, each having $\sigma^2 = 100$, are to be compared to a control treatment also having $\sigma^2 = 100$. It is desired to select the control treatment with probability 0.80 if $\mu_{[5]} \leq \mu_0$, and to select the best experimental treatment with probability 0.90 if $\mu_{[5]} \geq \max\{\mu_{[4]}, \mu_0\} + 7$. The yardstick for procedure $\mathcal{N}_{P/C}$ is given by $h = Z_{5,1/2}^{(0.20)} = 1.535$. The common sample size is the smallest integer n for which

$$P\left\{1.535 - \frac{7\sqrt{n}}{10\sqrt{2}} = 1.535 - 0.495\sqrt{n} \leq U_0; \right.$$
$$\left. -0.495\sqrt{n} \leq U_i \ (1 \leq i \leq 4)\right\} \geq 0.90.$$

After several iterations using Dunnett's program, we find that $n = 33$ observations are required. The procedure $\mathcal{N}_{P/C}$ selects the experimental treatment associated with $\bar{y}_{[5]}$ if

$$\bar{y}_{[5]} > \bar{y}_0 + 1.535(10)\sqrt{2/33} = \bar{y}_0 + 3.779$$

and selects the control otherwise.

5.3 SELECTING A SUBSET OF TREATMENTS

In problems involving a given standard μ_0 or a control treatment with mean μ_0, we provide procedures for selecting a subset of the t treatments containing *all* of the treatments with means μ_i that exceed μ_0. As usual in subset selection, the data on which the decision is to be based can come from a designed experiment or an observational study.

The number of treatments in the selected subset is a random variable that can vary from zero to t. The expected number of treatments in the selected subset depends on the true configuration of treatment means, the standard or control treatment mean μ_0, the common variance σ^2, and the common sample size n. Whether σ^2 is known or unknown, as $n \to \infty$, the expected number of treatments selected by the procedures described below converges to the true number of treatments with means larger than μ_0. Thus, in the case when the experimenter has control over the sample size, it is natural to choose n sufficiently large so that the selected subset will contain all of the treatments having means larger than μ_0, but only a small number of treatments having means less than μ_0.

The subset approach can be regarded as a screening device. The treatments in the selected subset can be subjected to further sampling with the objective of selecting all the "best" ones or a subset of the best ones using, say, the indifference-zone approach.

In Section 5.3.1 we consider subset selection with respect to a given standard μ_0. Section 5.3.2 is concerned with subset selection with respect to a control treatment having unknown mean μ_0.

5.3.1 Screening Involving a Standard

We now discuss the problem of selecting a subset of the treatments having means greater than a given standard.

Statistical Assumptions: Independent random samples of n observations $Y_{i1}, \ldots, Y_{in}$ $(1 \leq i \leq t)$ are taken from $t \geq 1$ normal treatments $\Pi_1, \ldots, \Pi_t$. Treatment Π_i has unknown mean μ_i $(1 \leq i \leq t)$; the Π_i have common known or common unknown variance σ^2. A known standard μ_0 is also given.

Let y_{ij} denote the observed value of Y_{ij} $(1 \leq i \leq t, 1 \leq j \leq n)$. The ordered μ_i-values are denoted by $\mu_{[1]} \leq \cdots \leq \mu_{[t]}$. The pairing of the μ_i with the $\mu_{[s]}$ $(1 \leq i, s \leq t)$ is assumed to be unknown.

The experimental objective and probability (design) requirement are stated in Goal 5.3.1 and (5.3.1), respectively.

Goal 5.3.1 To select a subset of the treatments that includes *all* those treatments having means $\mu_i > \mu_0$.

If Goal 5.3.1 is achieved, then we say that a *correct selection* (CS) has occurred.

Probability Requirement: For specified constant P^* with $1/t < P^* < 1$, we require that

$$P\{CS\} \geq P^* \tag{5.3.1}$$

for all $\boldsymbol{\mu}$ and σ^2, if σ^2 is unknown.

The following procedure, due to Gupta and Sobel (1958), can be used for both the common known and common unknown variance cases.

Procedure $\mathcal{N}_{GS}$ (Given Standard μ_0) Calculate the t sample means $\bar{y}_i = \sum_{j=1}^{n} y_{ij}/n$ $(1 \leq i \leq t)$. If σ^2 is unknown also calculate $s_\nu^2 = \sum_{i=1}^{t} \sum_{j=1}^{n} (y_{ij} - \bar{y}_i)^2/\nu$, the pooled unbiased estimate of σ^2 based on $\nu = t(n-1)$ d.f.

Case 1 (σ^2 Known): Include Π_i in the selected subset if and only if

$$\bar{y}_i > \mu_0 - h\sigma/\sqrt{n}$$

where $h = z^{(1-(P^*)^{1/t})}$ and $z^{(\gamma)}$ denotes the upper-γ quantile of the standard normal distribution.

Case 2 (σ^2 Unknown): Include Π_i in the selected subset if and only if

$$\bar{y}_i > \mu_0 - hs_\nu/\sqrt{n}$$

where $h = T_{t,\nu,0}^{(1-P^*)}$ is found in Table B.4.

Note that, in contrast to Goal 5.2.1, the subset selection Goal 5.3.1 can be achieved using a single-stage procedure whether or not σ^2 is known. Goal 5.2.1 can be achieved using a single-stage procedure only if σ^2 is known.

Procedure $\mathcal{N}_{GS}$ uses the constant h that accommodates the most difficult configuration of treatment means; specifically, h is chosen to achieve (5.3.1) when $\mu_1 = \cdots = \mu_t = \mu_0$. When σ^2 is known, h is related to the upper-$(1-P^*)$ one-sided equicoordinate point of the multivariate normal distribution whose components are independent, identically distributed $N(0, 1)$ random variables; specifically, $z^{(1-(P^*)^{1/t})} = Z_{t,0}^{(1-P^*)}$. As the amount of information increases, the σ^2 known and σ^2 unknown procedures become identical in that $T_{t,\nu,0}^{(1-P^*)} \to Z_{t,0}^{(1-P^*)}$ and S_ν converges almost surely to σ as $\nu \to \infty$.

If it is known that exactly (or at most) q treatments ($1 \leq q < t$) have means that are at least as large as μ_0, then a smaller constant h can be used. If σ^2 is known, then $h = z^{(1-(P^*)^{1/q})} = Z_{q,0}^{(1-P^*)}$ will guarantee (5.3.1), whereas if σ^2 is unknown, $h = T_{q,\nu,0}^{(1-P^*)}$ will guarantee (5.3.1).

Remark 5.3.1 As is true in general for subset selection procedures, the larger the value of n, the smaller the expected number of treatments in the selected subset.

Example 5.3.1 Recall Example 3.2.3 in which we studied the amounts of fat absorbed by doughnuts during cooking using one of $t = 4$ brands of shortening—A, B, C and D. The data for this experiment are listed in Table 3.4. Suppose that it is desired to select a subset containing all those brands whose mean fat absorption is at least 175 grams with confidence $P^* = 0.90$. To implement procedure $\mathcal{N}_{GS}$ we compute that $\nu = t(n-1) = 20$ d.f. From Table B.4 we find that $h = T_{t,\nu,0}^{(1-P^*)} = T_{4,20,0}^{(0.10)} = 2.06$. We select those brands for which

$$\bar{y}_i > \mu_0 - hs_\nu/\sqrt{n} = 175 - 2.06 \times \sqrt{100.9/6} = 166.55.$$

Hence we select brands A, B and C.

Unbalanced Experiments
Procedure $\mathcal{N}_{GS}$ can be modified to accommodate unbalanced data as follows: Suppose that n_i observations are available from the ith treatment ($1 \leq i \leq t$). Treatment Π_i is selected if and only if

$$\bar{y}_i \geq \mu_0 - hs_\nu/\sqrt{n_i} \qquad (5.3.2)$$

where s_ν^2 is the pooled unbiased estimator of σ^2 based on $\nu = \sum_{i=1}^{t}(n_i - 1)$ d.f. and $h = T_{t,\nu,0}^{(1-P^*)}$ is found in Table B.4. If σ^2 is known, then σ replaces s_ν in (5.3.2) and $z^{1-(P^*)^{1/t}}$ is used for h.

Blocking
Blocking designs *cannot* be used for comparisons with respect to a *standard* because the decision rule is *not* location invariant.

5.3.2 Screening Involving a Control Treatment

We consider the problem of selecting a subset of the treatments having means greater than that of a control treatment.

Statistical Assumptions: The assumptions here are the same as those of Section 5.3.1 except that there is now a control treatment Π_0 from which a random sample of n observations, $Y_{01}, \ldots, Y_{0n}$, is taken, independent of the other tn observations. The Y_{0j} are normal with *unknown* mean μ_0 and the same variance, σ^2, as the other treatments.

In addition to the notation of Section 5.3.1, we let y_{0j} denote the observed value of Y_{0j} ($1 \leq j \leq n$). The probability requirement is again given by (5.3.1), except that now μ_0 in Goal 5.3.1 is an *unknown* control treatment mean instead of a given standard. The following modification of procedure $\mathcal{N}_{GS}$ accommodates the control treatment.

Procedure $\mathcal{N}_{GS}$ (Control Treatment—Balanced Experiment) Calculate the $t + 1$ sample means $\bar{y}_i = \sum_{j=1}^{n} y_{ij}/n$ ($0 \leq i \leq t$). If σ^2 is unknown, also calculate $s_\nu^2 = \sum_{i=0}^{t} \sum_{j=1}^{n} (y_{ij} - \bar{y}_i)^2 / \nu$, the pooled unbiased estimate of σ^2 based on $\nu = (t + 1)(n - 1)$ d.f.

Case 1 (σ^2 Known): Include Π_i in the selected subset if and only if

$$\bar{y}_i > \bar{y}_0 - h\sigma\sqrt{2/n}$$

where $h = Z_{t,1/2}^{(1-P^*)}$ is found in Table B.1.

Case 2 (σ^2 Unknown): Include Π_i in the selected subset if and only if

$$\bar{y}_i > \bar{y}_0 - hs_\nu\sqrt{2/n}$$

where $h = T_{t,\nu,1/2}^{(1-P^*)}$ is found in Table B.3.

Unbalanced Experiments

It is straightforward to extend procedure $\mathcal{N}_{GS}$ to the following special case involving the comparison of treatments with a control. If the experimenter has allocated n_0 observations to the control treatment and n to each of the other t test treatments, then the rule above is modified by changing the constant in the yardstick to reflect the correct standard error of the difference $\bar{Y}_i - \bar{Y}_0$ and the correct value of h.

Procedure $\mathcal{N}_{GS}$ (Control Treatment—Unbalanced Experiments) Take a random sample of n observations $Y_{i1}, \ldots, Y_{in}$ from Π_i ($1 \leq i \leq t$) and a random sample of n_0 observations $Y_{01}, \ldots, Y_{0n_0}$ from the control treatment Π_0. Calculate the $t+1$ sample means $\bar{y}_i = \sum_{j=1}^{n} y_{ij}/n$ ($1 \leq i \leq t$) and $\bar{y}_0 = \sum_{j=1}^{n_0} y_{0j}/n_0$. If σ^2 is unknown, also calculate

$$s_\nu^2 = \left[\sum_{i=1}^{t}\sum_{j=1}^{n}(y_{ij} - \bar{y}_i)^2 + \sum_{j=1}^{n_0}(y_{0j} - \bar{y}_0)^2 \right] / \nu,$$

the pooled unbiased estimate of σ^2 based on $\nu = t(n-1) + (n_0 - 1)$ d.f.

Case 1 (σ^2 Known): Include Π_i in the selected subset if and only if

$$\bar{y}_i > \bar{y}_0 - h\sigma \sqrt{\frac{1}{n} + \frac{1}{n_0}}$$

where $h = Z_{t,\rho}^{(1-P^*)}$ with $\rho = n/(n + n_0)$ is found in Table B.1.

Case 2 (σ^2 Unknown): Include Π_i in the selected subset if and only if

$$\bar{y}_i > \bar{y}_0 - h s_\nu \sqrt{\frac{1}{n} + \frac{1}{n_0}}$$

where $h = T_{t,\nu,\rho}^{(1-P^*)}$ with $\rho = n/(n + n_0)$ is found in Table B.4.

Values of $T_{t,\nu,\rho}^{(1-P^*)}$ for ρ not listed in Table B.4 can be obtained by performing interpolation or using the FORTRAN program given in Dunnett (1989).

Blocking

Blocking designs *can* be used for comparisons with respect to a control because the decision rule is invariant with respect to *common* location shifts of all the treatment and control observations. For example, if observations are available in n complete blocks of size $t+1$ and Y_{ij} is the response on treatment i in the jth block ($0 \leq i \leq t$, $1 \leq j \leq n$), then procedure $\mathcal{N}_{GS}$ selects those treatments Π_i for which

$$\bar{y}_i > \bar{y}_0 - h\sigma \sqrt{2/n}$$

where $\bar{y}_i = \sum_{j=1}^{n} y_{ij}/n$ ($0 \leq i \leq t$) and $h = Z_{t,1/2}^{(1-P^*)}$. In a similar way, balanced incomplete block designs or Latin square designs can be used for this problem. (See Section 2.2.)

5.4 SIMULTANEOUS CONFIDENCE INTERVALS

Sections 5.2 and 5.3 considered problems of selecting a single treatment or subset of treatments that are *at least as good as* a standard or a control. However, in many medical studies and other applications, it is more relevant to quantify the differences between the (experimental) treatments and the standard or control mean.

This section describes methods for quantifying the differences between the set of treatment means and either the standard value or the mean of the control treatment. Specifically, we show how to form simultaneous confidence intervals and confidence bounds for the differences between t treatment means and μ_0, where μ_0 is given in the case of a standard and μ_0 is unknown in the case of a control. In the first case, described in Section 5.4.1, the problem has a simple solution; in the second case, described in Section 5.4.2, μ_0 must be estimated and the solution to the problem requires special multivariate student t-distribution tables. Section 5.4.3 describes comparisons with respect to *two* control treatments.

5.4.1 Comparison with a Standard

We first study the problem of comparing treatments with a known standard.

Statistical Assumptions: Independent random samples of n observations $Y_{i1}, \ldots, Y_{in}$ $(1 \leq i \leq t)$ are taken from $t \geq 1$ normal treatments $\Pi_1, \ldots, \Pi_t$. Treatment Π_i has unknown mean μ_i; the t treatments have common known or common unknown variance σ^2. A standard μ_0 is also given.

Let y_{ij} denote the observed value of Y_{ij} $(1 \leq i \leq t, 1 \leq j \leq n)$. If σ^2 is known, choose $n \geq 1$, while if σ^2 is unknown, choose $n \geq 2$ in order to calculate an estimate of σ^2. Calculate $\bar{y}_i = \sum_{j=1}^{n} y_{ij}/n$ $(1 \leq i \leq t)$ and $s_\nu^2 = \sum_{i=1}^{t} \sum_{j=1}^{n} (y_{ij} - \bar{y}_i)^2/\nu$, the unbiased pooled estimate of σ^2 based on $\nu = t(n-1)$ d.f.

The purpose of the experiment and the associated probability (design) requirement are stated in Goal 5.4.1 and (5.4.1), respectively.

Goal 5.4.1 To obtain joint confidence interval estimates for the t differences $\mu_i - \mu_0$ $(1 \leq i \leq t)$, where μ_0 is a given standard, using either two-sided interval estimates or simultaneous upper or lower confidence bounds (one-sided interval estimates).

Probability Requirement: For specified $0 < P^* < 1$, we require that

$$P\{\text{Confidence region covers } \mu_i - \mu_0 \ (1 \leq i \leq t) \mid (\boldsymbol{\mu}, \sigma^2)\} \geq P^* \qquad (5.4.1)$$

for all $\boldsymbol{\mu}$ and σ^2.

The problem of forming simultaneous confidence intervals for comparison with a given standard μ_0 is straightforward because the differences $\{\bar{Y}_i - \mu_0 \ (1 \leq i \leq t)\}$ are *independent* normal random variables.

> **Simultaneous Two-Sided Confidence Bounds (Comparison with a Given Standard μ_0)**
>
> *Case 1 (σ^2 Known):* Simultaneous two-sided confidence bounds for $\{\mu_i - \mu_0 \ (1 \leq i \leq t)\}$ with joint confidence coefficient P^* are given by
>
> $$\left\{ \mu_i - \mu_0 \in \bar{y}_i - \mu_0 \pm h\sigma/\sqrt{n} \ \ (1 \leq i \leq t) \right\} \qquad (5.4.2)$$
>
> where $h = z^{(q/2)}$ with $q = 1 - (P^*)^{1/t}$.
>
> *Case 2 (σ^2 Unknown):* Simultaneous two-sided confidence bounds are
>
> $$\left\{ \mu_i - \mu_0 \in \bar{y}_i - \mu_0 \pm hs_\nu/\sqrt{n} \ \ (1 \leq i \leq t) \right\} \qquad (5.4.3)$$
>
> where $h = |M|_{t,\nu}^{(1-P^*)}$, the upper-$(1 - P^*)$ critical point of the t-dimensional studentized maximum modulus distribution.

The studentized maximum modulus quantile, $|M|_{p,\nu}^{(\alpha)}$, was defined by Equation (4.2.1) in Section 4.2. The values of the studentized maximum modulus critical point can be found in Table B.7 or in Bechhofer and Dunnett (1988), or can be computed via the Dunnett (1989) MVTPRD program using the relationship $|M|_{p,\nu}^{(\alpha)} = |T|_{p,\nu,0}^{(\alpha)}$.

Remark 5.4.1 The quantity $z^{(q/2)}$ with $q = 1 - (P^*)^{1/t}$ is a special case of the two-sided upper-$(1 - P^*)$ equicoordinate point of the multivariate normal distribution. In particular, we have $z^{(q/2)} = |Z|_{t,0}^{(1-P^*)}$, a relationship demonstrated in Appendix A. Thus, the unknown variance procedure converges to the known variance procedure because $|M|_{t,\nu}^{(1-P^*)} \to |Z|_{t,0}^{(1-P^*)}$ as $\nu \to \infty$.

To form simultaneous lower or upper confidence bounds for the differences $\{\mu_i - \mu_0 \ (1 \leq i \leq t)\}$, one need only use the analogous one-sided quantile of the appropriate normal or t-distribution. For example, when the variance is unknown, we have

$$\left\{ \mu_i - \mu_0 \geq \bar{y}_i - \mu_0 - hs_\nu/\sqrt{n} \ \ (1 \leq i \leq t) \right\}$$

where $h = T_{t,\nu,0}^{(1-P^*)}$, are joint lower confidence bounds for $\{\mu_i - \mu_0 \ (1 \leq i \leq t)\}$ with joint confidence coefficient P^*.

Example 5.4.1 Recall Examples 3.2.3 and 5.3.1 concerning fat absorption by doughnuts during cooking using one of $t = 4$ brands of shortening—A, B, C and D (see Table 3.4). Suppose we wish to form 90% simultaneous two-sided confidence bounds for the differences between the mean fat absorption for each brand and the threshold value of 175 grams of fat. We require the critical value $h = |M|_{t,\nu}^{(1-P^*)} = |M|_{4,20}^{(0.10)} = 2.39$ from the Dunnett (1989) multivariate t-distribution critical point program. Using Equation (5.4.3), we have the following confidence bounds.

$$\mu_i - \mu_0 \in \bar{y}_i - \mu_0 \pm hs_\nu/\sqrt{n} = \bar{y}_i - 175 \pm 2.39 \times \sqrt{100.9/6} = \bar{y}_i - 175 \pm 9.80$$

for $i \in \{A,B,C,D\}$.

Fixed-Width Interval Estimates

As stated, the simultaneous confidence intervals (5.4.2) and (5.4.3) are basically an *analysis* tool that can be used with *any* sample size n. When the experimenter has control over n, then the choice of sample size allows the experimenter to determine the width of the confidence interval (or the length of the yardstick below $\bar{y}_i - \mu_0$ for the case of a confidence bound). For example, if σ^2 is *known*, then setting $n = \lceil (\sigma z^{(q/2)}/d)^2 \rceil$, where $q = 1 - (P^*)^{1/t}$, guarantees an interval having half-width equal to d. If σ^2 is *unknown*, then a two-stage procedure must be used to form fixed-width confidence intervals for the treatment minus control differences (see Healy 1956).

Unbalanced Data

If the data are unbalanced with n_i observations taken from the ith treatment ($1 \le i \le t$), then (5.4.2) and (5.4.3) require only minor modifications. Let y_{ij} denote the observed value of Y_{ij} ($1 \le i \le t$, $1 \le j \le n_i$). Furthermore, let $\bar{y}_i = \sum_{j=1}^{n_i} y_{ij}/n_i$ ($1 \le i \le t$), and let $s_\nu^2 = \sum_{i=1}^{t} \sum_{j=1}^{n_i} (y_{ij} - \bar{y}_i)^2/\nu$ be the unbiased pooled estimate of σ^2 based on $\nu = \sum_{i=1}^{t}(n_i - 1)$ d.f. Then (5.4.2) and (5.4.3) become

$$\{\mu_i - \mu_0 \in \bar{y}_i - \mu_0 \pm h\sigma/\sqrt{n_i} \quad (1 \le i \le t)\}$$

and

$$\{\mu_i - \mu_0 \in \bar{y}_i - \mu_0 \pm hs_\nu/\sqrt{n_i} \quad (1 \le i \le t)\}$$

where $h = z^{(q/2)}$, $q = 1 - (P^*)^{1/t}$ and $h = |T|_{t,\nu,0}^{(1-P^*)}$ for the σ^2 known and σ^2 unknown cases, respectively.

5.4.2 Comparison with a Control Treatment

In this section, μ_0 is assumed to be the unknown mean of the control treatment and must be estimated from a sample taken from that treatment. We consider primarily the case of common *unknown* σ^2; the case of common *known* σ^2 will be discussed only briefly because it can be handled as in Section 5.4.1. It is assumed that data are collected from a completely randomized design; experiments with blocking are mentioned at the end of this section.

Statistical Assumptions: The assumptions are the same as those of Section 5.4.1 except that there is now a control treatment Π_0 from which a random sample of n observations, $Y_{01}, \ldots, Y_{0n}$, is taken, independent of the other tn observations. The Y_{0j}

are normal with unknown mean μ_0 and have the same common known or common unknown variance, σ^2, as the other treatments.

We introduce some minor modifications of the notation of Section 5.4.1. Let y_{ij} ($0 \leq i \leq t$, $1 \leq j \leq n$) denote the observed value of Y_{ij}. Furthermore, let $\bar{y}_i = \sum_{j=1}^{n} y_{ij}/n$ ($0 \leq i \leq t$) and $s_\nu^2 = \sum_{i=0}^{t} \sum_{j=1}^{n} (y_{ij} - \bar{y}_i)^2/\nu$ denote the unbiased pooled estimate of σ^2 based on $\nu = (t+1)(n-1)$ d.f. The probability requirement is again given by (5.4.1), except that now μ_0 in Goal 5.4.1 is an unknown control treatment mean instead of a given standard.

As in Section 5.4.1, we start by considering an analysis procedure that guarantees the probability requirement (5.4.1) for an arbitrary common sample size $n \geq 2$. If the choice of the sample size is under the control of the experimenter, we describe an optimal method of choosing that sample size.

Dunnett (1955, 1964) proposed the following single-stage procedure for achieving Goal 5.4.1 and guaranteeing (5.4.1) when σ^2 is unknown.

> **Procedure $\mathcal{N}_D$ (Simultaneous Confidence Bounds—Comparison with a Control)** When σ^2 is unknown, simultaneous lower confidence bounds with joint confidence coefficient P^* are given by
>
> $$\left\{ \mu_i - \mu_0 \geq \bar{y}_i - \bar{y}_0 - hs_\nu\sqrt{2/n} \quad (1 \leq i \leq t) \right\} \quad (5.4.4)$$
>
> where $h = T_{t,\nu,1/2}^{(1-P^*)}$ is found in Table B.3. Joint two-sided intervals with confidence coefficient P^* are given by
>
> $$\left\{ \mu_i - \mu_0 \in \bar{y}_i - \bar{y}_0 \pm hs_\nu\sqrt{2/n} \quad (1 \leq i \leq t) \right\} \quad (5.4.5)$$
>
> where $h = |T|_{t,\nu,1/2}^{(1-P^*)}$ is found in Table B.5.

If σ^2 is *known*, then $100 \times P^*\%$ joint lower confidence bounds are obtained by replacing the product $T_{t,\nu,1/2}^{(1-P^*)} s_\nu$ in (5.4.4) by $Z_{t,1/2}^{(1-P^*)}\sigma$; similarly, $100 \times P^*\%$ joint two-sided intervals are constructed by replacing the product $|T|_{t,\nu,1/2}^{(1-P^*)} s_\nu$ in (5.4.5) by the product $|Z|_{t,1/2}^{(1-P^*)}\sigma$ where $|Z|_{p,\rho}^{(\alpha)}$ (the two-sided upper-$(1-P^*)$ equicoordinate point of the equicorrelated multivariate normal distribution) is defined implicitly by

$$P\left\{ \max_{1 \leq j \leq p} |W_j| \leq |Z|_{p,\rho}^{(\alpha)} \right\} = 1 - \alpha$$

where $W_1, \ldots, W_p$ are independent standard normal random variables.

Unbalanced Experiments and the Optimal Choice of Sample Size

The simplest method of relaxing the requirement that a common number of observations must be taken from each of the $t+1$ treatments, and the only one that we will discuss, is to permit the number of observations taken from the control treatment to be different from (and larger than) the common number of observations to be taken from each of the t test treatments. This type of unbalance is intuitively attractive in that one might suspect that it is desirable to take more observations from the control treatment because the control mean is used in *every* confidence interval statement whereas each test treatment mean is used only once.

Let n_0 denote the number of observations to be taken from the control treatment and let n denote the common number of observations to be taken from *each* of the t test treatments; thus $T = n_0 + tn$ is the *total* number of observations to be collected. There are two methods of formulating an optimal allocation problem that have been considered in the statistical literature. In the first formulation, T is fixed and the apportionment of the total number T of observations to the control, n_0, and experimental treatments, n, is determined to maximize the joint confidence coefficient; it can be shown that for large T, this optimal allocation of observations is $n_0 = n\sqrt{t}$ (Bechhofer 1969 and Bechhofer and Nocterne 1972). This allocation implies that $n = T/(t + \sqrt{t})$ and $n_0 = T\sqrt{t}/(t + \sqrt{t})$. If this allocation is used then $\sqrt{2/T}$ in (5.4.4) and (5.4.5) is replaced by $\sqrt{(1/n_0) + (1/n)}$ in the interval estimates, and the critical point is replaced by $T_{t,\nu,\rho}^{(1-P^*)}$ where the correlation is $\rho = n/(n_0 + n)$. Upper-α equicoordinate percentage points for selected common $\rho \neq 1/2$, ν and α can be found in Table B.4; a more complete listing can be found in Bechhofer and Dunnett (1988).

In the second formulation, an interval width (in the case of two-sided confidence intervals) or yardstick (in the case of confidence bounds) is fixed. For a given confidence level, the minimum total number of observations is determined to achieve that interval width or yardstick length together with the allocation of observations to the control and test treatments. Such an interval (or bound) can be constructed with a *single-stage* procedure if σ^2 is *known* and with a *two-stage* procedure if σ^2 is *unknown*. These two cases will now be considered.

Suppose that d is the desired half-width of a two-sided confidence interval or the desired length of the "yardstick" for a confidence bound.

Case 1 (Fixed-Width Confidence Intervals—σ^2 known): Given d, suppose that it is desired to obtain $100 \times P^*\%$ joint confidence intervals of the form:

One-Sided: $\quad \{\mu_i - \mu_0 \geq \bar{y}_i - \bar{y}_0 - d \quad (1 \leq i \leq t)\}$

Two-Sided: $\quad \{\mu_i - \mu_0 \in \bar{y}_i - \bar{y}_0 \pm d \quad (1 \leq i \leq t)\}$

Bechhofer and Tamhane (1983a) have prepared tables of optimal allocations of observations (n_0, n) to guarantee joint confidence coefficients of $P^* = 0.75, 0.90, 0.95$ and 0.99. These are given for both one-sided and two-sided intervals. Let $\hat{T} = \hat{n}_0 + t\hat{n}$ denote the minimum possible total number of observations taken. Bechhofer and Tamhane compute $\hat{\gamma}_0 = \hat{n}_0/\hat{T}$, the proportion of the total observations to be allocated to the control treatment, and the ratio $\hat{\lambda} = d\hat{T}^{1/2}/\sigma$ from which $\hat{T}$ can be calculated. A proportion $(1 - \hat{\gamma}_0)/t$ of $\hat{T}$ is then allocated to each of the test treatments. Table 5.2 contains a selection of optimizing constants $(\hat{\gamma}_0, \hat{\lambda})$ that are abstracted from the Bechhofer and Tamhane tables.

Table 5.2. Optimal Allocation on the Control $\hat{\gamma}_0$ and Associated $\hat{\lambda}$ to Achieve Joint Confidence Coefficient P^* When Comparing t Normal Treatments with a Normal Control

P^*	t	One-sided Intervals		Two-sided Intervals	
		$\hat{\gamma}_0$	$\hat{\lambda}$	$\hat{\gamma}_0$	$\hat{\lambda}$
0.75	2	0.352	2.482	0.387	3.541
	3	0.292	3.351	0.332	4.495
	4	0.257	4.094	0.298	5.309
	5	0.233	4.757	0.274	6.033
	6	0.215	5.363	0.255	6.693
	7	0.202	5.927	0.240	7.305
	8	0.190	6.456	0.228	7.879
	9	0.181	6.958	0.218	8.422
	10	0.173	7.436	0.209	8.938
0.90	2	0.389	3.838	0.400	4.651
	3	0.335	4.819	0.348	5.699
	4	0.301	5.654	0.314	6.590
	5	0.277	6.397	0.290	7.381
	6	0.258	7.074	0.271	8.101
	7	0.243	7.701	0.256	8.768
	8	0.231	8.290	0.244	9.393
	9	0.220	8.846	0.233	9.893
	10	0.211	9.375	0.224	10.544
0.95	2	0.399	4.651	0.405	5.360
	3	0.348	5.700	0.354	6.469
	4	0.314	6.591	0.321	7.411
	5	0.290	7.382	0.297	8.247
	6	0.271	8.103	0.278	9.007
	7	0.256	8.770	0.263	9.711
	8	0.243	9.395	0.250	10.369
	9	0.233	9.986	0.239	10.990
	10	0.223	10.547	0.230	11.581

From Bechhofer and Tamhane (1983a), pp. 90–91. Reprinted with permission from *Technometrics*. Copyright 1983 by the American Statistical Association and the American Society for Quality Control. All rights reserved.

Given t and joint confidence coefficient P^*, obtain $(\hat{\gamma}_0, \hat{\lambda})$ from Table 5.2 according to the experimenter's desire for one-sided or two-sided intervals. Determine the *minimum* total number of observations $\hat{T} = \hat{n}_0 + t\hat{n}$ to achieve a common interval width with a joint confidence coefficient of P^* by solving $\hat{\lambda} = d\hat{T}^{1/2}/\sigma$ to obtain

$$\hat{T} = \lceil (\hat{\lambda}\sigma/d)^2 \rceil.$$

Then

$$\hat{n}_0 = \lceil \hat{\gamma}_0 \hat{T} \rceil \quad \text{and} \quad \hat{n} = \lceil (\hat{T} - \hat{n}_0)/t \rceil$$

are the optimal numbers of observations to be taken on the control and on each test treatment, respectively. Although approximate, the samples sizes $(\hat{n}_0, \hat{n})$ will be very close to the exact optimal integer allocations if $\hat{T}$ is large.

Example 5.4.2 Suppose that $t = 4$ and $\sigma = 2$ and that it is desired to find 90% joint two-sided intervals of width $d = 1.5$. From Table 5.2 we find that $\hat{\gamma}_0 = 0.314$ and $\hat{\lambda} = 6.590$. Thus,

$$\hat{T} = \left\lceil \left(\frac{(6.590)2}{1.5} \right)^2 \right\rceil = \lceil 77.2 \rceil = 78$$

and

$$\hat{n}_0 = \lceil 0.314(78) \rceil = \lceil 24.49 \rceil = 25.$$

For each treatment we take

$$\hat{n} = \left\lceil \frac{78 - 25}{4} \right\rceil = \lceil 13.25 \rceil = 14$$

observations; thus the total number of observations required is $4(14) + 25 = 81$. The resulting joint two-sided 90% confidence intervals are $\{\mu_i - \mu_0 \in \bar{y}_i - \bar{y}_0 \pm 1.5\;(1 \le i \le 4)\}$.

Case 2 (Fixed-Width Confidence Intervals—σ^2 Unknown): If σ^2 is not known, then, for *design* purposes, replace σ^2 in the formula for $\hat{T}$ by a *conjectured* upper bound. The experimenter is cautioned that if this bound is too large, then the resulting $\hat{T}$ will be too large, causing the true confidence coefficient to be much greater than the desired confidence coefficient. However, for *analysis* purposes, *after* the experiment has been conducted, the assumed upper bound on σ^2 should be replaced by the unbiased pooled estimate of σ^2, that is,

$$s_\nu^2 = \left[\sum_{i=1}^{t} \sum_{j=1}^{\hat{n}} (y_{ij} - \bar{y}_i)^2 + \sum_{j=1}^{\hat{n}_0} (y_{0j} - \bar{y}_0)^2 \right] / \nu$$

where $\nu = t(\hat{n} - 1) + (\hat{n}_0 - 1)$. Then we obtain $100 \times P^*\%$ joint confidence intervals of the form:

One-Sided: $\left\{ \mu_i - \mu_0 \ge \bar{y}_i - \bar{y}_0 - T_{t,\nu,\rho}^{(1-P^*)} s_\nu \sqrt{(1/\hat{n}) + (1/\hat{n}_0)} \;\; (1 \le i \le t) \right\}$

Two-Sided: $\left\{ \mu_i - \mu_0 \in \bar{y}_i - \bar{y}_0 \pm |T|_{t,\nu,\rho}^{(1-P^*)} s_\nu \sqrt{(1/\hat{n}) + (1/\hat{n}_0)} \;\; (1 \le i \le t) \right\}$

where $\rho = \hat{n}_0/(\hat{n} + \hat{n}_0)$ in both of the above equations. A more theoretically sound method of forming fixed-width intervals for the common *unknown* variance case was developed by Healy (1956), who devised an analogue of the Stein (1945) two-stage procedure.

Balanced Treatment Incomplete Block Design

The procedure described at the beginning of Section 5.4.2 assumes that the experimenter will employ a completely randomized design or a randomized complete block design in which the block sizes are sufficiently large to accommodate the t test treatments and the control treatment as well. If the blocks available have a common size $k < t + 1$, that is, if the $t + 1$ test treatments are to be compared using incomplete blocks of size k, then entirely new considerations are required to determine the optimal incomplete block design. The problem of how to accomplish this is considered in Bechhofer and Tamhane (1981, 1983b); they introduce a new class of incomplete block designs called *balanced treatment incomplete block* (BTIB) designs. The optimal design tables for BTIB designs with blocks of common size $k = 2, 3$ for $t = k(1)6$ treatments are listed in Bechhofer and Tamhane (1983b).

5.4.3 Comparisons with Respect to Two Control Treatments

In certain situations, it is appropriate to compare simultaneously $t \geq 1$ treatments to *two* controls. For example, suppose that t new pain relievers are to be compared simultaneously with respect to hours of relief (or quality of relief) to two commonly used standard therapies such as aspirin and acetaminophen. (When $t = 1$, we can use the procedure in Section 5.4.2 where we compare two treatments to one control.)

This section concerns multiple comparisons with respect to two control treatments with unknown means $\mu_{0,1}$ and $\mu_{0,2}$.

Statistical Assumptions: The assumptions are the same as those of Section 5.4.1 except that there are now *two* control treatments, Π_{01} and Π_{02}. Independent random samples of n observations $Y_{(0,c)1}, \ldots, Y_{(0,c)n}$, independent of the other tn observations, are taken from Π_{0c} ($c = 1, 2$). The $Y_{(0,c)j}$ ($1 \leq j \leq n$) are normal with unknown means $\mu_{0,c}$ ($c = 1, 2$) and the same common known or common unknown variance, σ^2, as the other treatments.

Let y_{ij} ($1 \leq i \leq t, 1 \leq j \leq n$) and $y_{(0,c)j}$ ($c = 1, 2, 1 \leq j \leq n$) denote the observed values of the treatments and controls, respectively. The experimental objective and probability requirement are stated in Goal 5.4.2 and (5.4.6), respectively.

Goal 5.4.2 To obtain joint confidence interval estimates of the t differences $\mu_i - \mu_{0,1}$ *and* the t differences $\mu_i - \mu_{0,2}$ ($1 \leq i \leq t$) using either one-sided interval estimates (simultaneous upper or lower confidence bounds) or two-sided interval estimates.

Probability Requirement: For specified constant $P^\star$ ($0 < P^\star < 1$) we require that

$$P\{\text{Confidence region covers } \mu_i - \mu_{0,1} \text{ and}$$
$$\mu_i - \mu_{0,2} \quad (1 \leq i \leq t) \mid (\boldsymbol{\mu}, \mu_{0,1}, \mu_{0,2}, \sigma^2)\} \geq P^\star \quad (5.4.6)$$

for all $\boldsymbol{\mu}$ and σ^2, if σ^2 is unknown.

Hoover (1991) generalized the Dunnett (1955) *single*-control procedure to *two* controls.

Procedure $\mathcal{N}_{Hoo}$ Calculate $\bar{y}_i = \sum_{j=1}^{n} y_{ij}/n$ $(1 \leq i \leq t)$, $\bar{y}_{0,c} = \sum_{j=1}^{n} y_{(0,c)j}/n$ $(c = 1, 2)$, and

$$s_\nu^2 = \left[\sum_{i=1}^{t}\sum_{j=1}^{n}(y_{ij} - \bar{y}_i)^2 + \sum_{c=1}^{2}\sum_{j=1}^{n}(y_{(0,c)j} - \bar{y}_{0,c})^2 \right] / \nu,$$

the unbiased pooled estimate of σ^2 based on $\nu = (t+2)(n-1)$ d.f.

Case 1 (Simultaneous Lower Confidence Bounds): The bounds

$$\left\{ \mu_i - \mu_{0,c} \geq \bar{y}_i - \bar{y}_{0,c} - d_1 s_\nu \sqrt{2/n} \quad (1 \leq i \leq t, c = 1, 2) \right\}$$

satisfy (5.4.6), where $d_1 = d_1(t, P^\star, \nu)$ is found in Table 5.3.

Case 2 (Simultaneous Two-Sided Confidence Intervals): The intervals

$$\left\{ \mu_i - \mu_{0,c} \in \bar{y}_i - \bar{y}_{0,c} \pm d_2 s_\nu \sqrt{2/n} \quad (1 \leq i \leq t, c = 1, 2) \right\}$$

satisfy (5.4.6), where $d_2 = d_2(t, P^\star, \nu)$ is found in Table 5.3.

The constants $d_1(t, P^\star, \nu)$ and $d_2(t, P^\star, \nu)$ have been computed only for $t = 1(1)10$, $P^\star = 0.95$ and selected ν. In the special case of $t = 1$ treatment, notice that $d_1(1, P^\star, \nu) = T_{2,\nu,1/2}^{(1-P^\star)}$ and $d_2(1, P^\star, \nu) = |T|_{2,\nu,1/2}^{(1-P^\star)}$. This is because the problem of forming simultaneous confidence intervals for a *single* treatment and *two* controls is the same as that of forming simultaneous confidence intervals for *two* treatments and a *single* control. The procedures $\mathcal{N}_D$ and $\mathcal{N}_{Hoo}$ coincide in this case and, in particular, have identical critical points.

Remark 5.4.2 Hoover gave the asymptotically ($T \to \infty$) optimal allocation of observations when a total of $T = mn_0 + tn$ observations is to be divided among t treatments and m controls; here n_0 [n] is the number of observations to be taken from *each* of the m controls [t treatments]. This asymptotically optimal allocation is $n = T/\sqrt{t}(\sqrt{t} + \sqrt{m})$ and $n_0 = T/\sqrt{m}(\sqrt{t} + \sqrt{m})$, that is, $n_0 = n\sqrt{t/m}$. When $m = 1$ this allocation reduces to the optimal allocation result given earlier in this section for one control. Different constants replacing d_1 or d_2 are required in order to implement the procedure for $m > 2$; such constants have not yet been calculated.

Table 5.3. Values of $d_1 = d_1(t, P^\star, \nu)$ (top entries) and $d_2 = d_2(t, P^\star, \nu)$ (bottom entries) for $P^\star = 0.95$, Two Controls, t Treatments and ν d.f.

ν	\multicolumn{10}{c}{t}									
	1	2	3	4	5	6	7	8	9	10
5	2.441	2.896	3.154	3.333	3.469	3.579	3.670	3.748	3.816	3.876
	3.031	3.529	3.815	4.012	4.164	4.285	4.387	4.475	4.550	4.618
6	2.337	2.752	2.986	3.148	3.271	3.370	3.452	3.523	3.584	3.638
	2.863	3.305	3.557	3.732	3.866	3.972	4.062	4.139	4.206	4.265
7	2.268	2.656	2.874	3.025	3.139	3.231	3.307	3.372	3.429	3.480
	2.752	3.157	3.388	3.547	3.670	3.767	3.849	3.920	3.980	4.034
8	2.218	2.588	2.795	2.937	3.045	3.132	3.204	3.265	3.319	3.367
	2.673	3.053	3.268	3.417	3.531	3.622	3.698	3.764	3.821	3.871
9	2.180	2.536	2.735	2.871	2.974	3.057	3.126	3.185	3.237	3.282
	2.614	2.975	3.180	3.321	3.429	3.515	3.587	3.649	3.703	3.751
10	2.151	2.496	2.688	2.820	2.920	3.000	3.066	3.123	3.172	3.216
	2.569	2.916	3.112	3.247	3.350	3.432	3.501	3.561	3.612	3.657
11	2.128	2.464	2.651	2.779	2.876	2.954	3.018	3.073	3.121	3.164
	2.532	2.868	3.057	3.188	3.287	3.366	3.433	3.490	3.539	3.584
12	2.109	2.438	2.621	2.746	2.840	2.916	2.979	3.033	3.080	3.121
	2.503	2.829	3.013	3.140	3.236	3.313	3.378	3.433	3.481	3.523
13	2.093	2.417	2.596	2.718	2.811	2.885	2.946	2.999	3.045	3.085
	2.478	2.797	2.977	3.100	3.194	3.269	3.331	3.386	3.432	3.474
14	2.079	2.398	2.574	2.695	2.786	2.858	2.919	2.970	3.015	3.055
	2.457	2.770	2.946	3.066	3.158	3.232	3.293	3.346	3.391	3.432
15	2.067	2.382	2.556	2.675	2.764	2.836	2.895	2.946	2.990	3.029
	2.440	2.747	2.919	3.038	3.128	3.200	3.260	3.312	3.356	3.396
16	2.057	2.369	2.540	2.657	2.745	2.816	2.875	2.925	2.968	3.007
	2.424	2.727	2.897	3.013	3.102	3.172	3.231	3.282	3.326	3.365
17	2.048	2.357	2.526	2.642	2.729	2.799	2.857	2.906	2.949	2.987
	2.411	2.710	2.877	2.992	3.079	3.148	3.206	3.256	3.300	3.338
18	2.041	2.346	2.514	2.629	2.715	2.784	2.841	2.890	2.933	2.970
	2.397	2.694	2.859	2.973	3.059	3.127	3.185	3.234	3.276	3.314
19	2.034	2.337	2.503	2.617	2.702	2.770	2.827	2.876	2.918	2.955
	2.389	2.681	2.844	2.956	3.041	3.108	3.165	3.214	3.256	3.293
20	2.028	2.329	2.494	2.606	2.691	2.758	2.815	2.863	2.904	2.941
	2.379	2.669	2.830	2.941	3.025	3.092	3.148	3.196	3.237	3.274
24	2.008	2.303	2.463	2.573	2.655	2.721	2.775	2.822	2.863	2.898
	2.350	2.631	2.787	2.894	2.975	3.040	3.094	3.140	3.180	3.216
30	1.990	2.277	2.433	2.540	2.620	2.684	2.737	2.782	2.822	2.856
	2.321	2.594	2.745	2.848	2.926	2.989	3.041	3.085	3.124	3.158
40	1.971	2.252	2.404	2.508	2.586	2.648	2.699	2.743	2.781	2.815
	2.293	2.558	2.704	2.804	2.879	2.939	2.989	3.032	3.069	3.103
60	1.953	2.227	2.376	2.477	2.552	2.613	2.663	2.705	2.742	2.775
	2.266	2.522	2.664	2.760	2.833	2.891	2.939	2.980	3.016	3.048
120	1.935	2.203	2.348	2.446	2.519	2.578	2.626	2.668	2.703	2.735
	2.239	2.488	2.624	2.717	2.788	2.843	2.890	2.930	2.964	2.995
∞	1.917	2.179	2.320	2.415	2.487	2.544	2.591	2.631	2.665	2.696
	2.213	2.454	2.586	2.676	2.743	2.797	2.842	2.880	2.913	2.942

5.5 CHAPTER NOTES

In Section 5.2, Bechhofer and Turnbull (1978) used an indifference-zone formulation to study the problem of selecting the best treatment provided it is better than a given standard. Wilcox (1984) considered the same problem and probability requirement but allowed unknown and unequal variances for the normal treaments. He proposed a two-stage procedure for this problem based on the Rinott (1978) solution to the corresponding selection problem without a standard (see Section 2.8). Dunnett (1984) also devised procedures for the control version of the Bechhofer and Turnbull problem. They considered selection of the best of t normal test treatments having means $\mu_1, \ldots, \mu_t$ relative to that of a normal control treatment having unknown mean μ_0 when all treatments have a common known variance; they also used the same probability requirement as Bechhofer and Turnbull. Dunnett provided extensive tables and a FORTRAN program to implement the procedure. Paulson (1962) gave a sequential procedure for both the standard and control versions of this problem.

Section 5.3 studied the problem of selecting a subset of treatments better than a standard or control. Schafer (1977) reviewed procedures for selecting all treatments better than a standard for several other parametric families. Tong (1969) and Norell (1990), and the references therein, considered related problems of partitioning a set of treatments with respect to the means of one or more controls. This formulation differs from that in Section 5.3 in its probability requirement. The cited papers guaranteed that the experiment is designed on such a scale that all treatments whose means are sufficiently below a control value are correctly classified as inferior *and* all treatments whose means are sufficiently greater than the control value are correctly classified as superior. Lastly, we note that there have been several loss function and Bayesian formulations of the subset standard/control problem. Among these are Randles and Hollander (1971), Gupta and Kim (1980) and Gupta and Hsiao (1983).

Section 5.4 considered the problem of forming simultaneous confidence intervals for all normal treatment-control differences when the observations have a common known or common unknown variance. Bofinger and Lewis (1992) devised simultaneous confidence intervals for all normal treatment-control differences when the observations from different treatments are allowed to have unknown and unequal variances.

CHAPTER 6

Selection Problems in Two-Factor Normal Response Experiments

6.1 INTRODUCTION

Many of the selection problems posed for single-factor experiments have generalizations to two- and higher-way factorial experiments. This chapter will discuss these problems for additive two-factor experiments; the generalizations to three- or higher-factor experiments are straightforward.

Consider a two-factor experiment in which random samples of observations $Y_{ij1}, Y_{ij2}, \ldots$ are taken from $a \times b$ normal treatments with means μ_{ij} ($1 \leq i \leq a$, $1 \leq j \leq b$). The factor A has a levels and the factor B has b levels. This is the so-called means model. It is more customary to define the main effects of the row factor, α_i, the column factor, β_j, and two-way interactions, $(\alpha\beta)_{ij}$, by

$$\mu_{ij} = \mu + \alpha_i + \beta_j + (\alpha\beta)_{ij}$$

subject to the identifiability conditions $\sum_{i=1}^{a} \alpha_i = \sum_{j=1}^{b} \beta_j = 0$ and $\sum_{j=1}^{b}(\alpha\beta)_{ij} \stackrel{i}{=} 0 \stackrel{j}{=} \sum_{i=1}^{a}(\alpha\beta)_{ij}$. The factors A and B are said to be *additive* if $(\alpha\beta)_{ij} \equiv 0$ for all i and j and are said to *interact* otherwise, that is, if not all of the $(\alpha\beta)_{ij}$ are zero.

In general, the objectives, design and analysis of factorial experiments depend heavily on whether or not the factors *interact* in their effects on the response variable being studied. In the two-factor experiment where interaction is present, then, as noted below, many of the formulations discussed in Chapters 2–5 are relevant. For example, depending on the application, one of the following goals may be of interest.

1. Regard the problem as a single-factor experiment with $t = ab$ treatments. Select the treatment combination associated with $\max_{i,j} \mu_{ij}$ (Goal 2.1.1 in Section 2.1) or select a subset of the treatment combinations associated with $\max_{i,j} \mu_{ij}$ (Goal 3.1.1 in Section 3.1).
2. Regard the problem as a independent single-factor experiments with each factor having b levels. Select the level of factor B associated with $\mu_{i[b]} \equiv$

INTRODUCTION

$\max\{\mu_{i1}, \ldots, \mu_{ib}\}$ or select a subset of levels of factor B associated with $\mu_{i[b]}$. Thus for each level of A, the goal is again that of Section 2.1 or of Section 3.1. Of course, the problem could be symmetrically stated by reversing the roles of rows and columns (see Section 6.8).

3. Form simultaneous confidence intervals for all treatment differences $\{\mu_{ij} - \mu_{i'j'} : (i, j) \neq (i', j')\}$.

4. Form simultaneous confidence intervals for $\{\mu_{ij} - \max_{i',j'} \mu_{i'j'} : (i, j)\}$, that is, all comparisons with the mean of the best treatment combination as described in Section 4.4.

5. Select the treatment combination associated with the largest absolute interaction, $\max_{i,j}\{|(\alpha\beta)_{ij}|\}$ (see Section 6.8).

The procedures described in Chapters 2–5 can be used to achieve Goals 1–4 as well as other possible objectives.

This chapter will discuss selection problems when the factor effects are additive. The various goals and procedures for two-factor experiments are similar to those for single-factor experiments. When additivity holds, factorial experiments for selection problems can be less costly in terms of total sample size than their single-factor counterparts. (See Example 6.2.2.)

Before the assumption of additivity is either uncritically accepted by the unwary or rejected out of hand by skeptics, we wish to describe our attitude concerning its validity. Philosophically, it is perhaps unreasonable to ever assume that additivity holds *exactly*. Rather, the issue should be whether or not the magnitude of the interactions is sufficient to invalidate the conclusions obtained based on this assumption.

In some cases an experimenter may be prepared to assume additivity on the basis of previous experience or knowledge of the factors being studied. In many more practical situations, the experimenter will not know whether additivity holds. In this case we recommend that the experimenter carry out a preliminary $\mathcal{F}$-test of the hypothesis $H_O : (\alpha\beta)_{ij} \stackrel{i,j}{=} 0$ versus the global alternative, $H_A : (\text{not } H_O)$. If H_O is accepted, then proceed as in the additive case by adopting Goal 6.2.1. If H_O is rejected, then proceed with one of the options discussed above. However, the theoretical performance characteristics of this composite preliminary test procedure have been studied in depth only for the special case $a = b = 2$ (Taneja 1986 and Borowiak and De Los Reyes 1992). The experimenter is cautioned that even if the interaction effects are "small," the preliminary test will indicate their existence if the sample sizes are sufficiently large. And yet these effects may not be large enough to invalidate conclusions based on the *no-interaction* assumption. The recent paper by Fabian (1991) and the discussion following it summarize some of these issues for several problems, including that of selecting the treatment with the largest mean, and recommend an alternative to the testing strategy described above. Clearly, much *subjective* judgment must be used here.

This chapter assumes that

$$\mu_{ij} = \mu + \alpha_i + \beta_j \qquad (6.1.1)$$

($1 \leq i \leq a, 1 \leq j \leq b$) where $\sum_{i=1}^{a} \alpha_i = \sum_{j=1}^{b} \beta_j = 0$. We denote the ordered values of the unknown α_i and β_j by $\alpha_{[1]} \leq \cdots \leq \alpha_{[a]}$ and $\beta_{[1]} \leq \cdots \leq \beta_{[b]}$, respectively. The values of the α_i and β_j are unknown as is their pairing with the $\alpha_{[q]}$ and $\beta_{[r]}$ ($1 \leq i, q \leq a$, $1 \leq j, r \leq b$). Let y_{ijs} denote the observed value of Y_{ijs} ($1 \leq i \leq a$, $1 \leq j \leq b, s \geq 1$).

Notice that additivity implies that there are "best" levels of the A and B factors in the sense that for *any* level, j, of B, the *same* level of A (that corresponding to $\alpha_{[a]}$) is associated with the largest treatment mean among $\mu_{1j}, \ldots, \mu_{aj}$. Similarly, for any level, i, of A, the *same* level of B (that corresponding to $\beta_{[b]}$) is associated with the largest treatment mean among $\mu_{i1}, \ldots, \mu_{ib}$. If there is interaction between the two factors then the best level of A can depend on the level j of B and a similar statement holds for the best level of B.

Section 6.2 describes procedures using completely randomized designs that employ the indifference-zone approach for selecting the factor-level combination associated with $\alpha_{[a]}$ and $\beta_{[b]}$ in *additive* two-factor experiments when there is a *common known* variance. Section 6.2.1 presents a single-stage selection procedure. Sections 6.2.2 and 6.2.3 give two sequential procedures that can be used if circumstances permit. One procedure is closed but does not allow treatments to be eliminated during sampling. The other sequential procedure is also closed but does allow treatment combinations to be eliminated; however, the second procedure can require more stages to terminate sampling as the comparisons in Section 6.2.4 show. Section 6.3 presents selection procedures for the same goals as in Section 6.2 but for conducting *split-plot* experiments. Section 6.4 describes multi-stage procedures for selecting the best treatment combination when there is a *common unknown* variance. As in the case of single-factor experiments, unless the design requirement is stated in terms of ratios of main effects to the standard deviation of the responses, single-stage procedures cannot be used. (See Equation (2.7.1) regarding the single-factor case.) Section 6.5 considers procedures for split-plot designs that are appropriate for the unknown measurement error case. Lastly, Sections 6.6 and 6.7 present subset selection procedures for additive completely randomized and split-plot experiments, respectively.

6.2 INDIFFERENCE-ZONE SELECTION USING COMPLETELY RANDOMIZED DESIGNS (COMMON KNOWN VARIANCE)

This section is concerned with indifference-zone (IZ) procedures for selecting the best treatment combination in additive two-factor normal experiments in which all of the combinations have *known* variance σ^2.

Statistical Assumptions: Independent random samples $Y_{ij1}, Y_{ij2}, \ldots$ are taken from ab ($a, b \geq 2$) normal treatments Π_{ij} ($1 \leq i \leq a$, $1 \leq j \leq b$). The (i, j)th treatment combination has unknown mean $\mu_{ij} = \mu + \alpha_i + \beta_j$ and common known variance σ^2.

The purpose of the experiment is stated in Goal 6.2.1, and the associated probability (design) requirement is given by Equation (6.2.1).

Goal 6.2.1 To select the treatment combination associated with $\alpha_{[a]}$ and $\beta_{[b]}$.

Thus, we seek to find simultaneously the "best" levels of both treatments. A *correct selection* (CS) is said to be made if Goal 6.2.1 is achieved.

Probability Requirement: For specified constants $(\delta_\alpha^\star, \delta_\beta^\star, P^\star)$ with $0 < \delta_\alpha^\star, \delta_\beta^\star < \infty$ and $1/ab < P^\star < 1$, we require that

$$P\{CS \mid \boldsymbol{\mu}\} \geq P^\star \qquad (6.2.1)$$

whenever $\boldsymbol{\mu}$ satisfies $\alpha_{[a]} - \alpha_{[a-1]} \geq \delta_\alpha^\star$ and $\beta_{[b]} - \beta_{[b-1]} \geq \delta_\beta^\star$.

The preference-zone and the IZ that correspond to the design requirement (6.2.1) are depicted in Figure 6.1 for a two-factor experiment when the α_i and β_j are in the slippage configurations where $\alpha_{[a]} - \alpha_{[i]} = \delta_\alpha$ ($1 \leq i \leq a - 1$) and $\beta_{[b]} - \beta_{[j]} = \delta_\beta$ ($1 \leq j \leq b - 1$) with $\delta_\alpha, \delta_\beta \geq 0$. The IZ is comprised of three parts. In

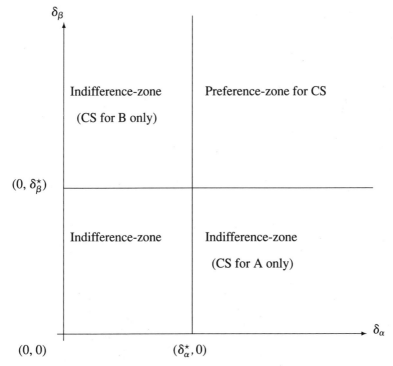

Figure 6.1. Preference- and indifference-zones for two-factor experiments when $\alpha_{[a]} - \alpha_{[i]} = \delta_\alpha$ ($1 \leq i \leq a - 1$) and $\beta_{[b]} - \beta_{[j]} = \delta_\beta$ ($1 \leq j \leq b - 1$).

the "southeast" corner (labeled "CS for A only") we have $\alpha_{[a]} - \alpha_{[a-1]} \geq \delta_\alpha^\star$ and $\beta_{[b]} - \beta_{[b-1]} < \delta_\beta^\star$; this part of the IZ might be thought of as the "B-main-effect-only IZ." The "northwest corner" of the indifference-zone (labeled "CS for B only") is the "A-main-effect-only IZ" because $\alpha_{[a]} - \alpha_{[a-1]} < \delta_\alpha^\star$ while $\beta_{[b]} - \beta_{[b-1]} \geq \delta_\beta^\star$. Lastly, in the "southwest" corner, we have $\alpha_{[a]} - \alpha_{[a-1]} < \delta_\alpha^\star$ and $\beta_{[b]} - \beta_{[b-1]} < \delta_\beta^\star$; neither the best level of the A main effect nor the best level of the B main effect is separated significantly from their respective second-best levels.

The $P\{CS \mid \boldsymbol{\mu}\}$ is controlled at level $P^\star$ in the preference-zone for a correct selection. For the procedures in Sections 6.2.1–6.2.3, the minimum of the $P\{CS \mid \boldsymbol{\mu}\}$ over the preference-zone defined by (6.2.1) occurs at the slippage configuration with $\delta_\alpha = \delta_\alpha^\star$ and $\delta_\beta = \delta_\beta^\star$. In the slippage configuration, the $P\{CS \mid \boldsymbol{\mu}\}$ approaches unity as δ_α and $\delta_\beta \to \infty$. If the true configuration is in the portion of the indifference-zone labeled "CS for A only," then the (marginal) probability that any of the procedures described in Sections 6.2.1–6.2.3 select the best level for factor A is greater than $P^\star$. However, the corresponding statement is not true for factor B in that region, and the probability of *simultaneous* correct selection of the best levels for both factors [the event CS in Equation (6.2.1)] may be less than $P^\star$. We now present three procedures, useful in different circumstances, that guarantee (6.2.1) when σ^2 is *known* and compare their performances in Section 6.2.4.

6.2.1 Single-Stage Procedure

Bechhofer (1954) proposed a single-stage procedure for the case of common known σ^2.

Procedure $\mathcal{N}_{B2}$ For the given (a, b) and specified $(\delta_\alpha^\star/\sigma, \delta_\beta^\star/\sigma, P^\star)$, determine n from Table 6.1 or as in Examples 6.2.1 and 6.2.2, below.

Sampling rule: Take a random sample of n observations Y_{ijs} $(1 \leq s \leq n)$ in a single stage from each Π_{ij} $(1 \leq i \leq a, 1 \leq j \leq b)$.

Terminal decision rule: Calculate the $a + b$ sample sums

$$A_{in} = \sum_{j=1}^{b}\sum_{s=1}^{n} y_{ijs} \quad (1 \leq i \leq a) \quad \text{and} \quad B_{jn} = \sum_{i=1}^{a}\sum_{s=1}^{n} y_{ijs} \quad (1 \leq j \leq b).$$

Select the factor-levels that yielded $A_{[a]n} = \max\{A_{1n}, \ldots, A_{an}\}$ and $B_{[b]n} = \max\{B_{1n}, \ldots, B_{bn}\}$ as the ones associated with $\alpha_{[a]}$ and $\beta_{[b]}$.

Table 6.1 is abstracted from Bechhofer and Goldsman (1988b); it is restricted to $P^\star = 0.9025 \, (= (0.95)^2)$ and $\delta_\alpha^\star/\sigma = 0.2 = \delta_\beta^\star/\sigma$. However, as Examples 6.2.1 and 6.2.2 show, it is straightforward to compute sample sizes to satisfy Equation (6.2.1) for other $P^\star$, $\delta_\alpha^\star/\sigma$ and $\delta_\beta^\star/\sigma$ given a table of $Z_{t,1/2}^{(1-P^\star)}$-values or a program that computes them (for example, the FORTRAN program PCSNB listed in Appendix C).

Table 6.1. Minimum Common Number n of Observations from Each of ab Treatment Combinations Necessary to Guarantee the Probability Requirement (6.2.1) Using Procedure $\mathcal{N}_{B2}$ with Selected (a, b) When $\delta_\alpha^\star/\sigma = \delta_\beta^\star/\sigma = 0.2$ and $P^\star = 0.9025$

a	\multicolumn{8}{c}{b}							
	1	2	3	4	5	6	7	8
2	84	68						
3	127	72	62					
4	153	79	60	54				
5	172	87	62	52	47			
6	187	94	64	51	45	42		
7	199	100	67	52	45	40	38	
8	209	105	70	54	45	40	37	35
9	218	109	73	55	45	40	36	34
10	226	113	76	57	46	40	36	33
11	233	117	78	59	47	40	36	33
12	239	120	80	60	48	41	36	32
13	245	123	82	62	49	42	36	33
14	250	125	84	63	50	42	37	33
15	255	128	85	64	51	43	37	33
16	260	130	87	65	52	44	38	33
17	264	132	88	66	53	44	38	34
18	268	134	90	67	54	45	39	34
19	272	136	91	68	55	46	39	35
20	275	138	92	69	55	46	40	35

Reprinted from Bechhofer and Goldsman (1988b), p. 111, by courtesy of Marcel Dekker, Inc.

Notice that sample size n depends only on the *ratios* $\delta_\alpha^\star/\sigma$ and $\delta_\beta^\star/\sigma$ rather than their individual values. The procedure $\mathcal{N}_{B2}$ uses $A_{1n} \ldots A_{an}$, each of which is based on $b \times n$ observations, for determining the best level of factor A; it uses $B_{1n} \ldots B_{bn}$, each of which is based on $a \times n$ observations, for selecting the best level of factor B.

To introduce the examples below, let $P\{CS(2) \mid (a, b; \delta_\alpha^\star/\sigma, \delta_\beta^\star/\sigma, n)\}$ denote the probability of a correct selection for the *two-factor* single-stage procedure $\mathcal{N}_{B2}$ when σ^2 is known, and suppose there are a [b] levels of the first [second] factor, a common number n of observations is taken from each of the ab treatment combinations, and the treatment effects are in a configuration for which $\alpha_{[a]} - \alpha_{[i]} = \delta_\alpha^\star$ ($1 \le i \le a - 1$) and $\beta_{[b]} - \beta_{[j]} = \delta_\beta^\star$ ($1 \le j \le b - 1$). The latter configuration is least-favorable (LF) for this two-factor problem. Similarly, let $P\{CS(1) \mid (t; \delta^\star/\sigma, n)\}$ denote the $P\{CS \mid \boldsymbol{\mu}\}$ using procedure $\mathcal{N}_B$ for Goal 2.1.1 when σ^2 is known and there are t levels of the factor, n observations are taken at each factor-level, and the treatment means are in a configuration for which $\mu_{[t]} - \mu_{[i]} = \delta^\star$ ($1 \le i \le t - 1$); this configuration is LF for the one-factor problem. From Section 2.2.1, the equation

$$n = \lceil 2(\sigma Z_{t-1,1/2}^{(1-P\{CS(1)|(t;\delta^\star/\sigma,n)\})}/\delta^\star)^2 \rceil$$

relates $P\{CS(1) \mid (t; \delta^*/\sigma, n)\}$, n, σ and δ^*. It can be shown that if there is no interaction between the two factors, then procedures $\mathcal{N}_B$ and $\mathcal{N}_{B2}$ have probabilities of CS in their respective LF-configurations that satisfy

$$P\{CS(2) \mid (a,b; \delta_\alpha^*/\sigma, \delta_\beta^*/\sigma, n)\}$$
$$= P\{CS(1) \mid (a; \delta_\alpha^*/\sigma, bn)\} \times P\{CS(1) \mid (b; \delta_\beta^*/\sigma, an)\} \quad (6.2.2)$$

(Bechhofer 1954, Equation (48)). This identity will enable us to determine the smallest n to guarantee (6.2.1) using procedure $\mathcal{N}_{B2}$. We illustrate the method in Examples 6.2.1 and 6.2.2.

Example 6.2.1 (Designing a Completely Randomized Two-Way Selection Experiment) Suppose that $\sigma = 0.9$ and $(a,b) = (3,5)$ and that the experimenter specifies $(\delta_\alpha^*, \delta_\beta^*) = (0.4, 0.6)$ and $P^* = 0.95$. For the two-way classification, the 3×5 matrix of cell sums $\sum_{s=1}^{n} y_{ijs}$ ($1 \leq i \leq 3$, $1 \leq j \leq 5$), row sums $A_{in} = \sum_{j=1}^{5} \sum_{s=1}^{n} y_{ijs}$ ($1 \leq i \leq 3$), and column sums $B_{jn} = \sum_{i=1}^{3} \sum_{s=1}^{n} y_{ijs}$ ($1 \leq j \leq 5$) can be depicted as follows:

$\sum_{s=1}^{n} y_{11s}$	$\cdots$	$\sum_{s=1}^{n} y_{15s}$	$A_{1n} = \sum_{j=1}^{5} \sum_{s=1}^{n} y_{1js}$
$\vdots$		$\vdots$	$\vdots$
$\sum_{s=1}^{n} y_{31s}$	$\cdots$	$\sum_{s=1}^{n} y_{35s}$	$A_{3n} = \sum_{j=1}^{5} \sum_{s=1}^{n} y_{3js}$
$B_{1n} = \sum_{i=1}^{3} \sum_{s=1}^{n} y_{i1s}$	$\cdots$	$B_{5n} = \sum_{i=1}^{3} \sum_{s=1}^{n} y_{i5s}$	

The selection decision employs the A_{in} ($1 \leq i \leq 3$) and B_{jn} ($1 \leq j \leq 5$), which are based on $5n$ and $3n$ observations, respectively. We calculate the smallest n such that the product (6.2.2) is at least 0.95. To accomplish this, choose a trial value of n, say $n = 9$. From Section 2.2.1, the individual factors $P_A = P\{CS(1) \mid (3; 0.4/0.9, 45)\}$ and $P_B = P\{CS(1) \mid (5; 0.6/0.9, 27)\}$ are implicitly defined by the equations

$$9(5) = \lceil 2(0.9 Z_{2,1/2}^{(1-P_A)}/0.4)^2 \rceil \quad \text{and} \quad 9(3) = \lceil 2(0.9 Z_{4,1/2}^{(1-P_B)}/0.6)^2 \rceil.$$

Solving the left-hand equation gives $Z_{2,1/2}^{(1-P_A)} = 0.4 \times \sqrt{45/2}/0.9 = 2.108$, which, by linear interpolation in Table B.1, corresponds to $P_A = 0.9676$. Similarly, the right-hand equation becomes $Z_{4,1/2}^{(1-P_B)} = 0.6 \times \sqrt{27/2}/0.9 = 2.449$, which corresponds to $P_B = 0.9747$. The product of these two probabilities is 0.9431, which is (barely) *less* than the specified $P^* = 0.95$. In the same way, $n = 10$ yields the values $P\{CS(1) \mid (3; 0.4/0.9, 50)\} = 0.9749$ and $P\{CS(1) \mid (5; 0.6/0.9, 30)\} = 0.9823$. The product of these two probabilities is 0.9576.

We conclude that the smallest common single-stage sample size that will guarantee $P^* = 0.95$ for the two-factor experiment is $n = 10$ observations (to be taken from each of the 15 treatment combinations). As a side benefit we notice that whenever $\alpha_{[3]} - \alpha_{[2]} \geq 0.4$, the best level of factor A is correctly selected with probability at least $= 0.9749$ no matter what the configuration of the β_j, and a similar statement holds for factor B.

This method of determining n can also be employed for more general goals. For example, suppose that an experimenter desires to select simultaneously the s_1 best levels of the first factor and the s_2 best levels of the second factor ($1 \leq s_1 < a$, $1 \leq s_2 < b$), both *without* regard to order, that is, a generalization of Goal 2.5.1. In order to implement the natural procedure based on the A_{in} and B_{jn}, the constants in Table 2.8 (expanded in Table I of Bechhofer 1954 or the Dunnett 1989 program as shown in Appendix C) are required.

Efficiency of Two-Factor Experiments
In addition to illustrating the determination of n, the following example shows, for single-stage procedures, the efficiency of a two-factor experiment relative to that of two independent single-factor experiments in the absence of interaction when both guarantee the same probability requirement (6.2.1).

Example 6.2.2 (Efficiency of the Two-Way Design) This example considers a problem that is *symmetric* in the number of levels of each factor and in the specified δ^*'s, that is, $a = b$ and $\delta_\alpha^* = \delta_\beta^*$. Suppose that $\sigma = 0.9$ and $a = b = 4$ and that the experimenter specifies $\delta_\alpha^* = \delta_\beta^* = 0.5$ and $P^* = 0.95$. Then the right-hand side of (6.2.2) becomes $[P\{CS(1) \mid (4; 0.5/0.9, 4n)\}]^2$, which we set equal to $P^* = 0.95$ to determine n, yielding a target value of

$$P\{CS(1) \mid (4; 0.5/0.9, 4n)\} = \sqrt{P^*} = 0.9747.$$

By interpolating in Table B.1 we obtain $Z_{3,1/2}^{(1-0.9747)} = 2.351$. Hence, the common number of observations per treatment combination is found by solving $4n = \lceil 2(0.9 \times 2.351/0.5)^2 \rceil = 36$ or $n = 9$. Thus a total of $4 \times 4 \times 9 = 144$ observations would be required for the two-factor experiment.

Alternatively, the probability requirement (6.2.1) could have been guaranteed using two *independent single-factor* experiments, each with specified $(\delta^*, P^*) = (0.5, 0.9747)$. For *each* experiment the sample size required per treatment would be found by computing $n = \lceil 2(0.9 \times 2.351/0.5)^2 \rceil = 36$ as in Section 2.2. Thus a total of $2 \times 4 \times 36 = 288$ observations would be required using two experiments.

For this example, we conclude that conducting the two-factor experiment instead of two independent single-factor experiments would require $100(144/288) = 50\%$ as many observations. This is so because, in the factorial experiment, one can think of each observation as "working twice"—once for rows and again for columns.

Remark 6.2.1 Bawa (1972) presents an asymptotic analysis (as $P^* \to 1$) of the ratio of the minimum sample sizes required for general multi-factor and multiple single-factor procedures (as illustrated in Example 6.2.2) when both attain the probability requirement (6.2.1). The choice of the symmetric setup in Example 6.2.2 is a special case of his setup. In general, the savings is less for nonsymmetric situations than for symmetric ones, but *the two-factor experiment always results in a savings in total sample size relative to two independent single-factor experiments* that guarantee the same probability requirement. This fact is also true for experiments with three or more factors.

Confidence Statement Formulation

Confidence statements for the effects in two-factor experiments are similar to those for single-factor experiments. Suppose $n = n(a, b; \delta_\alpha^*/\sigma, \delta_\beta^*/\sigma, P^*)$ is chosen to achieve the $P\{CS \mid \boldsymbol{\mu}\}$ according to procedure $\mathcal{N}_{B2}$. Let α_{S_1} denote the selected level of the first factor and β_{S_2} denote the selected level of the second factor. Then the experimenter can assert

$$P\left\{\alpha_{[a]} - \delta_\alpha^* \leq \alpha_{S_1} \leq \alpha_{[a]} \quad \text{and} \quad \beta_{[b]} - \delta_\beta^* \leq \beta_{S_2} \leq \beta_{[b]} \mid \boldsymbol{\mu}\right\} \geq P^*$$

whenever $\boldsymbol{\mu}$ satisfies $\mu_{ij} = \mu + \alpha_i + \beta_j$ for all i and j.

Randomized Complete Block Design

Suppose that there is inadequate homogeneous experimental material to conduct a completely randomized experiment, but that c blocks are available, each containing uniform experimental material and sufficiently large to accommodate one or more replications of the entire set of ab treatment combinations. Also assume that the relative treatment effects are the same in all blocks (that is, there is no treatment × block interaction). Thus the data satisfy the standard linear model with *no* treatment interaction and an *additive* block effect. Formally, assume that r replications of the entire experiment are possible in each block. Let Y_{ijks} denote the sth replication on the (i, j)th treatment combination in the kth block $(1 \leq i \leq a, 1 \leq j \leq b, 1 \leq k \leq c, 1 \leq s \leq r)$. The Y_{ijks} are assumed to satisfy $E\{Y_{ijks}\} = \mu + \alpha_i + \beta_j + \gamma_k$ with $\sum_{i=1}^{a} \alpha_i = \sum_{j=1}^{b} \beta_j = \sum_{k=1}^{c} \gamma_k = 0$. Here the α_i $(1 \leq i \leq a)$ and β_j $(1 \leq j \leq b)$ are treatment main effects, the γ_k $(1 \leq k \leq c)$ are block effects, and there is no interaction between the treatments or between the treatments and the blocks. Lastly, assume that $\text{Var}\{Y_{ijks}\} = \sigma^2$ for all i, j, k, s and that σ^2 is *known*.

Selection proceeds by first choosing the number of blocks c to satisfy the equation

$$P\{CS(2) \mid (a, b; \delta_\alpha^*/\sigma, \delta_\beta^*/\sigma, cr)\}$$
$$= P\{CS(1) \mid (a; \delta_\alpha^*/\sigma, bcr)\} \times P\{CS(1) \mid (b; \delta_\beta^*/\sigma, acr)\}.$$

Then replicate the $(a \times b)$-factor experiment r times in each of the c blocks. Set

$$A_i = \sum_{j=1}^{b}\sum_{k=1}^{c}\sum_{s=1}^{r} y_{ijks} \quad (1 \leq i \leq a) \quad \text{and} \quad B_j = \sum_{i=1}^{a}\sum_{k=1}^{c}\sum_{s=1}^{r} y_{ijks} \quad (1 \leq j \leq b)$$

and choose the levels of the observed $A_{[a]} = \max\{A_1, \ldots, A_a\}$ and $B_{[b]} = \max\{B_1, \ldots, B_b\}$ as the ones associated with $\alpha_{[a]}$ and $\beta_{[b]}$.

6.2.2 A Closed Sequential Procedure without Elimination

As in the single-factor case, it is possible to give a sequential procedure that usually outperforms the single-stage procedure $\mathcal{N}_{B2}$. We first describe a closed selection procedure for two-factor experiments with common known variance due to Bechhofer and Goldsman (1988b). In particular, this procedure can provide considerable savings in the expected total number of observations relative to the single-stage procedure if the treatment means are in a configuration favorable to the experimenter; yet it requires only a modest increase in observations should the equal-means (EM) configuration hold.

At each stage, one observation is taken from all treatment combinations, thus yielding a matrix of $a \times b$ data points. Let $A_{im} = \sum_{j=1}^{b} \sum_{s=1}^{m} y_{ijs}$ and $B_{jm} = \sum_{i=1}^{a} \sum_{s=1}^{m} y_{ijs}$ ($1 \le i \le a$, $1 \le j \le b$, $m \ge 1$) denote the row and columns sums, respectively, after m matrix-observations have been taken. Denote the ordered values of the A_{im} and B_{jm} by $A_{[1]m} < \cdots < A_{[a]m}$ and $B_{[1]m} < \cdots < B_{[b]m}$, respectively. Furthermore, define

$$u_m = \sum_{i=1}^{a-1} \exp\{-\delta_\alpha^\star (A_{[a]m} - A_{[i]m})/\sigma^2\},$$

$$v_m = \sum_{j=1}^{b-1} \exp\{-\delta_\beta^\star (B_{[b]m} - B_{[j]m})/\sigma^2\}$$

and

$$z_m = u_m + v_m + u_m \cdot v_m. \tag{6.2.3}$$

Procedure $\mathcal{N}_{BG2}$ For the given (a, b) and specified $(\delta_\alpha^\star/\sigma, \delta_\beta^\star/\sigma, P^\star)$, choose the *truncation number* $n_0 = n_0(a, b; \delta_\alpha^\star/\sigma, \delta_\beta^\star/\sigma, P^\star)$ from Table 6.2.

Sampling rule: At the mth stage of experimentation ($m \ge 1$), observe the matrix $\{Y_{ijm} \ (1 \le i \le a, 1 \le j \le b)\}$.

Stopping rule: After the mth matrix-observation has been observed, compute z_m in (6.2.3). Stop sampling when, for the first time, either

$$z_m \le (1 - P^\star)/P^\star \quad \text{or} \quad m = n_0,$$

whichever occurs first.

Terminal decision rule: Let N be the value of m at termination (N is a random variable). Select the factor-levels associated with $A_{[a]N}$ and $B_{[b]N}$ as the ones associated with $\alpha_{[a]}$ and $\beta_{[b]}$, respectively.

Table 6.2. Truncation Numbers Required to Implement Procedure $\mathcal{N}_{BG2}$ When $\delta_\alpha^\star/\sigma = \delta_\beta^\star/\sigma = 0.2$ and $P^\star = 0.9025$

	b				
a	2	3	4	5	6
2	86	92	102	110	120
3		76	74	76	80
4			64	61	61
5				55	53
6					48

Reprinted from Bechhofer and Goldsman (1988b), pp. 118–120, by courtesy of Marcel Dekker, Inc.

In the same way as for procedure $\mathcal{N}_{B2}$, the truncation numbers in Table 6.2 are available for a limited range of a and b but only for $\delta_\alpha^\star/\sigma = \delta_\beta^\star/\sigma = 0.2$ and $P^\star = 0.9025 = (0.95)^2$. Unlike procedure $\mathcal{N}_{B2}$, it is not easy to extend the range of truncation numbers to other cases because they require extensive simulation to estimate the amount by which the $P\{CS\}$ exceeds the nominal $P^\star$, a quantity called "the excess over the boundary."

Comparison of Tables 6.1 and 6.2 shows that the truncation numbers for the *sequential* procedure $\mathcal{N}_{BG2}$ are only slightly larger than the common sample sizes for the comparable *single-stage* procedure. Thus, if it is feasible to use the sequential procedure, the experimenter can gain most of the benefits of early stopping even if the treatment means are in a configuration in (or near) the EM-configuration; substantial savings in the expected number of matrix-observations are achieved in more favorable configurations.

If the truncation number is not available in Table 6.2, the experimenter might be tempted to use procedure $\mathcal{N}_{BG2}$ with the modified stopping rule that omits the stopping due to truncation; explicitly, the untruncated rule that stops sampling only when $z_m \leq (1 - P^\star)/P^\star$. The resulting procedure achieves the design requirement in Equation (6.2.1) for *any* $(a, b; \delta_\alpha^\star/\sigma, \delta_\beta^\star/\sigma, P^\star)$ but it is *conservative*. Experimenters should be warned that while this untruncated procedure guarantees the desired probability level, it may require an unacceptably large number of observations to terminate sampling if the means are "close" together (Bechhofer and Goldsman 1988b).

6.2.3 A Closed Sequential Procedure with Elimination

This section describes a competitor to procedure $\mathcal{N}_{BG2}$; it is also closed and sequential but it allows the experimenter to cease sampling from treatments no longer indicated as contending for $\alpha_{[a]}$ and $\beta_{[b]}$. The elimination process for this rule continues until a single treatment-factor combination remains. Despite its attractive attributes, we shall see in the next section that the rule sometimes requires more *stages* of sampling to stop than does the closed $\mathcal{N}_{BG2}$ procedure stated in the previous section. Thus we will conclude with recommendations regarding circumstances in which it is preferable to use each of the two rules.

IZ SELECTION—CR DESIGNS

To describe the sampling and operation of the rule, we introduce the notation A_m [B_m] to denote the number of levels remaining for factor A [factor B] *before* stage m, that is, the number of indices i [j] for which the associated treatment levels with α_i [β_j] have not been eliminated. For example, $A_1 = a$ and $B_1 = b$ prior to taking the first set of observations. Sampling is terminated when only one treatment-factor combination remains; formally this is when $A_m = B_m = 1$. We refer to noneliminated treatment combinations as being *active*. This procedure is due to Hartmann (1993) who generalized the work of Paulson (1964) and Fabian (1974) to multi-factor experiments in which there is no interaction between the factor-levels.

Procedure $\mathcal{N}_{H2}$ For specified $P^\star$, choose $P^\star_\alpha, P^\star_\beta > 0$ with $P^\star_\alpha \times P^\star_\beta \geq P^\star$. (We recommend the choice $P^\star_\alpha = P^\star_\beta = \sqrt{P^\star}$.) For the given $(\delta^\star_\alpha, \delta^\star_\beta)$ define

$$a_\alpha = \frac{2\sigma^2}{\delta^\star_\alpha} \ln\left\{\frac{a-1}{2(1-P^\star_\alpha)}\right\} \quad \text{and} \quad a_\beta = \frac{2\sigma^2}{\delta^\star_\beta} \ln\left\{\frac{b-1}{2(1-P^\star_\beta)}\right\}. \quad (6.2.4)$$

Sampling rule: At the mth stage ($m \geq 1$) take a single observation Y_{ijm} from each of the $A_m \times B_m$ *active* treatment combinations. Compute the row sum over the active levels of B for each active level i of factor A, that is, $U_{im} = \sum_{j \text{ active for B}} Y_{ijm}$. Also compute the column sum over the active levels of A for each active level j of factor B, that is, $V_{jm} = \sum_{i \text{ active for A}} Y_{ijm}$. Eliminate any active level i of factor A for which

$$\sum_{r=1}^m U_{ir} < \max_k \left(\sum_{r=1}^m U_{kr}\right) - \max\left(a_\alpha - \frac{\delta^\star_\alpha}{2}\sum_{r=1}^m B_r, 0\right)$$

where the leftmost maximum is taken over k corresponding to active A factor levels at stage m. Also eliminate any active level j of factor B for which

$$\sum_{r=1}^m V_{jr} < \max_l \left(\sum_{r=1}^m V_{\ell r}\right) - \max\left(a_\beta - \frac{\delta^\star_\beta}{2}\sum_{r=1}^m A_r, 0\right)$$

where the leftmost maximum is taken over l corresponding to the active B factor-levels.

Stopping rule: Stop sampling when $A_m = B_m = 1$.

Terminal decision rule: At stopping, select the single remaining treatment-factor combination as the one associated with the factor-levels $\alpha_{[a]}$ and $\beta_{[b]}$.

Example 6.2.3 We illustrate the calculations with a simple example. Consider a two-factor experiment involving normal observations having unit variance in which A has $a = 2$ levels and B has $b = 3$ levels. Suppose that we specify $P^\star_\alpha = P^\star_\beta = 0.90$

and $\delta_\alpha^\star = \delta_\beta^\star = 2.0$. Substitution into Equation (6.2.4) yields $a_\alpha = 1.6094$ and $a_\beta = 2.3025$. Suppose that the first-stage data are:

m	y_{11m}	y_{12m}	y_{13m}	y_{21m}	y_{22m}	y_{23m}
1	1.2	2.0	1.9	0.9	2.5	2.7

Thus, $(U_{11}, U_{21}) = (5.1, 6.1), (V_{11}, V_{21}, V_{31}) = (2.1, 4.5, 4.6)$ and $(A_1, B_1) = (2, 3)$. We eliminate treatment level 1 of factor A because

$$U_{11} = 5.1 < \max_i(U_{i1}) - \max\left(a_\alpha - \frac{\delta_\alpha^\star}{2}B_1, 0\right)$$

$$= 6.1 - \max(1.6094 - 1 \times 3, 0) = 6.1.$$

Similarly, level 1 of factor B is eliminated because

$$2.1 < 4.6 - \max(2.3025 - 1 \times 2, 0) = 4.2975.$$

Level 2 of factor A and levels 2 and 3 of factor B are active going into stage $m = 2$ and thus $A_2 = 1$ and $B_2 = 2$.

In stage 2, we sample from treatment combinations (2,2) and (2,3). Suppose that after the second stage, the observed data are:

m	y_{11m}	y_{12m}	y_{13m}	y_{21m}	y_{22m}	y_{23m}
1	1.2	2.0	1.9	0.9	2.5	2.7
2					2.1	2.9

Now $U_{22} = 5.0, (V_{22}, V_{32}) = (2.1, 2.9)$ and $(A_2, B_2) = (1, 2)$. We eliminate level 2 of factor B because

$$V_{21} + V_{22} = 4.5 + 2.1 = 6.6$$

$$< \max(4.5 + 2.1, 4.6 + 2.9) - \max(2.3025 - 1 \times (2+1), 0) = 7.5.$$

Thus $A_3 = 1 = B_3$ and the procedure stops after two stages and selects the treatment combination (2,3).

6.2.4 Comparison of the Performance Characteristics of the Procedures

Criteria for Comparisons

Sections 6.2.1–6.2.3 described three procedures ($\mathcal{N}_{B2}$, $\mathcal{N}_{BG2}$ and $\mathcal{N}_{H2}$), all of which require the same statistical assumptions (in particular, a *common known* variance σ^2), achieve Goal 6.2.1, and guarantee the same probability requirement (6.2.1). Procedure $\mathcal{N}_{B2}$ is a single-stage procedure, while procedures $\mathcal{N}_{BG2}$ and $\mathcal{N}_{H2}$ are closed multi-stage procedures with $\mathcal{N}_{H2}$ allowing elimination.

In this section, we use Monte Carlo (MC) simulation to compare the three procedures in terms of their achieved $P\{CS\}$, $E\{N\}$ = E{Number of *stages* to terminate sampling}, and E{T} = E{*Total* number of *observations* to terminate sampling} for various configurations of the treatment means. (See Section 2.4 for additional discussion of these terms.) As usual, the quantity E{N} can often be equated to the *duration* of the study, while E{T} is simply the number of experimental units one would expect to sample during the course of the study. Three configurations of the treatment means were studied: (i) equally-spaced main effects $\delta^\star$ apart in both factors (denoted by ES($\delta^\star$)): $\alpha_{[i+1]} - \alpha_{[i]} = \delta^\star$ ($1 \leq i \leq a - 1$) and $\beta_{[j+1]} - \beta_{[j]} = \delta^\star$ ($1 \leq j \leq b - 1$); (ii) least-favorable in both factors (denoted by LF): $\alpha_{[a]} - \alpha_{[i]} = \delta^\star$ ($1 \leq i \leq a - 1$) and $\beta_{[b]} - \beta_{[j]} = \delta^\star$ ($1 \leq j \leq b - 1$), and (iii) equal main effects in both factors (hence equal means and thus denoted by EM): $\alpha_{[a]} = \alpha_{[1]}$ and $\beta_{[b]} = \beta_{[1]}$.

Summary of Comparisons

As was the case in Section 2.4 with the comparison of selection procedures for the single-factor case, our two-factor MC studies showed that both of the multi-stage procedures $\mathcal{N}_{BG2}$ and $\mathcal{N}_{H2}$ usually required a substantially smaller expected total number of observations, E{T}, than did procedure $\mathcal{N}_{B2}$. Thus, if the experimental conditions permit the use of multi-stage procedures, the experimenter should consider procedures $\mathcal{N}_{BG2}$ or $\mathcal{N}_{H2}$. A comparison of procedures $\mathcal{N}_{BG2}$ and $\mathcal{N}_{H2}$ showed that neither minimized E{T} uniformly for *all* configurations of $\boldsymbol{\mu}$ but that procedure $\mathcal{N}_{BG2}$ *always yielded substantially lower total number of stages*, E{N}, *than did procedure* $\mathcal{N}_{H2}$. Indeed, we can make the following general statements concerning the relative performances of these two multi-stage procedures.

- To minimize the expected total number of stages, E{N}, use procedure $\mathcal{N}_{BG2}$.
- To minimize the expected total number of observations, E{T}, use procedure $\mathcal{N}_{H2}$ if it is believed that $\boldsymbol{\mu}$ is in or near the EM-configuration *or* if both a and b are ≥ 4; otherwise, use procedure $\mathcal{N}_{BG2}$.

Details

Tables 6.3 and 6.4 serve as partial guides to the relative performances of the procedures $\mathcal{N}_{B2}$, $\mathcal{N}_{BG2}$ and $\mathcal{N}_{H2}$. These tables are abstracted from a large simulation study in Bechhofer, Goldsman and Hartmann (1993). The values listed in Table 6.3 are MC estimates of the quantities of interest (except for procedure $\mathcal{N}_{B2}$ for which some of the calculations are exact); below each estimate (in parentheses) is the estimated standard error of the value above it. The tables include results for the special case $P^\star = 0.9025 = (0.95)^2$, $a = 2(1)6$, $b = a(1)6$, and $\delta^\star/\sigma = 0.2$ (which can be thought of as representative values of these quantities). All of the procedures guarantee the probability requirement (6.2.1) (although procedures $\mathcal{N}_{B2}$ and $\mathcal{N}_{H2}$ sometimes *overprotect* by a sizable amount, especially as a or b increase). In the tables, we have used the notation $\hat{p}$, $\hat{N}$ and $\hat{T}$ to denote our MC estimates of the P{CS}, E{N} and E{T}, respectively.

Table 6.4 lists the better of procedures $\mathcal{N}_{BG2}$ and $\mathcal{N}_{H2}$ in terms of the minimum estimated E{N} and minimum estimated E{T} for the "representative" $P^\star$, (a, b), $\delta^\star/\sigma$, and $\boldsymbol{\mu}$ listed above. Any rough insights that the experimenter may have concerning $\boldsymbol{\mu}$ can assist in making an informed choice concerning the procedure

Table 6.3. Estimated Performance Characteristics of Procedures $\mathcal{N}_{B2}$, $\mathcal{N}_{BG2}$ and $\mathcal{N}_{H2}$ When $P^\star = 0.9025$ and $\delta_\alpha^\star = 0.2\sigma = \delta_\beta^\star$

(a,b)	Procedure	ES(0.2σ) $\hat{p}$	$\hat{N}$	$\hat{T}$	LF $\hat{p}$	$\hat{N}$	$\hat{T}$	EM $\hat{N}$	$\hat{T}$
(2,2)	$\mathcal{N}_{B2}$	0.9034	1	272	0.9034	1	272	1	272
	$\mathcal{N}_{BG2}$ $n_0 = 86$	0.9040 (0.0027)	43.8 (0.2)	175.2 (0.4)	0.9032 (0.0003)	43.5 (<0.1)	173.9 (0.2)	65.9 (0.2)	263.4 (0.4)
	$\mathcal{N}_{H2}$	0.9067 (0.0027)	65.3 (0.3)	188.9 (0.6)	0.9105 (0.0018)	65.7 (0.2)	189.2 (0.4)	85.2 (0.3)	251.0 (0.6)
(2,3)	$\mathcal{N}_{B2}$	0.9338 (0.0023)	1	432	0.9034	1	432	1	432
	$\mathcal{N}_{BG2}$ $n_0 = 92$	0.9252 (0.0024)	39.7 (0.2)	238.1 (0.4)	0.9038 (0.0003)	47.3 (<0.1)	284.1 (0.2)	74.4 (0.2)	446.4 (0.5)
	$\mathcal{N}_{H2}$	0.9330 (0.0023)	80.6 (0.4)	271.5 (0.8)	0.9159 (0.0018)	93.2 (0.3)	319.7 (0.7)	129.4 (0.4)	440.7 (1.1)
(2,4)	$\mathcal{N}_{B2}$	0.9568 (0.0019)	1	632	0.9029	1	632	1	632
	$\mathcal{N}_{BG2}$ $n_0 = 102$	0.9308 (0.0023)	37.4 (0.2)	299.0 (0.5)	0.9037 (0.0003)	52.3 (0.1)	418.2 (0.3)	83.7 (0.2)	669.8 (0.6)
	$\mathcal{N}_{H2}$	0.9410 (0.0022)	97.1 (0.4)	351.2 (1.0)	0.9194 (0.0018)	123.3 (0.3)	480.3 (0.9)	174.9 (0.5)	668.9 (1.5)
(2,5)	$\mathcal{N}_{B2}$	0.9665 (0.0016)	1	870	0.9039	1	870	1	870
	$\mathcal{N}_{BG2}$ $n_0 = 110$	0.9313 (0.0023)	36.2 (0.2)	361.9 (0.5)	0.9027 (0.0001)	56.6 (<0.1)	565.6 (0.1)	90.8 (0.2)	907.6 (0.7)
	$\mathcal{N}_{H2}$	0.9441 (0.0021)	109.8 (0.4)	420.1 (1.1)	0.9257 (0.0017)	145.7 (0.3)	647.6 (1.2)	208.3 (0.5)	902.3 (1.8)
(2,6)	$\mathcal{N}_{B2}$	0.9745 (0.0014)	1	1128	0.9042	1	1128	1	1128
	$\mathcal{N}_{BG2}$ $n_0 = 120$	0.9362 (0.0022)	34.7 (0.2)	416.8 (0.6)	0.9037 (0.0003)	61.0 (0.1)	732.0 (0.4)	98.8 (0.3)	1186.0 (0.9)
	$\mathcal{N}_{H2}$	0.9466 (0.0021)	120.0 (0.5)	481.0 (1.1)	0.9222 (0.0017)	163.3 (0.4)	821.3 (1.4)	233.7 (0.3)	1138.2 (1.0)
(3,3)	$\mathcal{N}_{B2}$	0.9463 (0.0021)	1	558	0.9051	1	558	1	558
	$\mathcal{N}_{BG2}$ $n_0 = 76$	0.9288 (0.0023)	32.8 (0.1)	295.1 (0.4)	0.9053 (0.0003)	41.6 (<0.1)	374.6 (0.2)	66.1 (0.1)	594.9 (0.4)
	$\mathcal{N}_{H2}$	0.9576 (0.0018)	76.7 (0.3)	316.9 (0.8)	0.9225 (0.0017)	83.4 (0.2)	377.0 (0.7)	108.3 (0.3)	512.6 (0.9)
(3,4)	$\mathcal{N}_{B2}$	0.9582 (0.0018)	1	720	0.9038	1	720	1	720
	$\mathcal{N}_{BG2}$ $n_0 = 74$	0.9301 (0.0023)	29.4 (0.1)	352.8 (0.4)	0.9048 (0.0003)	41.2 (<0.1)	493.9 (0.3)	65.5 (0.1)	785.7 (0.5)
	$\mathcal{N}_{H2}$	0.9619 (0.0017)	83.9 (0.4)	373.1 (0.9)	0.9217 (0.0017)	99.2 (0.3)	498.8 (0.9)	134.2 (0.4)	686.3 (1.3)
(3,5)	$\mathcal{N}_{B2}$	0.9667 (0.0016)	1	930	0.9057	1	930	1	930
	$\mathcal{N}_{BG2}$ $n_0 = 76$	0.9338 (0.0023)	27.2 (0.1)	408.5 (0.5)	0.9047 (0.0002)	41.9 (<0.1)	628.4 (0.2)	67.7 (0.1)	1015.1 (0.5)
	$\mathcal{N}_{H2}$	0.9673 (0.0016)	93.7 (0.4)	431.0 (1.0)	0.9249 (0.0017)	121.4 (0.3)	652.8 (1.1)	172.6 (0.5)	909.7 (1.7)

Reprinted from Bechhofer, Goldsman and Hartmann (1993), pp. 216–218, by courtesy of Marcel Dekker, Inc.

Table 6.3. *Continued*

(a,b)	Procedure	ES(0.2σ) $\hat{p}$	$\hat{N}$	$\hat{T}$	LF $\hat{p}$	$\hat{N}$	$\hat{T}$	EM $\hat{N}$	$\hat{T}$
(3,6)	$\mathcal{N}_{B2}$	0.9710 (0.0015)	1	1152	0.9036	1	1152	1	1152
$n_0 = 80$	$\mathcal{N}_{BG2}$	0.9333 (0.0023)	25.9 (0.1)	465.7 (0.5)	0.9043 (0.0003)	43.5 (<0.1)	783.5 (0.4)	70.9 (0.1)	1275.8 (0.6)
	$\mathcal{N}_{H2}$	0.9719 (0.0015)	104.0 (0.4)	486.7 (1.1)	0.9292 (0.0017)	142.0 (0.4)	820.5 (1.4)	203.4 (0.5)	1140.3 (2.1)
(4,4)	$\mathcal{N}_{B2}$	0.9593 (0.0018)	1	864	0.9060	1	864	1	864
$n_0 = 64$	$\mathcal{N}_{BG2}$	0.9318 (0.0023)	25.2 (0.1)	403.6 (0.4)	0.9042 (0.0004)	37.3 (<0.1)	597.3 (0.3)	58.5 (<0.1)	936.2 (0.4)
	$\mathcal{N}_{H2}$	0.9673 (0.0016)	82.1 (0.4)	410.7 (0.9)	0.9258 (0.0017)	92.2 (0.3)	565.9 (0.9)	118.1 (0.4)	769.5 (1.2)
(4,5)	$\mathcal{N}_{B2}$	0.9673 (0.0016)	1	1040	0.9066	1	1040	1	1040
$n_0 = 61$	$\mathcal{N}_{BG2}$	0.9299 (0.0023)	23.0 (0.1)	459.8 (0.4)	0.9034 (0.0004)	36.1 (<0.1)	721.3 (0.3)	56.3 (<0.1)	1126.0 (0.4)
	$\mathcal{N}_{H2}$	0.9718 (0.0015)	86.5 (0.4)	454.6 (1.0)	0.9288 (0.0017)	102.4 (0.3)	681.4 (1.1)	136.5 (0.4)	936.3 (1.6)
(4,6)	$\mathcal{N}_{B2}$	0.9723 (0.0015)	1	1224	0.9027	1	1224	1	1224
$n_0 = 61$	$\mathcal{N}_{BG2}$	0.9386 (0.0022)	21.3 (<0.1)	511.9 (0.5)	0.9033 (0.0004)	35.9 (<0.1)	861.3 (0.3)	56.6 (<0.1)	1359.0 (0.4)
	$\mathcal{N}_{H2}$	0.9746 (0.0014)	94.2 (0.4)	502.2 (1.0)	0.9253 (0.0017)	121.0 (0.3)	835.9 (1.3)	167.7 (0.5)	1147.8 (2.0)
(5,5)	$\mathcal{N}_{B2}$	0.9722 (0.0015)	1	1175	0.9042	1	1175	1	1175
$n_0 = 55$	$\mathcal{N}_{BG2}$	0.9328 (0.0023)	20.3 (<0.1)	508.4 (0.4)	0.9037 (0.0004)	33.4 (<0.1)	834.2 (0.3)	51.5 (<0.1)	1288.7 (0.4)
	$\mathcal{N}_{H2}$	0.9771 (0.0014)	86.6 (0.4)	488.9 (1.0)	0.9262 (0.0017)	97.6 (0.3)	756.7 (1.0)	122.7 (0.4)	1024.2 (1.4)
(5,6)	$\mathcal{N}_{B2}$	0.9738 (0.0015)	1	1350	0.9037	1	1350	1	1350
$n_0 = 53$	$\mathcal{N}_{BG2}$	0.9374 (0.0022)	19.0 (<0.1)	568.6 (0.5)	0.9048 (0.0004)	32.1 (<0.1)	962.3 (0.3)	49.8 (<0.1)	1495.5 (0.4)
	$\mathcal{N}_{H2}$	0.9802 (0.0013)	89.7 (0.4)	526.1 (1.0)	0.9295 (0.0017)	105.6 (0.3)	871.6 (1.2)	137.4 (0.5)	1187.8 (1.7)
(6,6)	$\mathcal{N}_{B2}$	0.9789 (0.0013)	1	1512	0.9050	1	1512	1	1512
$n_0 = 48$	$\mathcal{N}_{BG2}$	0.9398 (0.0022)	17.3 (<0.1)	622.5 (0.5)	0.9029 (0.0002)	30.0 (<0.1)	1080.9 (0.1)	45.7 (<0.1)	1643.5 (0.3)
	$\mathcal{N}_{H2}$	0.9832 (0.0012)	89.8 (0.4)	555.7 (1.1)	0.9291 (0.0017)	101.7 (0.3)	949.6 (1.3)	126.5 (0.4)	1286.0 (1.6)

Table 6.4. The Multi-Stage Procedure $\mathcal{N}_{BG2}$ or $\mathcal{N}_{H2}$ Having Minimum Estimated $E\{N\}$ and Minimum Estimated $E\{T\}$ When $P^\star = 0.9025$ for the Configurations: ES(0.2σ), LF and EM

	ES(0.2σ)		LF		EM	
(a,b)	$E\{N\}$	$E\{T\}$	$E\{N\}$	$E\{T\}$	$E\{N\}$	$E\{T\}$
$a = 2,3$ and $b = a(1)6$	$\mathcal{N}_{BG2}$	$\mathcal{N}_{BG2}$	$\mathcal{N}_{BG2}$	$\mathcal{N}_{BG2}$	$\mathcal{N}_{BG2}$	$\mathcal{N}_{H2}$
$(4,4)$	$\mathcal{N}_{BG2}$	$\mathcal{N}_{BG2}$	$\mathcal{N}_{BG2}$	$\mathcal{N}_{H2}$	$\mathcal{N}_{BG2}$	$\mathcal{N}_{H2}$
$a = 4,5,6$ and $b = a(1)6$	$\mathcal{N}_{BG2}$	$\mathcal{N}_{H2}$	$\mathcal{N}_{BG2}$	$\mathcal{N}_{H2}$	$\mathcal{N}_{BG2}$	$\mathcal{N}_{H2}$

to use. Both procedures $\mathcal{N}_{BG2}$ and $\mathcal{N}_{H2}$ have strengths and weaknesses. A major strength is that both procedures are *closed* and generally require fewer observations than does the single-stage procedure $\mathcal{N}_{B2}$. Furthermore, both are adaptive procedures that, with high probability, will terminate in only a relatively small number of *stages* if the μ-configuration is favorable to the experimenter (that is, if the best treatment combination mean is much larger than the remaining means); if the μ-configuration is unfavorable (such as the EM-configuration), then the closed procedures will protect the experimenter against taking an excessive number of observations to detect small differences among the means.

One weakness of procedure $\mathcal{N}_{H2}$ is that it tends to take a prohibitively large number of *stages* compared to procedure $\mathcal{N}_{BG2}$. [See Hartmann (1993) for possible modifications to procedure $\mathcal{N}_{H2}$ that may make the expected number of stages more competitive with procedure $\mathcal{N}_{BG2}$.] On the other hand, truncation tables for procedure $\mathcal{N}_{BG2}$ are only available at present for (a, b) with $1 \leq a, b \leq 6$ when $\delta_\alpha^\star = \delta_\beta^\star = \delta^\star = 0.2(0.1)0.8$ and $P^\star = 0.9025$. As noted above, one should be careful when using the *untruncated* version of procedure $\mathcal{N}_{BG2}$ because N can with sizable probability take on excessively large values (particularly in the EM or contiguous configurations).

Remark 6.2.2 The procedures $\mathcal{N}_{BG2}$ and $\mathcal{N}_{H2}$ both allow the user to specify any probabilities $P_\alpha^\star$ and $P_\beta^\star$ so that $P_\alpha^\star \cdot P_\beta^\star = P^\star$. However, the reader should be warned that the symmetric choice $P_\alpha^\star = \sqrt{P^\star} = P_\beta^\star$ is *not* optimal when the problem is *not* symmetric in (a, b) and $(\delta_\alpha^\star, \delta_\beta^\star)$. To illustrate, Table 6.5 shows, for procedure $\mathcal{N}_{H2}$, the effect on the estimated $E\{N \mid \text{EM}\}$ and $E\{T \mid \text{EM}\}$ when $P_\alpha^\star$ is varied subject to $P_\alpha^\star \cdot P_\beta^\star = 0.9025$ when $a = 2, b = 6, \delta_\alpha^\star = \delta_\beta^\star = 0.2, \sigma^2 = 1$ and $P^\star = 0.9025$. (The number in parentheses below each $\hat{N}$ and $\hat{T}$ entry is the estimated standard error of that entry.) It can be seen that the optimal choice to minimize $E\{T \mid \text{EM}\}$ is approximately $P_\alpha^\star = 0.997$, while the optimal choice to minimize $E\{N \mid \text{EM}\}$ is approximately $P_\alpha^\star = 0.998$. It should be noted that a small change in $P_\alpha^\star$ (say) can lead

IZ SELECTION — SP DESIGNS   159

Table 6.5. For Procedure $\mathcal{N}_{H2}$, Effect on E{N | EM} and E{T | EM} of Varying $P_\alpha^\star$ and $P_\beta^\star$ Subject to $P_\alpha^\star \cdot P_\beta^\star = P^\star = 0.9025$ When $(a, b) = (2, 6)$ and $\delta_\alpha^\star = 0.2\sigma = \delta_\beta^\star$

$P_\alpha^\star$	$P_\beta^\star$	$\hat{N}$	$\hat{T}$
0.93000	0.97043	279.3 (0.6)	1345.3 (2.4)
0.95000	0.95000	233.9 (0.5)	1138.7 (2.1)
0.97043	0.93000	201.4 (0.5)	1007.2 (1.9)
0.99176	0.91000	166.1 (0.4)	916.2 (1.7)
0.99500	0.90703	159.0 (0.4)	913.2 (1.7)
0.99700	0.90522	153.8 (0.4)	912.4 (1.6)
0.99800	0.90431	152.6 (0.4)	921.4 (1.5)
0.99850	0.90386	153.3 (0.4)	932.4 (1.5)
0.99900	0.90340	157.1 (0.4)	951.1 (1.7)

Reprinted from Bechhofer, Goldsman and Hartmann (1993), p. 222, by courtesy of Marcel Dekker, Inc.

to large changes in $\hat{N}$ and $\hat{T}$. At the present time, we do not know how to *calculate* the optimal choice of $P_\alpha^\star$ for given (a, b) and $(\delta_\alpha^\star, \delta_\beta^\star)$; but we do recommend that greater probability be placed on the factor having fewer levels.

6.3 INDIFFERENCE-ZONE SELECTION USING SPLIT-PLOT DESIGNS (COMMON KNOWN VARIANCE)

6.3.1 Introduction

In many practical settings, experimenters find that a completely randomized experimental design is difficult or impossible to use. As an example, Anderson and McLean (1974) review an experiment conducted by Beeson (1965) in which it was desired to perform an *in vitro* study of four different heart valve designs under six circulatory conditions (corresponding to six pulse rates). The measured response was the maximum flow gradient measured in units of mm Hg. A tank of fluid was used to mimic the human circulatory system, and a pump was employed to vary the pulse rate. The experimenter chose to run all 24 treatment combinations (of valve type by pulse rate) twice in his evaluation of the designs.

To conduct such an experiment using a randomized complete block design with two blocks, the experimenter would first randomly select a valve type by pulse rate combination and then determine the maximum flow gradient. Then the experimenter would select a second valve type by pulse rate combination from among the remaining 23 combinations and continue in this manner until the 24 treatment combinations had been used. Then the entire process would be repeated with a new randomization.

The experimenter did not want to conduct the experiment in this manner because the equipment required a substantial setup time to insert each valve. Instead, Beeson proposed to randomly select a valve type, insert it, and then measure the flow gradients for all six pulse rates tested in a random order; changing the simulated pulse rate alone was simple to do because it required merely that the speed of the pump motor be varied. Then a second valve type was randomly selected from among the remaining three unused valve types, and a new randomization of pulse rates was determined. The entire process was repeated until the 24 treatment combinations had been used.

This second method of conducting the experiment is typical of many industrial and agricultural examples in which the experimenter finds it easier to restrict the randomization by first *fixing* a randomly chosen level of one factor and then experimenting with all the factor levels of the second factor (chosen by a separate randomization). Such a design is called a *split-plot* design. Historically, the factor whose levels are fixed in time (or, more commonly, in space) is called the *whole-plot factor*; the factor whose levels vary according to the separate randomization is called the *split-plot factor*. Split-plot designs were first used in agricultural field trials where it was often convenient to restrict the randomization of one factor so that all plots in each row of a test field would have the same level of one factor (say, the same fertilizer) and the levels of a second factor (say, variety) would be randomized over the plots in the field. Most books on experimental design describe hypothesis testing for split-plot designs (see Milliken and Johnson 1984, for example).

This section extends the methods presented in Section 6.2 for designing completely randomized two-factor experiments satisfying additivity (6.1.1) to allow for split-plot designs laid out in blocks.

Statistical Assumptions: Let Y_{ijk} denote the response on the (i, j)th treatment combination in the kth block for a split-plot experiment that is conducted in c complete blocks with the row factor as the whole-plot factor ($1 \leq i \leq a, 1 \leq j \leq b, 1 \leq k \leq c$). We assume

$$Y_{ijk} = \mu + \phi_k + \alpha_i + \beta_j + \omega_{ik} + \epsilon_{ijk} \qquad (6.3.1)$$

where the block effects ϕ_k, the whole-plot effects ω_{ik}, and the measurement errors ϵ_{ijk} are mutually independent with distributions

ϕ_k	i.i.d.	$N(0, \sigma_b^2)$	$(1 \leq k \leq c)$
ω_{ik}	i.i.d.	$N(0, \sigma_w^2)$	$(1 \leq i \leq a, 1 \leq k \leq c)$
ϵ_{ijk}	i.i.d.	$N(0, \sigma_\epsilon^2)$	$(1 \leq i \leq a, 1 \leq j \leq b, 1 \leq k \leq c)$

The interpretation of the model quantities is that ϕ_k is the effect of the kth block, ω_{ik} is the confounding effect due to lack of complete randomization in the kth block when the ith level of the row factor is fixed, and ϵ_{ijk} is the measurement error associated with the response Y_{ijk}. The whole-plot variance component σ_w^2 and the measurement error σ_ϵ^2 can either be known or unknown. Note that under the additive mean model (6.1.1) we have

$$Y_{ijk} = \mu + \phi_k + \alpha_i + \omega_{ik} \qquad (6.3.2)$$
$$+ \beta_j + \epsilon_{ijk}.$$

which shows that the A factor satisfies the assumptions of a randomized complete block design where ω_{ik} plays the role of the measurement error.

All the procedures described in this chapter are functionally *independent* of the block variance component σ_b^2. In this section we assume that the whole-plot variance component σ_w^2 and the measurement error σ_ϵ^2 variance are known. As usual, the notation in which a dot replaces a subscript means that an average has been computed with respect to that subscript; for example, $\overline{Y}_{\cdot \cdot k} = \sum_{i=1}^{a} \sum_{j=1}^{b} Y_{ijk}/ab$. Lastly, let

$$\overline{Y}_{[1] \cdot \cdot} \leq \cdots \leq \overline{Y}_{[a] \cdot \cdot} \quad \text{and} \quad \overline{Y}_{\cdot [1] \cdot} \leq \cdots \leq \overline{Y}_{\cdot [b] \cdot}.$$

denote the ordered $\overline{Y}_{i \cdot \cdot}$ and $\overline{Y}_{\cdot j \cdot}$, respectively.

As in Section 6.2, we study procedures that achieve Goal 6.2.1 of simultaneously identifying the "best" levels of both treatments; a *correct selection* (CS) is said to be made if Goal 6.2.1 is achieved. Our design requirement remains (6.2.1). In Section 6.5 we will allow various error components to be unknown and modify the probability requirement accordingly.

6.3.2 IZ Selection When $(\sigma_w^2, \sigma_\epsilon^2)$ Is Known

Because the experimental design calls for complete sets of treatment combinations to be laid out in blocks, the experiment must be designed so that an adequate number of blocks are used to guarantee the probability requirement. As in the case of the completely randomized design, the procedure proposed to solve this problem is independent of the block variance component, σ_b^2.

Pan and Santner (1993) proposed a single-stage procedure to solve this problem.

Procedure $\mathcal{N}_{PS2}$ (Indifference-Zone, Common Known Variance) For the given (a, b) and specified $(\delta_\alpha^*/\sigma_\epsilon, \delta_\beta^*/\sigma_\epsilon, P^*)$, determine c to satisfy Equation (6.3.4) below.

Sampling rule: Sample c blocks of data $\{Y_{ijk}(1 \leq i \leq a, 1 \leq j \leq b)\}$ $(1 \leq k \leq c)$, each using a split-plot design with the A factor as the whole-plot factor.

Terminal decision rule: Select the level of the A factor associated with $\overline{Y}_{[a] \cdot \cdot}$ as best and select the level of the B factor associated with $\overline{Y}_{\cdot [b] \cdot}$ as best.

Pan and Santner (1993) showed that the least favorable configuration for this problem is the slippage configuration

$$\alpha_{[1]} = \cdots = \alpha_{[a-1]} = \alpha_{[a]} - \delta_\alpha^\star \quad \text{and} \quad \beta_{[1]} = \cdots = \beta_{[b-1]} = \beta_{[b]} - \delta_\beta^\star. \quad (6.3.3)$$

The minimal number of blocks c required to satisfy the design requirement is the smallest integer c satisfying

$$P\left\{U_i \leq \frac{\sqrt{c}\delta_\alpha^\star}{\rho_1 \sigma_\epsilon} \; (1 \leq i < a)\right\} \times P\left\{V_j \leq \frac{\sqrt{c}\delta_\beta^\star}{\rho_2 \sigma_\epsilon} \; (1 \leq j < b)\right\} \geq P^\star \quad (6.3.4)$$

where

$$\rho_1 = \sqrt{2\left(\gamma_w + \frac{1}{b}\right)} \quad \text{and} \quad \rho_2 = \sqrt{\frac{2}{a}} \quad (6.3.5)$$

with

$$\gamma_w = \sigma_w^2 / \sigma_\epsilon^2,$$

and where $\boldsymbol{U} = (U_1, \ldots, U_{a-1})$ and $\boldsymbol{V} = (V_1, \ldots, V_{b-1})$ are $(a-1)$-dimensional and $(b-1)$-dimensional multivariate normal distributions, respectively, with zero mean vectors, unit variances, and common correlation $1/2$. Thus c can be computed by trial and error from standard tables such as Bechhofer and Dunnett (1988) or FORTRAN programs such as Dunnett (1989).

Remark 6.3.1 There are no confounding effects when $\gamma_w = 0$. In this case the expression (6.3.4) for the probability of correct selection reduces to that of a completely randomized experiment. The critical points in (6.3.4) become

$$\frac{\sqrt{c}\delta_\alpha^\star}{\rho_1 \sigma_\epsilon} = \frac{\sqrt{cb}\delta_\alpha^\star}{\sigma_\epsilon \sqrt{2}} \quad \text{and} \quad \frac{\sqrt{c}\delta_\beta^\star}{\rho_2 \sigma_\epsilon} = \frac{\sqrt{ca}\delta_\beta^\star}{\sigma_\epsilon \sqrt{2}}.$$

Substituting in (6.3.4) we obtain the expression in Bechhofer (1954) for the single-stage procedure based on a completely randomized design. Similar reductions hold for all other special cases of the split-plot procedures described in this book.

Example 6.3.1 (Designing a Selection Procedure) To illustrate the implementation of procedure $\mathcal{N}_{PS2}$, suppose that it is desired to select the treatment combination with the largest main effects in a factorial experiment having $a = 4$ levels for factor A and $b = 3$ levels for factor B. We specify that correct selection should occur with probability at least 0.90 when $\delta_\alpha^\star = \delta_\beta^\star$. If $\sigma_b^2 = 0.75$, $\sigma_w^2 = 0.25$ and $\sigma_\epsilon^2 = 1.0$, then it is straightforward to calculate that

$$\rho_1 = \sqrt{2\left(\gamma_w + \frac{1}{b}\right)} = \sqrt{2\left(\frac{0.25}{1} + \frac{1}{3}\right)} = 1.080$$

$$\text{and} \quad \rho_2 = \sqrt{\frac{2}{a}} = \sqrt{\frac{2}{4}} = 0.707.$$

Thus the coordinates in the product of probabilities (6.3.4) are:

$$\frac{\sqrt{c}\delta_\alpha^\star}{\rho_1 \sigma_\epsilon} = 1.0184 \times \sqrt{c} \quad \text{and} \quad \frac{\sqrt{c}\delta_\beta^\star}{\rho_2 \sigma_\epsilon} = 1.5556 \times \sqrt{c}.$$

Using the equicoordinate probability program MVNPRD of Dunnett (1989) (see Appendix C) we find, by trial and error, that the product (6.3.4) is approximately 0.82 for $c = 2$, it is 0.90 for $c = 3$, and it is 0.95 for $c = 4$. Therefore, the use of $c = 3$ blocks guarantees (6.3.2) at level $P^\star = 0.90$.

6.4 IZ SELECTION USING COMPLETELY RANDOMIZED DESIGNS (COMMON UNKNOWN VARIANCE)

In this section, we assume that the treatment combinations have *common unknown variance* σ^2. Recall that in Section 2.7 we stated that for single-factor experiments, there does not exist a single-stage procedure that will guarantee the probability requirement (2.1.1) when the common variance is unknown. The same conclusion applies with respect to the probability requirement (6.2.1) for multi-factor experiments.

We describe a (noneliminating) two-stage procedure that samples from all treatments at both stages. Although it has been proved that this two-stage and other sequential procedures guarantee the probability requirement (6.2.1), there have been no studies of the performance characteristics of any of these procedures. The notation in this section is the same as that given in Section 6.2. The purpose of the experiment is stated in Goal 6.2.1, and the associated probability (design) requirement is given by (6.4.1), which, the reader should notice, is required to hold for all σ^2.

Probability Requirement: For specified constants $(\delta^\star, P^\star)$ with $0 < \delta^\star < \infty$ and $1/ab < P^\star < 1$, we require that

$$P\{CS \mid (\boldsymbol{\mu}, \sigma^2)\} \geq P^\star \tag{6.4.1}$$

whenever $\boldsymbol{\mu}$ satisfies $\alpha_{[a]} - \alpha_{[a-1]} \geq \delta^\star$ and $\beta_{[b]} - \beta_{[b-1]} \geq \delta^\star$ and $\sigma^2 > 0$. (Notice that the indifference-zones for both factors must use the *same* $\delta^\star$ in order to insure the validity of the procedure to be discussed below.)

We describe a generalization of the two-stage procedure $\mathcal{N}_{BDS}$ that is due to Bechhofer and Dunnett (1986), who consider a multi-factor setting in which the factor-levels do not interact.

Procedure $\mathcal{N}_{BD2}$ For the given (a, b), specify (δ^*, P^*).

Sampling rule:

Stage 1: Take an arbitrary common number $n_1 \geq 2$ of observations from each of the ab treatments.

Stage 2: Calculate

$$s_\nu^2 = \sum_{i=1}^{a} \sum_{j=1}^{b} \sum_{s=1}^{n_1} \left(y_{ijs} - \sum_{s=1}^{n_1} y_{ijs}/n_1 \right)^2 / \nu,$$

the unbiased estimate of σ^2 based on $\nu = ab(n_1 - 1)$ d.f. Enter Table 6.6 with (a, b) (where $a \leq b$) and n_1 to obtain the constant h. Take a common number $N - n_1$ of additional observations from each of the ab treatments where $N = \max\{\lceil 2(hs_\nu/\delta^*)^2/a \rceil, n_1\}$. Calculate the $a + b$ overall row and column sums

$$A_{iN} = \sum_{j=1}^{b} \sum_{s=1}^{N} y_{ijs} \quad (1 \leq i \leq a) \quad \text{and} \quad B_{jN} = \sum_{i=1}^{a} \sum_{s=1}^{N} y_{ijs} \quad (1 \leq j \leq b)$$

based on the combined first- and second-stage observations.

Terminal decision rule: Select the treatment combination that yielded $A_{[a]N} = \max\{A_{1N}, \ldots, A_{aN}\}$ and $B_{[b]N} = \max\{B_{1N}, \ldots, B_{bN}\}$ as the one associated with $\alpha_{[a]}$ and $\beta_{[b]}$.

The h-values given in Table 6.6 are for $P^* = 0.90$. Table D of Bechhofer and Dunnett (1988) contains h-values for other choices of P^*. In Table D take $p_1 = a - 1$, $p_2 = b - 1$, $\nu = ab(n_1 - 1)$, and $1 - \alpha = P^*$ yielding $h = g_2$.

Remark 6.4.1 Because it is assumed that additivity holds, one could legitimately pool the sum of squares used in the sampling rule above with the residual sum of squares to obtain an unbiased estimate of σ^2 based on $\nu = (a - 1)(b - 1) + ab(n_1 - 1)$ d.f. However, if the no-interaction assumption is not true, this pooling can inflate the estimate of σ^2 and thus introduce an additional source of error. Therefore, we recommend the conservative choice of the unbiased estimator of σ^2 given in the procedure. In particular, Table 6.6 was determined under the assumption that $\nu = ab(n_1 - 1)$.

Table 6.6. Constants h Required to Implement Procedure $\mathcal{N}_{BD2}$ for $P^* = 0.90$

(a,b)	\multicolumn{7}{c}{n_1}						
	4	5	6	7	8	9	10
(2,2)	1.76	1.73	1.71	1.69	1.68	1.68	1.67
(2,3)	1.77	1.75	1.74	1.73	1.72	1.71	1.71
(2,4)	1.84	1.82	1.81	1.80	1.79	1.79	1.79
(2,5)	1.91	1.89	1.88	1.88	1.87	1.87	1.87
(2,6)	1.97	1.96	1.95	1.95	1.94	1.94	1.94
(3,3)	1.98	1.96	1.95	1.94	1.94	1.93	1.93
(3,4)	1.94	1.92	1.91	1.91	1.90	1.90	1.90
(3,5)	1.95	1.94	1.93	1.93	1.92	1.92	1.92
(3,6)	1.98	1.97	1.97	1.96	1.96	1.96	1.96
(4,4)	2.10	2.09	2.08	2.07	2.07	2.07	2.07
(4,5)	2.05	2.04	2.03	2.03	2.03	2.03	2.02
(4,6)	2.04	2.03	2.03	2.03	2.02	2.02	2.02

Reprinted from "Percentage points of multivariate Student t distributions," by R. E. Bechhofer and C. W. Dunnett (1988), *Selected Tables in Mathematical Statistics*, Volume 11, pp. 364–367, by permission of the American Mathematical Society.

Example 6.4.1 (Designing a Selection Experiment When σ^2 Is Unknown) This example is taken from Section 7 of Bechhofer and Dunnett (1988) and uses data from Smith (1969) concerning an experiment involving $a = 3$ catalysts and $b = 4$ reagents. The experimenter was interested in determining the combination of reagent and catalyst yielding the highest production rate; Smith assumed a no-interaction model. Initially, the experimenter used $n_1 = 2$ replicates from each of the $ab = 12$ treatment combinations, and obtained $s_\nu^2 = 7.33$. Suppose that we wish to determine the number of observations required from each treatment combination in a second stage to achieve a probability of at least $P^* = 0.90$ of simultaneously selecting the best catalyst *and* the best reagent for $\delta^* = 2.0$.

Since there is no $n_1 = 2$ column in Table 6.6, we use Table D of Bechhofer and Dunnett (1988). We take $p_1 = a - 1 = 2$, $p_2 = b - 1 = 3$, $\nu = ab(n_1 - 1) = 12$, and $1 - \alpha = 0.90$. These choices yield $h = g_2 = 2.06$. Then we calculate

$$\frac{2h^2 s_\nu^2}{a(\delta^*)^2} = \frac{2(2.06)^2(7.33)}{3(2.0)^2} = 5.18.$$

Therefore, we must take a *total* of $N = \max\{\lceil 5.18 \rceil, n_1\} = \max\{\lceil 5.18 \rceil, 2\} = 6$ observations from each treatment combination, that is, 4 *additional* observations on each treatment combination in the second stage.

Remark 6.4.2 As in the common known variance case, Monte Carlo simulations in the completely symmetric setup (that is, $a = b$) with no interaction indicate great efficiencies in the use of factorial experiments instead of multiple independent single-factor experiments. As $P^* \to 1$, the probability requirement can be guaranteed with approximately half as many total observations on the average as would be required by two independent single-factor experiments that guarantee the same probability requirement.

Confidence Statement Formulation
If procedure $\mathcal{N}_{BD2}$ is carried out at confidence level P^*, then the experimenter can assert that

$$P\{\alpha_{[a]} - \delta^* \leq \alpha_{S_1} \leq \alpha_{[a]} \text{ and } \beta_{[b]} - \delta^* \leq \beta_{S_2} \leq \beta_{[b]} \mid \boldsymbol{\mu}\} \geq P^*$$

whenever $\boldsymbol{\mu}$ satisfies $\mu_{ij} = \mu + \alpha_i + \beta_j$ for all i and j. Here α_{S_1} denotes the selected level of the A factor and β_{S_2} denotes the selected level of the B factor.

6.5 IZ SELECTION USING SPLIT-PLOT DESIGNS (COMMON UNKNOWN VARIANCE)

This section describes a procedure to select the levels of the A and B factors associated with $\alpha_{[a]}$ and $\beta_{[b]}$ (Goal 6.2.1) under the statistical assumptions of Section 6.3 when the measurement error σ_ϵ^2 is *unknown* but the ratio of the variance component for the confounding factor to the measurement error variance, $\gamma_w = \sigma_w^2/\sigma_\epsilon^2$, is *known*. We propose a single-stage procedure that selects the best treatment with prespecified probability under the following modification of the probability requirement (6.2.1).

Probability Requirement: For specified constants $(\delta_\alpha^*, \delta_\beta^*, P^*)$ $0 < \delta_\alpha^*, \delta_\beta^* < \infty$ and $1/ab < P^* < 1$, we require that

$$P\{CS \mid \boldsymbol{\mu}, \sigma_\epsilon^2\} \geq P^* \qquad (6.5.1)$$

whenever $\alpha_{[a]} - \alpha_{[a-1]} \geq \delta_\alpha^* \sigma_\epsilon$ and $\beta_{[b]} - \beta_{[b-1]} \geq \delta_\beta^* \sigma_\epsilon$.

This design requirement is analogous to power requirements for $\mathcal{F}$-tests in analysis of variance because it is stated in terms of the *number of measurement error standard deviations* that can be detected rather than in terms of the absolute differences of the main effects. [If the design requirement were stated in terms of differences of main effects, as in (2.7.3), then a procedure involving at least two stages, such as those given in Sections 2.7.1 and 2.7.2 for single-factor experiments, would be required; such a procedure has not yet been devised.]

In the statement of procedure $\mathcal{N}_{PS2}$, recall that Y_{ijk} denotes the response on the ith level of the whole-plot factor and the jth level of the split-plot factor in the kth block.

Procedure $\mathcal{N}_{PS2}$ (Indifference-Zone, Common Unknown Variance) For the given (a, b) and specified $(\delta_\alpha^*/\sigma_\epsilon, \delta_\beta^*/\sigma_\epsilon, P^*)$, determine c to satisfy Equation (6.5.2), below.

Sampling rule: Sample c blocks of data $\{Y_{ijk}\ (1 \leq i \leq a,\ 1 \leq j \leq b)\}\ (1 \leq k \leq c)$, each using a split-plot design with the A factor as the whole-plot factor.

Terminal decision rule: Select the level of the A factor associated with $\overline{Y}_{[a]\cdot\cdot}$ as best and select the level of the B factor associated with $\overline{Y}_{\cdot[b]\cdot}$ as best.

Pan and Santner (1993) showed that the LF-configuration for this procedure is the slippage configuration (6.3.3), which is the same LF-configuration as for the common known variance version of procedure $\mathcal{N}_{PS2}$. The minimal number of blocks c required to satisfy (6.5.1) is the smallest integer c satisfying

$$P\left\{U_i \leq \frac{\sqrt{c}\delta_\alpha^\star}{\rho_1}\ (1 \leq i < a)\right\} \times P\left\{V_j \leq \frac{\sqrt{c}\delta_\beta^\star}{\rho_2}\ (1 \leq j < b)\right\} \geq P^\star$$

(6.5.2)

where ρ_1, ρ_2, $\boldsymbol{U} = (U_1, \ldots, U_{a-1})$ and $\boldsymbol{V} = (V_1, \ldots, V_{b-1})$ are defined as in (6.3.5) and its subsequent discussion.

6.6 SUBSET SELECTION USING CR DESIGNS

This section proposes single-stage subset selection procedures for selecting the best treatment combination under the statistical assumptions of Section 6.2 when the normal measurement errors have *common known* or *common unknown* variance σ^2. The purpose of the experiment is stated in Goal 6.6.1, and the associated probability (design) requirement is given by Equation (6.6.1).

Goal 6.6.1 To select a subset of treatment combinations associated with $\alpha_{[a]}$ and $\beta_{[b]}$.

Thus, we seek to simultaneously find the "best" levels of both treatments. A *correct selection* (CS) is said to be made if Goal 6.6.1 is achieved.

Probability Requirement: For specified constant $P^\star$ with $1/ab < P^\star < 1$, we require that

$$P\{CS \mid (\boldsymbol{\mu}, \sigma^2)\} \geq P^\star \tag{6.6.1}$$

whenever $\boldsymbol{\mu}$ satisfies the additivity assumption (6.1.1) and $\sigma^2 > 0$.

Bechhofer and Dunnett (1987) analyzed the following two-factor generalization of the Gupta (1956, 1965) subset selection procedure $\mathcal{N}_G$. Let $\overline{y}_{[i]\cdot\cdot}$ and $\overline{y}_{\cdot[j]\cdot}$ denote the ordered $\overline{y}_{i\cdot\cdot}$ $(1 \leq i \leq a)$ and $\overline{y}_{\cdot j\cdot}$ $(1 \leq j \leq b)$, respectively.

Procedure $\mathcal{N}_{G2}$ (Completely Randomized Design) Calculate the sample means $\bar{y}_{i\cdot\cdot}$ and $\bar{y}_{\cdot j\cdot}$ ($1 \le i \le a$, $1 \le j \le b$). If σ^2 is unknown, calculate

$$s_v^2 = \sum_{i=1}^{a} \sum_{j=1}^{b} \sum_{k=1}^{n} (y_{ijk} - \bar{y}_{i\cdot\cdot} - \bar{y}_{\cdot j\cdot} + \bar{y}_{\cdot\cdot\cdot})^2 / v,$$

the unbiased (pooled) estimate of σ^2 based on $v = abn - a - b + 1$ d.f.

Case 1 (σ^2 Known): Include level i of the row factor in the selected subset if

$$\bar{y}_{i\cdot\cdot} \ge \bar{y}_{[a]\cdot\cdot} - h\sigma\sqrt{2/bn} \qquad (6.6.2)$$

and include level j of the column factor in the selected subset if

$$\bar{y}_{\cdot j\cdot} \ge \bar{y}_{\cdot[b]\cdot} - h\sigma\sqrt{2/an}. \qquad (6.6.3)$$

The constant h is selected to satisfy Equation (6.6.6).

Case 2 (σ^2 Unknown): Include level i of the row factor in the selected subset if

$$\bar{y}_{i\cdot\cdot} \ge \bar{y}_{[a]\cdot\cdot} - hs_v\sqrt{2/bn} \qquad (6.6.4)$$

and include level j of the column factor in the selected subset if

$$\bar{y}_{\cdot j\cdot} \ge \bar{y}_{\cdot[b]\cdot} - hs_v\sqrt{2/an}. \qquad (6.6.5)$$

The constant h is selected to satisfy Equation (6.6.7).

The value h required to implement (6.6.2)–(6.6.3), the σ^2 known case, is the solution of the equation

$$P\{U_i < h\,(1 \le i < a)\} \times P\{V_j < h\,(1 \le j < b)\} = P^\star \qquad (6.6.6)$$

where $(U_1, \ldots, U_{a-1})$ and $(V_1, \ldots, V_{b-1})$ have multivariate normal distributions with zero means, unit variances, and common correlation 1/2. Equation (6.6.6) can be solved by trial and error using either Table B.1 or the programs in Appendix C.

The value h required to implement (6.6.4)–(6.6.5), the σ^2 unknown case, is the solution of the equation

$$P\{U_i < h\,(1 \le i < a) \quad \text{and} \quad V_j < h\,(1 \le j < b)\} = P^\star \qquad (6.6.7)$$

where $(U_1, \ldots, U_{a-1}, V_1, \ldots, V_{b-1})$ has the multivariate student t-distribution with zero means, unit variances, $v = nab - a - b + 1$ d.f., constant correlation equal to

1/2 for all (U_i, U_j) with $i \neq j$ and for all (V_i, V_j) with $i \neq j$, and correlation equal to zero for all (U_i, V_j). This block covariance matrix for the joint distribution of (U, V) does not have the product form that the Dunnett (1989) program MVNPRD requires, nor do any of the tables presented in this book apply. However, the special tables C.1 to C.3 in Bechhofer and Dunnett (1988) solve (6.6.7) for h.

When σ^2 is unknown, a conservative choice of h can be computed that *does* use the tables or the FORTRAN programs in this book. It can be shown that

$$P\{U_i < h\, (1 \le i < a) \quad \text{and} \quad V_j < h\, (1 \le j < b)\}$$
$$\ge P\{U_i < h\, (1 \le i < a)\} \times P\{V_j < h\, (1 \le j < b)\} = P^* \quad (6.6.8)$$

where $(U_1, \ldots, U_{a-1})$ and $(V_1, \ldots, V_{b-1})$ have the marginal multivariate student t-distributions in (6.6.7). Substituting the value of h that solves the equality in (6.6.8) into (6.6.4)–(6.6.5) gives a conservative solution that satisfies (6.6.1).

Example 6.6.1 Smith (1969) describes an experiment involving four reagents (Factor A) and three catalysts (Factor B) to optimize the production of a chemical; two replications of each treatment combination were performed. The experimenter was interested in determining the treatment combination yielding the highest production rate of the chemical. Initially, Smith assumed a main effects model without interaction. The (coded) mean yields were 7, 9, 13 and 11 for the four reagents and 9, 12 and 9 for the three catalysts; the residual sum of squares for this model was $s_{18}^2 = 7.33$ based on $18\, (= 24 - 4 - 3 + 1)$ degrees of freedom.

To implement the exact procedure we determine $h = 2.475$ from Table C.3 of Bechhofer and Dunnett (1988). Thus the yardsticks of procedure $\mathcal{N}_{G2}$ are $hs_\nu\sqrt{2/bn} = 2.475 \times \sqrt{7.33} \times \sqrt{2/3(2)} = 3.87$ for selection of reagents and $hs_\nu\sqrt{2/an} = 2.475 \times \sqrt{7.33} \times \sqrt{2/4(2)} = 3.35$ for selection of catalysts. We select all reagents with means greater than

$$\bar{y}_{[4]\cdot\cdot} - hs_\nu\sqrt{2/bn} = 13 - 3.87 = 9.13$$

and we select all catalysts with means greater than

$$\bar{y}_{\cdot[3]\cdot} - hs_\nu\sqrt{2/an} = 12 - 3.35 = 8.65.$$

The selected treatment combinations have reagent means 11 and 13 and catalyst means 9 and 12.

6.7 SUBSET SELECTION USING SPLIT-PLOT DESIGNS

This section presents subset procedures for split-plot experiments. The purpose of the analysis is stated in Goal 6.6.1.

We adopt the notation and statistical assumptons of Section 6.3; in particular, y_{ijk} is the observation on the ith level of the whole-plot factor and the jth level of the split-plot factor in the kth block ($1 \leq i \leq a$, $1 \leq j \leq b$, $1 \leq k \leq c$). We also remind the reader of the notation $\gamma_w = \sigma_w^2/\sigma_\epsilon^2$ for the ratio of the confounding factor variance to the measurement error variance. As in previous sections, we base the yardsticks used to define our subset selection procedures on the variances of the differences of whole-plot and split-plot treatment sample means. In particular, it can be shown that for whole-plot treatment mean differences,

$$\text{Var}(\overline{Y}_{i_1\cdot\cdot} - \overline{Y}_{i_2\cdot\cdot}) = 2(\sigma_\epsilon^2 + b\sigma_w^2)/bc = 2\sigma_\epsilon^2(b\gamma_w + 1)/bc$$

for any $i_1 \neq i_2$ with $1 \leq i_1, i_2 \leq a$, and for split-plot treatment mean differences,

$$\text{Var}(\overline{Y}_{\cdot j_1 \cdot} - \overline{Y}_{\cdot j_2 \cdot}) = 2\sigma_\epsilon^2/ac$$

for any $j_1 \neq j_2$ with $1 \leq j_1, j_2 \leq b$. As usual for this design, we let $\overline{y}_{[1]\cdot\cdot} \leq \cdots \leq \overline{y}_{[a]\cdot\cdot}$ denote the ordered whole-plot means and $\overline{y}_{\cdot[1]\cdot} \leq \cdots \leq \overline{y}_{\cdot[b]\cdot}$ denote the ordered split-plot means.

None of the procedures in Section 6.7.1 or Section 6.7.2 depends on the block variance component σ_b^2. In Section 6.7.1 we assume that the variance ratio γ_w is known (operationally this ordinarily means that the variance component σ_w^2 is known) and the measurement error variance σ_ϵ^2 is known or unknown. Section 6.7.2 assumes that neither σ_w^2 nor σ_ϵ^2 is known.

6.7.1 Subset Selection When γ_w Is Known

The probability (design) requirement used in this section is (6.7.1).

Probability Requirement: For specified constant P^* with $1/ab < P^* < 1$, we require that

$$P\{CS \mid (\boldsymbol{\mu}, \sigma_b^2, \sigma_\epsilon^2)\} \geq P^* \qquad (6.7.1)$$

whenever $\boldsymbol{\mu}$ satisfies the additivity assumption (6.1.1) and $\sigma_b^2, \sigma_\epsilon^2 > 0$.

Santner and Pan (1994) proposed the following procedure to guarantee Equation (6.7.1).

Procedure $\mathcal{N}_{SP1}$ Suppose that c blocks have been used in a split-plot experiment with a levels of the whole-plot factor, A, and b levels of the split-plot factor, B. Calculate the whole-plot sample means $\overline{y}_{i\cdot\cdot}$ ($1 \leq i \leq a$) and the split-plot sample means $\overline{y}_{\cdot j\cdot}$ ($1 \leq j \leq b$). If σ_ϵ^2 is unknown, calculate the unbiased estimate

$$s_\epsilon^2 = \sum_{i=1}^{a}\sum_{j=1}^{b}\sum_{k=1}^{c}\left(y_{ijk} - \overline{y}_{i\cdot k} - \overline{y}_{\cdot j\cdot} + \overline{y}_{\cdot\cdot\cdot}\right)^2/\nu \qquad (6.7.2)$$

based on $\nu = (ac - 1)(b - 1)$ d.f.

Case 1 (σ_ϵ^2 Known): Include level i of the row factor in the selected subset if

$$\bar{y}_{i..} > \bar{y}_{[a]..} - h\sigma_\epsilon\sqrt{2(b\gamma_w + 1)/bc} \qquad (6.7.3)$$

and include level j of the column factor if

$$\bar{y}_{.j.} > \bar{y}_{.[b].} - h\sigma_\epsilon\sqrt{2/ac} \qquad (6.7.4)$$

where the constant h is selected to satisfy Equation (6.6.6).

Case 2 (σ_ϵ^2 Unknown): Include level i of the row factor if

$$\bar{y}_{i..} > \bar{y}_{[a]..} - hs_\epsilon\sqrt{2(b\gamma_w + 1)/bc} \qquad (6.7.5)$$

and include level j of the column factor if

$$\bar{y}_{.j.} > \bar{y}_{.[b].} - hs_\epsilon\sqrt{2/ac} \qquad (6.7.6)$$

where the constant h is selected to satisfy Equation (6.6.7) for an exact procedure or Equation (6.6.8) for a conservative procedure, in either case taking $\nu = (ac - 1)(b - 1)$ d.f.

The estimator s_ϵ^2 in Equation (6.7.2) is the usual unbiased ANOVA moment estimator of σ_ϵ^2 corresponding to the residual row for the fitted model Equation (6.3.1). When σ_ϵ^2 is *known*, the value of h required to implement procedure $\mathcal{N}_{SP1}$ is the *same* as that for corresponding two-factor case using the CR design (see Section 6.6); when σ_ϵ^2 is *unknown* only the d.f. changes from that used to determine h in the CR design.

Remark 6.7.1 The selection procedures for the split-plot factor given in (6.7.4) and (6.7.6) coincide with the selection procedures for the column factor in the completely randomized design [(6.6.3) and (6.6.5)]. When there is no whole-plot confounding effect, that is, $\sigma_w^2 = 0$, then the selection procedures for the whole-plot factor [(6.7.3) and (6.7.5)] coincide with those for the row factor in the CR design [(6.6.2) and (6.6.4)] because

$$h\sigma_\epsilon\sqrt{2(b\gamma_w + 1)/bc} = h\sigma_\epsilon\sqrt{2(0 + 1)/bc} = h\sigma_\epsilon\sqrt{2/bc}.$$

This quantity is the same yardstick as that used in (6.6.2) because σ_ϵ^2 corresponds to the measurement error in Section 6.6 and c is the number of replications of each treatment combination.

Example 6.7.1 (Screening in a Split-Plot Experiment) Beeson (1965) conducted an experiment to evaluate the performance of four prosthetic cardiac valve designs (see the detailed and more widely available account of the experiment in Anderson and McLean 1974). Beeson used two valves of each design in his experiment. Each of the eight valves was tested at the six pulse rates, 60, 80, ..., 160 beats per minute, in a tank of fluid that mimicked the circulatory system. The measured response for each valve design by pulse rate combination was the "maximum flow gradient." Suppose that it is desired to select the valve design by pulse rate combination for which the mean maximum flow gradient is greatest, based on Beeson's experimental data.

In more detail, Beeson conducted his experiment in $c = 2$ blocks by selecting a valve type at random and measuring the response for all $b = 6$ pulse rates run in a random order (using a different rate randomization for each valve). He forced all $a = 4$ valve types to be used once and then performed the entire experiment a second time based on a new randomization of the valve designs and new randomizations for pulse rates. The data are displayed in Table 6.7.

This experiment was conducted using a split-plot design in two blocks. There was a lack of randomization of the treatment combinations over time; once a valve type was selected it was used for six consecutive pulse rates. Thus valve design was the whole-plot treatment and pulse rate was the split-plot treatment. Anderson and McLean's analysis of these data showed no evidence of interaction between the valve design and pulse rate factors. We make the same assumption in our illustration of procedure $\mathcal{N}_{SP1}$ to select the best heart valve design by pulse rate combination. We take $P^* = 0.80$ and, in this section, assume that the ratio of the variance of the confounding factor to the measurement error is $\gamma_w = 0.25$.

Tables 6.8 and 6.9 list the sample means for the valves and the pulse rates, respectively. From standard statistical packages we determine that

$$s_\epsilon^2 = \sum_{i=1}^{a}\sum_{j=1}^{b}\sum_{k=1}^{c} \left(y_{ijk} - \bar{y}_{i\cdot k} - \bar{y}_{\cdot j\cdot} + \bar{y}_{\cdots}\right)^2 / \nu = 89.12 = (9.44)^2$$

with $\nu = (4 \times 2 - 1)(6 - 1) = 35$ d.f. The exact value h that solves (6.6.7) is $h = 1.85$; it is found in Table C.1 of Bechhofer and Dunnett (1988) [corresponding to $(1 - \alpha) = 0.80$ and to $(p_1, p_2) = (3, 5)$, in Bechhofer and Dunnett's notation]. Alternatively, the conservative constant satisfying (6.6.8) with $\nu = 35$ is also $h = 1.85$ to two decimal places.

Table 6.7. Maximum Flow Gradient for Beeson (1965) Two-Factor Split-Plot Experiment Involving Four Valve Designs and Six Pulse Rates

Block	Valve Design	Pulse Rate					
		60	80	100	120	140	160
1	1	27	42	38	43	39	60
	2	20	27	38	42	57	27
	3	18	20	30	27	43	38
	4	17	48	40	43	37	20
2	1	20	52	48	45	28	57
	2	27	32	38	47	52	37
	3	20	17	38	22	40	37
	4	20	10	37	25	30	33

Table 6.8. Average Maximum Flow Gradient $\bar{y}_{i..}$ for Four Valve Designs in Beeson (1965) Cardiac Valve Experiment

	Valve Design		
1	2	3	4
41.58	37.00	29.17	30.00

Table 6.9. Average Maximum Flow Gradient $\bar{y}_{.j.}$ for Six Pulse Rates in Beeson (1965) Cardiac Valve Experiment

		Pulse Rate			
60	80	100	120	140	160
21.13	31.00	38.38	36.75	40.75	38.63

Thus procedure $\mathcal{N}_{SP1}$ selects those valve designs for which

$$\bar{y}_{i..} > 41.58 - 1.85 \times 9.44 \times \sqrt{2(6 \times 0.25 + 1)/6(2)} = 41.58 - 11.27 = 30.31$$

and selects those pulse rates for which

$$\bar{y}_{.j.} > 40.75 - 1.85 \times 9.44 \times \sqrt{2/4(2)} = 40.75 - 8.73 = 32.02.$$

These criteria lead to the selection of valve designs 1 and 2 and pulse rates 100, 120, 140 and 160.

6.7.2 Subset Selection When σ_w^2 and σ_ϵ^2 Are Unknown

The probability (design) requirement used in this section is (6.7.7).

Probability Requirement: For specified constant P^* with $1/ab < P^* < 1$, we require that

$$P\{CS \mid (\boldsymbol{\mu}, \sigma_b^2, \sigma_w^2, \sigma_\epsilon^2)\} \geq P^* \quad (6.7.7)$$

whenever $\boldsymbol{\mu}$ satisfies the additivity assumption (6.1.1) and $\sigma_b^2, \sigma_w^2, \sigma_\epsilon^2 > 0$.

Santner and Pan (1994) proposed the following procedure to guarantee Equation (6.7.7).

Procedure $\mathcal{N}_{SP2}$ Suppose that c blocks have been used in a split-plot experiment with a levels of the whole-plot factor, A, and b levels of the split-plot factor, B. Calculate the whole-plot sample means $\bar{y}_{i..}$ ($1 \leq i \leq a$) and the split-plot sample means $\bar{y}_{.j.}$ ($1 \leq j \leq b$). Also calculate the estimator s_ϵ^2 of σ_ϵ^2 given by (6.7.2) and based on $\nu_\epsilon = (ac-1)(b-1)$ d.f. and the estimator of $\sigma_\epsilon^2 + b\sigma_w^2$

$$s_{\epsilon+w}^2 = b \times \sum_{i=1}^{a}\sum_{k=1}^{c}\left(\bar{y}_{i\cdot k} - \bar{y}_{i\cdot\cdot} - \bar{y}_{\cdot\cdot k} + \bar{y}_{\cdot\cdot\cdot}\right)^2 / \nu_w$$

based on $\nu_w = (a-1)(c-1)$ d.f.

Include level i of the row factor in the selected subset if

$$\bar{y}_{i\cdot\cdot} > \bar{y}_{[a]\cdot\cdot} - h_a s_{\epsilon+w}\sqrt{2/bc}$$

and include level j of the column factor if

$$\bar{y}_{\cdot j\cdot} > \bar{y}_{\cdot[b]\cdot} - h_b s_\epsilon \sqrt{2/ac}$$

where the constants h_a and h_b are chosen to satisfy Equation (6.7.8).

The estimators s_ϵ^2 and $s_{\epsilon+w}^2$ are the unbiased ANOVA moment estimators corresponding to the residual row and block by whole-plot treatment row, respectively, for the fitted model (6.3.1). The values of h_a and h_b required to implement procedure $\mathcal{N}_{SP2}$ are *any* solutions of the equation

$$P\{U_i < h_a\ (1 \leq i < a)\} \times P\{V_j < h_b\ (1 \leq j < b)\} = P^\star \qquad (6.7.8)$$

where $(U_1,\ldots,U_{a-1})$ has the $(a-1)$-dimensional multivariate t-distribution with zero mean vector, unit variances, common correlation $1/2$ and $(a-1)(c-1)$ d.f. and $(V_1,\ldots,V_{b-1})$ has the $(b-1)$-dimensional multivariate t-distribution with zero mean vector, unit variances, common correlation $1/2$ and $(ac-1)(b-1)$ d.f. Using the Dunnett (1989) quadrant probability program, two solutions h_a and h_b of Equation (6.7.8) are easily computed. One solution is obtained by selecting equal length row and column factors, $h_a = h_b$; the second solution is determined by requiring equal selection probabilities for the best row and column factors, that is,

$$P\{U_i < h_a\ (1 \leq i < a)\} = \sqrt{P^\star} = P\{V_j < h_b\ (1 \leq j < b)\}.$$

In either case the appropriate constants can be solved by trial and error.

Example 6.7.2 (Example 6.7.1 Continued) Suppose we again consider simultaneous selection of the best valve and pulse rate combination with probability at least $P^\star = 0.80$ but *without* the assumption that σ_w^2 is known. From an ANOVA corresponding to the model (6.3.1), we had previously obtained $s_\epsilon^2 = 89.12$ based on $(ac-1)(b-1) = 35$ d.f. and now also find $s_{\epsilon+w}^2 = 77.19$ based on $(a-1)(c-1) = 3$ d.f. By trial and error we find that $h_a = 2.35$ satisfies

$$P\{U_i < h_a\ (1 \leq i \leq 3)\} = \sqrt{0.80} = 0.894$$

where (U_1, U_2, U_3) has the multivariate t-distribution with 3 d.f., and $h_b = 1.94$ satisfies

$$P\{V_j < h_b \,(1 \le j \le 5)\} = \sqrt{0.80} = 0.894$$

where $(V_1, \ldots, V_5)$ has the multivariate t-distribution with 35 d.f. and both U and V have zero mean vectors, unit variances, and common correlations 1/2.

Hence we select those valve designs for which

$$\bar{y}_{i..} > 41.58 - 2.35 \times \sqrt{77.19} \times \sqrt{2/6(2)} = 41.58 - 8.42 = 33.16$$

and those pulse rates for which

$$\bar{y}_{.j.} > 40.75 - 1.94 \times \sqrt{89.12} \times \sqrt{2/4(2)} = 40.75 - 9.16 = 31.59.$$

Procedure $\mathcal{N}_{SP2}$ again selects valve designs 1 and 2 and pulse rates 100, 120, 140 and 160.

6.8 CHAPTER NOTES

Section 6.1 mentions several alternative formulations of selection, screening and multiple comparison problems that are appropriate for non-additive factorial experiments. Suppose that

$$E(Y_{ijk}) = \mu_{ij} = \mu + \alpha_i + \beta_j + (\alpha\beta)_{ij}$$

where the interaction parameters $(\alpha\beta)_{ij}$ are not all zero. Wu and Cheung (1994) presented a procedure for selecting, for each level i of the row factor, a subset of the levels of the column factor having the largest treatment mean, that is, the largest mean among $\mu_{i1}, \ldots, \mu_{ib}$; their procedure guarantees simultaneously that all subsets contain the best levels of the column factor. Santner and Pan (1994) presented the analogues of the Wu and Cheung (1994) subset procedure for split-plot experiments. Bechhofer, Santner and Turnbull (1977) proposed procedures for conducting a completely randomized experiment to select the treatment combination associated with the largest absolute interaction, $|(\alpha\beta)|_{[ab]}$ where $|(\alpha\beta)|_{[1]} \le |(\alpha\beta)|_{[2]} \le \cdots \le |(\alpha\beta)|_{[ab]}$ denote the ordered absolute interactions.

Federer and McCulloch (1984) derived simultaneous confidence intervals for comparisons of main effects with the best main effect in the split-plot model (6.3.1) or in the split-plot model having interacting factors. Specifically they provided formulas for simultaneous confidence intervals for the parameter sets $\{\alpha_i - \alpha_{[a]}\}_i$ and $\{\beta_j - \beta_{[b]}\}_j$.

In some sciences, two-factor experiments are conducted using types of randomization restricted designs other than the split-plot design. An example is the strip-plot design that is used in agricultural experiments. The strip-plot design has randomization restrictions in both row and column directions. Pan and Santner (1994) provided indifference-zone selection procedures for blocked strip-plot experiments that achieve Goal 6.2.1.

There are relatively few applications of two-factor selection procedures in the literature. Gupta and Hsu (1980) presented a case study of an application of a subset selection procedure to an interesting observational data set; they studied the motor vehicle traffic fatality rates for the 48 contiguous states and the District of Columbia over a 17-year period. While they were primarily interested in differences in the fatality rates across states, the problem of discerning temporal trends could also be of interest.

CHAPTER 7

Selecting Best Treatments in Single-Factor Bernoulli Response Experiments

7.1 INTRODUCTION

The Bernoulli (binomial) distribution has long been employed as an appropriate model for *acceptance sampling* problems where a manufacturing process produces independent items that are nondefective with probability p and defective with probability $1 - p$, and p is constant from item to item. For example, this model was used at the Bell Telephone System by Dodge and Romig (1944) for their acceptance sampling plans. The same model has been used successfully in clinical trials where p might represent the cure probability for a particular drug regimen; see Armitage (1960, 1975), Anscombe (1963) and Paulson (1969).

Sections 7.2–7.4 of this chapter discuss procedures for selecting the best treatment in a single-factor Bernoulli response experiment. The complementary screening problem using the subset approach is the subject of Section 7.5. Sections 7.2 and 7.4 adopt the indifference-zone approach, while Section 7.3 discusses both an adaptive sampling scheme for collecting data given that there is a bound on the maximum number of observations to be taken from each treatment and the corresponding terminal decision rule for selecting the best treatment. Throughout we make the following assumptions.

Statistical Assumptions: Independent random samples $X_{i1}, X_{i2}, \ldots$ ($1 \leq i \leq t$) are taken from $t \geq 2$ Bernoulli treatments $\Pi_1, \ldots, \Pi_t$, where Π_i has unknown "success" probability p_i. Here $X_{ij} = 1$ [0] if the jth observation from Π_i is a "success" ["failure"] ($1 \leq i \leq t$, $j \geq 1$).

We denote the ordered values of $p_1, \ldots, p_t$ by $p_{[1]} \leq \cdots \leq p_{[t]}$. The treatment associated with $p_{[t]}$ is referred to as the "best" treatment. Neither the values of the $p_{[s]}$ nor the pairing of the Π_i with the $p_{[s]}$ ($1 \leq i, s \leq t$) is assumed to be known. Let

x_{ij} ($1 \leq i \leq t$, $j \geq 1$) denote the observed value of X_{ij}. The indifference-zone and screening objectives are stated as Goals 7.1.1 and 7.1.2, respectively.

Goal 7.1.1 To select the treatment associated with $p_{[t]}$.

Goal 7.1.2 To select a (random-size) subset of the t treatments that contains the treatment associated with $p_{[t]}$.

A *correct selection* (CS) is said to be made for Goal 7.1.1 if the treatment associated with $p_{[t]}$ is selected. A CS is said to be made for Goal 7.1.2 if the selected subset contains the treatment associated with $p_{[t]}$.

In Sections 7.2 and 7.4 there are several intuitive ways of forming indifference-zones that embody the notion that $p_{[t]}$ and $p_{[t-1]}$ are "separated." One method is to use the *difference* between $p_{[t]}$ and $p_{[t-1]}$. An indifference-zone based on this measure requires that the difference $p_{[t]} - p_{[t-1]}$ be sufficiently large, that is,

$$p_{[t]} - p_{[t-1]} \geq \Delta^\star \tag{7.1.1}$$

where $\Delta^\star$ ($0 < \Delta^\star < 1$) is specified. Figure 7.1 illustrates this indifference-zone (as well as the accompanying preference-zone).

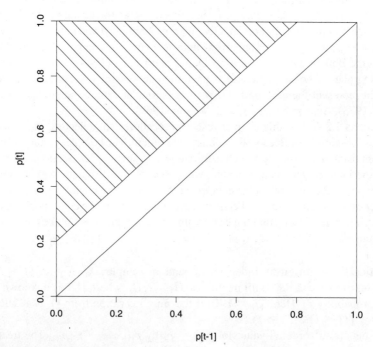

Figure 7.1. Shaded region is the portion of the ($p_{[t-1]}, p_{[t]}$) parameter space that is in the preference-zone defined by the difference (7.1.1) when $\Delta^\star = 0.2$.

INTRODUCTION

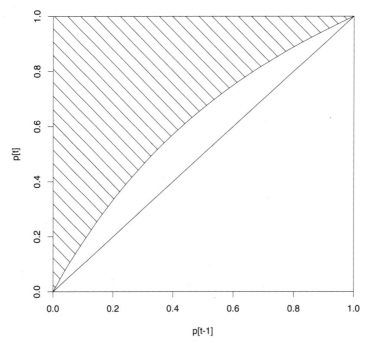

Figure 7.2. Shaded region is the portion of the $(p_{[t-1]}, p_{[t]})$ parameter space that is in the preference-zone defined by the odds ratio (7.1.2) when $\theta^\star = 2.0$.

A second measure, one that has often been used in the biomedical community, is based on the *odds of success*, $p/(1-p)$, associated with the Bernoulli success probability p. The resulting indifference-zone based on the odds of success requires that the *odds ratio* associated with $p_{[t]}$ and $p_{[t-1]}$ be sufficiently large, that is,

$$\frac{p_{[t]}(1 - p_{[t-1]})}{(1 - p_{[t]})p_{[t-1]}} \geq \theta^\star \qquad (7.1.2)$$

where $\theta^\star$ ($1 < \theta^\star < \infty$) is specified. Figure 7.2 illustrates this indifference-zone (as well as the accompanying preference-zone).

A third measure that embodies the notion that $p_{[t]}$ and $p_{[t-1]}$ are "separated" requires that the *relative risk* associated with the two best treatments be sufficiently large, that is,

$$\frac{p_{[t]}}{p_{[t-1]}} \geq \theta^\star \qquad (7.1.3)$$

where $\theta^\star$ ($1 < \theta^\star < \infty$) is specified. Figure 7.3 illustrates this method of defining an indifference-zone.

Both the odds ratio (7.1.2) and the relative risk (7.1.3) have the property that the probabilities $p_{[t]}$ and $p_{[t-1]}$ in the preference-zone (where it is desired to make a CS with high probability) can be arbitrarily close in the sense of Euclidean distance. For

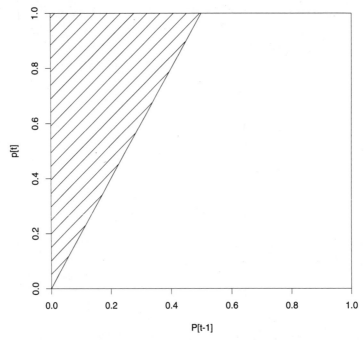

Figure 7.3. Shaded region is the portion of the $(p_{[t-1]}, p_{[t]})$ parameter space that is in the preference-zone defined by the relative risk (7.1.3) when $\theta^\star = 2.0$.

example, Figure 7.2 is a plot of $(p_{[t-1]}, p_{[t]})$ points in the preference-zone for the odds ratio (7.1.2) when $\theta^\star = 2.0$; correct selection for $(p_{[t-1]}, p_{[t]})$ pairs near $(0,0)$ or $(1,1)$ requires substantial sampling. Notice that the $(p_{[t-1]}, p_{[t]})$ parameter space includes only points *above* the line $p_{[t]} = p_{[t-1]}$ because $p_{[t]} \geq p_{[t-1]}$. As a second example, using (7.1.3) as the requirement with $\theta^\star = 2.0$, the $(p_{[t-1]}, p_{[t]})$ pairs $(0.25, 0.50)$ and $(1 \times 10^{-6}, 2 \times 10^{-6})$ are both on the boundary of the preference-zone. Clearly it will be much more difficult to make a CS when $(p_{[t-1]}, p_{[t]}) = (1 \times 10^{-6}, 2 \times 10^{-6})$ than when $(p_{[t-1]}, p_{[t]}) = (0.25, 0.50)$.

An indifference-zone defined by the difference $p_{[t]} - p_{[t-1]}$ does not have the same technical difficulty because points $(p_{[t-1]}, p_{[t]})$ in the preference-zone cannot be "close" to each other. For example, Figure 7.1 illustrates points $(p_{[t-1]}, p_{[t]})$ in the indifference-zone satisfying $p_{[t]} - p_{[t-1]} = \Delta^\star = 0.2$. In contrast to Figures 7.2 or 7.3, points $(p_{[t-1]}, p_{[t]})$ in the indifference-zone defined by (7.1.1) always remain separated.

Section 7.2 considers a fixed-sample-size selection procedure $\mathcal{B}_{SH}$ due to Sobel and Huyett (1957), who were the first to devise a single-stage procedure for the indifference-zone based on the difference (7.1.1). Section 7.3 assumes that the experimenter has a fixed upper bound n on the number of observations that they would be willing to sample from each treatment *and* that these $t \times n$ experimental units (for example, patients in a clinical trial) will become available one at a time. For each experimental unit, procedure $\mathcal{B}_{BK}$ specifies which treatment it is assigned in a

INTRODUCTION   181

manner that reduces the total number of failures experienced during the experiment subject to achieving the same $P\{CS\}$ as the Sobel and Huyett procedure $\mathcal{B}_{SH}$. Section 7.4 discusses sequential procedures where the indifference-zone is based on the odds ratio (7.1.2). In particular, we study an open noneliminating sequential selection procedure $\mathcal{B}_{BKS}$ in Section 7.4.1 and an open eliminating sequential procedure $\mathcal{B}_P$ in Section 7.4.2; we compare the performances of these procedures and make recommendations as to when to use each in Section 7.4.3. Lastly, Section 7.5 introduces a fixed-sample-size subset selection procedure $\mathcal{B}_{GS}$ due to Gupta and Sobel (1960) that selects a random size subset of the treatments containing the best treatment.

In some applications, experimenters will want to select the Bernoulli treatment having the *smallest* success probability. For example, one might be interested in determining which of a number of drugs produces the least onerous side effects. One can easily solve this problem by applying the procedures in Sections 7.2 and 7.4 to $1 - X_{ij}$ (for all i, j) (that is, interchanging "successes" and "failures") for the two types of indifference-zones discussed below. Using this interchange technique with the single-stage procedure $\mathcal{B}_{SH}$ of Section 7.2 permits one to assert that the probability of correct selection is at least P^* whenever

$$p_{[2]} - p_{[1]} \geq \Delta^\star. \tag{7.1.4}$$

This follows because the variables $1 - X_{ij}$ will then have "success" probabilities $q_i = 1 - p_i$, and hence (7.1.4) becomes

$$q_{[t]} - q_{[t-1]} \geq \Delta^\star \tag{7.1.5}$$

where $q_{[1]} \leq \cdots \leq q_{[t]}$ are the ordered q_i-values. Alternatively, the experimenter can use an indifference-zone defined by

$$\frac{p_{[2]}(1 - p_{[1]})}{(1 - p_{[2]})p_{[1]}} \geq \theta^\star \tag{7.1.6}$$

by applying the the open sequential procedure $\mathcal{B}_{BKS}$ to the variables $1 - X_{ij}$ because (7.1.6) then has "success" probabilities that satisfy

$$\frac{q_{[t]}(1 - q_{[t-1]})}{(1 - q_{[t]})q_{[t-1]}} \geq \theta^\star. \tag{7.1.7}$$

In addition, the procedure $\mathcal{B}_{GS}$ can be applied to the $1 - X_{ij}$ to give a subset selection procedure that selects a random-size subset containing the treatment associated with $p_{[1]}$ no matter what the true $\boldsymbol{p}$.

However, the reader is warned that this technique of interchanging "successes" and "failures" *does not* permit one to modify procedures for selecting the treatment associated with $p_{[t]}$ using an indifference-zone defined by the relative risk into a procedure for selecting the treatment associated with $p_{[1]}$ based on an indifference-zone defined by $p_{[2]}/p_{[1]}$.

7.2 A SINGLE-STAGE PROCEDURE FOR THE INDIFFERENCE-ZONE $p_{[t]} - p_{[t-1]} \geq \Delta^*$

This section considers a single-stage procedure that is designed to select the Bernoulli treatment having success probability $p_{[t]}$ and satisfies an indifference requirement defined in terms of the difference $p_{[t]} - p_{[t-1]}$.

Probability Requirement: For specified constants (Δ^*, P^*) with $0 < \Delta^* < 1$ and $1/t < P^* < 1$, we require

$$P\{CS\} \geq P^* \quad \text{whenever } p_{[t]} - p_{[t-1]} \geq \Delta^*. \tag{7.2.1}$$

The probability in (7.2.1) depends on the entire vector $\boldsymbol{p} = (p_1, p_2, \ldots, p_t)$ and on the common number n of independent observations taken from each of the t treatments. The constant Δ^* can be interpreted as the "smallest $p_{[t]} - p_{[t-1]}$ difference worth detecting."

7.2.1 Completely Unknown p-Values

The following *single-stage* procedure was proposed by Sobel and Huyett (1957) to guarantee (7.2.1).

Procedure $\mathcal{B}_{SH}$ For the given t and (Δ^*, P^*), find n from Table 7.1.

Sampling rule: Take a random sample of n observations X_{ij} $(1 \leq j \leq n)$ in a *single* stage from each Π_i $(1 \leq i \leq t)$.

Terminal decision rule: Calculate the t sample sums $y_{in} = \sum_{j=1}^{n} x_{ij}$. Denote the ordered values of the y_{in} $(1 \leq i \leq t)$ by $y_{[1]n} \leq \cdots \leq y_{[t]n}$. Select the treatment that yielded the largest sample sum, $y_{[t]n}$, as the one associated with $p_{[t]}$; in the case of ties, randomize.

Example 7.2.1 Suppose that an experimenter wishes to select the best of $t = 4$ treatments with probability at least $P^* = 0.95$ whenever $p_{[4]} - p_{[3]} \geq 0.10$ in probability requirement (7.2.1). Table 7.1 shows that $n = 212$ observations per treatment are required. Now suppose that, at the end of sampling, we have $y_{1,212} = 70$, $y_{2,212} = 145$, $y_{3,212} = 95$ and $y_{1,212} = 102$. Then we select Π_2 as the best treatment, and we can assert with confidence at least 0.95 that the selection is correct whenever $p_{[4]} - p_{[3]} \geq 0.10$.

One might guess that

$$p_{[1]} = p_{[t-1]} = (1 - \Delta^*)/2, \qquad p_{[t]} = (1 + \Delta^*)/2 \tag{7.2.2}$$

Table 7.1. Smallest Sample Size per Treatment Needed for Procedure $\mathcal{B}_{SH}$ to Guarantee the Probability Requirement (7.2.1) for Selecting the Binomial Treatment Having the Largest p-Value for Various Δ^*, t and P^*

t	Δ^*	P^*						
		0.60	0.75	0.80	0.85	0.90	0.95	0.99
2	0.05	14	92	142	215	329	541	1082
	0.10	4	23	36	54	83	135	270
	0.15	2	11	16	24	37	60	120
	0.20	1	6	9	14	21	34	67
	0.25	1	4	6	9	13	22	43
	0.30	1	3	4	6	9	15	29
	0.35	1	2	3	5	7	11	21
	0.40	1	2	3	4	5	9	16
	0.45	1	2	2	3	4	7	13
	0.50	1	1	2	3	4	5	10
3	0.05	79	206	273	364	498	735	1308
	0.10	20	52	69	91	125	184	327
	0.15	9	23	31	41	55	82	145
	0.20	5	13	17	23	31	46	81
	0.25	4	9	11	15	20	29	52
	0.30	3	6	8	10	14	20	35
	0.35	2	5	6	8	10	15	26
	0.40	2	4	5	6	8	11	20
	0.45	2	3	4	5	6	9	15
	0.50	2	3	3	4	5	7	12
4	0.05	134	283	359	458	601	850	1442
	0.10	34	71	90	114	150	212	360
	0.15	15	32	40	51	67	94	160
	0.20	9	18	23	29	38	53	89
	0.25	6	12	14	18	24	34	57
	0.30	4	8	10	13	17	23	39
	0.35	3	6	7	9	12	17	28
	0.40	3	5	6	7	9	13	21
	0.45	2	4	5	6	7	10	17
	0.50	2	3	4	5	6	8	13
10	0.05	314	513	606	725	890	1169	1803
	0.10	79	128	151	181	222	291	449
	0.15	35	57	67	80	98	129	198
	0.20	20	32	38	45	55	72	111
	0.25	13	20	27	29	35	46	70
	0.30	9	14	17	20	24	32	48
	0.35	7	11	13	15	18	23	35
	0.40	5	8	10	11	13	17	26
	0.45	4	6	8	9	11	14	20
	0.50	4	5	6	7	9	11	16

Reprinted from Gibbons, Olkin and Sobel (1977), pp. 425–426, by courtesy of John Wiley & Sons.

is the LF-configuration of the p_i $(1 \leq i \leq t)$ because the variance of the binomial distribution is maximized when $p = 1/2$. When $t = 2$ it can be shown that (7.2.2) is, indeed, the LF-configuration for all n (see Eaton and Gleser 1989, Section 4). However, in general for $t \geq 3$, the *true* LF-configuration depends on (t, Δ^*, n). (Example 7.2.2 shows how much the *true* LF-configuration differs from the conjectured LF-configuration for *small n* when $t = 3$.) Fortunately, for fixed (t, Δ^*), the $P\{CS\}$ calculated under (7.2.2), the conjectured LF-configuration, rapidly approaches the $P\{CS\}$ calculated under the *true* LF-configuration as $n \to \infty$. Thus (7.2.2) serves as a large-sample *approximation* to the LF-configuration of the p_i. The sample sizes in Table 7.1 are computed under the *true* LF-configuration.

Example 7.2.2 For $t = 3$ and $n = 2$, Sobel and Huyett (1957) showed that the *exact* LF-configuration is given by $p_{[1]} = p_{[2]} = p_{[3]} - \Delta^*$, where

$$p_{[3]} = \left[13\Delta^* + 12 - \sqrt{25(\Delta^*)^2 + 24\Delta^*}\right]/18.$$

In particular, when $\Delta^* = 0.1$, the LF-configuration is given by $p_{[1]} = p_{[2]} = 0.5484$ and $p_{[3]} = 0.6484$, rather than by $p_{[1]} = p_{[2]} = 0.45$ and $p_{[3]} = 0.55$ from (7.2.2).

7.2.2 An Alternative Design Specification

Sometimes the experimenter will have *a priori* reasons for believing that $p_{[t]}$ is in some particular region of $[0, 1]$ that is not "near" the limiting LF-configuration (7.2.2). This might be the situation if $p_{[t]}$ is the proportion of satisfactory items produced by a very reliable manufacturer, for example, $p_{[t]} > 0.8$. In such a case the experimenter can often guarantee a higher probability of achieving Goal 7.1.1 with a much smaller sample size than that given by Table 7.1.

In these situations, the following alternate probability requirement can be used.

Probability Requirement: For specified constants (p_L^*, p_U^*, P^*) with $0 \leq p_L^* \leq p_U^* - \Delta^* \leq 1$ and $1/t < P^* < 1$, we require

$$P\{CS\} \geq P^* \quad \text{whenever } p_{[t-1]} \leq p_L^* \text{ and } p_U^* \leq p_{[t]}. \tag{7.2.3}$$

For the sampling rule and terminal decision rule of procedure $\mathcal{B}_{SH}$, Sobel and Huyett (1957) give tables of the smallest common sample size to guarantee the probability requirement (7.2.3). Their tables are reproduced as Table 7.2 and are entered with the given t and specified constants (p_L^*, p_U^*, P^*).

Example 7.2.3 Suppose that $t = 4$ and that the experimenter adopts the alternative probability requirement (7.2.3) with $p_L^* = 0.80$, $p_U^* = 0.90$ (because it is believed that $p_{[3]} \leq 0.80$ and $p_{[4]} \geq 0.90$), and $P^* = 0.95$. How large must the common single-stage sample size n be to guarantee (7.2.3)? Table 7.2 shows that $n = 107$

A SINGLE-STAGE PROCEDURE FOR THE INDIFFERENCE-ZONE $p_{[t]} - p_{[t-1]} \geq \Delta^\star$

Table 7.2. Smallest Sample Size per Treatment Needed to Guarantee the $(p_L^\star, p_U^\star, P^\star)$ Requirement (7.2.3) for Selecting the Binomial Treatment Having the Largest p-Value for Selected $(p_L^\star, p_U^\star)$ and $P^\star$ When $t = 2, 4$

t	$P^\star$	$(p_L^\star, p_U^\star)$				
		(0.60, 0.75)	(0.80, 0.95)	(0.80, 0.90)	(0.80, 0.85)	(0.90, 0.95)
2	0.60	2	2	3	9	6
	0.75	10	6	13	53	27
	0.80	14	8	19	83	40
	0.85	21	11	28	124	60
	0.90	32	16	42	189	91
	0.95	53	25	68	312	149
	0.99	106	49	35	623	298
4	0.50	7	4	10	42	21
	0.60	14	8	18	79	39
	0.75	28	14	37	168	80
	0.80	35	18	46	211	101
	0.85	45	22	59	268	128
	0.90	59	28	77	350	171
	0.95	83	39	107	493	239
	0.99	139	65	182	831	399

Reprinted from Sobel and Huyett (1957), p. 557, by courtesy of the *Bell System Technical Journal*.

observations per treatment are required. Comparing the sample size required by this example to the 212 observations per treatment required by Example 7.2.1, we find that a substantial savings is achieved; this is a consequence of the additional information provided by the experimenter.

Optimality Property
Hall (1959) proved that procedure $\mathcal{B}_{SH}$ is *most economical* for indifference-zones defined by either (7.2.1) or (7.2.3) in the sense that there is no *single-stage* procedure requiring fewer observations per treatment that guarantees these probability requirements.

Confidence Statement Formulation
Let p_S be the p-value of the *selected* treatment. If the common sample size is chosen to satisfy (7.2.1), the experimenter can assert with confidence coefficient at least $P^\star$ that

$$p_{[t]} - \Delta^\star \leq p_S \leq p_{[t]}.$$

Curtailment
Once the choice of sample size is made, the experimenter may well be able to take fewer observations by sampling sequentially and then eliminating treatments for which the partial sample sums preclude the possibility of their being

selected as the best; this is referred to as *curtailment*. Section 7.3 discusses a closed adaptive, sequential procedure that employs curtailment and has many optimality properties.

7.3 A CLOSED ADAPTIVE SEQUENTIAL PROCEDURE

7.3.1 Introduction

This section also considers Goal 7.1.1, selection of the Bernoulli treatment having the largest "success" probability $p_{[t]}$. However, unlike the approach employed in Section 7.2, *no* probability requirement is specified. Rather, we suppose that a maximum common number, n, of observations per treatment has been determined using some criterion. For example, Section 7.2 provides two methods of determining such an n. Alternatively, economic considerations may dictate the choice of n. Under these circumstances, single-stage selection procedures can be inferior to sequential procedures for the following reason: If observations are taken sequentially from the treatments, the experimenter may learn early in the sampling process that particular treatments are noncontending for selection, and that it is a waste of resources to continue sampling from them. This can be especially striking for the Bernoulli selection problem, as demonstrated by the simple example below.

Example 7.3.1 Suppose that an experimenter has decided to take up to $n = 10$ observations from each of $t = 2$ treatments. If the experimenter were to carry out a single-stage procedure, a total of 20 observations must be collected before a selection is made. Instead, suppose that after taking six observations sequentially from each treatment, treatment Π_1 yielded six successes while treatment Π_2 yielded only one success. Clearly, treatment Π_2 cannot be selected even if Π_2 only has successes in its remaining four observations and Π_1 only has failures. Thus the experimenter can *cease* sampling and save the cost of taking the remaining eight observations required by the single-stage procedure.

The device used in Example 7.3.1 is called *weak curtailment*. Weak curtailment is characterized by termination of sampling when it is *certain that a particular treatment will be selected* over all other treatments.

This section requires a more sophisticated notation than that used in Sections 7.1 and 7.2. For ease of exposition we adopt the language of clinical trials and refer to the experimental units as patients (although the applications of the Bernoulli model are, of course, more general). For each patient ($1 \leq m \leq tn$) we will need to indicate the treatment the patient is to be assigned and the outcome of the Bernoulli experiment. For the mth patient, let S_i^m [F_i^m] denote that a success [failure] occurred using treatment Π_i ($1 \leq i \leq t$). The following example illustrates this notation.

Example 7.3.2 Suppose that $n = 1$ observation is taken from each of $t = 2$ treatments Π_1 and Π_2. The possible outcomes taking *order* into account are given by the following eight sequences.

A CLOSED ADAPTIVE SEQUENTIAL PROCEDURE

Sequence	Outcomes		Decision Using $\mathcal{B}_{SH}$
1	S_1^1	S_2^2	Randomize
2	S_1^1	F_2^2	Select Π_1
3	S_2^1	S_1^2	Randomize
4	S_2^1	F_1^2	Select Π_2
5	F_1^1	S_2^2	Select Π_2
6	F_1^1	F_2^2	Randomize
7	F_2^1	S_1^2	Select Π_1
8	F_2^1	F_1^2	Randomize

The first patient in Sequence 1 is assigned to treatment Π_1 and experiences a success; the second patient is assigned to Π_2 and also experiences a success. The first patient in Sequence 4 is assigned to treatment Π_2 and experiences a success; the second patient is assigned to Π_1 and experiences a failure.

Application of procedure $\mathcal{B}_{SH}$ would lead to the selection of Π_1 for either Sequence 2 or 7, the selection of Π_2 for either Sequence 4 or 5, and randomization between Π_1 and Π_2, each with probability 1/2, for all other sequences.

Example 7.3.3 Consider the same setup as in Example 7.3.2. Suppose now that the experimenter decides to make the selection after having taken only the *first* observation and uses the rule that selects the sampled treatment if it produces a success and selects the other (nonsampled) treatment if the sampled treatment produces a failure. Thus, the experimenter would select Π_1 [Π_2] after the first observation from Sequences 1, 2, 7 and 8 [3, 4, 5 and 6]. These are the *same* decisions as $\mathcal{B}_{SH}$ would make for Sequences 2, 4, 5 and 7; likewise, $\mathcal{B}_{SH}$ would randomize between the two treatments for the other four outcome sequences. Interestingly, it will be shown that this new rule and $\mathcal{B}_{SH}$ lead to exactly the same $P\{CS\}$, but the new rule always saves one observation (out of two) from each treatment.

The device used in Example 7.3.3 is called *strong curtailment*. Strong curtailment is characterized by termination of sampling when it is certain that a particular treatment can do no worse than *tie* for best (see Jennison 1983).

7.3.2 The Procedure

Bechhofer and Kulkarni (1982) capitalized on the above ideas and proposed a closed adaptive sequential procedure $\mathcal{B}_{BK}$ for Goal 7.1.1. They suppose that a maximum number, n, of observations per treatment has been determined using some criterion. The present section describes the procedure $\mathcal{B}_{BK}$. This procedure takes no more than n Bernoulli observations from each of the t treatments and permits elimination of noncontending treatments after each observation. The procedure takes observations *one-at-a-time* from the t treatments (instead of nt in a single stage as does procedure $\mathcal{B}_{SH}$). The sampling procedure is *adaptive* in the sense that the decision as to which treatment is to be sampled next is dictated by all of the earlier outcomes. For any given n and p-vector, procedure $\mathcal{B}_{BK}$ achieves *exactly the same* $P\{CS\}$ uniformly in p as does the corresponding single-stage Sobel–Huyett procedure $\mathcal{B}_{SH}$. Procedure $\mathcal{B}_{BK}$ has several *optimality* properties; these will be enumerated after the description of the procedure and a number of illustrative examples.

To describe procedure $\mathcal{B}_{BK}$, we must first define the following quantities. Let n_{im} denote the total number of observations taken from Π_i through the mth observation, and let w_{im} denote the total number of successes yielded by Π_i through the first m observations ($1 \leq i \leq t$, $1 \leq m \leq tn$).

Procedure $\mathcal{B}_{BK}$

Sampling rule: At the mth *observation* ($1 \leq m \leq tn - 1$), if sampling has not yet stopped, take the next observation from the treatment that has the smallest number of failures among all Π_i for which $n_{im} < n$ ($1 \leq i \leq t$). If there is a tie among such smallest-number-of-failures treatments, take the next observation from that one of them that has the largest number of successes. If there is a further tie among such smallest-number-of-failures treatments, select one of these tied treatments at random and take the next observation from it.

Stopping rule: Stop sampling at the first observation m at which there exists a treatment Π_i such that

$$w_{im} \geq w_{jm} + n - n_{jm} \quad \text{for all} \quad j \neq i \quad (1 \leq i, j \leq t). \tag{7.3.1}$$

Terminal decision rule: Suppose that there are q treatments that satisfy (7.3.1) when sampling stops. If $q = 1$, then select the treatment Π_i that satisfies (7.3.1) as the one associated with $p_{[t]}$; if $q > 1$, then randomize.

Procedure $\mathcal{B}_{BK}$ employs a so-called *least failures* sampling rule. In addition, its (strong curtailment) stopping criterion (7.3.1) has a very intuitive explanation. The left-hand side of (7.3.1) represents the current total number of successes from Π_i, while the right-hand side represents the current total number of successes from Π_j plus the total number of *potential* successes if all of the *remaining* $n - n_{j,m}$ observations from Π_j after stage m were to be successes. Hence (7.3.1) tells us to stop sampling as soon as there exists one or more treatment that has at least as many successes as the maximum possible number of successes at termination from any other treatment. In other words, sampling stops when there is a treatment Π_i for which the total number of successes can "do no worse" than *tie* for the greatest number of successes if sampling were to be completed with n observations from all treatments.

Remark 7.3.1 We point out that, regardless of the sequence observed, the stopping rule (7.3.1) would have led to stopping after one observation in Example 7.3.2.

We now give several examples to illustrate the working of procedure $\mathcal{B}_{BK}$.

Example 7.3.4 For $t = 3$ and $n = 1$, if we observe

$$\begin{array}{ccc} \Pi_1 & \Pi_2 & \Pi_3 \\ \hline & & S_3^1 \end{array} \quad \text{or} \quad \begin{array}{ccc} \Pi_1 & \Pi_2 & \Pi_3 \\ \hline F_1^1 & F_2^2 & \end{array} \quad \text{or} \quad \begin{array}{ccc} \Pi_1 & \Pi_2 & \Pi_3 \\ \hline F_1^2 & F_2^1 & \end{array}$$

stop sampling and select Π_3.

Example 7.3.5 For $t = 3$ and $n = 2$, if we observe

$$\begin{array}{ccc} \Pi_1 & \Pi_2 & \Pi_3 \\ \hline F_1^5 & S_2^1 & S_3^3 \\ & F_2^2 & F_3^4 \end{array}$$

stop sampling and randomize between Π_2 and Π_3 using probability 1/2 for each.

Example 7.3.6 For $t = 3$ and $n = 3$, if we observe

$$\begin{array}{ccc} \Pi_1 & \Pi_2 & \Pi_3 \\ \hline F_1^1 & S_2^3 & F_3^2 \\ & F_2^4 & \\ & S_2^5 & \end{array}$$

stop sampling and select Π_2.

Remark 7.3.2 Percus and Percus (1984) studied procedure $\mathcal{B}_{BK}$ for $t = 2$ and derived an exact closed-form expression for the $P\{CS\}$ and for the expected total number of observations when sampling stops.

As mentioned above, procedure $\mathcal{B}_{BK}$ guarantees exactly the *same* $P\{CS\}$ uniformly in p as does the single-stage procedure $\mathcal{B}_{SH}$. This result was proved in Bechhofer and Kulkarni (1982) and is a special case of a more general theorem proved in Jennison (1983).

7.3.3 Optimality Properties of Procedure $\mathcal{B}_{BK}$

The spirit of the optimality results stated below is that it is desirable to minimize the expected number of observations taken from the *inferior* treatments. This is the ethical consideration that occurs in the context of clinical trials or the economic one that occurs in sampling inspection settings.

In this section, the notations $\mathcal{R}^*$, $\mathcal{S}^*$ and $\mathcal{T}^*$ represent the sampling, stopping and terminal decision rules, respectively, of procedure $\mathcal{B}_{BK}$. Furthermore, the notation $\mathcal{R}$ represents an *arbitrary* sampling rule that takes no more than n observations from any one of the $t \geq 2$ treatments and which is used in conjunction with $\mathcal{S}^*$ and $\mathcal{T}^*$ of procedure $\mathcal{B}_{BK}$. We now state several optimality properties of the Bechhofer and Kulkarni sampling rule $\mathcal{R}^*$; more complete discussions are given in Bechhofer and Kulkarni (1982, 1984) and Kulkarni and Jennison (1984).

In the following, $T_{(i)}$ denotes the total number of observations taken from the treatment associated with $p_{[i]}$ ($1 \leq i \leq t$) at the time procedure $\mathcal{B}_{BK}$ terminates sampling, $T = \sum_{i=1}^{t} T_{(i)}$ is the total number of observations the procedure takes, and T^F is the *total number of failures* that have occurred (whether on the best or another treatment).

Optimality Properties When $t = 2$

a. A necessary and sufficient condition that procedure $\mathcal{B}_{BK} = (\mathcal{R}^\star, \mathcal{S}^\star, \mathcal{T}^\star)$ minimize $E\{T \mid (p_1, p_2)\}$ among all procedures $(\mathcal{R}, \mathcal{S}^\star, \mathcal{T}^\star)$ (that is, with arbitrary sampling rule and the same stopping and termination rules as $\mathcal{B}_{BK}$) is that $p_1 + p_2 \geq 1$.

b. A necessary and sufficient condition that procedure $\mathcal{B}_{BK} = (\mathcal{R}^\star, \mathcal{S}^\star, \mathcal{T}^\star)$ minimize $E\{T_{(1)} \mid (p_1, p_2)\}$ among all procedures $(\mathcal{R}, \mathcal{S}^\star, \mathcal{T}^\star)$ is that

$$p_{[2]} \geq \frac{3 - p_{[1]} - \sqrt{(3 - p_{[1]})^2 - 4}}{2}$$

(which is 81.6% of the parameter space).

c. A necessary and sufficient condition that procedure $\mathcal{B}_{BK} = (\mathcal{R}^\star, \mathcal{S}^\star, \mathcal{T}^\star)$ minimize $E\{T^F \mid (p_1, p_2)\}$ among all procedures $(\mathcal{R}, \mathcal{S}^\star, \mathcal{T}^\star)$ is

$$p_{[1]} + p_{[2]} \geq 1$$

or

$$p_{[2]} \geq \frac{2 - 4p_{[1]} + p_{[1]}^2 - (1 - p_{[1]})\sqrt{2 - 4p_{[1]} + p_{[1]}^2}}{1 - 2p_{[1]}}.$$

Optimality Properties When $t \geq 3$

a. A sufficient condition that procedure $\mathcal{B}_{BK} = (\mathcal{R}^\star, \mathcal{S}^\star, \mathcal{T}^\star)$ minimize $E\{T \mid \mathbf{p}\}$ among all procedures $(\mathcal{R}, \mathcal{S}^\star, \mathcal{T}^\star)$ is that $p_{[1]} + \sum_{i=2}^{t} p_{[i]}/(t-1) \geq 1$.

b. A sufficient condition that procedure $\mathcal{B}_{BK} = (\mathcal{R}^\star, \mathcal{S}^\star, \mathcal{T}^\star)$ minimize $\sum_{i=1}^{s} E\{T_{(i)} \mid \mathbf{p}\}$ for all s ($1 \leq s \leq t$) among all procedures $(\mathcal{R}, \mathcal{S}^\star, \mathcal{T}^\star)$ is that $p_{[1]} + p_{[2]} \geq 1$.

Monte Carlo estimates of $E\{T_{(i)}\}$ and other performance characteristics of procedure $\mathcal{B}_{BK}$ are given in Bechhofer and Frisardi (1983). Two of the most relevant tables are reproduced here as Tables 7.3 and 7.4. For $t = 5$, $n = 50$ and various $\mathbf{p}$, Table 7.3 gives estimates of the expected numbers of observations taken from the inferior treatments, $E\{T_{(i)}\}$ ($1 \leq i \leq 4$). Notice that most of the table entries are remarkably small relative to $n = 50$, particularly for the vector $(0.55, 0.65, 0.75, 0.85, 0.95)$. Because procedures $\mathcal{B}_{SH}$ and $\mathcal{B}_{BK}$ achieve *exactly the same* $P\{CS\}$ uniformly in $\mathbf{p}$ and because the former procedure requires 50 observations per treatment, the results are particularly striking.

Table 7.3. Monte Carlo Estimates of $E\{T_{(i)}\}$ ($1 \leq i \leq 4$) for Procedure $\mathcal{B}_{BK}$ When $t = 5$ and $n = 50$ for Selected p

$p = (p_{[1]}, \ldots, p_{[5]})$	$E\{T_{(1)}\}$	$E\{T_{(2)}\}$	$E\{T_{(3)}\}$	$E\{T_{(4)}\}$
(0.05, 0.15, 0.25, 0.35, 0.45)	28.56	31.94	35.99	41.96
(0.15, 0.25, 0.35, 0.45, 0.55)	25.74	29.33	34.03	39.96
(0.25, 0.35, 0.45, 0.55, 0.65)	22.33	25.90	30.61	37.57
(0.35, 0.45, 0.55, 0.65, 0.75)	18.99	22.16	27.02	34.91
(0.45, 0.55, 0.65, 0.75, 0.85)	13.26	15.86	20.40	29.50
(0.55, 0.65, 0.75, 0.85, 0.95)	5.28	6.55	9.39	17.17

Reprinted from Bechhofer and Frisardi (1983), p. 191, by courtesy of Gordon and Breach Science Publishers.

Table 7.4 gives estimates of the expected total number of observations, $E\{T\}$, for $t = 5$ and various n and p. The estimates in Table 7.4 compare very favorably to the total numbers of observations required by the single-stage procedure $\mathcal{B}_{SH}$, that is, $tn = 50, 150$ and 250 for the cases $n = 10, 30$ and 50, respectively. Note that the p-vectors $(0.45, \ldots, 0.85)$ and $(0.55, \ldots, 0.95)$ satisfy the sufficient condition (a) for the case $t \geq 3$, and hence no sampling rule can have a smaller expected total number of observations.

Summary

Very substantial savings in $E\{T\}$ can be realized if procedure $\mathcal{B}_{BK}$ is used in place of the single-stage procedure $\mathcal{B}_{SH}$ with both achieving exactly the same $P\{CS\}$; these savings increase as the p_i-values ($1 \leq i \leq t$) increase. In addition, procedure $\mathcal{B}_{BK}$ samples far more frequently from the treatments having the larger p_i-values than from the inferior treatments (that is, those having small p_i-values), thus making it particularly attractive for use in clinical trials (provided that the nature of the treatment is such that the experimenter can wait until the outcome of a trial is known before the next trial is conducted). From a practical point of view, procedure $\mathcal{B}_{BK}$ is very easy to carry out, and no special tables are needed for its implementation.

Remark 7.3.3 For $t = 2$, let $\bar{\mathcal{R}}^*$ denote the conjugate sampling rule in which $w_{jm} - n_{jm}$ and w_{im} of (7.3.1) are replaced by $-w_{jm}$ and $n_{im} - w_{im}$, respectively, for $1 \leq i, j \leq 2$. Intuitively, this rule tends to sample from treatments with high failure

Table 7.4. Monte Carlo Estimates of $E\{T\}$ for Procedure $\mathcal{B}_{BK}$ When $t = 5$ and $n = 10, 30, 50$ for Selected p

$p = (p_{[1]}, \ldots, p_{[5]})$	$n = 10$	$n = 30$	$n = 50$
(0.05, 0.15, 0.25, 0.35, 0.45)	34.48	110.28	187.58
(0.15, 0.25, 0.35, 0.45, 0.55)	31.54	104.81	178.16
(0.25, 0.35, 0.45, 0.55, 0.65)	29.57	98.54	165.27
(0.35, 0.45, 0.55, 0.65, 0.75)	26.15	89.75	151.87
(0.45, 0.55, 0.65, 0.75, 0.85)	22.09	75.04	127.76
(0.55, 0.65, 0.75, 0.85, 0.95)	17.46	54.40	87.38

Reprinted from Bechhofer and Frisardi (1983), p. 195, by courtesy of Gordon and Breach Science Publishers.

probabilities while searching for the treatment with the largest success probability. In this case, a necessary and sufficient condition that procedure $(\bar{\mathcal{R}}^\star, \mathcal{S}^\star, \mathcal{T}^\star)$ minimize $E\{T \mid (p_1, p_2)\}$ among all procedures $(\mathcal{R}, \mathcal{S}^\star, \mathcal{T}^\star)$ is that $p_1 + p_2 \leq 1$.

7.4 OPEN SEQUENTIAL PROCEDURES FOR THE ODDS RATIO INDIFFERENCE-ZONE

This section uses an indifference-zone defined by the odds ratio. As mentioned in Section 7.1, there are certain classes of problems for which the odds ratio has traditionally been used as a measure of distance. This is particularly the case in biomedical trials where p_i is the "success" probability for the ith treatment. An *open sequential* procedure must be employed for this problem since there do not exist single-stage procedures that guarantee the corresponding probability requirement, that is, we must use a procedure for which, prior to the start of experimentation, no fixed upper bound can be placed on the number of stages required to terminate sampling. This is in contrast to the probability requirement (7.2.1) for which a *fixed-sample-size* procedure can be used.

Probability Requirement: For specified constants $(\theta^\star, P^\star)$ with $1 < \theta^\star < \infty$ and $1/t < P^\star < 1$, we require

$$P\{CS\} \geq P^\star \quad \text{whenever} \quad \frac{p_{[t]}(1 - p_{[t-1]})}{(1 - p_{[t]})p_{[t-1]}} \geq \theta^\star. \quad (7.4.1)$$

We refer to the *p* satisfying

$$\frac{p_{[t]}(1 - p_{[t-1]})}{(1 - p_{[t]})p_{[t-1]}} = \theta^\star \quad \text{with } p_{[1]} = p_{[t-1]} \quad (7.4.2)$$

as a family of *least-favorable* (LF) configurations.

7.4.1 An Open Sequential Procedure without Elimination

The following open sequential procedure was proposed by Bechhofer, Kiefer and Sobel (1968) to guarantee (7.4.1).

Procedure $\mathcal{B}_{BKS}$ For the given t, specify $(\theta^\star, P^\star)$.

Sampling rule: At the mth stage of experimentation ($m \geq 1$), observe the random Bernoulli vector $(X_{1m}, \ldots, X_{tm})$.

Stopping rule: Let $y_{im} = \sum_{j=1}^{m} x_{ij}$ ($1 \leq i \leq t$) and denote the ordered y_{im}-values by $y_{[1]m} \leq \cdots \leq y_{[t]m}$. After the mth stage of experimentation, compute the statistic

$$z_m = \sum_{i=1}^{t-1} (1/\theta^\star)^{y_{[t]m} - y_{[i]m}}. \quad (7.4.3)$$

OPEN SEQUENTIAL PROCEDURES FOR THE ODDS RATIO INDIFFERENCE-ZONE   193

Stop at the first value of m (call it N) for which $z_m \leq (1 - P^*)/P^*$. Note that N is a random variable.

Terminal decision rule: Select the treatment that yielded the largest sample sum, $y_{[t]N}$, as the one associated with $p_{[t]}$; in the case of ties, randomize.

We illustrate the application of procedure $\mathcal{B}_{BKS}$ with the following (artificial) example.

Example 7.4.1 For $t = 3$ and $(\theta^*, P^*) = (2, 0.75)$, suppose that the following sequence of vector-observations is obtained using procedure $\mathcal{B}_{BKS}$:

m	x_{1m}	x_{2m}	x_{3m}	y_{1m}	y_{2m}	y_{3m}	$y_{[3]m} - y_{[2]m}$	$y_{[3]m} - y_{[1]m}$	z_m
1	1	0	1	1	0	1	0	1	1.5
2	0	1	1	1	1	2	1	1	1.0
3	0	1	1	1	2	3	1	2	0.75
4	0	0	1	1	2	4	2	3	0.375
5	1	1	1	2	3	5	2	3	0.375
6	1	0	1	3	3	6	3	3	0.25

Because $z_6 \leq (1 - P^*)/P^* = 1/3$, sampling stops at stage $N = 6$ and treatment Π_3 is selected as best.

Remark 7.4.1 The procedure $\mathcal{B}_{BKS}$ and the normal means procedure $\mathcal{N}_{BKS}$ of Section 2.3.2 have three similarities and one dissimilarity that we wish to emphasize. First, both procedures base stopping on large differences between the largest and the remaining cumulative sample sums. Second, both sample until the procedure can distinguish the best treatment with probability at least P^*. Third, both procedures require sampling from *all* treatments at *every* stage, that is, no treatments can be eliminated prior to stopping the experiment. However, the procedures differ greatly with respect to their stopping characteristics. While both procedures achieve $P\{CS\} \geq P^*$ for all LF-configurations, the procedure $\mathcal{N}_{BKS}$ takes approximately the same expected number of observations for all μ that are least favorable. In contrast, the expected number of observations that procedure $\mathcal{B}_{BKS}$ takes varies over the boundary of LF-configurations (see Example 7.4.2). The reason is that there are LF-configurations for $\mathcal{B}_{BKS}$ that can have arbitrarily small differences $p_{[t]} - p_{[t-1]}$ (as described in Section 7.1). In this case, the procedure must sample longer to attain $P\{CS\} \geq P^*$.

Example 7.4.2 For $t = 3$ and $(\theta^*, P^*) = (2.0, 0.90)$, Equation (7.4.2) says that any p satisfying

$$p_{[3]}(1 - p_{[1]})/(1 - p_{[3]})p_{[1]} = p_{[3]}(1 - p_{[2]})/(1 - p_{[3]})p_{[2]} = 2.0$$

Table 7.5. Estimated Achieved $P\{CS\}$ and $E\{N\}$ for Procedure $\mathcal{B}_{BKS}$ for Five Odds Ratio LF-Configurations, Each with $(\theta^\star, P^\star) = (2, 0.90)$

$p_{[1]}$	$p_{[2]}$	$p_{[3]}$	$\hat{p}$	$\hat{N}$
2/3	2/3	0.80	0.9266	37.0
			(0.0001)	(0.2)
17/23	17/23	0.85	0.9264	44.6
			(0.0001)	(0.3)
9/11	9/11	0.90	0.9262	60.6
			(0.0001)	(0.4)
19/21	19/21	0.95	0.9262	109.3
			(0.0001)	(0.7)
97/103	97/103	0.97	0.9264	176.8
			(0.0001)	(1.1)

is an LF-configuration. Table 7.5 lists Monte Carlo (MC) sampling results for procedure $\mathcal{B}_{BKS}$. The $P\{CS\}$ and expected number of stages to terminate sampling, $E\{N\}$, were estimated for five different LF-configurations $\boldsymbol{p}$. The notations $\hat{p}$ and $\hat{N}$ represent estimated values for $P\{CS\}$ and $E\{N\}$, respectively; the number in parentheses below a table entry is the estimated standard error of that entry (using the methods in Bechhofer, Kiefer and Sobel 1968). The MC results show how $E\{N\}$ depends on the unknown $\boldsymbol{p}$-vector even if $\boldsymbol{p}$ is in the LF-configuration. In fact, we see that although the achieved $P\{CS\}$ is approximately constant (and conservative), $E\{N\}$ increases as $p_{[3]} \to 1$.

In general, $E\{N\} \to \infty$ as $p_{[1]} \to 0$ or $p_{[t]} \to 1$ for $\boldsymbol{p}$ in the LF-configuration because the procedure cannot distinguish between the treatments for which $p_i \le p_{[t]}$ or $p_{[1]} \le p_i$. It is for this reason that an *open sequential* procedure is required in order to guarantee the probability requirement (7.2.3). Of course, if the experimenter has prior knowledge that $0 < p_L^\star \le p_{[1]}$ and $p_{[t]} \le p_U^\star < 1$ where $p_L^\star$ and $p_U^\star$ are known, then $\max_{\boldsymbol{p}} E\{N\}$ is finite and, in principle, can be calculated *before* experimentation. It is also possible to devise a single-stage procedure that achieves $P\{CS\} \ge P^\star$ under these circumstances, that is, whenever the odds ratio is at least $\theta^\star$ and $0 < p_L^\star \le p_{[1]}$ and $p_{[t]} \le p_U^\star < 1$.

Blocking

In most applications of the Bernoulli response model, blocking is called "stratification." For example, this is the case in the analysis of retrospective studies as described in the epidemiological literature. We will adopt this convention for the remainder of this subsection.

Procedure $\mathcal{B}_{BKS}$ can be used when the Bernoulli observations $(X_{1j}, \ldots, X_{tj})$ taken at the jth block of the experiment are independent with success probabilities $(p_{1j}, \ldots, p_{tj})$ that satisfy

$$\ln\left(\frac{p_{ij}}{1 - p_{ij}}\right) = \tau + \xi_i + \eta_j \tag{7.4.4}$$

($1 \leq i \leq t, 1 \leq j \leq b$) where b is the number of blocks sampled (see Bechhofer and Goldsman 1988a, Section 4.2.)

For example, in matched-pair case–control studies, the observed quantity X_{ij} is the occurrence (1) or not (0) of exposure to a given substance for the ith subject in the jth matched pair. Specifically, p_{1j} could represent the probability of exposure to a suspected etiological agent given that the subject is diseased (a case), and p_{2j} is the probability of exposure given that the subject is not diseased (a control). Model (7.4.4) states that there is a *common* log odds ratio of exposure for cases to controls in each stratum but that the baseline incidence of disease can vary across strata. It can be easily calculated that the value of the common log odds ratio is $\xi_1 - \xi_2$.

Another example of the use of model (7.4.4), proposed by Rasch (1960, 1980) and called the *Rasch model*, occurs in item analysis. The model is used to analyze the set of answers of t students who take a test consisting of b questions. The observed quantity X_{ij} is unity if the ith student correctly answers the jth question on the test. The "stratum" effect η_j is related to the difficulty of the jth question.

In the general case of (7.4.4), the *same* treatment is associated with the ith ordered probability in all strata, namely, it is the treatment associated with $\xi_{[i]}$ where $\xi_{[1]} \leq \cdots \leq \xi_{[t]}$ denote the ordered "treatment" effects. If $p_{[1]j} \leq \cdots \leq p_{[t]j}$ denote the ordered probabilities in the jth stratum, then the log odds ratio

$$\ln\left(\frac{p_{[t]j}(1 - p_{[t-1]j})}{(1 - p_{[t]j})p_{[t-1]j}}\right) = \xi_{[t]} - \xi_{[t-1]}$$

is *independent* of τ or the strata effects. If procedure $\mathcal{B}_{BKS}$ is used for data satisfying (7.4.4), then the procedure will guarantee the probability requirement (7.4.1).

7.4.2 An Open Sequential Procedure with Elimination

We now describe a sequential procedure, $\mathcal{B}_P$, due to Paulson (1993) that also guarantees the odds ratio indifference-zone probability requirement defined by (7.4.1). However, procedure $\mathcal{B}_P$ is *eliminating*, unlike procedure $\mathcal{B}_{BKS}$ which is *noneliminating*. Procedure $\mathcal{B}_P$ is the Bernoulli counterpart of Paulson's normal procedure $\mathcal{N}_P$ that was described in Section 2.3.3. The performances of procedures $\mathcal{B}_{BKS}$ and $\mathcal{B}_P$ are compared in Section 7.4.3, and recommendations are given as to when each should be used.

Procedure $\mathcal{B}_P$ For the given t, specify $(\theta^\star, P^\star)$.

Sampling rule: At the mth stage of experimentation ($m \geq 1$), observe the random Bernoulli vector ($X_{im} : i \in R_m$); here x_{im} is the mth observation from Π_i ($i \in R_m$), and R_m is the set of treatments that have not yet been eliminated.

Stopping rule: Let $y_{im} = \sum_{r=1}^{m} x_{ir}$ ($1 \leq i \leq t, m \geq 1$). Furthermore, let e_i ($1 \leq i \leq t$) denote the stage at which treatment i was eliminated, and define $n_{im} = \min\{m, e_i\}$ as the number of observations taken from treatment Π_i through the mth stage. For treatments Π_i still in contention, let

$$g_i(m) = \sum_{\substack{j=1 \\ j \neq i}}^{t} (\theta^\star)^{y_{j,n_{jm}} - y_{i,n_{jm}}} \qquad (i \in R_m).$$

After the mth stage of experimentation, eliminate from further consideration and sampling any remaining treatment Π_i for which

$$g_i(m) > \frac{t-1}{1-P^\star}.$$

Stop at the first value of m for which only one treatment remains.

Terminal decision rule: Select the remaining treatment as the one associated with $p_{[t]}$.

Notice that once a treatment Π_i is eliminated, $n_{im} = e_i$ remains constant for subsequent stages m. The intuition behind the procedure is that at each stage m, procedure $\mathcal{B}_P$ compares the total number of successes from Π_i until time n_{jm} ($y_{i,n_{jm}}$) with the total number of successes from each of the other treatments ($y_{j,n_{jm}}$ for $j \neq i$); it eliminates Π_i once there is sufficient cumulative evidence against the possibility that Π_i is best. We illustrate the application of procedure $\mathcal{B}_P$ with the following (artificial) example:

Example 7.4.3 For $t = 3$, $(\theta^\star, P^\star) = (2, 0.75)$, suppose that the sequence of vector-observations in Table 7.6 is obtained using procedure $\mathcal{B}_P$. We see that Π_1 is eliminated after Stage 3, for $g_1(3) = 12 > 8 = (t-1)/(1-P^\star)$. Similarly, Π_2 is eliminated after Stage 5, because $g_2(5) = 8.25 > 8$; at that point, Π_3 is the only remaining treatment, and it is declared the winner. The entire procedure required 5 stages and a total of 13 observations.

7.4.3 Comparison of Procedures $\mathcal{B}_{BKS}$ and $\mathcal{B}_P$

Criteria for Comparisons
Sections 7.4.1 and 7.4.2 described two procedures ($\mathcal{B}_{BKS}$ and $\mathcal{B}_P$), both of which require the same statistical assumptions, achieve Goal 7.1.1, and guarantee the same probability requirement (7.4.1). Procedure $\mathcal{B}_P$ permits elimination of treatments, whereas procedure $\mathcal{B}_{BKS}$ does not.

We carried out extensive Monte Carlo sampling experiments to compare the performance characteristics of the two procedures with respect to their $P\{CS\}$, $E\{N\} = E\{$Number of *stages* to terminate sampling$\}$, and $E\{T\} = E\{$Total number of *observations* to terminate sampling$\}$ for two configurations of means: the LF-configuration, given by (7.4.2), and the equal-probability (EP) configuration,

Table 7.6. Calculations to Implement Procedure $\mathcal{B}_P$ in Example 7.4.3.

	Stage				
m	1	2	3	4	5
x_{1m}	0	0	0	—	—
x_{2m}	0	1	1	0	0
x_{3m}	1	1	1	1	1
y_{1m}	0	0	0	—	—
y_{2m}	0	1	2	2	2
y_{3m}	1	2	3	4	5
n_{1m}	1	2	3	3	3
n_{2m}	1	2	3	4	5
n_{3m}	1	2	3	4	5
$(\theta^\star)^{y_{2,n_{2m}} - y_{1,n_{2m}}}$	1	2	4	—	—
$(\theta^\star)^{y_{3,n_{3m}} - y_{1,n_{3m}}}$	2	4	8	—	—
$(\theta^\star)^{y_{1,n_{1m}} - y_{2,n_{1m}}}$	1	0.5	0.25	0.25	0.25
$(\theta^\star)^{y_{3,n_{3m}} - y_{2,n_{3m}}}$	2	2	2	4	8
$(\theta^\star)^{y_{1,n_{1m}} - y_{3,n_{1m}}}$	0.5	0.25	0.125	0.125	0.125
$(\theta^\star)^{y_{2,n_{2m}} - y_{3,n_{2m}}}$	0.5	0.5	0.5	0.25	0.125
$g_1(m)$	3	6	12	—	—
$g_2(m)$	3	2.5	2.25	4.25	8.25
$g_3(m)$	1	0.75	0.625	0.375	0.25

$p_{[1]} = p_{[t]}$. Tables 7.7 and 7.8 present results for the cases $t = 3$ and 5, respectively, for $\theta^\star = 2$, $P^\star = 0.75$, 0.90, and 0.95, and a *sequence* of odds ratio LF-configurations (7.4.2) in which $p_{[t]} = 0.80, 0.85, 0.90$ and 0.95. In addition, we estimated $E\{N\}$ and $E\{T\}$ for both procedures under a sequence of EP-configurations in which $p_{[1]} = p_{[t]} = 0.50, 0.80$ and 0.90. These results are reported in Table 7.9. In each of the tables, the notations $\hat{p}$, $\hat{N}$ and $\hat{T}$ represent estimated values for $P\{CS\}$, $E\{N\}$ and $E\{T\}$, respectively; the number in parentheses below a table entry is the estimated standard error of that entry.

Summary of Comparisons

We remind the reader that $E\{N\}$ can often be equated to the *duration* of a study, while $E\{T\}$ may be more relevant if the cost of obtaining single observations is very high. (See the Criteria for Comparisons in Section 2.4 for relevant comments concerning the normal means selection problem.) Our general conclusions are as follows.

- If the main intent of the experiment is to minimize $E\{N\}$ while guaranteeing (7.4.1), then procedure $\mathcal{B}_{BKS}$ should be used.
- If the main intent of the experiment is to minimize $E\{T\}$ while guaranteeing (7.4.1), then procedure $\mathcal{B}_P$ should be used. As an added bonus, procedure $\mathcal{B}_P$ often achieves a higher $P\{CS \mid LF\}$ than does procedure $\mathcal{B}_{BKS}$.

Table 7.7. Estimated Achieved $P\{CS\}$, $E\{N\}$ and $E\{T\}$ Using the Odds Ratio LF-Configuration (7.4.2) for Procedures $\mathcal{B}_{BKS}$ and $\mathcal{B}_P$ When $t = 3$ and $\theta^* = 2$

P^*	$p_{[1]} = p_{[2]}$	$p_{[3]}$	Procedure $\mathcal{B}_{BKS}$			Procedure $\mathcal{B}_P$		
			$\hat{p}$	$\hat{N}$	$\hat{T}$	$\hat{p}$	$\hat{N}$	$\hat{T}$
0.75	2/3	0.80	0.7854 (0.0002)	18.2 (0.1)	54.7 (0.3)	0.8145 (0.0039)	22.7 (0.1)	55.6 (0.3)
	17/23	0.85	0.7838 (0.0002)	21.9 (0.1)	65.6 (0.4)	0.8193 (0.0038)	27.3 (0.2)	66.8 (0.4)
	9/11	0.90	0.7823 (0.0002)	29.8 (0.2)	89.3 (0.6)	0.8227 (0.0038)	37.9 (0.2)	92.0 (0.5)
	19/21	0.95	0.7808 (0.0002)	54.8 (0.3)	164.4 (1.0)	0.8227 (0.0038)	67.9 (0.4)	165.1 (1.0)
0.90	2/3	0.80	0.9266 (0.0001)	37.0 (0.2)	111.0 (0.7)	0.9391 (0.0024)	44.3 (0.3)	108.6 (0.6)
	17/23	0.85	0.9264 (0.0001)	44.6 (0.3)	133.8 (0.8)	0.9399 (0.0024)	53.7 (0.3)	130.7 (0.7)
	9/11	0.90	0.9262 (0.0001)	60.6 (0.4)	181.7 (1.1)	0.9381 (0.0024)	72.4 (0.4)	176.1 (1.0)
	19/21	0.95	0.9262 (0.0001)	109.3 (0.7)	328.0 (2.0)	0.9373 (0.0024)	134.6 (0.8)	324.9 (1.8)
0.95	2/3	0.80	0.9626 (<0.0001)	48.0 (0.3)	143.9 (0.9)	0.9659 (0.0018)	54.2 (0.3)	134.3 (0.7)
	17/23	0.85	0.9625 (<0.0001)	57.3 (0.3)	172.0 (1.0)	0.9683 (0.0018)	66.1 (0.4)	162.9 (0.9)
	9/11	0.90	0.9624 (<0.0001)	77.6 (0.5)	232.9 (1.4)	0.9689 (0.0017)	89.8 (0.5)	220.0 (1.2)
	19/21	0.95	0.9624 (<0.0001)	141.9 (0.8)	425.7 (2.5)	0.9677 (0.0018)	164.7 (1.0)	402.3 (2.1)

Details

No results have been reported heretofore in the literature for either procedure $\mathcal{B}_{BKS}$ or $\mathcal{B}_P$. All table entries are based on 10,000 independent experiments; the results for each entry are independent of all other entries. The number in parentheses below an entry is the estimated standard error of that entry. We also conducted MC sampling for other values of t and $p_{[t]}$, but these results are not reported here.

We focus first on the panel of Table 7.7 for which the specification is $t = 3$, $P^* = 0.75$, $\theta^* = 2$ and with true probabilities being LF-points $(p_{[1]} = p_{[2]}, p_{[3]})$ on the border of the preference-zone of Figure 7.2. We first observe that the achieved $P\{CS\}$ remains stable as $p_{[3]}$ increases; the differences between their values are *not* due to sampling error, but rather arise from the discreteness of the Bernoulli distribution. Second, the achieved $P\{CS\}$ for procedure $\mathcal{B}_P$ is greater than that of procedure $\mathcal{B}_{BKS}$ showing that procedure $\mathcal{B}_P$ provides greater protection against an incorrect selection than does procedure $\mathcal{B}_{BKS}$. Third, both $E\{N\}$ and $E\{T\}$ increase as $p_{[3]}$ increases for both procedures. However, $E\{N\}$ for procedure $\mathcal{B}_{BKS}$ is *less* than $E\{N\}$ for procedure

Table 7.8. Estimated Achieved $P\{CS\}$, $E\{N\}$ and $E\{T\}$ Using the Odds Ratio LF-Configuration (7.4.2) for Procedures $\mathcal{B}_{BKS}$ and $\mathcal{B}_P$ When $t = 5$ and $\theta^* = 2$

P^*	$p_{[1]} = p_{[4]}$	$p_{[5]}$	Procedure $\mathcal{B}_{BKS}$			Procedure $\mathcal{B}_P$		
			$\hat{p}$	$\hat{N}$	$\hat{T}$	$\hat{p}$	$\hat{N}$	$\hat{T}$
0.75	2/3	0.80	0.7812 (0.0003)	29.9 (0.2)	149.6 (0.8)	0.8104 (0.0039)	41.2 (0.2)	125.8 (0.6)
	17/23	0.85	0.7797 (0.0002)	36.0 (0.2)	179.8 (0.9)	0.8197 (0.0038)	50.0 (0.3)	151.5 (0.7)
	9/11	0.90	0.7773 (0.0002)	49.1 (0.3)	245.4 (1.3)	0.8151 (0.0039)	67.3 (0.4)	203.6 (0.9)
	19/21	0.95	0.7750 (0.0002)	89.7 (0.5)	448.4 (2.3)	0.8108 (0.0039)	123.4 (0.7)	370.1 (1.6)
0.90	2/3	0.80	0.9194 (0.0001)	50.5 (0.3)	252.4 (1.4)	0.9288 (0.0026)	65.7 (0.4)	208.7 (0.9)
	17/23	0.85	0.9190 (0.0001)	60.9 (0.3)	304.5 (1.6)	0.9291 (0.0026)	79.7 (0.4)	251.1 (1.1)
	9/11	0.90	0.9186 (0.0001)	83.3 (0.4)	416.6 (2.2)	0.9314 (0.0025)	108.7 (0.6)	340.6 (1.4)
	19/21	0.95	0.9180 (0.0001)	150.7 (0.8)	753.4 (3.9)	0.9282 (0.0026)	197.2 (1.0)	612.8 (2.6)
0.95	2/3	0.80	0.9603 (<0.0001)	62.5 (0.3)	312.5 (1.6)	0.9652 (0.0018)	77.7 (0.4)	259.6 (1.1)
	17/23	0.85	0.9600 (<0.0001)	76.7 (0.4)	383.7 (2.0)	0.9617 (0.0019)	94.4 (0.5)	313.1 (1.3)
	9/11	0.90	0.9596 (<0.0001)	103.0 (0.5)	514.9 (2.6)	0.9661 (0.0018)	129.4 (0.7)	424.5 (1.8)
	19/21	0.95	0.9594 (<0.0001)	188.0 (0.9)	940.1 (4.7)	0.9630 (0.0019)	234.3 (1.2)	764.3 (3.1)

$\mathcal{B}_P$ for *all* $p_{[3]}$, while the estimated $E\{T\}$ for procedure $\mathcal{B}_{BKS}$ is approximately the same as for procedure $\mathcal{B}_P$.

As we move from the panel for $P^* = 0.75$ to the panels for $P^* = 0.90$ and 0.95, $E\{N\}$ for procedure $\mathcal{B}_{BKS}$ is still less than that for procedure $\mathcal{B}_P$ (the differences *increasing* with increasing P^*) but there is a reversal in the comparison with respect to $E\{T\}$. Now the estimated $E\{T\}$ for procedure $\mathcal{B}_{BKS}$ is *greater* than the corresponding quantity for procedure $\mathcal{B}_P$ (the differences *increasing* with increasing P^*).

As we study the corresponding panels in Table 7.8 for $t = 5$, we find that all of the above-mentioned changes are greatly magnified: Specifically, the estimated $E\{N\}$ for procedure $\mathcal{B}_P$ is *much greater* that that for procedure $\mathcal{B}_{BKS}$, while the estimated $E\{T\}$ for procedure $\mathcal{B}_P$ is *much less* that that for procedure $\mathcal{B}_{BKS}$. Also, the achieved $P\{CS\}$ is *greater* for procedure $\mathcal{B}_P$ than for procedure $\mathcal{B}_{BKS}$.

The entries in Table 7.9 study the stopping abilities of procedures $\mathcal{B}_{BKS}$ and $\mathcal{B}_P$ in worst-case scenarios, that is, in the EP-configuration $p_{[1]} = p_{[t]}$. These values exhibit the same general relationship between $\mathcal{B}_{BKS}$ and $\mathcal{B}_P$ as do those in Tables 7.7 and 7.8 for the LF-configuration. Furthermore, $\min E\{N \mid \boldsymbol{p}\}$ and $\min E\{T \mid \boldsymbol{p}\}$ for both proce-

Table 7.9. Estimated Achieved $E\{N\}$ and $E\{T\}$ Using the EP-Configuration for Procedures $\mathcal{B}_{BKS}$ and $\mathcal{B}_P$ When $t = 3$ and 5 and $\theta^\star = 2$

		$t = 3$				$t = 5$			
		Procedure $\mathcal{B}_{BKS}$		Procedure $\mathcal{B}_P$		Procedure $\mathcal{B}_{BKS}$		Procedure $\mathcal{B}_P$	
$P^\star$	$p_{[1]} = p_{[t]}$	$\hat{N}$	$\hat{T}$	$\hat{N}$	$\hat{T}$	$\hat{N}$	$\hat{T}$	$\hat{N}$	$\hat{T}$
0.75	0.50	18.2 (0.1)	54.6 (0.4)	22.0 (0.2)	53.8 (0.3)	32.2 (0.2)	160.8 (1.0)	44.6 (0.3)	130.4 (0.7)
	0.80	29.5 (0.2)	88.4 (0.6)	36.5 (0.2)	88.3 (0.5)	54.5 (0.3)	272.4 (1.6)	75.1 (0.4)	217.1 (1.0)
	0.90	53.0 (0.4)	158.9 (1.1)	67.4 (0.5)	162.3 (1.0)	100.2 (0.6)	500.8 (2.9)	135.7 (0.8)	391.8 (1.9)
0.90	0.50	47.9 (0.3)	143.7 (1.0)	60.5 (0.4)	143.1 (0.9)	70.3 (0.5)	351.5 (2.3)	98.3 (0.6)	282.0 (1.5)
	0.80	78.1 (0.6)	234.3 (1.7)	96.2 (0.7)	226.1 (1.4)	118.3 (0.8)	591.6 (3.8)	159.6 (1.0)	455.4 (2.2)
	0.90	140.3 (1.0)	420.9 (3.0)	173.3 (1.2)	406.5 (2.5)	212.9 (1.3)	1064.4 (6.7)	287.3 (1.8)	818.3 (4.1)
0.95	0.50	73.0 (0.5)	219.0 (1.6)	85.7 (0.6)	202.5 (1.3)	103.8 (0.7)	518.9 (3.5)	135.6 (0.9)	396.4 (2.0)
	0.80	118.2 (0.9)	354.5 (2.6)	137.4 (1.0)	322.8 (2.1)	174.3 (1.1)	871.7 (5.6)	220.1 (1.4)	641.7 (3.2)
	0.90	214.0 (1.5)	642.0 (4.6)	249.3 (1.7)	584.0 (3.7)	313.3 (2.0)	1566.3 (9.9)	395.5 (2.4)	1153.3 (5.7)

dures occur at the particular EP-configuration $p_{[1]} = 1/2 = p_{[t]}$, while $\max E\{N \mid \boldsymbol{p}\}$ and $\max E\{T \mid \boldsymbol{p}\}$ occur when $\max\{p_1, \ldots, p_t\} \to 0$ or $\min\{p_1, \ldots, p_t\} \to 1$. Recall that it is for this reason that an open sequential procedure is required in order to guarantee the probability requirement (7.4.1).

7.5 A SINGLE-STAGE SUBSET SELECTION PROCEDURE

This section presents procedures that are designed to select a subset of treatments containing the Bernoulli treatment associated with $p_{[t]}$ (Goal 7.1.2). We desire a procedure that satisfies the following probability requirement while yielding subsets that are as small as possible. The statistical assumptions are the same as in Section 7.1.

Probability Requirement: For specified constant $P^\star$ ($1/t < P^\star < 1$), we require

$$P\{CS\} \geq P^\star \quad \text{for all } \boldsymbol{p} = (p_1, p_2, \ldots, p_t). \quad (7.5.1)$$

Here CS denotes the event of selecting a subset containing the treatment associated with $p_{[t]}$.

The probability on the left-hand side of (7.5.1) depends on the entire vector p and on the common number n of Bernoulli trials taken from each of the t treatments. The requirement (7.5.1) *does not* use a measure of distance as does the indifference-zone requirement (7.2.1). The subset selection approach can be used as an *analysis* tool even if the sample size was originally chosen by some auxiliary criterion such as budgetary considerations. The "cost" of using too "small" a sample size is that the number of treatments in the selected subset will tend to be large if $p_{[t]}$ is not sufficiently separated from the remaining treatment probabilities $p_{[1]}, \ldots, p_{[t-1]}$.

7.5.1 Screening in Balanced Experiments

Gupta and Sobel (1960) proposed the following single-stage procedure for this subset selection problem (see also Gupta 1965).

Procedure $\mathcal{B}_{GS}$ Calculate the t sample sums $y_i = \sum_{j=1}^{n} x_{ij}$. Let $y_{[1]} \leq \cdots \leq y_{[t]}$ denote the ordered sums.

Include Π_i in the selected subset if and only if

$$y_i \geq y_{[t]} - d \qquad (7.5.2)$$

where $d = d(n, t, P^*)$ is from Table 7.10.

Table 7.10 lists the constants d to implement procedure $\mathcal{B}_{GS}$ for $P^* = 0.75, 0.90, 0.95, 0.99$, $t = 2(1)20$ and selected n; d is the minimum value required to achieve (7.5.1). In the same way as for procedure $\mathcal{B}_{SH}$, the minimum of the $P\{CS\}$ over p can be expressed in closed form only for $t = 2$. When $t = 2$ the minimum occurs when $p_1 = p_2 = 1/2$; when $t > 2$ the minimizing d occurs when $p_{[1]} = p_{[t]}$, but the exact common value where the minimum of the $P\{CS\}$ occurs must be determined by a numerical search over those p for which $0 < p_{[1]} = p_{[t]} < 1$ (however, the point $p_{[1]} = 1/2 = p_{[t]}$ is the minimizer as $n \to \infty$). Table 7.10 is computed using the exact small-sample binomial probability expressions for $n \leq 10$ and a normal approximation [see Equation (7.5.6)] for $n > 10$.

Example 7.5.1 Suppose that a random sample of $n = 40$ observations is taken from each of $t = 6$ Bernoulli treatments, and that $P^* = 0.90$ is specified. From Table 7.10 we find that $d = d(40, 6, 0.90) = 9$. If $y_1 = 20$, $y_2 = 32$, $y_3 = 21$, $y_4 = 27$, $y_5 = 25$ and $y_6 = 19$ are observed, then $y_{[6]} - d = 32 - 9 = 23$ and Π_2, Π_4 and Π_5 are included in the selected subset because $y_i \geq 23$ for $i = 2, 4, 5$.

One basis for choosing the sample size n is to control the expected proportion of treatments retained by the selected subset. Table 7.11 lists the expected proportion of

Table 7.10. Constants $d = d(n, t, P^\star)$ Required to Implement Procedure $\mathcal{B}_{GS}$ for Selected (t, n) and $P^\star = 0.75$

										$P^\star = 0.75$									
$n \backslash t$	2	3	4	5	6	7	8	9	10	11	12	13	14	15	16	17	18	19	20
1	0	1	1	1	1	1	1	1	1	1	1	1	1	1	1	1	1	1	1
2	1	1	1	1	1	1	2	2	2	2	2	2	2	2	2	2	2	2	2
3	1	1	1	2	2	2	2	2	2	2	2	2	2	2	2	2	2	2	2
4	1	1	2	2	2	2	2	2	2	2	2	2	2	2	2	2	2	3	3
5	1	2	2	2	2	2	2	2	3	3	3	3	3	3	3	3	3	3	3
6	1	2	2	2	2	3	3	3	3	3	3	3	3	3	3	3	3	3	3
7	1	2	2	2	3	3	3	3	3	3	3	3	3	3	3	3	3	3	3
8	1	2	2	3	3	3	3	3	3	3	3	3	3	3	3	4	4	4	4
9	1	2	3	3	3	3	3	3	3	3	4	4	4	4	4	4	4	4	4
10	2	2	3	3	3	3	3	3	4	4	4	4	4	4	4	4	4	4	4
11	2	2	3	3	3	3	4	4	4	4	4	4	4	4	4	4	4	4	4
12	2	2	3	3	3	4	4	4	4	4	4	4	4	4	4	4	4	4	4
13	2	3	3	3	4	4	4	4	4	4	4	4	4	4	4	5	5	5	5
14	2	3	3	3	4	4	4	4	4	4	4	4	5	5	5	5	5	5	5
15	2	3	3	4	4	4	4	4	4	4	5	5	5	5	5	5	5	5	5
16	2	3	3	4	4	4	4	4	5	5	5	5	5	5	5	5	5	5	5
17	2	3	3	4	4	4	4	5	5	5	5	5	5	5	5	5	5	5	5
18	2	3	4	4	4	4	5	5	5	5	5	5	5	5	5	5	5	5	5
19	2	3	4	4	4	4	5	5	5	5	5	5	5	5	5	5	6	6	6
20	2	3	4	4	4	5	5	5	5	5	5	5	5	5	6	6	6	6	6
25	2	4	5	5	5	5	5	6	6	6	6	6	6	6	6	6	6	6	6
30	3	4	5	5	5	6	6	6	6	6	6	7	7	7	7	7	7	7	7
35	3	4	5	5	6	6	6	7	7	7	7	7	7	7	7	7	8	8	8
40	3	5	5	6	6	7	7	7	7	7	7	8	8	8	8	8	8	8	8
45	3	5	6	6	7	7	7	7	8	8	8	8	8	8	8	8	9	9	9
50	3	5	6	7	7	7	8	8	8	8	8	8	9	9	9	9	9	9	9
60	4	6	7	7	8	8	8	9	9	9	9	9	9	10	10	10	10	10	10
70	4	6	7	8	8	9	9	9	9	10	10	10	10	10	10	11	11	11	11
80	4	6	8	8	9	9	10	10	10	10	11	11	11	11	11	11	11	12	12
90	5	7	8	9	9	10	10	10	11	11	11	11	12	12	12	12	12	12	12
100	5	7	9	9	10	10	11	11	11	12	12	12	12	12	12	13	13	13	13
125	5	8	9	10	11	12	12	12	13	13	13	13	14	14	14	14	14	14	15
150	6	9	10	11	12	13	13	14	14	14	14	15	15	15	15	15	16	16	16
175	6	9	11	12	13	14	14	15	15	15	16	16	16	16	17	17	17	17	17
200	7	10	12	13	14	15	15	16	16	16	17	17	17	17	18	18	18	18	18
250	8	11	13	15	16	16	17	17	18	18	19	19	19	20	20	20	20	20	21
300	8	12	15	16	17	18	19	19	20	20	20	21	21	21	22	22	22	22	23
350	9	13	16	17	18	19	20	21	21	22	22	22	23	23	23	24	24	24	24
400	10	14	17	18	20	21	21	22	23	23	24	24	24	25	25	25	26	26	26
450	10	15	18	20	22	22	23	23	24	25	25	26	26	26	26	27	27	27	28
500	11	16	19	21	22	23	24	25	25	26	26	27	27	28	28	28	29	29	29

Reprinted from *Contributions to Probability and Statistics: Essays in Honor of Harold Hotelling*, edited by Ingram Olkin and others, with the permission of the publishers, Stanford University Press. © 1960 by the Board of Trustees of the Leland Stanford Junior University.

A SINGLE-STAGE SUBSET SELECTION PROCEDURE

Table 7.10. *Continued,* $P^* = 0.90$

$n\backslash t$	2	3	4	5	6	7	8	9	10	11	12	13	14	15	16	17	18	19	20
1	1	1	1	1	1	1	1	1	1	1	1	1	1	1	1	1	1	1	1
2	1	2	2	2	2	2	2	2	2	2	2	2	2	2	2	2	2	2	2
3	2	2	2	2	2	2	2	2	2	3	3	3	3	3	3	3	3	3	3
4	2	2	2	3	3	3	3	3	3	3	3	3	3	3	3	3	3	3	3
5	2	2	3	3	3	3	3	3	3	3	3	3	3	3	3	3	4	4	4
6	2	3	3	3	3	3	3	4	4	4	4	4	4	4	4	4	4	4	4
7	2	3	3	3	4	4	4	4	4	4	4	4	4	4	4	4	4	4	4
8	3	3	3	4	4	4	4	4	4	4	4	4	4	4	4	4	5	5	5
9	3	3	4	4	4	4	4	4	4	4	5	5	5	5	5	5	5	5	5
10	3	4	4	4	4	4	4	5	5	5	5	5	5	5	5	5	5	5	5
11	3	4	4	4	4	5	5	5	5	5	5	5	5	5	5	5	5	5	5
12	3	4	4	5	5	5	5	5	5	5	5	5	5	5	5	6	6	6	6
13	3	4	4	5	5	5	5	5	5	5	5	6	6	6	6	6	6	6	6
14	3	4	5	5	5	5	5	5	6	6	6	6	6	6	6	6	6	6	6
15	4	4	5	5	5	5	6	6	6	6	6	6	6	6	6	6	6	6	6
16	4	4	5	5	5	6	6	6	6	6	6	6	6	6	6	6	6	6	7
17	4	5	5	5	6	6	6	6	6	6	6	6	6	6	7	7	7	7	7
18	4	5	5	6	6	6	6	6	6	6	7	7	7	7	7	7	7	7	7
19	4	5	5	6	6	6	6	6	6	7	7	7	7	7	7	7	7	7	7
20	4	5	5	6	6	6	7	7	7	7	7	7	7	7	7	7	7	7	7
25	5	6	6	6	7	7	7	7	7	8	8	8	8	8	8	8	8	8	8
30	5	6	7	7	7	8	8	8	8	8	8	8	9	9	9	9	9	9	9
35	5	7	7	8	8	8	8	9	9	9	9	9	9	9	9	10	10	10	10
40	6	7	8	8	9	9	9	9	9	10	10	10	10	10	10	10	10	10	10
45	6	7	8	9	9	9	10	10	10	10	10	10	11	11	11	11	11	11	11
50	6	8	9	9	10	10	10	10	11	11	11	11	11	11	11	11	12	12	12
60	7	9	10	10	10	11	11	11	12	12	12	12	12	12	12	13	13	13	13
70	8	9	10	10	11	12	12	12	12	13	13	13	13	13	13	14	14	14	14
80	8	10	11	12	12	13	13	13	13	14	14	14	14	14	14	14	15	15	15
90	9	11	12	12	13	13	14	14	14	15	15	15	15	15	15	15	16	16	16
100	9	11	12	13	14	14	14	15	15	15	15	15	16	16	16	16	16	16	16
125	10	12	14	15	15	16	16	16	17	17	17	17	18	18	18	18	18	18	18
150	11	14	15	16	17	17	18	18	18	19	19	19	19	19	20	20	20	20	20
175	12	15	16	17	18	19	19	19	20	20	20	20	21	21	21	21	22	22	22
200	13	16	17	18	19	20	20	21	21	22	22	22	22	22	23	23	23	23	23
250	14	18	19	21	21	22	23	23	24	24	24	24	25	25	25	26	26	26	26
300	16	19	21	23	23	24	25	25	26	26	27	27	27	27	28	28	28	29	29
350	17	21	23	24	25	26	27	27	28	28	29	29	29	30	30	30	31	31	31
400	18	22	25	26	27	28	29	29	30	30	31	31	31	32	32	32	33	33	33
450	19	24	26	28	29	30	30	31	32	32	33	33	33	34	34	34	35	35	35
500	20	25	27	29	30	31	32	33	33	34	34	34	35	35	36	36	37	37	37

continued

Table 7.10. Continued, $P^* = 0.95$

	$P^* = 0.95$																		
$n\backslash t$	2	3	4	5	6	7	8	9	10	11	12	13	14	15	16	17	18	19	20
1	1	1	1	1	1	1	1	1	1	1	1	1	1	1	1	1	1	1	1
2	2	2	2	2	2	2	2	2	2	2	2	2	2	2	2	2	2	2	2
3	2	2	2	3	3	3	3	3	3	3	3	3	3	3	3	3	3	3	3
4	2	3	3	3	3	3	3	3	3	3	3	3	3	3	3	3	3	3	4
5	3	3	3	3	3	4	4	4	4	4	4	4	4	4	4	4	4	4	4
6	3	3	3	4	4	4	4	4	4	4	4	4	4	4	4	4	4	4	4
7	3	4	4	4	4	4	4	4	4	4	5	5	5	5	5	5	5	5	5
8	3	4	4	4	4	4	5	5	5	5	5	5	5	5	5	5	5	5	5
9	3	4	4	5	5	5	5	5	5	5	5	5	5	5	5	5	5	5	5
10	4	4	5	5	5	5	5	5	5	5	5	6	6	6	6	6	6	6	6
11	4	4	5	5	5	5	5	5	6	6	6	6	6	6	6	6	6	6	6
12	4	5	5	5	5	6	6	6	6	6	6	6	6	6	6	6	6	6	6
13	4	5	5	5	6	6	6	6	6	6	6	6	6	6	6	6	7	7	7
14	4	5	5	6	6	6	6	6	6	6	6	7	7	7	7	7	7	7	7
15	5	5	6	6	6	6	6	6	7	7	7	7	7	7	7	7	7	7	7
16	5	5	6	6	6	6	7	7	7	7	7	7	7	7	7	7	7	7	7
17	5	6	6	6	6	7	7	7	7	7	7	7	7	7	7	7	7	8	8
18	5	6	6	6	7	7	7	7	7	7	7	7	7	8	8	8	8	8	8
19	5	6	6	7	7	7	7	7	7	7	8	8	8	8	8	8	8	8	8
20	5	6	6	7	7	7	7	7	8	8	8	8	8	8	8	8	8	8	8
25	6	7	7	8	8	8	8	8	8	9	9	9	9	9	9	9	9	9	9
30	6	7	8	8	9	9	9	9	9	9	10	10	10	10	10	10	10	10	10
35	7	8	9	9	9	10	10	10	10	10	10	10	11	11	11	11	11	11	11
40	7	9	9	10	10	10	10	11	11	11	11	11	11	11	11	11	12	12	12
45	8	9	10	10	11	11	11	11	11	12	12	12	12	12	12	12	12	12	12
50	8	10	11	11	11	12	12	12	12	12	13	13	13	13	13	13	13	13	13
60	9	10	11	12	12	12	13	13	13	13	14	14	14	14	14	14	14	14	14
70	10	11	12	13	13	14	14	14	14	14	15	15	15	15	15	15	15	15	16
80	10	12	13	14	14	14	15	15	15	15	16	16	16	16	16	16	16	17	17
90	11	13	14	14	15	15	16	16	16	16	17	17	17	17	17	17	17	18	18
100	12	14	15	15	16	16	17	17	17	17	18	18	18	18	18	18	18	18	19
125	13	15	16	17	18	18	19	19	19	19	20	20	20	20	20	20	21	21	21
150	14	17	18	19	19	20	20	21	21	21	21	22	22	22	22	22	23	23	23
175	15	18	19	20	21	21	22	22	23	23	23	23	24	24	24	24	24	24	25
200	16	19	21	22	22	23	23	24	24	24	25	25	25	25	26	26	26	26	26
250	18	21	23	24	25	26	26	27	27	27	28	28	28	28	29	29	29	29	29
300	20	23	25	26	27	28	29	29	30	30	30	31	31	31	31	31	32	32	32
350	22	25	27	29	30	30	31	32	32	32	33	33	33	34	34	34	34	35	35
400	23	27	29	31	32	32	33	34	34	35	35	35	36	36	36	37	37	37	37
450	25	29	31	32	33	34	35	36	36	37	37	38	38	38	38	39	39	39	39
500	25	30	33	34	35	36	37	38	38	39	39	40	40	40	41	41	41	41	42

treatments, $E\{S\}/t$, selected by procedure $\mathcal{B}_{GS}$ under the slippage configuration

$$p_{[1]} = p_{[t-1]} = p \quad \text{and} \quad p_{[t]} = p - \Delta^\star \qquad (7.5.3)$$

for selected $\Delta^\star > 0$, p, t and n; Gupta and McDonald (1986) give more complete tables.

Example 7.5.2 An experimenter wishes to use procedure $\mathcal{B}_{GS}$ for $t = 10$ and $P^\star = 0.90$. The common sample size n to be taken from each treatment is to be determined. The experimenter decides to use the smallest value of n such that if the true configuration is given by

$$p_{[1]} = p_{[9]} = 0.50 \quad \text{and} \quad p_{[10]} = 0.75, \qquad (7.5.4)$$

then the procedure can be expected to eliminate *at least* three of the ten treatments. Equation (7.5.4) corresponds to using $\Delta^\star = 0.25$ and $p = 0.50$. From Table 7.11 we find that the value of n corresponding to an expected proportion of retained treatments of *at most* 0.7, lies between 20 and 25; linear interpolation gives the estimated number as $n = 24$. Table 7.10 shows that $d = 8$ corresponds to $n = 24$.

Example 7.5.3 This example illustrates the use of Table 7.11 to determine the sample size for a more demanding design requirement. Suppose that the experimenter wishes to determine the smallest value of n such that for *any* true configuration satisfying

$$0 < p_{[1]} = p_{[4]} = p \leq 0.75 \quad \text{and} \quad p_{[5]} = p + \Delta^\star, \qquad (7.5.5)$$

then the procedure can be expected to eliminate *at least* two of the five treatments, that is, $E\{S\} \leq 3$. Note the difference between this probability requirement and the one in Example 7.5.2. In the present example we wish to bound the expected proportion of treatments selected from above *simultaneously* for a *set* of slippage configurations corresponding to *different* p and a fixed $\Delta^\star$; in Example 7.5.2 we bounded the expected proportion of treatments selected for the single slippage configuration (7.5.4). In principle, to solve this design problem one needs to know the expected proportion selected for all $p \in (0.0, 0.75)$. Table 7.11 gives the expected proportion selected for a set of four p-values and thus provides a (very) approximate solution to the design problem. From the panel for $P^\star = 0.75$ and $n = 20$ of Table 7.11 we see that no more than the following proportions will be selected for $p = 0.25, 0.50, 0.70$ and 0.75.

p	Proportion Selected
0.25	0.528
0.50	0.527
0.70	0.531
0.75	0.532

Table 7.11. Expected Proportion of Treatments Retained in the Selected Subset by Procedure $\mathcal{B}_{GS}$ When $p_{[1]} = p_{[t-1]} = p$ and $p_{[t]} = p + \Delta^\star$ for Selected $(t, n, p, P^\star)$ and $\Delta^\star = 0.00, 0.10$

$P^\star$		0.75				0.90				0.95			
	t	2	3	5	10	2	3	5	10	2	3	5	10
n	p						$\Delta^\star = 0.00$						
5	0.50	0.828	0.904	0.845	0.935	0.945	0.904	0.964	0.935	0.989	0.980	0.964	0.992
	0.75	0.868	0.950	0.926	0.986	0.969	0.950	0.989	0.986	0.996	0.993	0.989	0.999
	0.95	0.982	0.999	0.999	1.000	0.999	0.999	1.000	1.000	1.000	1.000	1.000	1.000
	1.00	1.000	1.000	1.000	1.000	1.000	1.000	1.000	1.000	1.000	1.000	1.000	1.000
10	0.50	0.868	0.785	0.838	0.886	0.942	0.962	0.934	0.961	0.979	0.962	0.979	0.961
	0.75	0.904	0.846	0.912	0.958	0.966	0.984	0.974	0.990	0.991	0.984	0.994	0.990
	0.95	0.993	0.990	0.999	1.000	0.999	1.000	1.000	1.000	1.000	1.000	1.000	1.000
	1.00	1.000	1.000	1.000	1.000	1.000	1.000	1.000	1.000	1.000	1.000	1.000	1.000
15	0.50	0.819	0.832	0.858	0.771	0.951	0.913	0.932	0.949	0.979	0.961	0.972	0.981
	0.75	0.855	0.887	0.923	0.880	0.972	0.952	0.971	0.985	0.990	0.983	0.991	0.996
	0.95	0.981	0.996	0.999	0.999	1.000	1.000	1.000	1.000	1.000	1.000	1.000	1.000
	1.00	1.000	1.000	1.000	1.000	1.000	1.000	1.000	1.000	1.000	1.000	1.000	1.000
20	0.50	0.785	0.781	0.792	0.804	0.923	0.928	0.939	0.947	0.960	0.965	0.972	0.977
	0.75	0.820	0.838	0.869	0.900	0.951	0.962	0.974	0.984	0.978	0.985	0.991	0.995
	0.95	0.966	0.990	0.998	1.000	0.999	1.000	1.000	1.000	1.000	1.000	1.000	1.000
	1.00	1.000	1.000	1.000	1.000	1.000	1.000	1.000	1.000	1.000	1.000	1.000	1.000
25	0.50	0.760	0.830	0.833	0.834	0.941	0.942	0.902	0.905	0.968	0.970	0.974	0.951
	0.75	0.794	0.883	0.901	0.919	0.964	0.971	0.951	0.963	0.984	0.988	0.992	0.985
	0.95	0.951	0.996	0.999	1.000	1.000	1.000	1.000	1.000	1.000	1.000	1.000	1.000
	1.00	1.000	1.000	1.000	1.000	1.000	1.000	1.000	1.000	1.000	1.000	1.000	1.000
							$\Delta^\star = 0.10$						
5	0.40	0.817	0.895	0.835	0.929	0.938	0.895	0.960	0.969	0.987	0.977	0.960	0.992
	0.65	0.838	0.922	0.881	0.963	0.952	0.922	0.976	0.963	0.991	0.985	0.976	0.997
	0.85	0.928	0.984	0.979	0.998	0.989	0.984	0.998	0.998	0.999	0.999	0.998	1.000
	0.90	0.959	0.993	0.993	1.000	0.996	0.994	1.000	1.000	1.000	1.000	1.000	1.000
10	0.40	0.846	0.761	0.819	0.874	0.925	0.951	0.923	0.956	0.970	0.951	0.975	0.956
	0.65	0.864	0.791	0.857	0.915	0.939	0.965	0.946	0.974	0.978	0.965	0.984	0.974
	0.85	0.937	0.908	0.967	0.992	0.983	0.995	0.994	0.999	0.997	0.995	0.999	0.999
	0.90	0.965	0.953	0.990	0.999	0.994	0.999	0.999	0.999	1.000	0.999	0.999	1.000
15	0.40	0.786	0.797	0.830	0.747	0.926	0.885	0.914	0.939	0.963	0.943	0.962	0.977
	0.65	0.786	0.797	0.830	0.747	0.926	0.885	0.914	0.939	0.963	0.943	0.962	0.977
	0.85	0.869	0.923	0.966	0.955	0.982	0.975	0.991	0.997	0.995	0.993	0.998	1.000
	0.90	0.908	0.963	0.990	0.989	0.994	0.992	0.992	0.998	1.000	0.999	0.999	1.000
20	0.40	0.743	0.733	0.750	0.774	0.884	0.894	0.914	0.934	0.930	0.941	0.957	0.970
	0.65	0.754	0.756	0.785	0.821	0.898	0.913	0.935	0.955	0.942	0.955	0.970	0.981
	0.85	0.806	0.856	0.916	0.959	0.954	0.976	0.990	0.997	0.983	0.993	0.997	0.999
	0.90	0.838	0.911	0.965	0.990	0.978	0.992	0.998	1.000	0.994	0.998	1.000	1.000
25	0.40	0.711	0.772	0.783	0.800	0.896	0.902	0.861	0.879	0.934	0.942	0.955	0.934
	0.65	0.719	0.793	0.814	0.841	0.909	0.919	0.887	0.910	0.946	0.956	0.968	0.954
	0.85	0.752	0.885	0.931	0.965	0.962	0.979	0.973	0.988	0.985	0.993	0.997	0.996
	0.90	0.769	0.935	0.973	0.991	0.983	0.994	0.992	0.998	0.995	0.998	1.000	1.000

Reprinted from *Contributions to Probability and Statistics: Essays in Honor of Harold Hotelling*, edited by Ingram Olkin and others, with the permission of the publishers, Stanford University Press. © 1960 by the Board of Trustees of the Leland Stanford Junior University.

A SINGLE-STAGE SUBSET SELECTION PROCEDURE

Table 7.11. *Continued,* $\Delta^* = 0.25, 0.50$

P^*		0.75				0.90				0.95			
t		2	3	5	10	2	3	5	10	2	3	5	10
n	p					$\Delta^* = 0.25$							
5	0.25	0.762	0.850	0.796	0.925	0.896	0.850	0.946	0.925	0.971	0.959	0.946	0.992
	0.50	0.762	0.849	0.794	0.916	0.896	0.849	0.941	0.916	0.971	0.958	0.941	0.988
	0.70	0.798	0.908	0.883	0.974	0.934	0.908	0.978	0.974	0.988	0.983	0.978	0.998
	0.75	0.816	0.931	0.917	0.986	0.948	0.931	0.987	0.986	0.992	0.990	0.997	0.999
10	0.25	0.742	0.643	0.715	0.816	0.838	0.883	0.853	0.926	0.915	0.883	0.939	0.926
	0.50	0.742	0.643	0.713	0.808	0.838	0.882	0.849	0.919	0.915	0.882	0.936	0.919
	0.70	0.754	0.668	0.783	0.895	0.870	0.928	0.912	0.968	0.947	0.928	0.973	0.968
	0.75	0.763	0.684	0.821	0.930	0.888	0.948	0.937	0.982	0.961	0.948	0.984	0.982
15	0.25	0.651	0.628	0.671	0.607	0.805	0.735	0.792	0.864	0.875	0.821	0.884	0.935
	0.50	0.651	0.628	0.670	0.604	0.805	0.734	0.789	0.858	0.875	0.830	0.882	0.931
	0.70	0.632	0.635	0.719	0.678	0.827	0.768	0.850	0.924	0.908	0.877	0.934	0.973
	0.75	0.618	0.641	0.749	0.718	0.843	0.791	0.881	0.949	0.926	0.901	0.955	0.984
20	0.25	0.596	0.528	0.528	0.580	0.714	0.705	0.747	0.812	0.781	0.792	0.837	0.892
	0.50	0.596	0.529	0.527	0.577	0.714	0.705	0.745	0.808	0.781	0.792	0.836	0.888
	0.70	0.566	0.503	0.531	0.628	0.709	0.728	0.798	0.877	0.796	0.832	0.892	0.943
	0.75	0.546	0.483	0.532	0.655	0.707	0.745	0.849	0.908	0.809	0.857	0.919	0.963
25	0.25	0.562	0.527	0.516	0.557	0.703	0.682	0.615	0.669	0.763	0.761	0.798	0.770
	0.50	0.562	0.527	0.516	0.555	0.703	0.682	0.614	0.666	0.763	0.761	0.797	0.766
	0.70	0.533	0.497	0.509	0.587	0.694	0.696	0.634	0.721	0.772	0.795	0.851	0.832
	0.75	0.516	0.476	0.503	0.605	0.689	0.707	0.639	0.754	0.781	0.818	0.880	0.866
						$\Delta^* = 0.50$							
5	0.00	0.594	0.667	0.600	0.831	0.750	0.667	0.850	0.831	0.906	0.875	0.850	0.972
	0.25	0.610	0.646	0.568	0.769	0.737	0.646	0.801	0.769	0.878	0.836	0.801	0.948
	0.45	0.600	0.657	0.586	0.806	0.744	0.657	0.831	0.806	0.896	0.860	0.831	0.964
	0.50	0.594	0.667	0.600	0.831	0.750	0.667	0.850	0.831	0.906	0.875	0.850	0.972
10	0.00	0.527	0.370	0.338	0.439	0.586	0.585	0.502	0.661	0.688	0.585	0.698	0.661
	0.25	0.551	0.400	0.369	0.440	0.607	0.588	0.505	0.625	0.691	0.588	0.667	0.625
	0.45	0.537	0.383	0.351	0.439	0.595	0.585	0.502	0.645	0.689	0.585	0.685	0.645
	0.50	0.527	0.370	0.338	0.439	0.586	0.585	0.502	0.661	0.688	0.585	0.698	0.661
15	0.00	0.502	0.345	0.247	0.153	0.530	0.373	0.321	0.373	0.575	0.434	0.443	0.550
	0.25	0.511	0.367	0.284	0.194	0.553	0.404	0.357	0.393	0.598	0.464	0.461	0.537
	0.45	0.505	0.353	0.263	0.170	0.539	0.385	0.336	0.381	0.585	0.447	0.450	0.543
	0.50	0.502	0.345	0.247	0.153	0.530	0.373	0.321	0.378	0.575	0.434	0.443	0.550
20	0.00	0.500	0.334	0.205	0.119	0.503	0.347	0.246	0.218	0.510	0.372	0.305	0.327
	0.25	0.502	0.341	0.221	0.149	0.513	0.370	0.283	0.261	0.527	0.402	0.343	0.356
	0.45	0.501	0.336	0.210	0.130	0.506	0.356	0.261	0.236	0.517	0.384	0.321	0.339
	0.50	0.500	0.334	0.205	0.119	0.503	0.347	0.246	0.218	0.510	0.372	0.305	0.327
25	0.00	0.500	0.334	0.202	0.107	0.501	0.338	0.206	0.119	0.504	0.348	0.243	0.148
	0.25	0.501	0.338	0.211	0.126	0.507	0.352	0.223	0.150	0.514	0.370	0.279	0.188
	0.45	0.500	0.335	0.205	0.113	0.503	0.343	0.212	0.131	0.507	0.356	0.257	0.165
	0.50	0.500	0.334	0.202	0.107	0.501	0.338	0.206	0.119	0.504	0.348	0.243	0.148

Thus $n = 20$ is a value that guarantees (on the average) that *no more* than 60% of the treatments will be selected for $p \in (0.25, 0.75)$. However, since the proportion selected increases as p decreases, we may wish to use a few additional observations, say a total of $n = 25$ per treatment. From Table 7.10 we find that $d = 5$ will achieve the basic probability requirement (7.5.1) for $n = 25$.

Large-Sample Procedure

When $n_1 = n_2 = \cdots = n_t = n$, say, is large, the following approximate procedure can be used in place of procedure $\mathcal{B}_{GS}$. Let

$$w_i = \arcsin\sqrt{\frac{y_i}{n+1}} + \arcsin\sqrt{\frac{y_i + 1}{n+1}}$$

$(1 \leq i \leq t)$ and let $w_{[1]} \leq \cdots \leq w_{[t]}$ denote the ordered w_i-values. Then Π_i is included in the selected subset if and only if

$$w_i \geq w_{[t]} - d \qquad (7.5.6)$$

where $d = \sqrt{2} Z_{t-1,1/2}^{(1-P^*)}/\sqrt{n}$.

Example 7.5.4 In Example 7.5.1 suppose that the large-sample procedure (7.5.6) is used. Then we have $w_1 = 1.571$, $w_2 = 2.196$, $w_3 = 1.620$, $w_4 = 1.919$, $w_5 = 1.818$ and $w_6 = 1.522$. The yardstick is $d = \sqrt{2} Z_{5,1/2}^{(0.10)}/\sqrt{40} = 0.428$. The approximate procedure selects all treatments with $w_i \geq 2.196 - 0.428 = 1.768$; thus it chooses the treatments Π_2, Π_4 and Π_5 (as does the exact procedure $\mathcal{B}_{GS}$).

The screening capabilities of procedure $\mathcal{B}_{GS}$ are *not* as effective as those of the normal means subset selection procedure $\mathcal{N}_G$ (Sanchez 1987a). For example, Table 7.11 shows that unless $\Delta^* \geq 0.5$, very little screening takes place for small $n \leq 25$ when p is in a slippage configuration (7.5.3). For instance, if $t = 10$, $P^* = 0.90$, $n = 15$ and the unknown p satisfies $p_{[1]} = p_{[9]} = 0.75$ and $p_{[10]} = 1.0$, then $E\{S\} = 9.84$. Therefore if we collect *no data*, and randomly select a subset of 9 of the 10 treatments, we still guarantee the probability requirement $P\{CS\} \geq 0.90$ but achieve (constant) $E\{S\} = 9.00$. Clearly this no-data procedure dominates the procedure $\mathcal{B}_{GS}$ with respect to sample size and expected subset size, but *only at this particular p-configuration*. Of course, procedure $\mathcal{B}_{GS}$ satisfies $E\{S\} < 9.00$ for other configurations in which the best treatment is larger than the remaining ones. Still, it should be a warning that use of procedure $\mathcal{B}_{GS}$ with too small a sample size can lead to disappointing results.

To obtain an intuitive understanding of the difficulty that procedure $\mathcal{B}_{GS}$ has in distinguishing between treatments having substantially different p_i's, it is useful to compare the binomial and normal subset selection procedures. Both procedures are based on sample sums. In the normal case, because the individual observations are unbounded, a given treatment's sample sum can be dominated by a single observation; furthermore, the sample sums of treatments having greatly different treatment μ_i-

values will, with high probability, be arbitrarily far apart. In the binomial case, each Bernoulli outcome is either zero or unity, so for small sample sizes, the sample numbers of successes from the t treatments cannot be too "far apart" no matter how different the success probabilities. Thus, the procedure has a more difficult time in identifying inferior treatments.

There is a silver lining to this gray cloud. Because no single observation can substantially change the sample sum of n Bernoulli trials, in certain circumstances it is possible to curtail sampling for the procedure because no single observation or small number of observations can greatly alter the final decision arrived at by the procedure if the first "few" observations indicate that a particular treatment definitely will or will not be in the selected subset. This fact has been exploited in the indifference-zone version of the Bernoulli problem when discussing adaptive sampling in conjunction with the problem of determining the single best treatment. Sanchez (1987b) studies an analogue of the adaptive sampling procedure in Section 7.3 for Bernoulli subset selection.

7.5.2 Screening in Unbalanced Experiments

Let

$$w_i = \arcsin\sqrt{\frac{y_i}{n_i + 1}} + \arcsin\sqrt{\frac{y_i + 1}{n_i + 1}}$$

$(1 \leq i \leq t)$ and $w_{[1]} \leq \cdots \leq w_{[t]}$ denote the ordered w_i values. Then Π_i is included in the selected subset if and only if

$$w_i \geq w_{[t]} - d$$

where d solves the equation

$$\min_{1 \leq i \leq t} P\left\{V_j^i \leq d\sqrt{\frac{n_j n_i}{n_j + n_i}} \text{ for } j \neq i\right\} = P^\star \qquad (7.5.7)$$

and $(V_1^i, \ldots, V_{i-1}^i, V_{i+1}^i, \ldots, V_t^i)$ have a joint multivariate normal distribution with mean vector zero, unit variances, and

$$\mathrm{Cor}(V_j^i, V_\ell^i) = \sqrt{\frac{n_j n_\ell}{(n_j + n_i)(n_\ell + n_i)}} \quad \text{for } i, j \text{ and } \ell \text{ all not equal}.$$

The minimum in (7.5.7) arises because of the fact that one has to pick the worst-case assignment of sample size n_i $(1 \leq i \leq t)$ to the treatment associated with $p_{[t]}$.

In principle, each probability in Equation (7.5.7) can be exactly computed using the FORTRAN program MVNPRD in Dunnett (1989) because the V_j^i have a correlation structure of the form $\rho_{j,\ell} = b_j \times b_\ell$ (i is fixed). Alternatively, an easily computed but conservative value of d can be determined by solving

$$\min_{1 \leq i \leq t} P\left\{U_j^i \leq d\sqrt{\frac{n_j n_i}{n_j + n_i}} \text{ for } j \neq i\right\} = P^\star \qquad (7.5.8)$$

where $(U_1^i, \ldots, U_{i-1}^i, U_{i+1}^i, \ldots, U_t^i)$ have a joint multivariate normal distribution with mean vector zero, unit variances, and a *common correlation*

$$\rho = \sqrt{\frac{n_{[1]}^i n_{[2]}^i}{(n_{[1]}^i + n_i)(n_{[2]}^i + n_i)}}$$

where the $n_{[j]}^i$ ($1 \leq j \leq t-1$) are the ordered values of $\{n_1, \ldots, n_{i-1}, n_{i+1}, \ldots, n_t\}$.

Example 7.5.5 Suppose that an experimenter wishes to select a subset of $t = 3$ treatments containing the best treatment based on the following data.

i	n_i	y_i	w_i
1	20	8	1.38
2	24	13	1.65
3	27	6	1.00

For fixed i ($1 \leq i \leq 3$), the probability

$$Q_i = P\left\{V_j^i \leq d\sqrt{\frac{n_j n_i}{n_j + n_i}} \quad \text{for } j \neq i\right\}$$

is a bivariate quadrant probability with correlations and northeast corner point given by

i	ρ	$d\sqrt{\frac{n_j n_i}{n_j + n_i}}$	$d\sqrt{\frac{n_\ell n_i}{n_\ell + n_i}}$
1	0.56	$d \times 3.303$	$d \times 3.390$
2	0.53	$d \times 3.303$	$d \times 3.565$
3	0.45	$d \times 3.390$	$d \times 3.565$

Suppose that we desire to select a subset containing the best treatment with probability 0.80. By trial and error we compute d from

d	Q_1	Q_2	Q_3	$\min_{1 \leq i \leq 3} Q_i$
1	0.999	0.999	0.999	0.999
0.50	0.919	0.924	0.926	0.919
0.35	0.805	0.812	0.811	0.805

Thus we select all those treatments with

$$w_i \geq w_{[3]} - 0.35 = 1.65 - 0.35 = 1.30$$

which leads to selection of treatments 1 and 2.

7.5.3 An Alternative Design Specification

There are occasions when an experimenter knows *a priori* that the p_i-values have to be at least a certain minimum (greater than 1/2). Formally this corresponds to the following probability requirement.

Probability Requirement: For specified constants $(p_g, P^\star)$ with $1/t < P^\star < 1$ and $1/2 < p_g$, we require

$$P\{CS\} \geq P^\star \quad \text{for all } \boldsymbol{p} = (p_1, p_2, \ldots, p_t) \text{ with } 1/2 < p_g < p_{[1]}. \quad (7.5.9)$$

Here CS denotes the event of (correctly) selecting the treatment associated with $p_{[t]}$.

In this case one can sharpen the procedure $\mathcal{B}_{GS}$ by replacing d by a (*smaller*) integer $d^\star$. Gupta and McDonald (1986) show that when n is large, $d^\star$ is given approximately by

$$d^\star = (2d + 1) \times \sqrt{p_g(1 - p_g)} - 0.5. \quad (7.5.10)$$

It is easy to see that $d^\star < d$. Thus the modified procedure will result in a smaller value of $E\{S\}$ for given n and $P^\star$. Alternatively, given $P^\star$, the experimenter can consider the cost in terms of $E\{S\}$ for decreasing n.

7.6 CHAPTER NOTES

This chapter has considered the goal of indifference-zone selection of the best treatment and the goal of selection of a subset of the t treatments that contains the best treatment. For the first goal, Sections 7.2 and 7.4 restrict attention to indifference-zones defined by the difference and the odds ratio, respectively. To complete the specification of possible indifference-zone procedures, we mention Taheri and Young (1974), who proposed two closed sequential selection procedures for $t = 2$ that employ the *relative risk* indifference-zone (7.1.3). A comprensive bibliographic source for Bernoulli selection procedures, as of the date of its publication, is Büringer, Martin and Schriever (1980). Sanchez and Higle (1992) present a case study of subset selection for treatments with "small" p_i.

The problem of allocating observations among treatments when the total number of observations available is given (the *fixed patient horizon setup*) with the objective of assigning a higher proportion of the patient population to the treatment with larger success probability has great practical importance. This problem has been studied in the medical setting by Armitage (1960, 1975), Colton (1963), Cornfield, Halperin and Greenhouse (1969), Zelen (1969), Canner (1970) and Sobel and Weiss (1972), among others. The procedures that have been proposed for adaptive sampling are intuitively appealing but ad hoc. For example, one intuitive strategy that has been considered is the the so-called "play-the-winner" principle that dictates if a success is observed from a particular treatment, then the next observation must be taken from that treatment. (Procedure $\mathcal{B}_{BK}$ does *not* employ a play-the-winner sampling rule, as can be seen from observation S_2^5 of Example 7.3.6.) The monograph by Büringer, Martin and Schriever (1980) presented detailed descriptions of a large number of Bernoulli sequential selection procedures intended to guarantee the probability requirement (7.1.1).

A related class of statistical procedures was developed to solve the so-called two-armed (or multi-armed) bandit problem (the *infinite patient horizon setup*). These problems are concerned with minimizing (or maximizing) appropriate objective functions, the principal mathematical tool used being dynamic programming. This is not the Bernoulli selection problem that we have formulated. A number of researchers have considered problems of this type, for example, Robbins (1952, 1956), Bradt, Johnson and Karlin (1956), Berry (1972, 1978), and Rodman (1978); these references will provide the interested reader with a starting place to investigate further.

There are a number of important alternative goals that have been considered in the literature. Gupta (1965) gave a single-stage procedure for selecting all binomial treatments better than a standard or a control; Gupta and Liang (1986) proposed an empirical Bayes procedure for this same goal. For a large class of selection rules, Sanchez (1989) presented unbiased estimators of any given treatment probability p_i following termination of the rule. Hochberg and Tamhane (1987, pp. 275–277) presented large-sample confidence intervals for all pairwise differences $\{p_i - p_j\}_{i \neq j}$ among the treatment means; for $t = 2$, Santner and Snell (1980), Santner and Yamagami (1993) and Coe and Tamhane (1993a, 1993b) gave small-sample confidence intervals for $p_1 - p_2$, but there are no small-sample solutions for $t \geq 3$.

We have focused on completely randomized experimental designs for selection. The literature contains a number of papers that consider alternatives. An important example is the use of the matched pairs design for dependent Bernoulli trials as discussed in Bhapkar and Somes (1976) and Tamhane (1980, 1985).

The procedures presented in this chapter considered only one-way layouts. There has been some work on the case where several qualitative factors affect the response probabilities. Bechhofer and Goldsman (1988a) proposed an open, noneliminating, one-matrix-at-a-time, sequential procedure for selecting the factor-level combination associated with $\alpha_{[a]}$ and $\beta_{[b]}$ where the treatment probabilities p_{ij} satisfy

$$\ln\left(\frac{p_{ij}}{1 - p_{ij}}\right) = \mu + \alpha_i + \beta_j$$

($1 \leq i \leq a, 1 \leq j \leq b$) subject to the identifiability conditions $\sum_{i=1}^{a} \alpha_i = \sum_{j=1}^{b} \beta_j = 0$. Here μ is the average log odds (logit) over the treatment combinations, and the main effect α_i is the difference between the average logit for the probabilities at the ith level of the row factor and the overall mean (the main effect β_j has a similar meaning). The largest ordered main effects, $\alpha_{[a]}$ and $\beta_{[b]}$, are associated with largest success probability, that is, $p_{[a][b]} = \max_{i,j} p_{ij}$. Their probability requirement is stated in terms of the odds ratios $[\alpha_{[a]}/(1 - \alpha_{[a]})]/[\alpha_{[a-1]}/(1 - \alpha_{[a-1]})]$ and $[\beta_{[b]}/(1 - \beta_{[b]})]/[\beta_{[b-1]}/(1 - \beta_{[b-1]})]$; this is analogous to the probability requirement (7.4.1) for single-factor Bernoulli experiments in Section 7.4.

In this text we have adopted a frequentist viewpoint. However, there is a large literature on Bayesian procedures for selecting the best treatment or a subset containing the best treatment. These papers can be distinguished by

- The set of actions they assume available to the statistician
- The information that they assume known regarding the prior distribution
- The loss function, if any, that they adopt

For example, the action space can specify that a *single* treatment be selected or that a *subset* of treatments be selected. The prior can be completely specified up to a known parameter (vector), or it can be nonparametric. Finally, if a loss is not specified, then the emphasis can be on the calculation of the posterior probability that each treatment is best (without a specified loss), or it can be to minimize the expected loss for a specified loss.

Several authors have developed (known prior) Bayesian procedures for the binomial selection problem. Among these are Bratcher and Bland (1975), who studied selection problems for a prior on $(p_1, \ldots, p_t)$ that has independent components with known and possibly different beta priors. More recently, Kulkarni and Kulkarni (1987) considered procedure $\mathcal{B}_{BK}$ for the special case $t = 2$ with independent *uniform* prior densities for p_1 and p_2. Deely and Gupta (1988) analyzed a hierarchical Bayesian model that extends the normal theory model of Berger and Deely (1988) to binomial responses (see Section 2.9).

In contrast, another line of research has pursued the unknown prior, empirical Bayes formulation of Robbins (1956, 1964) in which the statistician uses the evidence in accumulating data to estimate the prior. Among the works which adopt this approach are those of Deely (1965) for general selection problems including the binomial case with unknown parameters $(p_1, \ldots, p_t)$ that he assumed to be independent beta with unknown parameters. Other empirical Bayes models, both parametric and nonparametric, have been considered in the literature. These include Liang (1984), Gupta and Liang (1986), Liang (1990, 1991) (a nonparametric empirical Bayes approach in which there are independent p_i's with unknown nonparametric prior) and Gupta and Liang (1989a) (a parametric empirical Bayes approach).

CHAPTER 8

Selection Problems for Categorical Response Experiments

8.1 INTRODUCTION

This chapter presents statistical procedures for selecting the category ("cell") that has the largest (or smallest) associated probability of occurrence using the indifference-zone and subset selection approaches. Two settings are considered—experiments with a univariate categorical response and experiments with bivariate categorical responses.

To begin, recall that the multinomial distribution for n independent replications of a univariate categorical response is specified in the following manner. Suppose that there are t outcomes (or categories) possible. Let p_i ($0 < p_i < 1$, $\sum_{i=1}^{t} p_i = 1$) denote the single-trial probability of the event associated with the ith category C_i ($1 \leq i \leq t$). Let Y_i denote the number of outcomes falling in C_i ($1 \leq i \leq t$) after n observations have been taken. Then $0 \leq Y_i \leq n$ ($1 \leq i \leq t$) and $\sum_{i=1}^{t} Y_i = n$. The t-variate discrete vector random variable $Y = (Y_1, Y_2, \ldots, Y_t)$ has the probability mass function

$$P\{Y_1 = y_1, Y_2 = y_2, \ldots, Y_t = y_t\} = \frac{n!}{\prod_{i=1}^{t} y_i!} \prod_{i=1}^{t} p_i^{y_i},$$

and we say that Y has a *multinomial* distribution with parameters n and $p = (p_1, \ldots, p_t)$. The multinomial distribution generalizes the binomial, for which there are $t = 2$ outcomes.

The multinomial distribution has been used as a model in preference studies, for example, when a voter is asked to name that one of $t \geq 2$ candidates that the voter believes is best qualified for a particular office—or in a telephone survey when a consumer is asked to name that one of $t \geq 2$ product brands (or television programs) that the consumer most prefers. The multinomial distribution also arises in many simulation applications. For example, a company might be interested in using simulation to determine which of $t \geq 2$ manufacturing layouts is most economical.

INTRODUCTION

Example 8.1.1 Suppose that three of the faces of a fair die are colored red, two are colored blue, and the remaining face is green, that is, the probability vector associated with red, blue and green is $p = (3/6, 2/6, 1/6)$. Now suppose that this die is tossed $n = 5$ times. Then the probability of observing exactly three reds, no blues and two greens is given by

$$P\{Y = (3, 0, 2)\} = \frac{5!}{3!\,0!\,2!}(3/6)^3(2/6)^0(1/6)^2 = 0.03472.$$

Example 8.1.2 (Example 8.1.1 continued) Suppose that we did not know the probabilities for red, blue and green in the previous example and that we wanted to select the color having the largest probability of occurring on a single trial. (Of course, in this example, that color is red.) The rule that we shall adopt is to select the color that occurs the most frequently during the five trials, using randomization to break ties. Let $Y = (Y_r, Y_b, Y_g)$ denote the number of occurrences of (red, blue, green) in five trials. The probability that we *correctly* select red is given by

$$P\{\text{red wins in 5 trials}\} = P\{Y_r > Y_b \text{ and } Y_g\} + 0.5P\{Y_r = Y_b, Y_r > Y_g\}$$
$$+ 0.5P\{Y_r > Y_b, Y_r = Y_g\}$$
$$= P\{Y = (5,0,0), (4,1,0), (4,0,1), (3,2,0), (3,1,1), (3,0,2)\}$$
$$+ 0.5P\{Y = (2,2,1)\} + 0.5P\{Y = (2,1,2)\}.$$

The following table lists the outcomes favorable to a *correct selection* of red, along with the associated probabilities of these outcomes, incorporating randomization when ties occur.

Outcome (red, blue, green)	Contribution to $P\{\text{red wins in 5 trials}\}$
(5, 0, 0)	0.03125
(4, 1, 0)	0.10417
(4, 0, 1)	0.05208
(3, 2, 0)	0.13889
(3, 1, 1)	0.13889
(3, 0, 2)	0.03472
(2, 2, 1)	(0.5)(0.13889)
(2, 1, 2)	(0.5)(0.06944)
	0.60416

Thus, the probability of correctly selecting red as the most probable color based on $n = 5$ trials is 0.6042. This probability can be increased by increasing the sample size n.

The remainder of this chapter is organized as follows. Sections 8.2 and 8.3 consider procedures for the case when there is a single multinomial response (a "univariate" multinomial response experiment). Section 8.2 presents three procedures that employ the indifference-zone approach for selecting the most probable category, and Section 8.3 discusses the subset approach for this same problem. Section 8.4 considers independent bivariate responses; it presents an indifference-zone procedure for selecting the most probable bivariate outcome.

8.2 INDIFFERENCE-ZONE PROCEDURES FOR MULTINOMIAL DATA

This section discusses indifference-zone (IZ) procedures for selecting the cell of a multinomial distribution having the largest probability.

Statistical Assumptions: Independent observations $X_j = (X_{1j}, \ldots, X_{tj})$ ($j \geq 1$) are taken from a single trial multinomial distribution having $t \geq 2$ categories with associated unknown probabilities $p = (p_1, \ldots, p_t)$. Let p_i ($0 < p_i < 1$, $\sum_{i=1}^{t} p_i = 1$) be the probability of the event E_i associated with the ith category C_i ($1 \leq i \leq t$). The C_i are mutually exclusive and exhaustive; $X_{ij} = 1$ [0] according to whether E_i does [does not] occur on the jth observation ($1 \leq i \leq t$, $j \geq 1$).

We denote the ordered values of $p_1, \ldots, p_t$ by $p_{[1]} \leq \cdots \leq p_{[t]}$. Neither the values of the $p_{[s]}$ nor the pairing of the C_i with the $p_{[s]}$ ($1 \leq i, s \leq t$) is assumed to be known. The category associated with $p_{[t]}$ is referred to as the most probable or "best" category. Here we use the ratio $p_{[t]}/p_{[t-1]}$ as the "measure of distance" between the largest and second largest cell probabilities to define the IZ. We denote the observed values of X_j by $x_j = (x_{1j}, \ldots, x_{tj})$ ($j \geq 1$). The cumulative sum for category C_i after $m \geq 1$ multinomial observations have been taken is denoted by $y_{im} = \sum_{j=1}^{m} x_{ij}$ ($1 \leq i \leq t$). The associated ordered values of the y_{im} are given by $y_{[1]m} \leq \cdots \leq y_{[t]m}$.

The purpose of the experiment is stated in Goal 8.2.1 and the associated probability (design) requirement is given in (8.2.1).

Goal 8.2.1 To select the category associated with $p_{[t]}$.

A *correct selection* (CS) is said to have been made if Goal 8.2.1 is achieved.

Probability Requirement: For specified constants $(\theta^\star, P^\star)$ with $1 < \theta^\star < \infty$ and $1/t < P^\star < 1$, we require

$$P\{\text{CS} \mid p\} \geq P^\star \quad \text{whenever } p_{[t]}/p_{[t-1]} \geq \theta^\star. \tag{8.2.1}$$

The probability in (8.2.1) depends on the entire vector p and on the number n of independent multinomial observations to be taken. The constant $\theta^\star$ can be interpreted as the "smallest $p_{[t]}/p_{[t-1]}$ ratio worth detecting."

In Section 8.2.1, we consider a single-stage procedure that guarantees the probability requirement (8.2.1). A curtailed sequential procedure related to the single-stage

INDIFFERENCE-ZONE PROCEDURES FOR MULTINOMIAL DATA

procedure is described in Section 8.2.2. As demonstrated in Section 2.3.2 for the normal means model with common known variance, when sequential procedures are feasible for a particular application, they can often produce substantial savings in the total number of observations to terminate sampling relative to single-stage procedures that guarantee the same probability requirement. Section 8.2.3 describes a closed sequential procedure that guarantees (8.2.1). Finally, Section 8.2.4 provides several additional applications of the multinomial selection problem.

8.2.1 A Single-Stage Procedure

The following *single-stage* procedure was proposed by Bechhofer, Elmaghraby and Morse (1959) to guarantee (8.2.1).

Procedure $\mathcal{M}_{BEM}$ For the given t and specified (θ^*, P^*), find n from Table 8.1 or Tables 8.2–8.5.

Sampling rule: Take a random sample of n multinomial observations $X_j = (X_{1j}, \ldots, X_{tj})$ ($1 \leq j \leq n$) in a *single* stage.

Terminal decision rule: Calculate the ordered sample sums $y_{[1]n} \leq \cdots \leq y_{[t]n}$. Select the category that yielded the largest sample sum, $y_{[t]n}$, as the one associated with $p_{[t]}$, randomizing to break ties.

The n-values found in Table 8.1 (some of which are abstracted from Gibbons, Olkin and Sobel 1977) are given for $t = 2(1)8, 10$ and selected (θ^*, P^*). They are computed so that procedure $\mathcal{M}_{BEM}$ achieves the nominal probability of correct selection P^* when the cell probabilities p are in the (least-favorable) configuration

$$p_{[1]} = p_{[t-1]} = 1/(\theta^* + t - 1) \quad \text{and} \quad p_{[t]} = \theta^*/(\theta^* + t - 1) \qquad (8.2.2)$$

as first determined by Kesten and Morse (1959). Alternatively, Tables 8.2–8.5 list the $P\{CS \mid LF\}$ for $t = 2(1)5$, $\theta^* = 1.2(0.2)2.0$, and a very detailed range of n. Using these tables, the experimenter can enter a row with a fixed sample size n and see the effect of increasing θ^* on the $P\{CS \mid LF\}$; equivalently the experimenter can enter a column with fixed θ^* and see the effect of increasing n on the $P\{CS \mid LF\}$.

Example 8.2.1 Suppose that a soft drink producer wishes to determine which of three proposed cola formulations will be the most popular. The company decides to give a taste test to n individuals, asking each person to state the brand they prefer. The company will declare as best that cola corresponding to the largest observed proportion of positive responses. The sample size n is to be chosen in such a way that a correct selection will be made with probability at least 0.95 whenever the ratio of the largest to second largest true (but unknown) proportions is at least 1.4. Entering Table 8.1 with $t = 3$, $P^* = 0.95$ and the ratio $\theta^* = 1.4$, we find that $n = 186$ individuals must be polled.

Table 8.1. Smallest Sample Size n Needed for the Single-Stage Multinomial Procedure $\mathcal{M}_{BEM}$ to Guarantee (8.2.1)

t	θ^*	P^* 0.75	P^* 0.90	P^* 0.95	P^* 0.99	t	θ^*	P^* 0.75	P^* 0.90	P^* 0.95	P^* 0.99
2	3.0	1	7	9	19	6	3.0	14	26	36	58
	2.8	3	7	11	21		2.8	16	30	41	68
	2.6	3	7	13	25		2.6	19	36	49	80
	2.4	3	9	15	29		2.4	23	44	60	97
	2.2	3	11	19	37		2.2	29	56	76	122
	2.0	5	15	23	47		2.0	38	74	101	163
	1.8	5	19	33	65		1.8	56	106	144	233
	1.6	9	31	49	101		1.6	90	172	233	377
	1.4	17	59	97	193		1.4	184	349	475	766
	1.2	55	199	327	653		1.2	658	1249	1697	2737
3	3.0	5	11	17	29	7	3.0	17	31	42	68
	2.8	6	13	19	33		2.8	20	36	49	79
	2.6	6	15	22	39		2.6	23	43	59	94
	2.4	7	18	26	46		2.4	28	53	72	114
	2.2	9	22	32	58		2.2	36	68	91	145
	2.0	12	29	42	75		2.0	48	90	121	193
	1.8	17	40	59	106		1.8	70	130	174	277
	1.6	26	64	94	167		1.6	114	211	283	450
	1.4	52	126	186	330		1.4	234	430	578	918
	1.2	181	437	645	1148		1.2	840	1545	2075	3297
4	3.0	8	16	23	39	8	3.0	20	37	49	78
	2.8	9	19	26	45		2.8	23	43	58	91
	2.6	10	22	31	53		2.6	28	51	69	108
	2.4	12	26	37	63		2.4	34	63	84	132
	2.2	15	33	46	79		2.2	43	80	106	167
	2.0	20	43	61	104		2.0	59	107	142	224
	1.8	29	61	87	147		1.8	86	154	205	322
	1.6	46	98	139	235		1.6	140	251	334	525
	1.4	92	196	278	471		1.4	286	514	684	1074
	1.2	326	692	979	1660		1.2	1030	1851	2464	3869
5	3.0	11	21	29	48	10	3.0	27	47	64	99
	2.8	12	24	34	56		2.8	31	57	74	115
	2.6	14	29	40	66		2.6	37	67	89	137
	2.4	17	35	48	80		2.4	46	83	109	168
	2.2	22	44	61	100		2.2	60	105	138	214
	2.0	29	58	81	133		2.0	81	141	186	287
	1.8	41	83	115	189		1.8	118	204	269	415
	1.6	68	134	185	305		1.6	192	335	440	679
	1.4	137	271	374	616		1.4	396	688	903	1394
	1.2	486	964	1331	2191		1.2	1433	2489	3269	5043

Reprinted from Gibbons, Olkin and Sobel (1977), p. 445–447, by courtesy of John Wiley & Sons.

Confidence Statement Formulation

If the sample size is chosen to guarantee (8.2.1), then the experimenter can assert that the probability p_S associated with the *selected* category satisfies

$$P\{p_{[t]}/\theta^* \leq p_S \leq p_{[t]} \mid \boldsymbol{p}\} \geq P^* \quad \text{for all } \boldsymbol{p}.$$

8.2.2 Use of Curtailment When the Maximum Number of Observations Is Specified

Bechhofer and Kulkarni (1984) proposed a closed sequential procedure for selecting the multinomial category associated with $p_{[t]}$ when the maximum total number of observations is specified to be n, say. Their procedure $\mathcal{M}_{BK}$ employs *curtailment* (as does procedure $\mathcal{B}_{BK}$ for the Bernoulli selection problem) and achieves the same probability of CS as does $\mathcal{M}_{BEM}$, that is, it has the property

$$P\{\text{CS using procedure } \mathcal{M}_{BK} \mid \boldsymbol{p}\} = P\{\text{CS using procedure } \mathcal{M}_{BEM} \mid \boldsymbol{p}\}$$

uniformly in $\boldsymbol{p}$. However, the special virtue of procedure $\mathcal{M}_{BK}$ is that

$$E\{N \text{ using procedure } \mathcal{M}_{BK} \mid \boldsymbol{p}\} \leq n \text{ using procedure } \mathcal{M}_{BEM}$$

uniformly in $\boldsymbol{p}$, where N is the (random) number of multinomial observations to the termination of sampling. Furthermore, these savings in $E\{N\}$ increase dramatically as $p_{[t]}$ approaches unity.

Unlike procedure $\mathcal{M}_{BEM}$, the distance measure θ^* does *not* play a role. The choice of the maximum number n of multinomial observations permitted can be made using criteria such as cost or availability of observations. Of course, one could also choose n to guarantee the probability requirement (8.2.1).

Procedure $\mathcal{M}_{BK}$ For the given t, specify n prior to the start of sampling.

Sampling rule: At the mth stage of experimentation ($m \geq 1$), take the random multinomial observation $X_m = (X_{1m}, \ldots, X_{tm})$.

Stopping rule: Calculate the sample sums y_{im} through stage m ($1 \leq i \leq t$). Stop sampling at the first stage m for which there exists a category C_i satisfying

$$y_{im} \geq y_{jm} + n - m \quad \text{for all } j \neq i \ (1 \leq i, j \leq t). \tag{8.2.3}$$

Terminal decision rule: Let N (a random variable) denote the value of m at the termination of sampling. If $N < n$, the procedure must terminate with a single category; select this category as the one associated with $p_{[t]}$. If $N = n$ and several categories simultaneously satisfy (8.2.3), then randomize to break ties.

The left-hand side of (8.2.3) represents the current total number of occurrences C_i; the right-hand side represents the current total number of occurrences C_j plus

Table 8.2. $P\{CS \mid LF\}$ for the Single-Stage Multinomial Procedure $\mathcal{M}_{BEM}$ as a Function of n and $\theta^\star$ for $t = 2$

		$\theta^\star$						$\theta^\star$			
n	1.2	1.4	1.6	1.8	2.0	n	1.2	1.4	1.6	1.8	2.0
1	0.5455	0.5833	0.6154	0.6429	0.6667	51	0.7434	0.8858	0.9534	0.9819	0.9931
2	0.5455	0.5833	0.6154	0.6429	0.6667	52	0.7434	0.8858	0.9534	0.9819	0.9931
3	0.5680	0.6238	0.6700	0.7085	0.7407	53	0.7474	0.8902	0.9565	0.9836	0.9940
4	0.5680	0.6238	0.6700	0.7085	0.7407	54	0.7474	0.8902	0.9565	0.9836	0.9940
5	0.5848	0.6534	0.7088	0.7536	0.7901	55	0.7513	0.8944	0.9593	0.9852	0.9947
6	0.5848	0.6534	0.7088	0.7536	0.7901	56	0.7513	0.8944	0.9593	0.9852	0.9947
7	0.5986	0.6773	0.7394	0.7882	0.8267	57	0.7552	0.8984	0.9620	0.9865	0.9954
8	0.5986	0.6773	0.7394	0.7882	0.8267	58	0.7552	0.8984	0.9620	0.9865	0.9954
9	0.6106	0.6977	0.7647	0.8160	0.8552	59	0.7589	0.9023	0.9644	0.9878	0.9959
10	0.6106	0.6977	0.7647	0.8160	0.8552	60	0.7589	0.9023	0.9644	0.9878	0.9959
11	0.6214	0.7155	0.7863	0.8390	0.8779	61	0.7625	0.9060	0.9667	0.9889	0.9964
12	0.6214	0.7155	0.7863	0.8390	0.8779	62	0.7625	0.9060	0.9667	0.9889	0.9964
13	0.6311	0.7314	0.8051	0.8583	0.8965	63	0.7661	0.9095	0.9689	0.9900	0.9969
14	0.6311	0.7314	0.8051	0.8583	0.8965	64	0.7661	0.9095	0.9689	0.9900	0.9969
15	0.6401	0.7457	0.8216	0.8748	0.9118	65	0.7695	0.9128	0.9708	0.9909	0.9973
16	0.6401	0.7457	0.8216	0.8748	0.9118	66	0.7695	0.9128	0.9708	0.9909	0.9973
17	0.6485	0.7588	0.8362	0.8890	0.9245	67	0.7729	0.9161	0.9727	0.9917	0.9976
18	0.6485	0.7588	0.8362	0.8890	0.9245	68	0.7729	0.9161	0.9727	0.9917	0.9976
19	0.6563	0.7707	0.8493	0.9013	0.9352	69	0.7762	0.9191	0.9744	0.9925	0.9979
20	0.6563	0.7707	0.8493	0.9013	0.9352	70	0.7762	0.9191	0.9744	0.9925	0.9979
21	0.6637	0.7818	0.8610	0.9120	0.9443	71	0.7794	0.9221	0.9760	0.9932	0.9981
22	0.6637	0.7818	0.8610	0.9120	0.9443	72	0.7794	0.9221	0.9760	0.9932	0.9981
23	0.6706	0.7921	0.8716	0.9215	0.9520	73	0.7826	0.9249	0.9776	0.9938	0.9984
24	0.6706	0.7921	0.8716	0.9215	0.9520	74	0.7826	0.9249	0.9776	0.9938	0.9984
25	0.6773	0.8017	0.8813	0.9297	0.9585	75	0.7857	0.9276	0.9790	0.9944	0.9986
26	0.6773	0.8017	0.8813	0.9297	0.9585	76	0.7857	0.9276	0.9790	0.9944	0.9986
27	0.6836	0.8106	0.8901	0.9371	0.9641	77	0.7887	0.9302	0.9803	0.9949	0.9987
28	0.6836	0.8106	0.8901	0.9371	0.9641	78	0.7887	0.9302	0.9803	0.9949	0.9987
29	0.6896	0.8190	0.8981	0.9435	0.9689	79	0.7917	0.9327	0.9815	0.9954	0.9989
30	0.6896	0.8190	0.8981	0.9435	0.9689	80	0.7917	0.9327	0.9815	0.9954	0.9989
31	0.6954	0.8269	0.9054	0.9493	0.9730	81	0.7946	0.9351	0.9827	0.9958	0.9990
32	0.6954	0.8269	0.9054	0.9493	0.9730	82	0.7946	0.9351	0.9827	0.9958	0.9990
33	0.7010	0.8344	0.9121	0.9544	0.9765	83	0.7974	0.9375	0.9837	0.9962	0.9991
34	0.7010	0.8344	0.9121	0.9544	0.9765	84	0.7974	0.9375	0.9837	0.9962	0.9991
35	0.7064	0.8414	0.9183	0.9590	0.9796	85	0.8002	0.9397	0.9847	0.9965	0.9992
36	0.7064	0.8414	0.9183	0.9590	0.9796	86	0.8002	0.9397	0.9847	0.9965	0.9992
37	0.7115	0.8480	0.9240	0.9630	0.9822	87	0.8029	0.9418	0.9857	0.9968	0.9993
38	0.7115	0.8480	0.9240	0.9630	0.9822	88	0.8029	0.9418	0.9857	0.9968	0.9993
39	0.7165	0.8543	0.9293	0.9667	0.9845	89	0.8056	0.9438	0.9866	0.9971	0.9994
40	0.7165	0.8543	0.9293	0.9667	0.9845	90	0.8056	0.9438	0.9866	0.9971	0.9994
41	0.7214	0.8602	0.9341	0.9699	0.9865	91	0.8082	0.9458	0.9874	0.9974	0.9995
42	0.7214	0.8602	0.9341	0.9699	0.9865	92	0.8082	0.9458	0.9874	0.9974	0.9995
43	0.7260	0.8659	0.9386	0.9729	0.9882	93	0.8108	0.9477	0.9882	0.9976	0.9995
44	0.7260	0.8659	0.9386	0.9729	0.9882	94	0.8108	0.9477	0.9882	0.9976	0.9995
45	0.7306	0.8712	0.9427	0.9755	0.9897	95	0.8133	0.9495	0.9889	0.9978	0.9996
46	0.7306	0.8712	0.9427	0.9755	0.9897	96	0.8133	0.9495	0.9889	0.9978	0.9996
47	0.7350	0.8763	0.9466	0.9779	0.9910	97	0.8158	0.9513	0.9896	0.9980	0.9996
48	0.7350	0.8763	0.9466	0.9779	0.9910	98	0.8158	0.9513	0.9896	0.9980	0.9996
49	0.7392	0.8812	0.9501	0.9800	0.9921	99	0.8182	0.9530	0.9902	0.9982	0.9997
50	0.7392	0.8812	0.9501	0.9800	0.9921	100	0.8182	0.9530	0.9902	0.9982	0.9997

INDIFFERENCE-ZONE PROCEDURES FOR MULTINOMIAL DATA

Table 8.2. *Continued*, $t = 2$

	θ^*						θ^*				
n	1.2	1.4	1.6	1.8	2.0	n	1.2	1.4	1.6	1.8	2.0
102	0.8206	0.9546	0.9908	0.9984	0.9997	204	0.9031	0.9917	0.9996	1.0000	1.0000
104	0.8230	0.9561	0.9913	0.9985	0.9998	208	0.9053	0.9922	0.9996	1.0000	1.0000
106	0.8253	0.9576	0.9919	0.9986	0.9998	212	0.9074	0.9927	0.9997	1.0000	1.0000
108	0.8276	0.9591	0.9924	0.9988	0.9998	216	0.9094	0.9931	0.9997	1.0000	1.0000
110	0.8298	0.9605	0.9928	0.9989	0.9998	220	0.9114	0.9936	0.9997	1.0000	1.0000
112	0.8320	0.9618	0.9932	0.9990	0.9999	224	0.9134	0.9940	0.9998	1.0000	1.0000
114	0.8341	0.9631	0.9937	0.9991	0.9999	228	0.9153	0.9943	0.9998	1.0000	1.0000
116	0.8362	0.9644	0.9940	0.9991	0.9999	232	0.9171	0.9947	0.9998	1.0000	1.0000
118	0.8383	0.9656	0.9944	0.9992	0.9999	236	0.9189	0.9950	0.9998	1.0000	1.0000
120	0.8403	0.9667	0.9947	0.9993	0.9999	240	0.9207	0.9953	0.9999	1.0000	1.0000
122	0.8423	0.9678	0.9950	0.9994	0.9999	244	0.9224	0.9956	0.9999	1.0000	1.0000
124	0.8443	0.9689	0.9953	0.9994	0.9999	248	0.9241	0.9959	0.9999	1.0000	1.0000
126	0.8463	0.9700	0.9956	0.9995	0.9999	252	0.9257	0.9961	0.9999	1.0000	1.0000
128	0.8482	0.9710	0.9959	0.9995	0.9999	256	0.9273	0.9964	0.9999	1.0000	1.0000
130	0.8501	0.9719	0.9961	0.9996	1.0000	260	0.9289	0.9966	0.9999	1.0000	1.0000
132	0.8519	0.9729	0.9963	0.9996	1.0000	264	0.9304	0.9968	0.9999	1.0000	1.0000
134	0.8537	0.9738	0.9966	0.9996	1.0000	268	0.9319	0.9970	0.9999	1.0000	1.0000
136	0.8555	0.9746	0.9968	0.9997	1.0000	272	0.9333	0.9972	0.9999	1.0000	1.0000
138	0.8573	0.9755	0.9970	0.9997	1.0000	276	0.9347	0.9973	0.9999	1.0000	1.0000
140	0.8590	0.9763	0.9971	0.9997	1.0000	280	0.9361	0.9975	1.0000	1.0000	1.0000
142	0.8607	0.9771	0.9973	0.9997	1.0000	284	0.9375	0.9976	1.0000	1.0000	1.0000
144	0.8624	0.9778	0.9975	0.9998	1.0000	288	0.9388	0.9978	1.0000	1.0000	1.0000
146	0.8641	0.9785	0.9976	0.9998	1.0000	292	0.9401	0.9979	1.0000	1.0000	1.0000
148	0.8657	0.9792	0.9978	0.9998	1.0000	296	0.9413	0.9980	1.0000	1.0000	1.0000
150	0.8673	0.9799	0.9979	0.9998	1.0000	300	0.9426	0.9982	1.0000	1.0000	1.0000
152	0.8689	0.9806	0.9980	0.9998	1.0000	304	0.9437	0.9983	1.0000	1.0000	1.0000
154	0.8705	0.9812	0.9981	0.9999	1.0000	308	0.9449	0.9984	1.0000	1.0000	1.0000
156	0.8720	0.9818	0.9982	0.9999	1.0000	312	0.9461	0.9985	1.0000	1.0000	1.0000
158	0.8735	0.9824	0.9983	0.9999	1.0000	316	0.9472	0.9986	1.0000	1.0000	1.0000
160	0.8750	0.9830	0.9984	0.9999	1.0000	320	0.9483	0.9987	1.0000	1.0000	1.0000
162	0.8765	0.9835	0.9985	0.9999	1.0000	324	0.9493	0.9987	1.0000	1.0000	1.0000
164	0.8780	0.9841	0.9986	0.9999	1.0000	328	0.9504	0.9988	1.0000	1.0000	1.0000
166	0.8794	0.9846	0.9987	0.9999	1.0000	332	0.9514	0.9989	1.0000	1.0000	1.0000
168	0.8808	0.9851	0.9988	0.9999	1.0000	336	0.9524	0.9989	1.0000	1.0000	1.0000
170	0.8822	0.9856	0.9988	0.9999	1.0000	340	0.9534	0.9990	1.0000	1.0000	1.0000
172	0.8836	0.9860	0.9989	0.9999	1.0000	344	0.9543	0.9991	1.0000	1.0000	1.0000
174	0.8849	0.9865	0.9990	0.9999	1.0000	348	0.9553	0.9991	1.0000	1.0000	1.0000
176	0.8863	0.9869	0.9990	0.9999	1.0000	352	0.9562	0.9992	1.0000	1.0000	1.0000
178	0.8876	0.9873	0.9991	0.9999	1.0000	356	0.9571	0.9992	1.0000	1.0000	1.0000
180	0.8889	0.9877	0.9991	1.0000	1.0000	360	0.9579	0.9993	1.0000	1.0000	1.0000
182	0.8901	0.9881	0.9992	1.0000	1.0000	364	0.9588	0.9993	1.0000	1.0000	1.0000
184	0.8914	0.9885	0.9992	1.0000	1.0000	368	0.9596	0.9994	1.0000	1.0000	1.0000
186	0.8926	0.9889	0.9993	1.0000	1.0000	372	0.9604	0.9994	1.0000	1.0000	1.0000
188	0.8939	0.9892	0.9993	1.0000	1.0000	376	0.9612	0.9994	1.0000	1.0000	1.0000
190	0.8951	0.9896	0.9994	1.0000	1.0000	380	0.9620	0.9995	1.0000	1.0000	1.0000
192	0.8963	0.9899	0.9994	1.0000	1.0000	384	0.9628	0.9995	1.0000	1.0000	1.0000
194	0.8975	0.9902	0.9994	1.0000	1.0000	388	0.9635	0.9995	1.0000	1.0000	1.0000
196	0.8986	0.9905	0.9995	1.0000	1.0000	392	0.9643	0.9996	1.0000	1.0000	1.0000
198	0.8998	0.9908	0.9995	1.0000	1.0000	396	0.9650	0.9996	1.0000	1.0000	1.0000
200	0.9009	0.9911	0.9995	1.0000	1.0000	400	0.9657	0.9996	1.0000	1.0000	1.0000

Table 8.3. $P\{CS \mid LF\}$ for the Single-Stage Multinomial Procedure $\mathcal{M}_{BEM}$ as a Function of n and $\theta^\star$ for $t = 3$

		$\theta^\star$						$\theta^\star$			
n	1.2	1.4	1.6	1.8	2.0	n	1.2	1.4	1.6	1.8	2.0
1	0.3750	0.4118	0.4444	0.4737	0.5000	51	0.5651	0.7494	0.8679	0.9344	0.9686
2	0.3750	0.4118	0.4444	0.4737	0.5000	52	0.5675	0.7529	0.8711	0.9367	0.9702
3	0.3896	0.4403	0.4856	0.5262	0.5625	53	0.5695	0.7560	0.8740	0.9389	0.9716
4	0.3988	0.4570	0.5085	0.5538	0.5938	54	0.5718	0.7593	0.8770	0.9411	0.9730
5	0.4040	0.4674	0.5237	0.5734	0.6172	55	0.5741	0.7626	0.8800	0.9432	0.9743
6	0.4120	0.4826	0.5449	0.5992	0.6465	56	0.5761	0.7655	0.8827	0.9452	0.9755
7	0.4190	0.4957	0.5629	0.6210	0.6709	57	0.5783	0.7687	0.8855	0.9471	0.9767
8	0.4234	0.5043	0.5751	0.6360	0.6880	58	0.5805	0.7718	0.8883	0.9490	0.9778
9	0.4298	0.5162	0.5914	0.6554	0.7094	59	0.5824	0.7747	0.8908	0.9508	0.9788
10	0.4355	0.5269	0.6057	0.6721	0.7275	60	0.5846	0.7777	0.8934	0.9525	0.9799
11	0.4395	0.5345	0.6163	0.6848	0.7414	61	0.5867	0.7807	0.8959	0.9542	0.9808
12	0.4449	0.5445	0.6295	0.7001	0.7577	62	0.5886	0.7834	0.8982	0.9558	0.9817
13	0.4498	0.5536	0.6415	0.7138	0.7721	63	0.5907	0.7863	0.9007	0.9573	0.9826
14	0.4535	0.5605	0.6509	0.7246	0.7835	64	0.5928	0.7891	0.9030	0.9588	0.9834
15	0.4582	0.5692	0.6622	0.7372	0.7965	65	0.5946	0.7917	0.9052	0.9602	0.9842
16	0.4627	0.5772	0.6725	0.7486	0.8080	66	0.5966	0.7944	0.9074	0.9617	0.9850
17	0.4660	0.5836	0.6808	0.7580	0.8175	67	0.5986	0.7971	0.9096	0.9630	0.9857
18	0.4703	0.5913	0.6906	0.7686	0.8280	68	0.6004	0.7996	0.9116	0.9643	0.9863
19	0.4744	0.5985	0.6997	0.7783	0.8375	69	0.6024	0.8022	0.9137	0.9655	0.9870
20	0.4776	0.6044	0.7072	0.7864	0.8455	70	0.6044	0.8048	0.9157	0.9667	0.9876
21	0.4815	0.6113	0.7158	0.7954	0.8541	71	0.6061	0.8071	0.9175	0.9678	0.9882
22	0.4852	0.6179	0.7239	0.8038	0.8620	72	0.6081	0.8096	0.9195	0.9690	0.9887
23	0.4882	0.6234	0.7306	0.8109	0.8687	73	0.6100	0.8121	0.9213	0.9701	0.9893
24	0.4919	0.6297	0.7383	0.8187	0.8758	74	0.6117	0.8144	0.9231	0.9711	0.9898
25	0.4954	0.6358	0.7455	0.8259	0.8824	75	0.6136	0.8168	0.9249	0.9721	0.9903
26	0.4982	0.6408	0.7517	0.8321	0.8881	76	0.6154	0.8191	0.9266	0.9730	0.9907
27	0.5016	0.6467	0.7586	0.8389	0.8941	77	0.6171	0.8213	0.9282	0.9740	0.9912
28	0.5049	0.6523	0.7651	0.8452	0.8996	78	0.6189	0.8236	0.9299	0.9749	0.9916
29	0.5076	0.6571	0.7707	0.8507	0.9044	79	0.6207	0.8258	0.9315	0.9757	0.9920
30	0.5108	0.6625	0.7769	0.8566	0.9094	80	0.6224	0.8279	0.9330	0.9765	0.9923
31	0.5140	0.6677	0.7828	0.8621	0.9141	81	0.6242	0.8301	0.9345	0.9774	0.9927
32	0.5166	0.6722	0.7879	0.8670	0.9182	82	0.6259	0.8323	0.9360	0.9781	0.9930
33	0.5196	0.6773	0.7935	0.8722	0.9224	83	0.6276	0.8343	0.9374	0.9789	0.9934
34	0.5226	0.6821	0.7989	0.8770	0.9264	84	0.6293	0.8364	0.9389	0.9796	0.9937
35	0.5251	0.6864	0.8036	0.8813	0.9299	85	0.6310	0.8384	0.9403	0.9803	0.9940
36	0.5280	0.6911	0.8087	0.8859	0.9335	86	0.6326	0.8403	0.9416	0.9810	0.9943
37	0.5308	0.6957	0.8136	0.8902	0.9368	87	0.6343	0.8423	0.9429	0.9816	0.9945
38	0.5332	0.6997	0.8180	0.8940	0.9398	88	0.6360	0.8443	0.9442	0.9823	0.9948
39	0.5360	0.7041	0.8227	0.8980	0.9429	89	0.6375	0.8462	0.9454	0.9829	0.9950
40	0.5387	0.7084	0.8271	0.9018	0.9457	90	0.6392	0.8481	0.9467	0.9834	0.9953
41	0.5410	0.7122	0.8311	0.9052	0.9483	91	0.6409	0.8500	0.9479	0.9840	0.9955
42	0.5437	0.7164	0.8354	0.9088	0.9509	92	0.6424	0.8517	0.9490	0.9845	0.9957
43	0.5463	0.7205	0.8395	0.9121	0.9533	93	0.6440	0.8536	0.9502	0.9851	0.9959
44	0.5485	0.7240	0.8432	0.9152	0.9555	94	0.6456	0.8554	0.9513	0.9856	0.9961
45	0.5511	0.7280	0.8471	0.9183	0.9578	95	0.6471	0.8571	0.9523	0.9861	0.9963
46	0.5536	0.7318	0.8509	0.9213	0.9598	96	0.6487	0.8589	0.9534	0.9865	0.9964
47	0.5558	0.7353	0.8543	0.9240	0.9617	97	0.6503	0.8606	0.9545	0.9870	0.9966
48	0.5582	0.7390	0.8579	0.9268	0.9636	98	0.6517	0.8623	0.9555	0.9874	0.9968
49	0.5606	0.7426	0.8614	0.9295	0.9654	99	0.6533	0.8640	0.9565	0.9879	0.9969
50	0.5628	0.7459	0.8645	0.9319	0.9670	100	0.6549	0.8656	0.9574	0.9883	0.9970

Table 8.3. *Continued*, $t = 3$

n	θ^*					n	θ^*				
	1.2	1.4	1.6	1.8	2.0		1.2	1.4	1.6	1.8	2.0
102	0.6578	0.8688	0.9593	0.9891	0.9973	204	0.7728	0.9600	0.9956	0.9996	1.0000
104	0.6607	0.8719	0.9611	0.9898	0.9976	208	0.7763	0.9617	0.9960	0.9997	1.0000
106	0.6637	0.8751	0.9628	0.9905	0.9978	212	0.7796	0.9634	0.9963	0.9997	1.0000
108	0.6666	0.8780	0.9644	0.9911	0.9980	216	0.7829	0.9650	0.9966	0.9998	1.0000
110	0.6694	0.8809	0.9660	0.9917	0.9982	220	0.7861	0.9666	0.9969	0.9998	1.0000
112	0.6723	0.8838	0.9675	0.9922	0.9983	224	0.7893	0.9680	0.9971	0.9998	1.0000
114	0.6750	0.8865	0.9689	0.9927	0.9985	228	0.7924	0.9694	0.9974	0.9998	1.0000
116	0.6777	0.8892	0.9702	0.9932	0.9986	232	0.7955	0.9708	0.9976	0.9999	1.0000
118	0.6805	0.8918	0.9715	0.9937	0.9987	236	0.7984	0.9720	0.9978	0.9999	1.0000
120	0.6831	0.8943	0.9727	0.9941	0.9989	240	0.8014	0.9733	0.9980	0.9999	1.0000
122	0.6857	0.8968	0.9739	0.9945	0.9990	244	0.8043	0.9744	0.9981	0.9999	1.0000
124	0.6884	0.8993	0.9751	0.9948	0.9990	248	0.8072	0.9755	0.9983	0.9999	1.0000
126	0.6910	0.9016	0.9761	0.9952	0.9991	252	0.8100	0.9766	0.9984	0.9999	1.0000
128	0.6935	0.9039	0.9771	0.9955	0.9992	256	0.8127	0.9776	0.9985	0.9999	1.0000
130	0.6961	0.9062	0.9781	0.9958	0.9993	260	0.8154	0.9786	0.9987	0.9999	1.0000
132	0.6985	0.9083	0.9791	0.9961	0.9993	264	0.8181	0.9795	0.9988	1.0000	1.0000
134	0.7010	0.9104	0.9800	0.9963	0.9994	268	0.8207	0.9804	0.9989	1.0000	1.0000
136	0.7035	0.9125	0.9808	0.9966	0.9995	272	0.8233	0.9813	0.9990	1.0000	1.0000
138	0.7059	0.9146	0.9817	0.9968	0.9995	276	0.8258	0.9821	0.9990	1.0000	1.0000
140	0.7082	0.9165	0.9824	0.9970	0.9995	280	0.8283	0.9828	0.9991	1.0000	1.0000
142	0.7106	0.9185	0.9832	0.9972	0.9996	284	0.8308	0.9836	0.9992	1.0000	1.0000
144	0.7129	0.9204	0.9839	0.9974	0.9996	288	0.8332	0.9843	0.9993	1.0000	1.0000
146	0.7152	0.9222	0.9846	0.9975	0.9997	292	0.8356	0.9850	0.9993	1.0000	1.0000
148	0.7175	0.9240	0.9853	0.9977	0.9997	296	0.8379	0.9856	0.9994	1.0000	1.0000
150	0.7198	0.9257	0.9859	0.9979	0.9997	300	0.8402	0.9862	0.9994	1.0000	1.0000
152	0.7220	0.9274	0.9865	0.9980	0.9997	304	0.8425	0.9868	0.9995	1.0000	1.0000
154	0.7242	0.9291	0.9871	0.9981	0.9998	308	0.8447	0.9874	0.9995	1.0000	1.0000
156	0.7264	0.9307	0.9876	0.9982	0.9998	312	0.8469	0.9879	0.9996	1.0000	1.0000
158	0.7285	0.9323	0.9881	0.9984	0.9998	316	0.8491	0.9884	0.9996	1.0000	1.0000
160	0.7307	0.9339	0.9886	0.9985	0.9998	320	0.8512	0.9889	0.9996	1.0000	1.0000
162	0.7328	0.9354	0.9891	0.9986	0.9998	324	0.8533	0.9894	0.9997	1.0000	1.0000
164	0.7349	0.9368	0.9896	0.9987	0.9999	328	0.8553	0.9899	0.9997	1.0000	1.0000
166	0.7370	0.9383	0.9900	0.9987	0.9999	332	0.8573	0.9903	0.9997	1.0000	1.0000
168	0.7391	0.9397	0.9905	0.9988	0.9999	336	0.8593	0.9907	0.9997	1.0000	1.0000
170	0.7411	0.9410	0.9909	0.9989	0.9999	340	0.8613	0.9911	0.9998	1.0000	1.0000
172	0.7431	0.9424	0.9912	0.9990	0.9999	344	0.8632	0.9915	0.9998	1.0000	1.0000
174	0.7451	0.9437	0.9916	0.9990	0.9999	348	0.8651	0.9919	0.9998	1.0000	1.0000
176	0.7471	0.9449	0.9920	0.9991	0.9999	352	0.8670	0.9922	0.9998	1.0000	1.0000
178	0.7491	0.9462	0.9923	0.9992	0.9999	356	0.8688	0.9925	0.9998	1.0000	1.0000
180	0.7510	0.9474	0.9926	0.9992	0.9999	360	0.8707	0.9928	0.9998	1.0000	1.0000
182	0.7529	0.9486	0.9929	0.9993	0.9999	364	0.8724	0.9932	0.9998	1.0000	1.0000
184	0.7548	0.9498	0.9932	0.9993	0.9999	368	0.8742	0.9934	0.9999	1.0000	1.0000
186	0.7567	0.9509	0.9935	0.9994	0.9999	372	0.8759	0.9937	0.9999	1.0000	1.0000
188	0.7585	0.9520	0.9938	0.9994	1.0000	376	0.8777	0.9940	0.9999	1.0000	1.0000
190	0.7604	0.9531	0.9941	0.9994	1.0000	380	0.8793	0.9942	0.9999	1.0000	1.0000
192	0.7622	0.9541	0.9943	0.9995	1.0000	384	0.8810	0.9945	0.9999	1.0000	1.0000
194	0.7640	0.9552	0.9945	0.9995	1.0000	388	0.8826	0.9947	0.9999	1.0000	1.0000
196	0.7658	0.9562	0.9948	0.9995	1.0000	392	0.8842	0.9949	0.9999	1.0000	1.0000
198	0.7676	0.9571	0.9950	0.9996	1.0000	396	0.8858	0.9952	0.9999	1.0000	1.0000
200	0.7693	0.9581	0.9952	0.9996	1.0000	400	0.8874	0.9954	0.9999	1.0000	1.0000

Table 8.4. $P\{CS \mid LF\}$ for the Single-Stage Multinomial Procedure $\mathcal{M}_{BEM}$ as a Function of n and θ^* for $t = 4$

	θ^*						θ^*				
n	1.2	1.4	1.6	1.8	2.0	n	1.2	1.4	1.6	1.8	2.0
1	0.2857	0.3182	0.3478	0.3750	0.4000	51	0.4489	0.6344	0.7760	0.8706	0.9284
2	0.2857	0.3182	0.3478	0.3750	0.4000	52	0.4510	0.6380	0.7800	0.8742	0.9311
3	0.2954	0.3379	0.3774	0.4141	0.4480	53	0.4530	0.6416	0.7840	0.8776	0.9337
4	0.3047	0.3558	0.4031	0.4466	0.4864	54	0.4550	0.6451	0.7878	0.8810	0.9362
5	0.3100	0.3662	0.4182	0.4659	0.5094	55	0.4571	0.6486	0.7917	0.8842	0.9386
6	0.3147	0.3759	0.4328	0.4850	0.5325	56	0.4591	0.6521	0.7954	0.8874	0.9409
7	0.3208	0.3880	0.4504	0.5073	0.5586	57	0.4610	0.6555	0.7990	0.8905	0.9432
8	0.3260	0.3983	0.4652	0.5258	0.5801	58	0.4630	0.6588	0.8026	0.8935	0.9453
9	0.3303	0.4070	0.4777	0.5417	0.5986	59	0.4649	0.6621	0.8062	0.8964	0.9474
10	0.3345	0.4155	0.4902	0.5574	0.6168	60	0.4668	0.6654	0.8096	0.8992	0.9494
11	0.3391	0.4245	0.5031	0.5734	0.6350	61	0.4687	0.6687	0.8130	0.9020	0.9513
12	0.3432	0.4328	0.5148	0.5878	0.6512	62	0.4706	0.6718	0.8163	0.9046	0.9531
13	0.3470	0.4403	0.5256	0.6010	0.6661	63	0.4725	0.6750	0.8196	0.9072	0.9549
14	0.3507	0.4477	0.5361	0.6139	0.6805	64	0.4744	0.6782	0.8227	0.9097	0.9565
15	0.3545	0.4552	0.5468	0.6267	0.6945	65	0.4762	0.6812	0.8259	0.9122	0.9582
16	0.3581	0.4624	0.5567	0.6386	0.7076	66	0.4781	0.6843	0.8289	0.9146	0.9597
17	0.3615	0.4691	0.5661	0.6498	0.7197	67	0.4799	0.6873	0.8320	0.9169	0.9613
18	0.3648	0.4756	0.5753	0.6606	0.7313	68	0.4817	0.6903	0.8349	0.9191	0.9627
19	0.3682	0.4823	0.5844	0.6713	0.7427	69	0.4835	0.6932	0.8378	0.9213	0.9641
20	0.3714	0.4887	0.5932	0.6814	0.7533	70	0.4853	0.6961	0.8406	0.9234	0.9654
21	0.3745	0.4947	0.6015	0.6910	0.7633	71	0.4870	0.6990	0.8434	0.9255	0.9667
22	0.3775	0.5006	0.6095	0.7003	0.7729	72	0.4888	0.7019	0.8462	0.9275	0.9680
23	0.3806	0.5066	0.6176	0.7094	0.7823	73	0.4905	0.7047	0.8489	0.9295	0.9692
24	0.3836	0.5124	0.6254	0.7181	0.7911	74	0.4923	0.7074	0.8515	0.9314	0.9703
25	0.3864	0.5180	0.6328	0.7264	0.7994	75	0.4940	0.7102	0.8541	0.9332	0.9714
26	0.3892	0.5234	0.6400	0.7345	0.8074	76	0.4957	0.7129	0.8566	0.9350	0.9725
27	0.3921	0.5289	0.6472	0.7424	0.8152	77	0.4974	0.7156	0.8591	0.9367	0.9735
28	0.3948	0.5342	0.6542	0.7499	0.8225	78	0.4991	0.7183	0.8616	0.9384	0.9745
29	0.3975	0.5394	0.6609	0.7572	0.8295	79	0.5008	0.7209	0.8640	0.9401	0.9755
30	0.4001	0.5444	0.6674	0.7642	0.8362	80	0.5025	0.7235	0.8663	0.9417	0.9764
31	0.4028	0.5495	0.6739	0.7711	0.8427	81	0.5041	0.7261	0.8687	0.9433	0.9772
32	0.4054	0.5544	0.6802	0.7777	0.8489	82	0.5057	0.7286	0.8709	0.9448	0.9781
33	0.4079	0.5592	0.6863	0.7840	0.8548	83	0.5074	0.7311	0.8732	0.9463	0.9789
34	0.4104	0.5639	0.6923	0.7902	0.8605	84	0.5090	0.7336	0.8754	0.9477	0.9797
35	0.4129	0.5687	0.6981	0.7962	0.8659	85	0.5106	0.7361	0.8775	0.9491	0.9804
36	0.4154	0.5733	0.7039	0.8020	0.8712	86	0.5122	0.7385	0.8796	0.9504	0.9812
37	0.4178	0.5778	0.7094	0.8076	0.8761	87	0.5138	0.7409	0.8817	0.9518	0.9819
38	0.4202	0.5822	0.7148	0.8131	0.8809	88	0.5154	0.7433	0.8837	0.9531	0.9825
39	0.4225	0.5866	0.7202	0.8184	0.8856	89	0.5170	0.7457	0.8857	0.9543	0.9832
40	0.4249	0.5909	0.7254	0.8235	0.8900	90	0.5186	0.7480	0.8877	0.9555	0.9838
41	0.4272	0.5951	0.7305	0.8284	0.8942	91	0.5201	0.7503	0.8896	0.9567	0.9844
42	0.4294	0.5993	0.7355	0.8332	0.8983	92	0.5217	0.7526	0.8915	0.9579	0.9850
43	0.4317	0.6034	0.7404	0.8379	0.9022	93	0.5232	0.7549	0.8934	0.9590	0.9855
44	0.4340	0.6075	0.7452	0.8425	0.9059	94	0.5247	0.7571	0.8952	0.9601	0.9861
45	0.4361	0.6115	0.7498	0.8468	0.9095	95	0.5263	0.7593	0.8970	0.9611	0.9866
46	0.4383	0.6154	0.7544	0.8511	0.9130	96	0.5278	0.7615	0.8988	0.9622	0.9871
47	0.4405	0.6193	0.7589	0.8552	0.9163	97	0.5293	0.7637	0.9005	0.9632	0.9876
48	0.4426	0.6232	0.7633	0.8593	0.9195	98	0.5308	0.7658	0.9022	0.9641	0.9880
49	0.4447	0.6269	0.7676	0.8631	0.9226	99	0.5323	0.7680	0.9039	0.9651	0.9885
50	0.4468	0.6306	0.7718	0.8669	0.9255	100	0.5338	0.7701	0.9055	0.9660	0.9889

Table 8.4. *Continued*, $t = 4$

n	$\theta^\star$ 1.2	1.4	1.6	1.8	2.0	n	$\theta^\star$ 1.2	1.4	1.6	1.8	2.0
102	0.5367	0.7742	0.9087	0.9678	0.9897	204	0.6564	0.9077	0.9836	0.9978	0.9998
104	0.5396	0.7783	0.9118	0.9695	0.9904	208	0.6602	0.9108	0.9847	0.9980	0.9998
106	0.5425	0.7823	0.9148	0.9711	0.9911	212	0.6639	0.9138	0.9857	0.9982	0.9998
108	0.5454	0.7862	0.9176	0.9726	0.9918	216	0.6676	0.9167	0.9866	0.9984	0.9998
110	0.5482	0.7900	0.9204	0.9740	0.9924	220	0.6712	0.9195	0.9874	0.9986	0.9999
112	0.5510	0.7937	0.9231	0.9754	0.9929	224	0.6748	0.9222	0.9883	0.9987	0.9999
114	0.5537	0.7974	0.9257	0.9766	0.9934	228	0.6784	0.9248	0.9890	0.9988	0.9999
116	0.5565	0.8010	0.9282	0.9779	0.9939	232	0.6818	0.9273	0.9897	0.9990	0.9999
118	0.5592	0.8045	0.9306	0.9790	0.9943	236	0.6853	0.9298	0.9904	0.9991	0.9999
120	0.5619	0.8080	0.9329	0.9801	0.9947	240	0.6887	0.9321	0.9910	0.9992	0.9999
122	0.5645	0.8114	0.9351	0.9811	0.9951	244	0.6920	0.9344	0.9916	0.9992	0.9999
124	0.5672	0.8147	0.9373	0.9821	0.9955	248	0.6953	0.9366	0.9921	0.9993	1.0000
126	0.5698	0.8180	0.9394	0.9830	0.9958	252	0.6986	0.9387	0.9926	0.9994	1.0000
128	0.5724	0.8212	0.9414	0.9839	0.9961	256	0.7018	0.9407	0.9931	0.9994	1.0000
130	0.5749	0.8244	0.9434	0.9848	0.9964	260	0.7050	0.9427	0.9935	0.9995	1.0000
132	0.5775	0.8275	0.9453	0.9855	0.9966	264	0.7081	0.9446	0.9939	0.9995	1.0000
134	0.5800	0.8305	0.9471	0.9863	0.9969	268	0.7112	0.9464	0.9943	0.9996	1.0000
136	0.5825	0.8335	0.9489	0.9870	0.9971	272	0.7143	0.9482	0.9947	0.9996	1.0000
138	0.5850	0.8364	0.9506	0.9877	0.9973	276	0.7173	0.9499	0.9950	0.9997	1.0000
140	0.5874	0.8392	0.9522	0.9883	0.9975	280	0.7203	0.9516	0.9953	0.9997	1.0000
142	0.5899	0.8420	0.9538	0.9889	0.9977	284	0.7233	0.9532	0.9956	0.9997	1.0000
144	0.5923	0.8448	0.9553	0.9895	0.9978	288	0.7262	0.9547	0.9959	0.9998	1.0000
146	0.5947	0.8475	0.9568	0.9900	0.9980	292	0.7291	0.9562	0.9962	0.9998	1.0000
148	0.5970	0.8501	0.9582	0.9905	0.9981	296	0.7319	0.9577	0.9964	0.9998	1.0000
150	0.5994	0.8527	0.9596	0.9910	0.9983	300	0.7347	0.9591	0.9966	0.9998	1.0000
152	0.6017	0.8553	0.9610	0.9915	0.9984	304	0.7375	0.9604	0.9969	0.9998	1.0000
154	0.6040	0.8578	0.9622	0.9919	0.9985	308	0.7402	0.9617	0.9971	0.9999	1.0000
156	0.6063	0.8603	0.9635	0.9923	0.9986	312	0.7429	0.9630	0.9972	0.9999	1.0000
158	0.6086	0.8627	0.9647	0.9927	0.9987	316	0.7456	0.9642	0.9974	0.9999	1.0000
160	0.6109	0.8651	0.9659	0.9931	0.9988	320	0.7482	0.9654	0.9976	0.9999	1.0000
162	0.6131	0.8674	0.9670	0.9935	0.9989	324	0.7509	0.9665	0.9977	0.9999	1.0000
164	0.6153	0.8697	0.9681	0.9938	0.9990	328	0.7534	0.9676	0.9979	0.9999	1.0000
166	0.6175	0.8719	0.9691	0.9941	0.9990	332	0.7560	0.9687	0.9980	0.9999	1.0000
168	0.6197	0.8741	0.9701	0.9944	0.9991	336	0.7585	0.9697	0.9981	0.9999	1.0000
170	0.6219	0.8763	0.9711	0.9947	0.9992	340	0.7610	0.9707	0.9983	0.9999	1.0000
172	0.6240	0.8784	0.9721	0.9950	0.9992	344	0.7635	0.9717	0.9984	0.9999	1.0000
174	0.6262	0.8805	0.9730	0.9952	0.9993	348	0.7659	0.9726	0.9985	0.9999	1.0000
176	0.6283	0.8826	0.9739	0.9955	0.9993	352	0.7683	0.9735	0.9986	1.0000	1.0000
178	0.6304	0.8846	0.9747	0.9957	0.9994	356	0.7707	0.9744	0.9987	1.0000	1.0000
180	0.6325	0.8866	0.9756	0.9959	0.9994	360	0.7730	0.9752	0.9987	1.0000	1.0000
182	0.6346	0.8885	0.9764	0.9961	0.9995	364	0.7754	0.9760	0.9988	1.0000	1.0000
184	0.6366	0.8904	0.9772	0.9963	0.9995	368	0.7777	0.9768	0.9989	1.0000	1.0000
186	0.6387	0.8923	0.9779	0.9965	0.9995	372	0.7799	0.9776	0.9990	1.0000	1.0000
188	0.6407	0.8941	0.9786	0.9967	0.9996	376	0.7822	0.9783	0.9990	1.0000	1.0000
190	0.6427	0.8959	0.9793	0.9969	0.9996	380	0.7844	0.9790	0.9991	1.0000	1.0000
192	0.6447	0.8977	0.9800	0.9970	0.9996	384	0.7866	0.9797	0.9991	1.0000	1.0000
194	0.6467	0.8995	0.9807	0.9972	0.9997	388	0.7888	0.9804	0.9992	1.0000	1.0000
196	0.6487	0.9012	0.9813	0.9973	0.9997	392	0.7909	0.9810	0.9993	1.0000	1.0000
198	0.6506	0.9029	0.9819	0.9975	0.9997	396	0.7930	0.9816	0.9993	1.0000	1.0000
200	0.6526	0.9045	0.9825	0.9976	0.9997	400	0.7951	0.9822	0.9993	1.0000	1.0000

Table 8.5. $P\{CS \mid LF\}$ for the Single-Stage Multinomial Procedure $\mathcal{M}_{BEM}$ as a Function of n and θ^* for $t = 5$

	θ^*						θ^*				
n	1.2	1.4	1.6	1.8	2.0	n	1.2	1.4	1.6	1.8	2.0
1	0.2308	0.2593	0.2857	0.3103	0.3333	51	0.3692	0.5427	0.6908	0.8020	0.8784
2	0.2308	0.2593	0.2857	0.3103	0.3333	52	0.3710	0.5463	0.6952	0.8064	0.8821
3	0.2376	0.2735	0.3076	0.3399	0.3704	53	0.3728	0.5498	0.6996	0.8107	0.8858
4	0.2455	0.2893	0.3310	0.3704	0.4074	54	0.3746	0.5533	0.7039	0.8149	0.8893
5	0.2509	0.3001	0.3469	0.3909	0.4321	55	0.3763	0.5567	0.7081	0.8190	0.8927
6	0.2547	0.3080	0.3589	0.4068	0.4516	56	0.3781	0.5601	0.7122	0.8230	0.8960
7	0.2589	0.3167	0.3722	0.4246	0.4734	57	0.3798	0.5635	0.7163	0.8269	0.8992
8	0.2635	0.3262	0.3864	0.4429	0.4954	58	0.3816	0.5668	0.7204	0.8307	0.9023
9	0.2676	0.3347	0.3989	0.4591	0.5146	59	0.3833	0.5701	0.7243	0.8344	0.9054
10	0.2712	0.3420	0.4099	0.4734	0.5317	60	0.3850	0.5734	0.7282	0.8380	0.9083
11	0.2746	0.3491	0.4206	0.4873	0.5482	61	0.3867	0.5766	0.7320	0.8416	0.9111
12	0.2780	0.3563	0.4314	0.5012	0.5645	62	0.3883	0.5798	0.7358	0.8451	0.9138
13	0.2815	0.3635	0.4420	0.5146	0.5802	63	0.3900	0.5830	0.7396	0.8485	0.9165
14	0.2847	0.3702	0.4519	0.5271	0.5948	64	0.3917	0.5861	0.7432	0.8518	0.9190
15	0.2877	0.3765	0.4612	0.5390	0.6085	65	0.3933	0.5892	0.7468	0.8551	0.9215
16	0.2907	0.3826	0.4704	0.5506	0.6218	66	0.3950	0.5923	0.7504	0.8582	0.9239
17	0.2937	0.3888	0.4794	0.5619	0.6347	67	0.3966	0.5953	0.7539	0.8613	0.9263
18	0.2966	0.3949	0.4883	0.5729	0.6471	68	0.3982	0.5983	0.7573	0.8644	0.9285
19	0.2994	0.4007	0.4967	0.5834	0.6588	69	0.3998	0.6013	0.7607	0.8673	0.9307
20	0.3021	0.4063	0.5049	0.5935	0.6700	70	0.4014	0.6043	0.7641	0.8702	0.9328
21	0.3047	0.4118	0.5130	0.6033	0.6809	71	0.4030	0.6072	0.7674	0.8730	0.9349
22	0.3074	0.4173	0.5209	0.6129	0.6914	72	0.4045	0.6101	0.7706	0.8758	0.9369
23	0.3100	0.4227	0.5286	0.6222	0.7016	73	0.4061	0.6130	0.7738	0.8785	0.9388
24	0.3125	0.4279	0.5360	0.6312	0.7113	74	0.4077	0.6159	0.7769	0.8812	0.9407
25	0.3149	0.4330	0.5433	0.6399	0.7206	75	0.4092	0.6187	0.7800	0.8838	0.9425
26	0.3174	0.4380	0.5505	0.6484	0.7296	76	0.4108	0.6215	0.7831	0.8863	0.9443
27	0.3198	0.4430	0.5575	0.6567	0.7383	77	0.4123	0.6243	0.7861	0.8888	0.9460
28	0.3221	0.4479	0.5644	0.6647	0.7468	78	0.4138	0.6271	0.7891	0.8912	0.9476
29	0.3245	0.4527	0.5711	0.6725	0.7549	79	0.4153	0.6298	0.7920	0.8935	0.9492
30	0.3268	0.4574	0.5776	0.6801	0.7627	80	0.4168	0.6325	0.7949	0.8959	0.9508
31	0.3290	0.4620	0.5841	0.6875	0.7702	81	0.4183	0.6352	0.7977	0.8981	0.9523
32	0.3312	0.4666	0.5904	0.6947	0.7775	82	0.4198	0.6378	0.8005	0.9003	0.9537
33	0.3335	0.4711	0.5966	0.7017	0.7846	83	0.4213	0.6405	0.8033	0.9025	0.9551
34	0.3356	0.4755	0.6027	0.7086	0.7914	84	0.4227	0.6431	0.8060	0.9046	0.9565
35	0.3378	0.4799	0.6086	0.7152	0.7980	85	0.4242	0.6457	0.8087	0.9067	0.9578
36	0.3399	0.4842	0.6145	0.7217	0.8044	86	0.4257	0.6482	0.8113	0.9087	0.9591
37	0.3420	0.4884	0.6202	0.7280	0.8105	87	0.4271	0.6508	0.8139	0.9107	0.9604
38	0.3441	0.4927	0.6259	0.7342	0.8165	88	0.4286	0.6533	0.8165	0.9126	0.9616
39	0.3461	0.4968	0.6314	0.7402	0.8222	89	0.4300	0.6558	0.8190	0.9145	0.9627
40	0.3481	0.5009	0.6368	0.7461	0.8278	90	0.4314	0.6583	0.8215	0.9163	0.9639
41	0.3501	0.5049	0.6422	0.7518	0.8332	91	0.4328	0.6608	0.8239	0.9182	0.9650
42	0.3521	0.5089	0.6474	0.7574	0.8384	92	0.4342	0.6632	0.8264	0.9199	0.9660
43	0.3541	0.5128	0.6526	0.7628	0.8435	93	0.4356	0.6656	0.8287	0.9217	0.9671
44	0.3560	0.5167	0.6576	0.7682	0.8483	94	0.4370	0.6680	0.8311	0.9233	0.9681
45	0.3580	0.5206	0.6626	0.7734	0.8531	95	0.4384	0.6704	0.8334	0.9250	0.9690
46	0.3599	0.5244	0.6675	0.7784	0.8576	96	0.4398	0.6728	0.8357	0.9266	0.9700
47	0.3618	0.5281	0.6723	0.7834	0.8621	97	0.4412	0.6751	0.8379	0.9282	0.9709
48	0.3636	0.5318	0.6771	0.7882	0.8663	98	0.4426	0.6774	0.8402	0.9298	0.9718
49	0.3655	0.5355	0.6817	0.7929	0.8705	99	0.4439	0.6797	0.8423	0.9313	0.9726
50	0.3673	0.5391	0.6863	0.7975	0.8745	100	0.4453	0.6820	0.8445	0.9328	0.9735

Table 8.5. *Continued, t = 5*

		$\theta^\star$						$\theta^\star$			
n	1.2	1.4	1.6	1.8	2.0	n	1.2	1.4	1.6	1.8	2.0
102	0.4480	0.6865	0.8487	0.9356	0.9751	204	0.5618	0.8455	0.9621	0.9929	0.9989
104	0.4507	0.6910	0.8528	0.9384	0.9765	208	0.5655	0.8496	0.9641	0.9935	0.9990
106	0.4533	0.6953	0.8568	0.9410	0.9779	212	0.5692	0.8536	0.9660	0.9940	0.9991
108	0.4559	0.6996	0.8607	0.9435	0.9793	216	0.5729	0.8576	0.9677	0.9945	0.9992
110	0.4585	0.7039	0.8645	0.9459	0.9805	220	0.5765	0.8614	0.9694	0.9950	0.9993
112	0.4611	0.7080	0.8681	0.9482	0.9817	224	0.5801	0.8651	0.9710	0.9954	0.9994
114	0.4637	0.7121	0.8717	0.9504	0.9828	228	0.5836	0.8687	0.9725	0.9958	0.9995
116	0.4662	0.7162	0.8751	0.9525	0.9838	232	0.5871	0.8722	0.9740	0.9961	0.9995
118	0.4687	0.7202	0.8785	0.9545	0.9848	236	0.5906	0.8756	0.9753	0.9964	0.9996
120	0.4712	0.7241	0.8818	0.9565	0.9857	240	0.5940	0.8789	0.9766	0.9967	0.9996
122	0.4737	0.7279	0.8850	0.9583	0.9865	244	0.5974	0.8822	0.9779	0.9970	0.9997
124	0.4762	0.7317	0.8881	0.9601	0.9873	248	0.6008	0.8853	0.9790	0.9972	0.9997
126	0.4786	0.7355	0.8911	0.9618	0.9881	252	0.6041	0.8884	0.9801	0.9975	0.9997
128	0.4810	0.7392	0.8940	0.9634	0.9888	256	0.6074	0.8913	0.9811	0.9977	0.9998
130	0.4834	0.7428	0.8969	0.9650	0.9895	260	0.6106	0.8942	0.9821	0.9979	0.9998
132	0.4858	0.7463	0.8996	0.9664	0.9901	264	0.6139	0.8970	0.9831	0.9980	0.9998
134	0.4882	0.7499	0.9023	0.9679	0.9907	268	0.6170	0.8998	0.9839	0.9982	0.9998
136	0.4906	0.7533	0.9049	0.9692	0.9912	272	0.6202	0.9024	0.9848	0.9984	0.9999
138	0.4929	0.7567	0.9075	0.9705	0.9918	276	0.6233	0.9050	0.9856	0.9985	0.9999
140	0.4952	0.7601	0.9100	0.9718	0.9922	280	0.6264	0.9075	0.9863	0.9986	0.9999
142	0.4975	0.7634	0.9124	0.9730	0.9927	284	0.6295	0.9100	0.9870	0.9987	0.9999
144	0.4998	0.7667	0.9147	0.9741	0.9931	288	0.6325	0.9124	0.9877	0.9988	0.9999
146	0.5020	0.7699	0.9170	0.9752	0.9936	292	0.6355	0.9147	0.9884	0.9989	0.9999
148	0.5043	0.7731	0.9192	0.9763	0.9939	296	0.6385	0.9170	0.9890	0.9990	0.9999
150	0.5065	0.7762	0.9214	0.9773	0.9943	300	0.6415	0.9192	0.9895	0.9991	0.9999
152	0.5088	0.7793	0.9235	0.9782	0.9946	304	0.6444	0.9213	0.9901	0.9992	0.9999
154	0.5110	0.7823	0.9256	0.9792	0.9950	308	0.6473	0.9234	0.9906	0.9992	1.0000
156	0.5132	0.7853	0.9275	0.9800	0.9953	312	0.6501	0.9254	0.9911	0.9993	1.0000
158	0.5153	0.7882	0.9295	0.9809	0.9955	316	0.6530	0.9274	0.9916	0.9994	1.0000
160	0.5175	0.7911	0.9314	0.9817	0.9958	320	0.6558	0.9293	0.9920	0.9994	1.0000
162	0.5196	0.7940	0.9332	0.9825	0.9961	324	0.6586	0.9312	0.9924	0.9995	1.0000
164	0.5218	0.7968	0.9350	0.9832	0.9963	328	0.6613	0.9330	0.9928	0.9995	1.0000
166	0.5239	0.7996	0.9367	0.9839	0.9965	332	0.6640	0.9348	0.9932	0.9995	1.0000
168	0.5260	0.8023	0.9384	0.9846	0.9967	336	0.6668	0.9365	0.9935	0.9996	1.0000
170	0.5281	0.8050	0.9400	0.9852	0.9969	340	0.6694	0.9382	0.9939	0.9996	1.0000
172	0.5302	0.8077	0.9416	0.9859	0.9971	344	0.6721	0.9398	0.9942	0.9996	1.0000
174	0.5322	0.8103	0.9432	0.9865	0.9973	348	0.6747	0.9414	0.9945	0.9997	1.0000
176	0.5343	0.8129	0.9447	0.9870	0.9974	352	0.6773	0.9429	0.9948	0.9997	1.0000
178	0.5363	0.8154	0.9462	0.9876	0.9976	356	0.6799	0.9444	0.9950	0.9997	1.0000
180	0.5384	0.8179	0.9476	0.9881	0.9977	360	0.6825	0.9459	0.9953	0.9997	1.0000
182	0.5404	0.8204	0.9490	0.9886	0.9979	364	0.6850	0.9473	0.9955	0.9998	1.0000
184	0.5424	0.8229	0.9504	0.9891	0.9980	368	0.6875	0.9487	0.9958	0.9998	1.0000
186	0.5444	0.8253	0.9517	0.9895	0.9981	372	0.6900	0.9501	0.9960	0.9998	1.0000
188	0.5463	0.8276	0.9530	0.9900	0.9982	376	0.6925	0.9514	0.9962	0.9998	1.0000
190	0.5483	0.8300	0.9542	0.9904	0.9983	380	0.6949	0.9527	0.9964	0.9998	1.0000
192	0.5503	0.8323	0.9554	0.9908	0.9984	384	0.6974	0.9539	0.9966	0.9998	1.0000
194	0.5522	0.8346	0.9566	0.9912	0.9985	388	0.6998	0.9551	0.9968	0.9999	1.0000
196	0.5541	0.8368	0.9578	0.9916	0.9986	392	0.7022	0.9563	0.9969	0.9999	1.0000
198	0.5561	0.8390	0.9589	0.9919	0.9987	396	0.7045	0.9575	0.9971	0.9999	1.0000
200	0.5580	0.8412	0.9600	0.9923	0.9988	400	0.7069	0.9586	0.9972	0.9999	1.0000

the additional number of potential occurrences of C_j if all of the $(n - m)$ remaining outcomes after stage m were also to be associated with C_j. Thus, *curtailment* takes place when one of the categories has sufficiently more successes than all of the other categories so that even if the remaining trials experience a "reversal of fortune" with *all* of the outcomes occurring from one of the losing categories, that losing category could still not defeat the current leader (at best, it could only *tie* the leader). The following examples illustrate this phenomenon.

Remark 8.2.1 Unlike procedure $\mathcal{B}_{BK}$ where both the adaptive sampling rule and the stopping rule are important, the single crucial issue for procedure $\mathcal{M}_{BK}$ is the stopping rule. There are not several treatments in this type of experiment, and hence no issue of switching among them.

Example 8.2.2 For $t = 3$ and $n = 2$, if we observe

m	x_{1m}	x_{2m}	x_{3m}	y_{1m}	y_{2m}	y_{3m}
1	1	0	0	1	0	0

stop sampling and select category C_1 because $y_{1m} = 1 \geq y_{jm} + n - m = 0 + 2 - 1 = 1$ for $j = 2$ and 3.

Example 8.2.3 For $t = 3$ and $n = 3$ or 4, if we observe

m	x_{1m}	x_{2m}	x_{3m}	y_{1m}	y_{2m}	y_{3m}
1	0	1	0	0	1	0
2	0	1	0	0	2	0

stop sampling and select category C_2 because $y_{2m} = 2 \geq y_{jm} + n - m = 0 + n - 2$ for $n = 3$ or $n = 4$ and both $j = 1$ and 3.

Example 8.2.4 For $t = 3$ and $n = 3$ suppose that

m	x_{1m}	x_{2m}	x_{3m}	y_{1m}	y_{2m}	y_{3m}
1	1	0	0	1	0	0
2	0	0	1	1	0	1
3	0	1	0	1	1	1

Because $y_{13} = y_{23} = y_{33} = 1$ we stop sampling and randomize among the three categories using probability $1/3$ for each.

Remark 8.2.2 Bechhofer and Kulkarni (1984) showed how procedure $\mathcal{M}_{BK}$ can be generalized to accommodate goals such as selecting the s $(1 \leq s \leq t - 1)$ best categories *without regard to order*. They also gave tables of $E\{N\}$ for procedure $\mathcal{M}_{BK}$ when $s = 1$, $t = 2(1)6$, and $n = 2(2)20$ with the p_i $(1 \leq i \leq t)$ in the LF-configuration ($p_{[1]} = p_{[s-1]} = 1/(\theta^* + s - 1)$, and $p_{[s]} = \theta^*/(\theta^* + s - 1)$) with selected θ^*. Procedure $\mathcal{M}_{BK}$ can also be generalized to accommodate cross-classified experiments of the type to be described in Section 8.4.

8.2.3 A Closed Sequential Procedure

This section describes a *closed* sequential procedure due to Bechhofer and Goldsman (1985, 1986) that incorporates curtailment. The basis of this procedure is the open sequential sampling procedure of Bechhofer, Kiefer and Sobel (1968), denoted $\mathcal{M}_{BKS}$ hereafter. Bechhofer and Goldsman (1985) studied the performance characteristics of procedure $\mathcal{M}_{BKS}$ and found that its achieved $P\{CS \mid LF\}$ exceeds the nominal lower bound of P^* by a *substantial* amount (it always "overprotects"). Furthermore, the distribution of the random number of multinomial observations N required by procedure $\mathcal{M}_{BKS}$ is highly skewed to the right. These phenomena result in a sizable proportion of experiments that terminate with excessively large values of N. Thus, procedure $\mathcal{M}_{BKS}$ yields large values of $E\{N \mid p\}$ and $Var\{N \mid p\}$, particularly when all of the p_i ($1 \leq i \leq t$) are equal or almost equal; these effects are magnified because of the $P\{CS \mid LF\}$ "overprotection."

These findings led Bechhofer and Goldsman to study a procedure that *truncates* $\mathcal{M}_{BKS}$ in such a way as to maintain $P\{CS \mid LF\} \geq P^*$, while *reducing* $E\{N \mid p\}$ and $Var\{N \mid p\}$, not only in the LF-configuration, but also *uniformly* in p. In addition to providing a closed procedure, Bechhofer and Goldsman studied a second improvement over the open sequential procedure $\mathcal{M}_{BKS}$ by incorporating curtailment, as described in Section 8.2.2. Curtailment permits early termination of the procedure with no loss in $P\{CS \mid p\}$.

Procedure $\mathcal{M}_{BG}$ For the given t, and specified (θ^*, P^*), find the truncation number n_0 from Table 8.6.

Sampling rule: At the mth stage of experimentation ($m \geq 1$), take the random multinomial observation $X_m = (X_{1m}, \ldots, X_{tm})$.

Stopping rule: At stage m, calculate the ordered category totals $y_{[1]m} \leq \cdots \leq y_{[t]m}$ and

$$z_m = \sum_{i=1}^{t-1} (1/\theta^*)^{(y_{[t]m} - y_{[i]m})}.$$

Stop sampling at the first stage when either

$$z_m \leq (1 - P^*)/P^* \quad \text{or} \quad m = n_0 \quad \text{or} \quad y_{[t]m} - y_{[t-1]m} \geq n_0 - m, \qquad (8.2.4)$$

whichever occurs first.

Terminal decision rule: Let N denote the value of m at the termination of sampling. Select the category that yielded $y_{[t]N}$ as the one associated with $p_{[t]}$; randomize in the case of ties.

Table 8.6. Truncation Numbers n_0 for the Sequential Multinomial Procedure $\mathcal{M}_{BG}$ to Guarantee (8.2.1), Along with Accompanying Performance Characteristics.[a]

t	$P^\star$	$\theta^\star$	$\mathcal{M}_{BG}$			t	$P^\star$	$\theta^\star$	$\mathcal{M}_{BG}$		
			n_0	$E\{N \mid \text{LF}\}$	$E\{N \mid \text{EP}\}$				n_0	$E\{N \mid \text{LF}\}$	$E\{N \mid \text{EP}\}$
2	0.75	3.0	1	1.00	1.00	3	0.75	3.0	5	3.24	3.48
		2.8	3	2.39	2.50			2.8	6	3.70	4.15
		2.6	3	2.40	2.50			2.6	7	3.94	4.38
		2.4	3	2.42	2.50			2.4	8	5.40	5.94
		2.2	3	2.43	2.50			2.2	10	6.00	6.68
		2.0	5	3.09	3.25			2.0	13	7.97	8.93
		1.8	7	3.44	3.63			1.8	18	11.34	12.74
		1.6	9	5.96	6.26			1.6	32	17.60	20.25
		1.4	19	11.35	12.06			1.4	71	34.02	39.84
		1.2	67	36.75	39.28			1.2	*285	117.89	140.85
	0.90	3.0	∞	3.20	4.00		0.90	3.0	12	6.97	8.93
		2.8	7	4.63	5.34			2.8	15	7.77	10.36
		2.6	9	5.23	6.26			2.6	16	9.17	11.83
		2.4	11	5.72	6.94			2.4	22	10.43	14.25
		2.2	15	6.33	7.84			2.2	25	13.30	17.77
		2.0	15	8.90	10.59			2.0	34	17.17	23.30
		1.8	27	11.04	13.91			1.8	50	23.71	32.89
		1.6	41	17.00	21.48			1.6	83	37.26	52.61
		1.4	79	32.92	41.84			1.4	*170	73.42	105.39
		1.2	267	112.28	143.50			1.2	*670	254.85	369.78
	0.95	3.0	11	5.25	6.94		0.95	3.0	20	8.90	13.57
		2.8	15	5.65	7.84			2.8	22	10.48	15.79
		2.6	13	7.54	9.66			2.6	25	12.27	18.27
		2.4	17	8.47	11.38			2.4	31	14.48	22.09
		2.2	23	9.43	13.13			2.2	41	17.56	27.76
		2.0	27	13.09	17.90			2.0	52	23.03	35.97
		1.8	35	18.03	24.30			1.8	71	32.63	50.41
		1.6	59	26.56	37.09			1.6	125	50.32	81.43
		1.4	151	48.31	72.36			1.4	*266	98.88	165.90
		1.2	455	166.54	245.31			1.2	*960	346.42	577.82

[a]Results are exact except for those denoted by an asterisk (*), which are obtained via Monte Carlo simulation.

Procedure $\mathcal{M}_{BG}$ has the same LF-configuration as does procedure $\mathcal{M}_{BEM}$; this determines the truncation numbers given in Table 8.6. The table lists truncation values for $t = 2(1)6$ and selected $(\theta^\star, P^\star)$.

Example 8.2.5 For this example, and Examples 8.2.6–8.2.8, suppose that $t = 3$, $P^\star = 0.75$ and $\theta^\star = 3.0$. Table 8.6 tells us to truncate sampling at $n_0 = 5$ observations. For the data

m	x_{1m}	x_{2m}	x_{3m}	y_{1m}	y_{2m}	y_{3m}
1	0	1	0	0	1	0
2	0	1	0	0	2	0

Table 8.6. *Continued*

t	$P^\star$	$\theta^\star$	n_0	$\mathcal{M}_{BG}$ $E\{N \mid LF\}$	$E\{N \mid EP\}$	t	$P^\star$	$\theta^\star$	n_0	$\mathcal{M}_{BG}$ $E\{N \mid LF\}$	$E\{N \mid EP\}$
4	0.75	3.0	9	4.91	5.75	5	0.75	3.0	*12	7.44	8.95
		2.8	9	6.00	6.79			2.8	*13	8.39	9.96
		2.6	11	7.05	8.16			2.6	*17	9.80	11.93
		2.4	15	8.29	9.91			2.4	*20	11.91	14.55
		2.2	17	10.44	12.26			2.2	*25	14.99	18.20
		2.0	24	13.78	16.45			2.0	*34	19.81	24.35
		1.8	35	19.42	23.34			1.8	*50	28.44	34.88
		1.6	57	31.11	37.65			1.6	*86	45.68	57.14
		1.4	*124	62.31	76.01			1.4	*184	92.68	117.62
		1.2	*495	219.69	270.89			1.2	*730	329.36	421.47
	0.90	3.0	19	9.84	13.85		0.90	3.0	*24	13.13	18.76
		2.8	22	11.29	15.94			2.8	*28	15.01	21.87
		2.6	26	13.20	18.87			2.6	*34	17.43	25.69
		2.4	31	15.93	22.77			2.4	*42	21.17	31.66
		2.2	39	19.79	28.39			2.2	*52	26.63	39.37
		2.0	53	25.71	37.31			2.0	*71	35.16	52.75
		1.8	*75	36.94	53.77			1.8	*104	50.34	75.69
		1.6	*126	58.69	86.83			1.6	*172	80.93	123.96
		1.4	*274	116.89	176.56			1.4	*374	163.55	252.55
		1.2	*1050	413.68	627.68			1.2	*1460	585.00	923.61
	0.95	3.0	26	12.97	20.34		0.95	3.0	*34	16.42	27.19
		2.8	30	14.74	23.29			2.8	*39	19.19	31.50
		2.6	36	17.19	27.36			2.6	*46	22.55	36.87
		2.4	44	20.68	33.38			2.4	*58	27.04	45.37
		2.2	56	25.75	42.16			2.2	*74	34.05	57.80
		2.0	*74	33.86	55.47			2.0	*98	45.01	76.28
		1.8	*106	47.80	79.05			1.8	*142	64.37	109.09
		1.6	*180	76.06	127.76			1.6	*240	103.53	180.02
		1.4	*380	152.72	264.25			1.4	*510	209.64	365.49
		1.2	*1500	537.10	962.45			1.2	*2000	741.70	1350.73

we stop sampling by the first criterion in (8.2.4) because $z_2 = (1/3)^2 + (1/3)^2 = 2/9 \leq (1 - P^\star)/P^\star = 1/3$, and we select category C_2.

Example 8.2.6 Consider the same set-up as in Example 8.2.5. For the data

m	x_{1m}	x_{2m}	x_{3m}	y_{1m}	y_{2m}	y_{3m}
1	0	1	0	0	1	0
2	1	0	0	1	1	0
3	0	1	0	1	2	0
4	1	0	0	2	2	0
5	1	0	0	3	2	0

we stop sampling by the second criterion in (8.2.4) because $m = n_0 = 5$ observations, and we select category C_1.

Example 8.2.7 Again consider the set-up of Example 8.2.5. For the data

m	x_{1m}	x_{2m}	x_{3m}	y_{1m}	y_{2m}	y_{3m}
1	0	1	0	0	1	0
2	1	0	0	1	1	0
3	0	1	0	1	2	0
4	1	0	0	2	2	0
5	0	0	1	2	2	1

we stop according to the second criterion in (8.2.4) because $m = n_0 = 5$. However, we now have a tie between y_{15} and y_{25} and thus randomly select between categories C_1 and C_2.

Example 8.2.8 In this last example using the specifications of Example 8.2.5, suppose that we observe the data

m	x_{1m}	x_{2m}	x_{3m}	y_{1m}	y_{2m}	y_{3m}
1	0	1	0	0	1	0
2	1	0	0	1	1	0
3	0	1	0	1	2	0
4	0	0	1	1	2	1

Because C_1 and C_3 can do no better than tie C_2 (if we were to take the potential remaining $n_0 - m = 5 - 4 = 1$ observation), the third criterion in (8.2.4) tells us to stop; we select category C_2.

Remark 8.2.3 The open Bechhofer, Kiefer and Sobel (1968) procedure $\mathcal{M}_{BKS}$ is exactly procedure $\mathcal{M}_{BG}$ with the modification that it uses only the first stopping criterion, $z_m \leq (1 - P^\star)/P^\star$, in (8.2.4). In addition to this first criterion, procedure $\mathcal{M}_{BG}$ uses the stopping criterion, $m = n_0$, to stop sampling if n_0 multinomial observations have been taken (truncation) and uses a third criterion, $y_{[t]m} - y_{[t-1]m} \geq n_0 - m$, to stop sampling if the cell currently in second place can do no better than tie the cell currently in first place (curtailment).

By construction, procedure $\mathcal{M}_{BG}$ is *superior* to the open procedure $\mathcal{M}_{BKS}$ in terms of $E\{N \mid \boldsymbol{p}\}$ and $\text{Var}\{N \mid \boldsymbol{p}\}$ uniformly in $\boldsymbol{p}$; the improvement is greatest when $p_{[1]} = p_{[t]}$. Procedures $\mathcal{M}_{BG}$ and $\mathcal{M}_{BKS}$ terminate sampling very quickly when $p_{[t]} \to 1$. If the $\boldsymbol{p}$ configuration is *very* favorable to the experimenter, then procedure $\mathcal{M}_{BG}$ (or procedure $\mathcal{M}_{BKS}$ if truncation numbers are not available) should *always* be used. In fact, in the *most favorable* configuration wherein $p_{[t]} = 1$, it is straightforward to show that procedure $\mathcal{M}_{BG}$ (or procedure $\mathcal{M}_{BKS}$) will require

$$N_{\min} = \left\lceil \ln\left\{\frac{(t-1)P^\star}{1 - P^\star}\right\} / \ln\{\theta^\star\} \right\rceil$$

observations. In this configuration, the savings in sample size relative to procedure $\mathcal{M}_{BEM}$ is considerable. For example, when $t = 4$, $\theta^* = 1.6$, $P^* = 0.75$, the single-stage procedure $\mathcal{M}_{BEM}$ requires exactly 46 observations to guarantee (8.2.1), whereas procedures $\mathcal{M}_{BG}$ and $\mathcal{M}_{BKS}$ require only $N_{\min} = 5$ observations.

Table 8.6 gives performance characteristics of procedure $\mathcal{M}_{BG}$ for various $(t; \theta^*, P^*)$ and configurations of p, namely, the LF-configuration, (8.2.2), and the equal-probability (EP) configuration, $p_{[1]} = p_{[t]}$. Procedure $\mathcal{M}_{BG}$ almost always dominates the single-stage procedure $\mathcal{M}_{BEM}$ in terms of smaller $E\{N \mid \text{LF}\}$ and $E\{N \mid \text{EP}\}$. In fact, to the best of our knowledge, procedure $\mathcal{M}_{BG}$ is superior, over a broad range of practical $(t; \theta^*, P^*)$-values, to all procedures that have thus far been proposed for the multinomial selection problem under consideration; we recommend its use when sequential sampling is an option of the experimenter and appropriate truncation numbers are available.

Unlike the normal means problem (see Goal 2.5.1), there does not exist a *single-stage* procedure for the problem of selecting the s ($1 < s \leq t$) categories associated with $p_{[t-s+1]}, \ldots, p_{[t]}$ *without regard to order* if one wishes to guarantee the probability requirement

$$P\{CS\} \geq P^* \quad \text{whenever } p_{[t-s+1]}/p_{[t-s]} \geq \theta^*.$$

However, this requirement can be guaranteed by a generalization of the *truncated sequential* procedure $\mathcal{M}_{BG}$. In particular, it can be guaranteed for $s = t - 1$, that is, selecting the category associated with the *smallest p_i-value*, $p_{[1]}$.

8.2.4 Applications

This section discusses two applications of the multinomial selection problem. The first concerns the selection of the treatment having the largest location parameter; the second is a nonparametric application intended to select the treatment having the highest probability of producing the "most desirable" observation from a given vector-observation of the competing treatments (see also Bechhofer and Sobel 1958).

Suppose that $W_{i1}, W_{i2}, \ldots$ is the ith of t mutually independent random samples. The form of the probability density or mass function (p.d.f. or p.m.f.) of the W_{ij} may be different for each i ($1 \leq i \leq t$) and is assumed to be unknown.

Selection of the Treatment Having the Largest Location Parameter

Suppose that the W_{ij} ($j \geq 1$) are *continuous* random variables with p.d.f. $f(w - \mu_i)$ where the μ_i ($1 \leq i \leq t$) are unknown location parameters. We consider the goal of selecting the p.d.f. $f(w - \mu_i)$ that has the highest probability of producing the largest observation from the vector $(W_{1j}, \ldots, W_{tj})$.

Define $X_{ij} = 1$ if $W_{ij} > \max_{i' \neq i} W_{i'j}$ on trial j and $X_{ij} = 0$ if not ($1 \leq i \leq t$, $j \geq 1$). Then $(X_{1j}, \ldots, X_{tj})$ ($j \geq 1$) has a multinomial distribution with probability vector p, where

$$p_i = P\{W_{i1} > W_{i'1} \ (i' \neq i; \ 1 \leq i' \leq t)\} \quad (1 \leq i \leq t).$$

Let $p_{[1]} \leq \cdots \leq p_{[t]}$ denote the ordered p_i-values. The goal of selecting the category associated with $p_{[t]}$ can be investigated using the multinomial selection procedures described in this chapter.

The p.d.f. having the highest probability, $p_{[t]}$, of producing the largest observation is the p.d.f. with the largest μ_i-value. However, the reader should be warned that the multinomial probability requirement (8.2.1) involving the p_i ($1 \leq i \leq t$) does not translate into any easily interpretable probability requirement involving the μ_i ($1 \leq i \leq t$). Hence, procedures that guarantee (8.2.1) may be difficult to explain in terms of the μ_i. Dudewicz (1971) studied the efficiency, in terms of sample size, of the single-stage multinomial selection procedure relative to the best competing *parametric* single-stage procedure for selecting the p.d.f. associated with the largest μ_i-value. As expected, he found that the multinomial procedure always required much larger sample sizes. However, the multinomial procedure does have the virtue that it is nonparametric—it requires no knowledge of the form of the underlying density functions.

Example 8.2.9 Suppose that random samples of mutually independent observations W_{ij} ($1 \leq i \leq t$, $j \geq 1$) are taken from $t = 3$ treatments. The p.d.f. of the W_{ij} are assumed to be of the same unknown form $f(w - \mu_i)$ ($1 \leq i \leq 3$), differing only with respect to the unknown location parameters μ_i. It is desired to select the treatment associated with $\mu_{[3]} = \max\{\mu_1, \mu_2, \mu_3\}$. By the introductory remarks, this problem is equivalent to selecting the multinomial category associated with $p_{[3]} = \max\{p_1, p_2, p_3\}$. Suppose that we specify $P^* = 0.75$ and $\theta^* = 3.0$. The truncation number $n_0 = 5$ from Table 8.6 for procedure $\mathcal{M}_{BG}$ applied to the following (artificial) data tells us to stop sampling if

m	w_{1m}	w_{2m}	w_{3m}	x_{1m}	x_{2m}	x_{3m}	y_{1m}	y_{2m}	y_{3m}
1	15	17	9	0	1	0	0	1	0
2	21	7	6	1	0	0	1	1	0
3	7	11	8	0	1	0	1	2	0
4	16	6	2	1	0	0	2	2	0
5	14	13	9	1	0	0	3	2	0

and select category C_1 (i.e., treatment 1). In doing so we are guaranteed that $P\{CS\} \geq 0.75$ whenever $p_{[3]} \geq 3p_{[2]}$. We again emphasize that the preference-zone $p_{[3]} \geq 3p_{[2]}$ does *not* have a straightforward interpretation in terms of the μ_i or their differences.

A General Nonparametric Interpretation

Suppose that we take i.i.d. vector-observations $W_j = (W_{1j}, \ldots, W_{tj})$ ($j \geq 1$), where the W_{ij} can be either discrete or continuous random variables. For a particular vector-observation W_j, suppose that the experimenter can determine which of the t observations W_{ij} ($1 \leq i \leq t$) is the "most desirable." The term "most desirable" is based on some criterion of goodness designated by the experimenter, and it can be quite general. For example, the "most desirable" observation might correspond to:

- The largest crop yield based on a vector-observation of t agricultural plots using competing fertilizers
- The smallest sample average customer waiting time based on a simulation run of each of t competing queueing strategies
- The smallest estimated variance of customer waiting times (from the above simulations)
- The smallest sample proportion of customer waiting times (from the above simulations) that are greater than some designated bound w

For a particular vector-observation $\boldsymbol{W}_j$, suppose that $X_{ij} = 1$ or 0 according as W_{ij} ($1 \le i \le t$) is the "most desirable" of the components of $\boldsymbol{W}_j$ or not. Then $(X_{1j}, \ldots, X_{tj})$ ($j \ge 1$) is a multinomial trial with probability vector $\boldsymbol{p}$, where

$$p_i = P\{W_{i1} \text{ is the "most desirable" component of } \boldsymbol{W}_1\} \qquad (1 \le i \le t).$$

Let $p_{[1]} \le \cdots \le p_{[t]}$ denote the ordered p_i-values. The problem of finding the category corresponding to the largest p_i can be thought of as that of finding the component having the highest probability of yielding the "most desirable" observation of those from a particular vector-observation; this latter problem can be approached using the multinomial selection methods described in this chapter.

Example 8.2.10 Suppose that it is desired to determine which of $t = 3$ job shop configurations is most likely to give the shortest expected times-in-system for a certain manufactured product. Because of the complicated configurations of the candidate job shops, it is necessary to simulate the three competitors. Suppose that the jth simulation run of configuration i yields W_{ij} ($1 \le i \le 3$, $j \ge 1$), the proportion of 1000 times-in-system greater than 20 minutes. Management has decided that the "most desirable" component of $\boldsymbol{W}_j = (W_{1j}, W_{2j}, W_{3j})$ will be that component corresponding to $\min_{1 \le i \le 3} W_{ij}$. If p_i denotes the probability that configuration i yields the smallest component of $\boldsymbol{W}_j$, then we seek to select the job shop configuration that corresponds to $p_{[3]}$. By the above remarks, this problem is equivalent to that of selecting the multinomial category associated with $p_{[3]} = \max\{p_1, p_2, p_3\}$. Suppose that we specify $P^\star = 0.75$ and $\theta^\star = 3.0$. The truncation number from Table 8.6 for procedure $\mathcal{M}_{BG}$ is $n_0 = 5$. We apply the procedure to the data found in Table 8.7 and select category C_2 (i.e., shop configuration 2).

Section 8.5 lists several nonparametric procedures for problems that have IZs with interpretations in terms of location parameters.

Table 8.7. Simulated Times-in-System for Products Produced by Three Job Shops

m	w_{1m}	w_{2m}	w_{3m}	x_{1m}	x_{2m}	x_{3m}	y_{1m}	y_{2m}	y_{3m}
1	0.13	0.09	0.14	0	1	0	0	1	0
2	0.24	0.10	0.07	0	0	1	0	1	1
3	0.17	0.11	0.12	0	1	0	0	2	1
4	0.13	0.08	0.02	0	0	1	0	2	2
5	0.14	0.13	0.15	0	1	0	0	3	2

8.3 SUBSET PROCEDURES FOR MULTINOMIAL DATA

At times, statisticians may not have the luxury of designing an experiment on a sufficiently large scale to select the single best multinomial cell but may be required to perform an analysis of multinomial response data with a sample size chosen according to economic criteria, convenience, or other reasons. In these cases, the statistician may suggest that an appropriate objective is to identify a subset of the multinomial cells that includes the cell associated with $p_{[t]}$. This section discusses a curtailed single-stage procedure and a sequential procedure to achieve this goal with prespecified $P\{CS\}$.

The statistical assumptions and notation of Section 8.2 are also used here. In particular, $y_{im} = \sum_{j=1}^{m} x_{ij}$ denotes the total number of outcomes of the ith category ($1 \leq i \leq t$) after $m \geq 1$ mulitinomial trials have been taken, and $p_{[1]} \leq \cdots \leq p_{[t]}$ denote the ordered values of the t cell probabilities in $\boldsymbol{p} = (p_1, \ldots, p_t)$. The purpose of the experiment is stated as Goal 8.3.1, and the associated probability (design) requirement is given in (8.3.1).

Goal 8.3.1 To select a (random-size) subset of the t categories that contains the category associated with $p_{[t]}$.

A *correct selection* (CS) is said to have been made if Goal 8.3.1 is achieved.

Probability Requirement: For specified constant P^* with $1/t < P^* < 1$, we require that

$$P\{CS\} \geq P^* \quad \text{for all } \boldsymbol{p} = (p_1, \ldots, p_t). \tag{8.3.1}$$

The probability on the left-hand side of (8.3.1) depends on the entire vector $\boldsymbol{p}$ and on the number n of multinomial trials; (8.3.1) does not involve a *measure of distance* as does the IZ probability requirement (8.2.1).

Sections 8.3.1 and 8.3.2 consider a single-stage procedure and its curtailed version, respectively. Both the original procedure and the curtailed version guarantee the *same* probability requirement (8.3.1) when used with the same implementing constant, and they select the *same* subset of the categories. The sequential procedure accomplishes these objectives with a smaller expected number of vector-observations, $E\{N\}$, than does the corresponding single-stage procedure. In Section 8.3.3 we describe a sequential procedure that is the best one thus far proposed for this problem; however, in order to implement the procedure, the user would have to undertake a tedious search to obtain certain constants necessary to implement the procedure.

8.3.1 A Single-Stage Procedure

The following single-stage procedure, proposed by Gupta and Nagel (1967), is designed to select a random-size subset of the t categories that contains the category with cell probability $p_{[t]}$. Their procedure was the first to address this problem, and

SUBSET PROCEDURES FOR MULTINOMIAL DATA   237

it is the standard against which the performances of later procedures for this problem
are to be compared.

Procedure $\mathcal{M}_{GN}$ Take a random sample of n multinomial vectors $X_j = (X_{1j}, \ldots, X_{tj})$ ($1 \leq j \leq n$) in a *single* stage.

Include category C_i in the selected subset if and only if

$$y_{in} \geq y_{[t]n} - d$$

where $d = d(t, n, P^*)$ is from Table 8.8.

Table 8.8 is from Gupta and Nagel (1967); the constants d were calculated so that procedure $\mathcal{M}_{GN}$ guarantees (8.3.1). This was accomplished by determining the configuration p at which the $P\{CS\}$ of procedure $\mathcal{M}_{GN}$ attained its global minimum. Gupta and Nagel showed that this LF-configuration is of the form

$$p = (0, 0, \ldots, 0, s, p, p, \ldots, p) \quad (s \leq p) \tag{8.3.2}$$

where $s = 1 - (r-1)p$ and r is the number of positive cell probabilities in (8.3.2). Thus, the global minimum of the $P\{CS\}$ is computed as

$$\min_{p} P\{CS \mid t, n, d; p\} = \min_{2 \leq r \leq t} \left[\min_{\frac{1}{r} \leq p \leq \frac{1}{r-1}} P\{CS \mid (0, 0, \ldots, 0, s, p, p, \ldots, p)\} \right].$$

Gupta and Nagel found that this minimum usually occurs at one end of the p-interval in question, that is, for $p = 1/r$ or for $p = 1/(r-1)$.

8.3.2 Curtailment of Procedure $\mathcal{M}_{GN}$

Like the single-stage IZ procedure $\mathcal{M}_{BEM}$, the subset procedure $\mathcal{M}_{GN}$ can be improved by observing that if one category C_i has sufficiently greater counts than the remaining

Table 8.8. Minimum d Such That $\min_p P\{CS\} \geq P^*$ for Procedure $\mathcal{M}_{GN}$ for Selected (t, n, P^*)

P^*	t	\multicolumn{14}{c}{n}
		2	3	4	5	6	7	8	9	10	11	12	13	14	15
0.75	2	0	1	2	1	2	1	2	3	2	3	2	3	2	3
	3(1)10	1	2	2	2	2	2	3	3	3	3	3	3	3	3
0.90	2	2	3	2	3	4	3	4	3	4	5	4	5	4	5
	3(1)10	2	3	3	3	4	4	4	4	4	5	5	5	5	5

Reprinted from Gupta and Nagel (1967), p. 17, by courtesy of the Indian Statistical Institute.

categories, then C_i must be associated with $y_{[t]n}$ and that no other categories can be selected. Bechhofer and Chen (1991) described the following curtailed version of procedure $\mathcal{M}_{GN}$ that employs this fact.

Procedure $\mathcal{M}_{GNBC}$ Let n denote the maximum number of multinomial observations to be taken. Find $d = d(t, n, P^*)$ from Table 8.8 corresponding to the given (t, n) and the specified P^*. If sampling has not yet stopped at stage m, take an additional multinomial observation $X_m = (X_{1m}, \ldots, X_{tm})$. Calculate the ordered sample cell totals, $y_{[1]m} \leq \cdots \leq y_{[t]m}$. Stop sampling at the first stage m such that

$$y_{[t]m} > y_{[t-1]m} + n - m + d \quad \text{or} \quad m = n$$

Having stopped at stage $m = N$, include category C_i in the selected subset if

$$y_{iN} \geq y_{[t]N} - d.$$

The procedures $\mathcal{M}_{GN}$ and $\mathcal{M}_{GNBC}$ select the *same* subset of the t categories. To see this, first notice that the constant d used by the curtailed procedure is that of Gupta and Nagel (1967). If sampling stops with $N < n$, then the cell identified as having the largest cell total at stage N will *exceed* all other cell counts by *more* than d and hence the terminal decision rule of procedure $\mathcal{M}_{GNBC}$ will select only that category corresponding to $y_{[t]N}$. If sampling stops with $N = n$, then the terminal decision rule is exactly that of $\mathcal{M}_{GN}$. In either case, both procedures select the same subset of the t categories. Consequently, the $P\{CS\}$ of both procedures is the same (uniformly in p). In particular, both procedure $\mathcal{M}_{GN}$ and procedure $\mathcal{M}_{GNBC}$ have the same LF-configuration p which is of the form in (8.3.2).

Example 8.3.1   Suppose that $t = 3$, $P^* = 0.75$ and no more than $n = 7$ trials are to be taken. Table 8.8 tells us to use $d = 2$ as the yardstick in the terminal decision rule. For the data

m	x_{1m}	x_{2m}	x_{3m}	y_{1m}	y_{2m}	y_{3m}
1	0	1	0	0	1	0
2	1	0	0	1	1	0
3	0	1	0	1	2	0
4	0	1	0	1	3	0
5	0	1	0	1	4	0
6	0	1	0	1	5	0

we stop sampling at stage $m = 6$ because $y_{26} = 5 > y_{j6} + n - m + d = y_{j6} + 7 - 6 + 2 = y_{j6} + 3$ for $j = 1, 3$ and we select category 2.

We illustrate the effectiveness of curtailment in procedure $\mathcal{M}_{GNBC}$ by comparing n when using procedure $\mathcal{M}_{GN}$ with the expected number of vector-observations, $E\{N\}$, for the procedure to terminate sampling. Table 8.9, abstracted from Bechhofer and Chen (1991), lists the value of $E\{N\}$ for selected t, n and given $\theta \geq 1$ under the *slippage configuration*, $\boldsymbol{p} = (p, p, \ldots, p, \theta p)$. Note that the requirement $\sum_{i=1}^{t} p_i = 1$ determines p for this configuration. When $\theta = 1$, all the cells have the same probability $1/t$, and intuition suggests that there will be minimal savings benefits in using the curtailed rule. As θ increases, the p-value of the best cell becomes more separated from those of the remaining cells and intuition suggests that the savings increases. Examination of Table 8.9 (and the complete set of $E\{N\}$ tables not given here) confirms the intuition suggested in this paragraph as well as several other patterns of behavior in the expected savings, $n - E\{N\}$; these are summarized as:

Table 8.9. $E\{N\}$ for Procedure $\mathcal{M}_{GNBC}$ with $d = 2$ Under the Slippage Configuration $(p, p, \ldots, p, \theta p)$ for Selected θ, t, and n

θ	n	t						
		2	3	4	5	6	7	8
1.0	5	4.875	4.963	4.984	4.992	4.995	4.997	4.998
	6	5.938	5.988	5.996	5.998	5.999	6.000	6.000
	9	8.555	8.916	8.956	8.974	8.984	8.990	8.993
	10	9.734	9.864	9.952	9.973	9.982	9.988	9.992
	13	12.229	12.781	12.903	12.944	12.967	12.980	12.987
	14	13.496	13.759	13.892	12.937	13.960	13.974	13.982
	17	15.920	16.670	16.839	16.906	16.944	16.962	16.973
	18	17.251	17.653	17.826	17.896	17.935	17.958	17.972
3.0	5	4.680	4.876	4.935	4.965	4.979	4.987	4.991
	6	5.762	5.922	5.968	5.985	5.992	5.996	5.998
	9	7.693	8.476	8.725	8.836	8.896	8.932	8.954
	10	8.897	9.336	9.645	9.791	9.867	9.910	9.937
	13	10.514	11.775	12.281	12.550	12.707	12.804	12.865
	14	11.781	12.593	13.153	13.461	13.642	13.755	13.829
	17	13.255	14.934	15.687	16.135	16.413	16.589	16.704
	18	14.555	15.708	16.527	17.015	17.323	17.524	17.656
5.0	5	4.517	4.739	4.847	4.904	4.937	4.957	4.970
	6	5.598	5.814	5.904	5.947	5.969	5.980	5.987
	9	7.130	7.918	8.314	8.535	8.672	8.762	8.824
	10	8.294	8.772	9.154	9.409	9.574	9.685	9.762
	13	9.581	10.748	11.381	11.807	12.106	12.322	12.481
	14	10.771	11.495	12.136	12.596	12.931	13.176	13.360
	17	11.995	13.437	14.261	14.871	15.328	15.677	15.945
	18	13.192	14.126	14.979	15.619	16.112	16.493	16.790

Reprinted from Bechhofer and Chen (1991), pp. 321–323. Copyright © 1991 by the American Sciences Press, Inc., Columbus, Ohio 43221-0161. Reprinted by permission.

- For fixed (n, t, d), the savings increase as θ increases. When $\theta \to \infty$, the configuration $\boldsymbol{p}$ approaches $(0, 0, \ldots, 0, 1)$; in this case, $E\{N \mid \boldsymbol{p} = (0, 0, \ldots, 0, 1)\} = \lceil (n + d)/2 \rceil$ for any t, and thus the savings is $n - \lceil (n + d)/2 \rceil$.
- For fixed (n, θ, d), the savings decreases as t increases.
- For fixed (t, θ, d), the savings increases as n increases.

8.3.3 A Curtailed Sequential Procedure

Chen and Hsu (1991) proposed a sequential procedure that achieves Goal 8.3.1 and guarantees (8.3.1). Their procedure employs a stopping rule that is a composite of the Gupta and Nagel (1967) fixed-sample-size rule, an inverse sampling rule due to Panchapakesan (1971) that stops sampling when the cell with the largest cell count reaches a certain value (see Section 8.5), a stopping rule of Ramey and Alam (1979) based on the difference between the two top cells, together with curtailment as in Bechhofer and Chen (1991). Four constants (n, d, M, r) are required to implement this composite procedure for given t and specified P^*. The procedure has the same LF-configuration to guarantee (8.3.1) as do procedures $\mathcal{M}_{GN}$ and $\mathcal{M}_{GNBC}$.

Procedure $\mathcal{M}_{CH}$ Let n denote the maximum number of multinomial observations to be taken. Find (d, M, r) in Table 8.10 corresponding to (t, n) and the specified P^*.

If sampling has not yet stopped at stage m, take a multinomial vector-observation $X_m = (X_{1m}, \ldots, X_{tm})$. Calculate the ordered sample cell totals, $y_{[1]m} \leq \cdots \leq y_{[t]m}$. Stop sampling at the first stage m such that:

1. $y_{[t]m} > y_{[t-1]m} + n - m + d$ or
2. $y_{[t]m} = M$ or
3. $y_{[t]m} - y_{[t-1]m} = r$ or
4. $m = n$.

If sampling stops at stage $m = N$ according to any of Criteria 1–3, include the single category associated with $y_{[t]N}$ in the selected subset. If sampling stops according to Criterion 4, include in the selected subset any category C_i such that

$$y_{iN} \geq y_{[t]N} - d.$$

Table 8.10, from Chen and Hsu (1991), gives (d, M, r)-values for $t = 3, 4$, $n = 10(5)30$ and $P^* = 0.75, 0.90, 0.95$. Concerning the performance of procedure $\mathcal{M}_{CH}$, notice that when $N = n$, the selection rule is the same as that of Gupta and Nagel, and the selected subset may contain one or more categories. Let S denote the number of cells selected. Chen and Hsu compare the performances of procedures $\mathcal{M}_{CH}$ and

Table 8.10. Values of (d, M, r) for Procedure $\mathcal{M}_{CH}$ for Given (t, n) and Specified $P^\star$

		$P^\star$		
t	n	0.75	0.90	0.95
	10	(3, 6, 4)	(5, 6, 5)	(5, 7, 5)
	15	(4, 8, 4)	(7, 9, 5)	(7, 9, 6)
3	20	(4, 10, 5)	(7, 11, 6)	(9, 11, 8)
	25	(4, 13, 6)	(7, 13, 7)	(8, 15, 8)
	30	(5, 15, 6)	(10, 16, 7)	(9, 19, 9)
	10	(4, 5, 3)	(4, 6, 4)	(5, 6, 5)
	15	(3, 8, 5)	(5, 8, 5)	(7, 9, 5)
4	20	(5, 8, 6)	(7, 10, 5)	(7, 11, 5)
	25	(4, 11, 6)	(7, 11, 6)	(7, 13, 8)
	30	(6, 12, 5)	(9, 13, 6)	(10, 13, 9)

Reprinted from Chen and Hsu (1991), pp. 408–409, by courtesy of the British Psychological Society.

$\mathcal{M}_{GN}$ by giving exact results for $P\{CS\}$, $E\{N\}$ and $E\{S\}$ (procedure $\mathcal{M}_{GN}$ always takes n observations). They provide a FORTRAN program to calculate $P\{CS\}$, $E\{N\}$ and $E\{S\}$ for procedure $\mathcal{M}_{CH}$ under the slippage configuration $p = (p, p, \ldots, p, \theta p)$ for given $\theta \geq 1$, $t \leq 10$, $n \leq 50$ and any particular (d, M, r). Taking $\theta = 1$, this program enables the experimenter to search for the optimal (d, M, r) for any specified $P^\star$; however, the search is exceedingly tedious for large t and n and for high $P^\star$. For comparison purposes the same program can also be used to calculate the $P\{CS\}$ and $E\{S\}$ for $t \leq 10$ and $n \leq 50$ for procedure $\mathcal{M}_{GN}$.

Based on their performance characteristics, Chen and Hsu concluded that the achieved $P\{CS\}$ of procedure $\mathcal{M}_{CH}$ may not be as large as that of procedure $\mathcal{M}_{GN}$ due to early termination of sampling. However, by choosing appropriate constants, procedure $\mathcal{M}_{CH}$ will always guarantee (8.3.1) with smaller $E\{N\}$ and $E\{S\}$ than will procedure $\mathcal{M}_{GN}$; the savings in $E\{N\}$ relative to that of the n of procedure $\mathcal{M}_{GN}$ is *very substantial* for large θ and n (although the decrease in $E\{S\}$ for these same constants is modest).

Concluding Remarks

If sequential sampling is an option for the experimenter, we recommend the use of procedure $\mathcal{M}_{CH}$ if the required constants (d, M, r) are either available in Table 8.10 or can be computed using the Chen and Hsu FORTRAN program. Otherwise, we recommend the curtailed version of the Gupta and Nagel (1967) rule, procedure $\mathcal{M}_{GNBC}$.

8.4 INDIFFERENCE-ZONE PROCEDURES FOR CROSS-CLASSIFIED DATA

In this section we consider selection procedures appropriate when a researcher collects data from two independent multinomial responses and is interested in identifying *simultaneously* the outcome of each that has the greatest probability. For example, suppose that a market research organization is planning a phone survey in which the characteristics of two products are to be surveyed simultaneously (to increase efficiency). In particular, it might be desired to identify the most popular brands of soap and cereal using a single survey. Alternatively, a political party might be interested in identifying that national candidate who is most popular at a given point in time and the voter opinion regarding a local school bond issue. Because of the nature of the characteristics, the planners of such a survey believe that the responses to the two questions will be *independent*. This problem is a generalization of the single-factor problem discussed in Sections 8.1 and 8.2.

Statistical Assumptions: Two sets of independent and identically distributed multinomial trials are observed: A_k ($k \geq 1$) with values $1, 2, \ldots, a$; and B_k ($k \geq 1$) with values $1, 2, \ldots, b$. Let $p_{ij} = P\{A_k = i, B_k = j\}$ ($1 \leq i \leq a, 1 \leq j \leq b$) be the joint probability of the outcome i on the A response and j on the B response (the same for all k).

Notice that the probabilities $\{p_{ij}\ (1 \leq i \leq a, 1 \leq j \leq b)\}$ satisfy $\sum_{i=1}^{a} \sum_{j=1}^{b} p_{ij} = 1$. Furthermore, $p_{i+} = \sum_{j=1}^{b} p_{ij}$ ($1 \leq i \leq a$) are the marginal probabilities for each A_k and $p_{+j} = \sum_{i=1}^{a} p_{ij}$ ($1 \leq j \leq b$) are the marginal probabilities for each B_k. Lastly, $p_{ij} = p_{i+} \times p_{+j}$ by independence of the A_k and B_k for each k.

The multinomial character of cross-classified data can be seen by defining $X_{ijk} = 1$ or 0 according to whether the result of the kth trial is $A_k = i$ and $B_k = j$ or not. Then $\{X_{ijk}\ (1 \leq i \leq a, 1 \leq j \leq b)\}$ is a single-trial (cross-classified) multinomial summary of the result of the kth trial and has corresponding matrix of cell probabilities $\{p_{ij}\ (1 \leq i \leq a, 1 \leq j \leq b)\}$. Throughout we say that category (or cell) C_{ij} has occurred on the kth observation if $X_{ijk} = 1$. We denote the ordered values of $p_{1+}, \ldots, p_{a+}$ and $p_{+1}, \ldots, p_{+b}$ by

$$p_{[1+]} \leq \cdots \leq p_{[(a-1)+]} \leq p_{[a+]} \quad \text{and} \quad p_{[+1]} \leq \cdots \leq p_{[+(b-1)]} \leq p_{[+b]},$$

respectively. Neither the values of the (p_{i+}, p_{+j}) nor their pairing with the $(p_{[s+]}, p_{[+t]})$ ($1 \leq i, s \leq a, 1 \leq j, t \leq b$) is assumed to be known.

The purpose of the experiment is stated in Goal 8.4.1, and the associated probability (design) requirement is given in Equation (8.4.1).

Goal 8.4.1 To select the category associated with $p_{[a+]}$ and the category associated with $p_{[+b]}$.

A *correct selection* (CS) is said to be made if Goal 8.4.1 is achieved. The category associated with $p_{[a+]}$ and $p_{[+b]}$ is referred to as the "best" category.

Probability Requirement: For specified constants $(\theta_\alpha^*, \theta_\beta^*, P^*)$ with $1 < \theta_\alpha^*, \theta_\beta^* < \infty$ and $1/ab < P^* < 1$, we require

$$P\{CS\} \geq P^* \quad \text{whenever } p_{[a+]} \geq \theta_\alpha^*\, p_{[(a-1)+]} \text{ and } p_{[+b]} \geq \theta_\beta^*\, p_{[+(b-1)]}.$$

(8.4.1)

The probability in Equation (8.4.1) depends on the (p_{i+}, p_{+j}) $(1 \leq i \leq a, 1 \leq j \leq b)$ and on the number n of multinomial trials. The constants $(\theta_\alpha^\star, \theta_\beta^\star)$ can be interpreted as the smallest $p_{[a+]}/p_{[(a-1)+]}$ and $p_{[+b]}/p_{[+(b-1)]}$ ratios "worth detecting." The following *single-stage* procedure was proposed by Bechhofer, Goldsman and Jennison (1989) to guarantee (8.4.1).

> **Procedure $\mathcal{M}_{BGJ}$** For the given (a, b) and specified $(\theta_\alpha^\star, \theta_\beta^\star, P^\star)$, determine n as in Example 8.4.1.
>
> *Sampling rule:* Take a random sample of n matrix-observations $\{X_{ijk}$ $(1 \leq i \leq a, 1 \leq j \leq b)\}$ $(1 \leq k \leq n)$ in a *single* stage. Let $R_{in} = \sum_{j=1}^{b} \sum_{k=1}^{n} X_{ijk}$ denote the total number of outcomes for row category i among the n trials, and let $C_{jn} = \sum_{i=1}^{a} \sum_{k=1}^{n} X_{ijk}$ denote the total number of outcomes in column category j among the n trials. Denote the ordered values of the R_{in} and C_{jn} by $R_{[1]n} \leq \cdots \leq R_{[a]n}$ and $C_{[1]n} \leq \cdots \leq C_{[b]n}$. Equal R_{in}-values and equal C_{jn}-values can occur with positive probabilities.
>
> *Terminal decision rule:* Select the categories that yielded $R_{[a]n}$ and $C_{[b]n}$ as the ones associated with $p_{[a]+}$ and $p_{+[b]}$, randomizing among tied categories for each variable, if several categories yield $R_{[a]n}$ or $C_{[b]n}$.

Determination of Sample Size

Let $P_n\{CS(2) \mid (a, b; \theta_\alpha^\star, \theta_\beta^\star)\}$ denote the $P\{CS\}$ for procedure $\mathcal{M}_{BGJ}$ when there are a [b] levels of the first [second] factor, n matrix-observations are taken, and the treatment effects are in the LF-configuration, that is, $p_{[a+]} = \theta_\alpha^\star p_{[i+]}$ $(1 \leq i \leq a - 1)$ and $p_{[+b]} = \theta_\beta^\star p_{[+j]}$ $(1 \leq j \leq b - 1)$. Similarly, let $P_n\{CS(1) \mid (t, \theta^\star)\}$ denote the $P\{CS\}$ for the univariate multinomial procedure $\mathcal{M}_{BEM}$ to achieve Goal 8.2.1 when there are t categories, n vector-observations are taken, and p is in the LF-configuration, that is, $p_{[t]} = \theta^\star \times p_{[i]}$ $(1 \leq i \leq t - 1)$. Then it can be shown that

$$P_n\{CS(2) \mid (a, b; \theta_\alpha^\star, \theta_\beta^\star)\} = P_n\{CS(1) \mid (a, \theta_\alpha^\star)\} \times P_n\{CS(1) \mid (b, \theta_\beta^\star)\} \qquad (8.4.2)$$

when there is independence between the row and column variables, that is, $p_{ij} = p_{i+} \times p_{+j}$ for all i and j (Bechhofer, Goldsman and Jennison 1989, Equation 4.10). This identity will enable us to determine the smallest n that will guarantee (8.4.1) if we use the single-stage procedure for given (a, b) and specified $(\theta_\alpha^\star, \theta_\beta^\star, P^\star)$. Tables 8.2–8.5 give $P_n\{CS(1) \mid (t, \theta^\star)\}$ for $t = 2(1)5$, respectively, and for various $\theta^\star$ and n. We give the following example to illustrate the method.

Example 8.4.1 Consider the phone survey problem described at the beginning of this section. Suppose that $(a, \theta_\alpha^\star) = (3, 1.4)$ and $(b, \theta_\beta^\star) = (4, 2.0)$ are specified. We are to determine n for procedure $\mathcal{M}_{BGJ}$ to guarantee (8.4.1). If we also specify

Table 8.11. $P\{CS\}$-Values for the Cross-Classified Multinomial Example with $(a, \theta_\alpha^\star) = (3, 1.4)$, $(b, \theta_\beta^\star) = (4, 2.0)$, and Selected n

n	$P_n\{CS(1) \mid (a, \theta_\alpha^\star)\}$[†]	$P_n\{CS(1) \mid (b, \theta_\beta^\star)\}$[‡]	$P_n\{CS(2) \mid (a, b; \theta_\alpha^\star, \theta_\beta^\star)\}$
126	0.9016	0.9958	0.8978
128	0.9039	0.9961	0.9004
130	0.9062	0.9964	0.9029
184	0.9498	0.9995	0.9493
186	0.9509	0.9996	0.9504
188	0.9520	0.9996	0.9516

[†]Probabilities obtained from Table 8.3
[‡]Probabilities obtained from Table 8.4

$P^\star = 0.90$ [0.95], then Table 8.11 tells us that approximately $n = 128$ [186] phone calls will guarantee (8.4.1). Clearly, the number of required phone calls will be very large if $\theta_\alpha^\star$ or $\theta_\beta^\star \to 1$ or $P^\star \to 1$.

Another (far less efficient) way of conducting the two-factor experiment is to handle it as two *separate* single-factor experiments, each using procedure $\mathcal{M}_{BEM}$, the first with $(a, \theta_\alpha^\star, P_1^\star)$ and the second with $(b, \theta_\beta^\star, P_2^\star)$, where $P_1^\star \cdot P_2^\star = P^\star$. In this situation, each person is asked one question (and the problem of independence of the responses does not arise). For convenience, one might take $P_1^\star = P_2^\star = \sqrt{P^\star}$. In particular, when $P^\star = 0.90$ we have $\sqrt{P^\star} = \sqrt{0.90} = 0.94868$ and it is necessary to phone $n_1 = 183$ respondents to find their preferences for the first characteristic and a further $n_2 = 60$ respondents to find their preferences for the second characteristic. This total of 243 contacts can be reduced by a more judicious choice of $(P_1^\star, P_2^\star)$ subject to their satisfying $P_1^\star \cdot P_2^\star = P^\star = 0.90$. The optimal pair $(P_1^\star, P_2^\star) = (0.9239, 0.9742)$ to minimize $n_1 + n_2$ yields $n_1 = 148$ and $n_2 = 78$, for a total of 226 calls. Comparison of this total number of 226 with the original two-factor number of 128 illustrates the advantage of asking each contact *both* questions when independence holds.

Remark 8.4.1 Bechhofer and Goldsman (1988a) showed how the open, noneliminating, sequential procedure $\mathcal{M}_{BKS}$ of Bechhofer, Kiefer and Sobel (1968) for single-factor experiments involving responses from exponential family populations can be generalized to multi-factor experiments in which *no interaction* is present. As a special case, their results apply to cross-classified multinomial data in which the classification variables are independent.

8.5 CHAPTER NOTES

The composite subset selection rule of Chen and Hsu (1991), procedure $\mathcal{M}_{CH}$, uses an "inverse sampling" criterion, among others, to stop sampling. Inverse sampling means that the number of multinomial trials is not determined *a priori* but that sampling continues until some event occurs. This type of stopping technique was first proposed for subset selection by Panchapakesan (1971), who developed a procedure that achieves Goal 8.3.1 and guarantees (8.3.1). His procedure takes multinomial trials X_m ($m \geq 1$) one-at-a-time until one of the cumulative cell totals equals a predetermined value M, that is, until the first stage m such that $y_{[t]m} = M$. Inverse sampling has also been used to terminate sampling for IZ procedures that select the most probable cell (Cacoullos and Sobel 1966, Alam 1971, Ramey and Alam 1979, Chen 1992) and for IZ procedures that select the *least* probable cell (Chen 1985). The four papers listed above that select the most probable cell use an IZ formulation based on the ratio "distance" ($p_{[t]}/p_{[t-1]}$). For example, Cacoullos and Sobel (1966) proposed an inverse sampling procedure that stops sampling when the total count in any cell reaches a preassigned number; Alam (1971) used a sequential procedure that stops when the difference between the largest and second-largest total cell counts reaches a preassigned number; Ramey and Alam (1979) proposed a stopping rule based on the simultaneous use of *both* of the above-mentioned criteria. Chen (1992) employed a stopping rule that combines Ramey and Alam's stopping rule with the Bechhofer and Kulkarni (1984) stopping rule, the latter stopping when the frequency of any cell is large enough to guarantee the selection of a particular cell.

There have been fewer proposals for selecting the multinomial cell with the *smallest* probability, that is, that associated with $p_{[1]}$. Unlike the binomial problem where selection of the smallest event probability $p_{[1]}$ can often be handled by symmetry using the transformed variables $1 - X_{ij}$, where X_{ij} is the jth Bernoulli outcome on the ith treatment, the multinomial problem does not have an analogous symmetry. Gupta and Nagel (1967) proposed a single-stage subset procedure that chooses a random-size subset of the t categories that contains the category associated with the smallest probability. All procedures that adopt the IZ approach for this problem use the distance measure $p_{[2]} - p_{[1]}$. (The ratio measure *cannot* be used for selecting the "worst" cell.) Alam and Thompson (1972) proposed a single-stage IZ procedure that correctly selects the least probable cell with probability at least P^* whenever $p_{[2]} - p_{[1]} \geq \Delta^*$. They provide tables to implement their procedure for selected t, Δ^* and P^*. In the same paper referred to above, Chen (1992) gave a sequential IZ procedure for selecting the "worst" category while satisfying the same probability requirement as Alam and Thompson (1972); he also provided constants to implement his rule.

Chen (1988) proposed a procedure for selecting the best category provided that it is better than a *control* category having *unknown* associated probability. His formulation parallels that of Bechhofer and Turnbull (1978) for the normal means (common known variance) problems as described in Section 5.2.1. Chen's procedure uses an inverse sampling rule; tables to implement the procedure are provided.

Both the minimax and Bayes approaches have been used for the multinomial problem. For example, Berger (1980) devised a class of minimax subset selection procedures for selecting the least probable cell that satisfy the probability requirement (8.3.1) when the loss is measured by the size of the selected subset. (See also Berger (1982) and the references therein.) Both Bayes and empirical Bayes rules have been developed for selecting the most probable multinomial category; see Gupta and Liang (1989b) for a guide to the literature using these approaches.

Several authors have proposed simultaneous confidence intervals for the t multinomial probabilities. Tortora (1978) and Thompson (1987) gave one set of intervals and determined the smallest sample size such that the intervals hold with joint confidence coefficient at least P^*. Bromaghm (1993) studied the properties of these intervals and proposed a new procedure to achieve the same goal; his procedure requires smaller recommended sample sizes than those of Tortora or Thompson.

APPENDIX A

Relationships Among Critical Points and Notation

We review the definitions of the various critical points used in this book and state their relationships. Let $W = (W_1, \ldots, W_p)$ have a multivariate normal distribution with mean vector zero, unit variances and common correlation ρ. Let

$$W_{[1]} \leq \cdots \leq W_{[p]}$$

denote the ordered components of W. Assume W is independent of V which has a χ_ν^2 distribution.

Definition A.1 Suppose that $\rho = 0$ in the specification of W. The upper-α critical point of the p-dimensional Studentized maximum modulus distribution with ν degrees of freedom, $|M|_{p,\nu}^{(\alpha)}$, is defined implicitly by

$$P\left\{\max_{1 \leq j \leq p} |W_j| \leq |M|_{p,\nu}^{(\alpha)} \sqrt{V/\nu}\right\} = 1 - \alpha.$$

Definition A.2 Suppose that $\rho = 0$ in the specification of W. The upper-α critical point of the Studentized range distribution, $Q_{p,\nu}^{(\alpha)}$, is defined implicitly by

$$P\left\{W_{[p]} - W_{[1]} \leq Q_{p,\nu}^{(\alpha)} \sqrt{V/\nu}\right\} = 1 - \alpha.$$

Definition A.3 The one-sided upper-α equicoordinate point of the equicorrelated multivariate central t-distribution, $T_{p,\nu,\rho}^{(\alpha)}$, is defined implicitly by

$$P\left\{\max_{1 \leq j \leq p} W_j \leq T_{p,\nu,\rho}^{(\alpha)} \sqrt{V/\nu}\right\} = 1 - \alpha.$$

Definition A.4 The two-sided upper-α equicoordinate point of the equicorrelated multivariate central t-distribution, $|T|^{(\alpha)}_{p,\nu,\rho}$, is defined implicitly by

$$P\left\{\max_{1\leq j\leq p} |W_j| \leq |T|^{(\alpha)}_{p,\nu,\rho}\sqrt{V/\nu}\right\} = 1 - \alpha.$$

Definition A.5 The one-sided upper-α equicoordinate point of the equicorrelated multivariate normal distribution, $Z^{(\alpha)}_{p,\rho}$, is defined implicitly by

$$P\left\{\max_{1\leq j\leq p} W_j \leq Z^{(\alpha)}_{p,\rho}\right\} = 1 - \alpha.$$

Definition A.6 The two-sided upper-α equicoordinate point of the equicorrelated multivariate normal distribution, $|Z|^{(\alpha)}_{p,\rho}$, is defined implicitly by

$$P\left\{\max_{1\leq j\leq p} |W_j| \leq |Z|^{(\alpha)}_{p,\rho}\right\} = 1 - \alpha.$$

Definition A.7 The upper-α quantile of the standard normal distribution, $z^{(\alpha)}$, is defined implicitly by

$$P\{N(0,1) \leq z^{(\alpha)}\} = 1 - \alpha.$$

Relationships

For any p and α, we have

$$|T|^{(\alpha)}_{p,\nu,\rho} \to |Z|^{(\alpha)}_{p,\rho} \quad \text{as } \nu \to \infty \quad \text{and} \quad T^{(\alpha)}_{p,\nu,\rho} \to Z^{(\alpha)}_{p,\rho} \quad \text{as } \nu \to \infty.$$

For any α, the following hold exactly:

$$|Z|^{(\alpha)}_{p,0} = z^{(\alpha^*/2)}$$

where $\alpha^* = 1 - (1-\alpha)^{1/p}$ and

$$|M|^{(\alpha)}_{p,\nu} = |T|^{(\alpha)}_{p,\nu,0}.$$

LIST OF NOTATION

$(a)^+$	$\max\{a, 0\}$
$(a)^-$	$\min\{a, 0\}$
$\lceil a \rceil$	Smallest integer greater than or equal to a
$\mathcal{B}_{BK}$	Bechhofer and Kulkarni (1982) adaptive closed curtailed procedure for selecting the Bernoulli treatment with the largest success probability (Section 7.3)

LIST OF NOTATION

$\mathcal{B}_{BKS}$	Bechhofer, Kiefer and Sobel (1968) open noneliminating sequential procedure for selecting the Bernoulli treatment with the largest success probability (Section 7.4)
$\mathcal{B}_{GS}$	Gupta and Sobel (1960) single-stage procedure for selecting a subset of Bernoulli treatments containing the treatment with the largest success probability (Section 7.5)
$\mathcal{B}_P$	Paulson (1993) open sequential procedure with elimination for selecting the Bernoulli treatment with the largest success probability (Section 7.4)
$\mathcal{B}_{SH}$	Sobel and Huyett (1957) single-stage procedure for selecting the Bernoulli treatment with the largest success probability (Section 7.2)
CS	Correct selection
d.f.	Degrees of freedom
EM	Equal means (configuration)
EP	Equal probability (configuration)
ES	Equally-spaced (configuration)
IZ	Indifference-zone (formulation)
LF	Least favorable (configuration)
$\ln(\cdot)$	natural logarithm
$\mu_{[i]}$	ith ordered population mean
$\lvert M \rvert_{k,\nu}^{(\alpha)}$	Two-sided upper-α critical point of the Studentized maximum modulus distribution with parameter k and ν degrees of freedom (Section 4.2)
$\mathcal{M}_{BEM}$	Bechhofer, Elmaghraby and Morse (1959) single-stage procedure for selecting the multinomial cell with the largest cell probability (Section 8.2)
$\mathcal{M}_{BG}$	Bechhofer and Goldsman (1985, 1986) closed sequential procedure for selecting the multinomial cell with the largest cell probability (Section 8.2)
$\mathcal{M}_{BGJ}$	Bechhofer, Goldsman and Jennison (1989) single-stage procedure for selecting the cell with the largest cell probability in a bivariate cross-classified setup where independence holds (Section 8.4)
$\mathcal{M}_{BK}$	Bechhofer and Kulkarni (1984) closed sequential procedure for selecting the multinomial cell with the largest cell probability (Section 8.2)
$\mathcal{M}_{CH}$	Chen and Hsu (1991) curtailed sequential subset selection procedure for selecting a subset of multinomial cells containing the cell with the largest cell probability (Section 8.3)
$\mathcal{M}_{GN}$	Gupta and Nagel (1967) single-stage subset selection procedure for selecting a subset of multinomial cells containing the cell with the largest cell probability (Section 8.3)
$\mathcal{M}_{GNBC}$	Bechhofer and Chen (1991) curtailed sequential version of the Gupta and Nagel (1967) procedure $\mathcal{M}_{GN}$ for selecting a subset of multinomial cells containing the cell with the largest cell probability (Section 8.3)
$\mathcal{N}_B$	Bechhofer (1954) single-stage procedure for selecting the treatment with the largest mean in a single-factor normal experiment when observations from all treatments have a common known variance (Section 2.2)

$\mathcal{N}_{B2}$ Bechhofer (1954) single-stage procedure for selecting simultaneously the treatment combination with the largest main effects in an additive two-factor normal experiment when observations from all treatment combinations have a common known variance (Section 6.2)

$\mathcal{N}_{BDS}$ Bechhofer, Dunnett and Sobel (1954) open noneliminating sequential procedure for selecting the treatment with the largest mean in a single-factor normal experiment when observations from all treatments have a common unknown variance (Section 2.7)

$\mathcal{N}_{BD2}$ Bechhofer and Dunnett (1986) two-stage procedure for selecting simultaneously the treatment combination with the largest main effects in an additive two-factor normal experiment when observations from all treatment combinations have a common unknown variance (Section 6.4)

$\mathcal{N}_{BKS}$ Bechhofer, Kiefer and Sobel (1968) open noneliminating sequential procedure for selecting the treatment with the largest mean in a single-factor normal experiment when observations from all treatments have a common known variance (Section 2.3)

$\mathcal{N}_{BG}$ Bechhofer and Goldsman (1987) truncated version of procedure $\mathcal{N}_{BKS}$ for selecting the treatment with the largest mean in a single-factor normal experiment when observations from all treatments have a common known variance (Section 2.3)

$\mathcal{N}_{BG2}$ Bechhofer and Goldsman (1988b) truncated non-eliminating sequential procedure for selecting the treatment combination with the largest main effects in an additive two-factor normal experiment when observations from all treatment combinations have a common known variance (Section 6.2)

$\mathcal{N}_{BT}$ Bechhofer and Turnbull (1978) single-stage procedure for selecting the best treatment relative to a standard in a single-factor normal experiment when observations from all treatments have a common known variance (Section 5.2)

$\mathcal{N}_{CGHBM}$ Carroll, Gupta and Huang (1975) single-stage procedure for selecting a subset of treatments containing the s largest means as modified by Bofinger and Mengersen (1986) when observations from all treatments have a common known or unknown variance (Section 3.4)

$\mathcal{N}_D$ Dunnett (1955) one- and two-sided simultaneous confidence intervals for comparing each treatment mean with a control when observations from all treatments have a common unknown variance (Section 5.4)

$\mathcal{N}_G$ Gupta (1956, 1965) single-stage procedure for selecting a subset of treatments containing the largest mean in a single-factor normal experiment when observations from all treatments have a common known or common unknown variance (Section 3.2)

$\mathcal{N}_{G2}$ Bechhofer and Dunnett (1987) single-stage procedure for selecting a subset of treatment combinations containing the treatment combination associated with the largest main effects in an additive two-factor normal

	experiment when observations from all treatment combinations have a common known or common unknown variance (Section 6.6)
$\mathcal{N}_{GS}$	Gupta and Sobel (1958) single-stage procedure for selecting a subset containing all treatments better than a standard or control in a single-factor normal experiment when observations from all treatments have a common known or unknown variance (Section 5.3)
$\mathcal{N}_{GSa}$	Gupta and Santner (1973) single-stage procedure for selecting a bounded subset of treatments containing the largest mean in a single-factor normal experiment when observations from all treatments have a common known variance (Section 3.4)
$\mathcal{N}_H$	Hartmann (1991) open multi-stage procedure for selecting the treatment with the largest mean in a single-factor normal experiment when observations from all treatments have a common unknown variance (Section 2.7)
$\mathcal{N}_{H2}$	Hartmann (1993) closed multi-stage procedure for selecting simultaneously the treatment combination with the largest main effects in an additive two-factor normal experiment when observations from all treatment combinations have a common known variance (Section 6.2)
$\mathcal{N}_{Ha}$	Hayter (1989) single-stage procedure for selecting the treatment with the largest mean in a single-factor normal experiment when observations from different treatments have unknown, possibly unequal, but bounded variances (Section 2.8)
$\mathcal{N}_{HL}$	Hochberg and Lachenbruch (1976) fixed-width simultaneous confidence intervals for differences of treatment means in a single-factor normal experiment when observations from all treatments have a common known variance (Section 4.3)
$\mathcal{N}_{Hoo}$	Hoover (1991) simultaneous confidence intervals for comparing treatment means with two controls in a single-factor normal experiment when observations from all treatments have a common unknown variance (Section 5.4)
$\mathcal{N}_{JH}$	Hsu (1984a) two-sided simultaneous confidence intervals for comparing each treatment mean with the best of the remaining means in a single-factor normal experiment when observations from all treatments have a common unknown variance (Section 4.4)
$\mathcal{N}_P$	Paulson (1993) closed eliminating sequential procedure for selecting the treatment with the largest mean in a single-factor normal experiment when observations from all treatments have a common known variance (Section 2.3)
$\mathcal{N}_{P/C}$	Paulson (1952) single-stage procedure for selecting the treatment with the largest mean with respect to a control treatment in a single-factor normal experiment when observations from all treatments have a common known variance (Section 5.2)
$\mathcal{N}_{PS2}$	Pan and Santner (1993) single-stage procedure for selecting simultaneously the treatment combination with the largest main effects in an addi-

	tive two-factor normal split-plot experiment when observations from all treatment combinations have a common known variance (Section 6.3)
$\mathcal{N}_R$	Rinott (1978) open two-stage procedure for selecting the treatment with the largest mean in a single-factor normal experiment when observations from all treatments have a common unknown variance (Section 2.8)
$\mathcal{N}_{SP1}$	Santner and Pan (1994) single-stage procedure for selecting a subset of treatments associated with the treatment combinations having the largest main effects in an additive two-factor normal split-plot experiment when observations from all treatment combinations have a common unknown confounding factor variance and a common unknown measurement error variance (Section 6.7)
$\mathcal{N}_{SP2}$	Santner and Pan (1994) single-stage procedure for selecting a subset of treatments associated with the treatment combinations having the largest main effects in an additive two-factor normal split-plot experiment when observations from all treatment combinations have a common known confounding factor variance and a common (known or unknown) measurement error variance (Section 6.7)
$\mathcal{N}_T$	Tukey (1953) two-sided simultaneous confidence intervals for all pairwise treatments in a single-factor normal experiment when observations from all treatments have a common unknown variance (Section 4.3)
$\mathcal{N}_{TB}$	Tamhane and Bechhofer (1977, 1979) two-stage procedure for selecting the treatment with the largest mean in a single-factor normal experiment when observations from all treatments have a common known variance (Section 2.3)
$P^\star$	Specified minimum probability of correct decision for a particular experimenal goal
$\phi(\cdot)$	Standard normal density function
$\Phi(\cdot)$	Standard normal distribution function
$Q_{k,\nu,\rho}^{(\alpha)}$	Upper-α critical point of the studentized range distribution with parameter k and ν degrees of freedom (Section 4.3)
SS	Subset selection (formulation)
$t_\nu^{(\alpha)}$	Upper-α critical point of student's t-distribution with ν degrees of freedom
$T_{k,\nu,\rho}^{(\alpha)}$	One-sided upper-α equicoordinate point of the k-variate equicorrelated t-distribution with common correlation ρ and ν d.f. (Section 2.7)
$\lvert T \rvert_{k,\nu,\rho}^{(\alpha)}$	Two-sided upper-α equicoordinate point of the k-variate equicorrelated t-distribution with common correlation ρ and ν degrees of freedom (Section 5.4)
$z^{(\alpha)}$	Upper-α critical point of the standard normal distribution
$Z_{k,\rho}^{(\alpha)}$	One-sided upper-α equicoordinate point of the k-variate equicorrelated multivariate normal distribution with common correlation ρ (Section 2.2)
$\lvert Z \rvert_{k,\rho}^{(\alpha)}$	Two-sided upper-α equicoordinate point of the k-variate equicorrelated multivariate normal distribution with common correlation ρ (Section 2.2)

APPENDIX B

Tables

The tables presented in this appendix complement the FORTRAN programs of Appendix C and will be particularly useful for those lacking access to a computer. They list various critical points for the equicorrelated normal distribution, the equicorrelated t-distribution, the studentized range distribution, and the studentized maximum modulus distribution. By groups, the tables are organized as follows:

- Tables B.1 and B.2 concern the equicorrelated normal distribution, with Table B.2 containing the information of Table B.1, but the latter is simpler to use for the very important and widely used case of common correlation 1/2.
- Tables B.3 and B.4 bear the same relationship as do Tables B.1 and B.2 but concern the equicorrelated t-distribution. Table B.5 contains the two-sided percentiles for the case of common correlation 1/2 and thus complements Table B.3.
- Tables B.6 and B.7 are critical points of quantities defined in terms of independent normal random variables and an independent "studentizing" chi-square variate.

Tables B.1 and B.3 were computed by the authors from the Dunnett (1989) program and checked using their own programs. Table B.2 of $Z_{p,\rho}^{(\alpha)}$ is from Table 2 of Hochberg and Tamhane (1987) (H–T, hereafter), who adapted it from Gupta, Nagel and Panchapakesan (1973). Table B.4 was computed by the authors from Dunnett's MVTPRD program for $\rho = 0$ and is taken from Table 4 of H–T for $\rho > 0$. Table B.5 is Table 5 from H–T with some rearrangement of the values. Tables 4 and 5 of H–T were prepared by Charles Dunnett using the program later published in Dunnett (1989). Table B.6 is Table 8 of H–T, who adapted it from Harter (1960, 1969). Lastly, Table B.7 is Table 7 from H–T, also prepared by Dunnett.

As recommended by H-T, the following rules should be used to interpolate in Tables B.2, B.4, B.5 and B.6.

- Interpolation with respect to the d.f. ν in Tables B.4, B.5 and B.6 should be done linearly in $1/\nu$.
- Interpolation with respect to the upper tail probability α should be done linearly in $\ln(\alpha)$.
- Interpolation with respect to the dimension p should be done linearly in $\ln(p)$.
- Interpolation with respect to the common correlation ρ in Tables B.2, B.4 and B.5 should be done linearly in $1/(1 - \rho)$.

Table B.1. Upper-α Equicoordinate Point $Z^{(\alpha)}_{p,1/2}$ of the p-Variate Standard Normal Distribution with Unit Variances and Common Correlation Coefficient 1/2 for $\alpha = 0.01(0.01)0.25$ and $p = 1(1)10$

α	p									
	1	2	3	4	5	6	7	8	9	10
0.01	2.326	2.558	2.685	2.772	2.837	2.889	2.933	2.970	3.002	3.031
0.02	2.054	2.300	2.435	2.526	2.595	2.649	2.695	2.733	2.767	2.797
0.03	1.881	2.138	2.277	2.371	2.442	2.498	2.544	2.584	2.618	2.649
0.04	1.751	2.016	2.158	2.255	2.327	2.384	2.432	2.472	2.507	2.538
0.05	1.645	1.916	2.062	2.160	2.234	2.292	2.341	2.381	2.417	2.448
0.06	1.555	1.832	1.981	2.080	2.155	2.214	2.263	2.305	2.340	2.372
0.07	1.476	1.758	1.909	2.010	2.086	2.146	2.195	2.237	2.273	2.306
0.08	1.405	1.692	1.845	1.947	2.024	2.084	2.134	2.177	2.214	2.246
0.09	1.341	1.632	1.787	1.891	1.968	2.029	2.079	2.122	2.159	2.192
0.10	1.282	1.577	1.734	1.838	1.916	1.978	2.029	2.072	2.109	2.142
0.11	1.227	1.526	1.684	1.790	1.868	1.931	1.982	2.025	2.063	2.096
0.12	1.175	1.478	1.638	1.744	1.824	1.886	1.938	1.982	2.019	2.053
0.13	1.126	1.433	1.594	1.701	1.781	1.845	1.896	1.941	1.979	2.012
0.14	1.080	1.390	1.553	1.661	1.741	1.805	1.857	1.902	1.940	1.974
0.15	1.036	1.349	1.513	1.622	1.703	1.767	1.820	1.865	1.903	1.937
0.16	0.994	1.310	1.475	1.586	1.667	1.731	1.784	1.829	1.868	1.902
0.17	0.954	1.273	1.439	1.550	1.632	1.697	1.750	1.795	1.834	1.868
0.18	0.915	1.237	1.405	1.516	1.599	1.664	1.717	1.763	1.802	1.836
0.19	0.878	1.202	1.371	1.483	1.566	1.632	1.686	1.731	1.770	1.805
0.20	0.842	1.168	1.339	1.452	1.535	1.601	1.655	1.701	1.740	1.775
0.21	0.806	1.136	1.307	1.421	1.505	1.571	1.625	1.671	1.711	1.745
0.22	0.772	1.104	1.277	1.391	1.475	1.542	1.596	1.642	1.682	1.717
0.23	0.739	1.073	1.247	1.362	1.446	1.513	1.568	1.614	1.654	1.689
0.24	0.706	1.043	1.218	1.333	1.418	1.486	1.541	1.587	1.627	1.662
0.25	0.674	1.014	1.189	1.306	1.391	1.458	1.514	1.560	1.601	1.636

Table B.2. Upper-α Equicoordinate Point $Z_{p,\rho}^{(\alpha)}$ of the p-Variate Standard Normal Distribution with Unit Variances and Common Correlation ρ for $\alpha = 0.01, 0.05, 0.10, 0.25$, $\rho = 0.1(0.2)0.7$, and $p = 2(1)10(2)20(4)40, 50$

	$\rho = 0.1$					$\rho = 0.3$			
$p\backslash\alpha$	0.25	0.10	0.05	0.01	$p\backslash\alpha$	0.25	0.10	0.05	0.01
2	1.09	1.63	1.95	2.57	2	1.06	1.61	1.94	2.57
3	1.31	1.81	2.12	2.71	3	1.28	1.78	2.10	2.70
4	1.46	1.93	2.23	2.80	4	1.39	1.90	2.20	2.80
5	1.56	2.02	2.31	2.88	5	1.49	1.98	2.28	2.86
6	1.65	2.10	2.38	2.93	6	1.57	2.05	2.35	2.92
7	1.72	2.16	2.43	2.98	7	1.63	2.11	2.40	2.97
8	1.78	2.21	2.48	3.02	8	1.69	2.16	2.45	3.01
9	1.83	2.25	2.52	3.06	9	1.73	2.20	2.49	3.04
10	1.87	2.29	2.56	3.09	10	1.77	2.24	2.52	3.07
12	1.95	2.36	2.62	3.14	12	1.84	2.30	2.58	3.12
14	2.01	2.41	2.67	3.18	14	1.90	2.35	2.63	3.17
16	2.06	2.46	2.72	3.22	16	1.95	2.40	2.67	3.20
18	2.11	2.50	2.75	3.26	18	1.99	2.43	2.71	3.24
20	2.15	2.54	2.79	3.29	20	2.03	2.47	2.74	3.27
24	2.22	2.60	2.85	3.34	24	2.09	2.53	2.79	3.32
28	2.27	2.65	2.89	3.38	28	2.14	2.57	2.84	3.36
32	2.32	2.70	2.94	3.42	32	2.18	2.61	2.88	3.39
36	2.36	2.73	2.97	3.45	36	2.22	2.65	2.91	3.42
40	2.40	2.77	3.00	3.48	40	2.26	2.68	2.94	3.45
50	2.48	2.84	3.07	3.54	50	2.33	2.75	3.01	3.51
	$\rho = 0.5$					$\rho = 0.7$			
$p\backslash\alpha$	0.25	0.10	0.05	0.01	$p\backslash\alpha$	0.25	0.10	0.05	0.01
2	1.01	1.58	1.92	2.56	2	0.95	1.53	1.88	2.53
3	1.19	1.73	2.06	2.68	3	1.09	1.66	2.00	2.64
4	1.31	1.84	2.16	2.77	4	1.18	1.75	2.08	2.72
5	1.39	1.92	2.23	2.84	5	1.25	1.81	2.14	2.78
6	1.46	1.98	2.29	2.89	6	1.30	1.86	2.19	2.82
7	1.51	2.03	2.34	2.93	7	1.35	1.90	2.23	2.86
8	1.56	2.07	2.38	2.97	8	1.39	1.94	2.27	2.89
9	1.60	2.11	2.42	3.00	9	1.42	1.97	2.30	2.92
10	1.64	2.14	2.45	3.03	10	1.45	1.99	2.32	2.94
12	1.70	2.20	2.50	3.08	12	1.49	2.04	2.37	2.98
14	1.75	2.24	2.55	3.12	14	1.53	2.08	2.40	3.02
16	1.79	2.28	2.58	3.15	16	1.56	2.11	2.43	3.05
18	1.82	2.32	2.62	3.18	18	1.59	2.13	2.46	3.07
20	1.85	2.35	2.64	3.21	20	1.62	2.16	2.48	3.09
24	1.91	2.40	2.69	3.26	24	1.66	2.20	2.52	3.13
28	1.95	2.44	2.73	3.29	28	1.70	2.23	2.56	3.16
32	1.99	2.48	2.77	3.33	32	1.72	2.26	2.58	3.19
36	2.02	2.51	2.80	3.36	36	1.75	2.29	2.61	3.21
40	2.05	2.53	2.83	3.38	40	1.77	2.31	2.63	3.23
50	2.11	2.59	2.88	3.43	50	1.82	2.35	2.67	3.28

Reprinted from Hochberg and Tamhane (1987), p. 381, by courtesy of John Wiley & Sons.

Table B.3. Upper-α Equicoordinate Point $T^{(\alpha)}_{p,\nu,1/2}$ of the p-Variate t-Distribution with Unit Variances, Common Correlation 1/2, and ν Degrees of Freedom for $\alpha = 0.05, 0.10, 0.25$, $p = 1(1)9$, and $\nu = 1(1)20(5)30, 60, 120$

	$\alpha = 0.05$								
	\multicolumn{9}{c}{p}
ν	1	2	3	4	5	6	7	8	9
1	6.31	9.51	11.58	13.10	14.27	15.23	16.04	16.73	17.34
2	2.92	3.80	4.34	4.71	5.00	5.24	5.44	5.60	5.75
3	2.35	2.94	3.28	3.52	3.70	3.85	3.97	4.08	4.17
4	2.13	2.61	2.88	3.08	3.22	3.34	3.44	3.52	3.59
5	2.02	2.44	2.68	2.85	2.98	3.08	3.16	3.24	3.30
6	1.94	2.34	2.56	2.71	2.83	2.92	3.00	3.06	3.12
7	1.89	2.27	2.48	2.62	2.73	2.82	2.89	2.95	3.00
8	1.86	2.22	2.42	2.55	2.66	2.74	2.81	2.87	2.92
9	1.83	2.18	2.37	2.50	2.60	2.68	2.75	2.81	2.86
10	1.81	2.15	2.34	2.47	2.56	2.64	2.70	2.76	2.81
11	1.80	2.13	2.31	2.43	2.53	2.60	2.67	2.72	2.77
12	1.78	2.11	2.29	2.41	2.50	2.58	2.64	2.69	2.74
13	1.77	2.09	2.27	2.39	2.48	2.55	2.61	2.66	2.71
14	1.76	2.08	2.25	2.37	2.46	2.53	2.59	2.64	2.69
15	1.75	2.07	2.24	2.36	2.44	2.52	2.57	2.62	2.67
16	1.75	2.06	2.23	2.34	2.43	2.50	2.56	2.61	2.65
17	1.74	2.05	2.22	2.33	2.42	2.49	2.54	2.59	2.64
18	1.73	2.04	2.21	2.32	2.41	2.48	2.53	2.58	2.62
19	1.73	2.03	2.20	2.31	2.40	2.47	2.52	2.57	2.61
20	1.72	2.03	2.19	2.30	2.39	2.46	2.51	2.56	2.60
25	1.71	2.00	2.17	2.27	2.36	2.42	2.48	2.52	2.56
30	1.70	1.99	2.15	2.25	2.34	2.40	2.45	2.50	2.54
35	1.69	1.98	2.13	2.24	2.32	2.38	2.44	2.48	2.52
40	1.68	1.97	2.13	2.23	2.31	2.37	2.42	2.47	2.51
45	1.68	1.96	2.12	2.22	2.30	2.36	2.41	2.46	2.50
50	1.68	1.96	2.11	2.22	2.29	2.36	2.41	2.45	2.49
55	1.67	1.96	2.11	2.21	2.29	2.35	2.40	2.44	2.48
60	1.67	1.95	2.10	2.21	2.28	2.35	2.40	2.44	2.48
120	1.66	1.93	2.08	2.18	2.26	2.32	2.37	2.41	2.45

Table B.3. *Continued*

	$\alpha = 0.10$								
					p				
ν	1	2	3	4	5	6	7	8	9
1	3.08	4.70	5.74	6.50	7.09	7.57	7.97	8.32	8.62
2	1.89	2.54	2.92	3.20	3.40	3.57	3.71	3.83	3.94
3	1.64	2.13	2.41	2.61	2.76	2.87	2.97	3.06	3.13
4	1.53	1.96	2.20	2.37	2.50	2.60	2.68	2.75	2.82
5	1.48	1.87	2.09	2.25	2.36	2.45	2.53	2.59	2.65
6	1.44	1.82	2.02	2.17	2.27	2.36	2.43	2.49	2.54
7	1.41	1.78	1.98	2.11	2.22	2.30	2.37	2.42	2.47
8	1.40	1.75	1.94	2.08	2.17	2.25	2.32	2.38	2.42
9	1.38	1.73	1.92	2.05	2.14	2.22	2.28	2.34	2.39
10	1.37	1.71	1.90	2.02	2.12	2.19	2.26	2.31	2.35
11	1.36	1.70	1.88	2.01	2.10	2.17	2.23	2.29	2.33
12	1.36	1.69	1.87	1.99	2.08	2.16	2.22	2.27	2.31
13	1.35	1.68	1.86	1.98	2.07	2.14	2.20	2.25	2.29
14	1.35	1.67	1.85	1.97	2.06	2.13	2.19	2.24	2.28
15	1.34	1.67	1.84	1.96	2.05	2.12	2.18	2.23	2.27
16	1.34	1.66	1.83	1.95	2.04	2.11	2.17	2.22	2.26
17	1.33	1.65	1.83	1.94	2.03	2.10	2.16	2.21	2.25
18	1.33	1.65	1.82	1.94	2.02	2.09	2.15	2.20	2.24
19	1.33	1.65	1.82	1.93	2.02	2.09	2.14	2.19	2.23
20	1.33	1.64	1.81	1.93	2.01	2.08	2.14	2.19	2.23
25	1.32	1.63	1.80	1.91	1.99	2.06	2.11	2.16	2.20
30	1.31	1.62	1.79	1.90	1.98	2.05	2.10	2.15	2.19
35	1.31	1.61	1.78	1.89	1.97	2.04	2.09	2.14	2.18
40	1.30	1.61	1.77	1.88	1.96	2.03	2.08	2.13	2.17
45	1.30	1.61	1.77	1.88	1.96	2.02	2.08	2.12	2.16
50	1.30	1.60	1.76	1.87	1.95	2.02	2.07	2.12	2.16
55	1.30	1.60	1.76	1.87	1.95	2.01	2.07	2.11	2.15
60	1.30	1.60	1.76	1.87	1.95	2.01	2.06	2.11	2.15
120	1.29	1.59	1.75	1.85	1.93	1.99	2.05	2.09	2.13

continued

Table B.3. *Continued*

	$\alpha = 0.25$								
	p								
ν	1	2	3	4	5	6	7	8	9
1	1.00	1.71	2.14	2.46	2.70	2.89	3.06	3.20	3.32
2	0.82	1.29	1.56	1.74	1.88	1.99	2.08	2.16	2.23
3	0.76	1.19	1.41	1.57	1.69	1.78	1.86	1.92	1.98
4	0.74	1.14	1.35	1.50	1.60	1.69	1.76	1.82	1.87
5	0.73	1.11	1.32	1.45	1.56	1.64	1.70	1.76	1.81
6	0.72	1.09	1.29	1.43	1.53	1.60	1.67	1.72	1.77
7	0.71	1.08	1.28	1.41	1.51	1.58	1.65	1.70	1.75
8	0.71	1.07	1.27	1.39	1.49	1.57	1.63	1.68	1.73
9	0.70	1.07	1.26	1.38	1.48	1.55	1.61	1.67	1.71
10	0.70	1.06	1.25	1.38	1.47	1.54	1.60	1.66	1.70
11	0.70	1.06	1.24	1.37	1.46	1.53	1.60	1.65	1.69
12	0.70	1.05	1.24	1.36	1.46	1.53	1.59	1.64	1.68
13	0.69	1.05	1.24	1.36	1.45	1.52	1.58	1.63	1.68
14	0.69	1.05	1.23	1.35	1.45	1.52	1.58	1.63	1.67
15	0.69	1.04	1.23	1.35	1.44	1.51	1.57	1.62	1.67
16	0.69	1.04	1.23	1.35	1.44	1.51	1.57	1.62	1.66
17	0.69	1.04	1.22	1.35	1.44	1.51	1.57	1.61	1.66
18	0.69	1.04	1.22	1.34	1.43	1.50	1.56	1.61	1.65
19	0.69	1.04	1.22	1.34	1.43	1.50	1.56	1.61	1.65
20	0.69	1.04	1.22	1.34	1.43	1.50	1.56	1.61	1.65
25	0.68	1.03	1.21	1.33	1.42	1.49	1.55	1.60	1.64
30	0.68	1.03	1.21	1.33	1.42	1.49	1.54	1.59	1.63
35	0.68	1.03	1.21	1.32	1.41	1.48	1.54	1.59	1.63
40	0.68	1.03	1.20	1.32	1.41	1.48	1.54	1.58	1.62
45	0.68	1.02	1.20	1.32	1.41	1.48	1.53	1.58	1.62
50	0.68	1.02	1.20	1.32	1.41	1.47	1.53	1.58	1.62
55	0.68	1.02	1.20	1.32	1.40	1.47	1.53	1.58	1.62
60	0.68	1.02	1.20	1.32	1.40	1.47	1.53	1.58	1.62
120	0.68	1.02	1.19	1.31	1.40	1.47	1.52	1.57	1.61

Table B.4. Upper-α Equicoordinate Point $T^{(\alpha)}_{p,\nu,\rho}$ of the p-Variate t-distribution with Unit Variances, Common Correlation Coefficient ρ, and ν Degrees of Freedom for $\alpha = 0.01, 0.05, 0.10, 0.20$, $\rho = 0.0, 0.1(0.2)0.7$, $p = 2(1)10(2)20$ and $\nu = 2(1)10, 12(4)24, 30, 40, 60, 120, \infty$

							$\rho = 0.0$							
ν	$\alpha\backslash p$	2	3	4	5	6	7	8	9	10	12	14	16	18	20
2	0.01	9.46	11.10	12.4	13.4	14.2	14.9	15.4	15.9	16.4	17.2	17.9	18.4	18.9	19.3
	0.05	4.08	4.83	5.40	5.84	6.21	6.52	6.78	7.01	7.22	7.57	7.87	8.12	8.34	8.53
	0.10	2.74	3.30	3.70	4.02	4.28	4.50	4.69	4.85	5.00	5.25	5.46	5.64	5.79	5.93
	0.20	1.73	2.15	2.45	2.68	2.87	3.02	3.16	3.28	3.38	3.56	3.70	3.83	3.94	4.03
3	0.01	5.71	6.46	7.02	7.46	7.83	8.14	8.41	8.64	8.86	9.23	9.54	9.80	10.00	10.20
	0.05	3.09	3.55	3.89	4.15	4.37	4.56	4.72	4.86	4.98	5.20	5.38	5.54	5.67	5.79
	0.10	2.27	2.65	2.93	3.14	3.32	3.47	3.60	3.71	3.81	3.98	4.13	4.25	4.36	4.45
	0.20	1.54	1.87	2.10	2.28	2.43	2.55	2.66	2.75	2.83	2.97	3.08	3.18	3.26	3.34
4	0.01	4.55	5.04	5.41	5.70	5.94	6.15	6.33	6.48	6.63	6.87	7.08	7.25	7.41	7.55
	0.05	2.72	3.08	3.34	3.54	3.71	3.85	3.97	4.08	4.18	4.34	4.48	4.60	4.71	4.80
	0.10	2.07	2.39	2.62	2.80	2.94	3.06	3.17	3.26	3.34	3.49	3.60	3.71	3.79	3.87
	0.20	1.46	1.75	1.96	2.12	2.24	2.35	2.44	2.52	2.59	2.71	2.82	2.90	2.98	3.04
5	0.01	4.00	4.39	4.67	4.90	5.08	5.24	5.38	5.50	5.61	5.80	5.95	6.09	6.21	6.32
	0.05	2.53	2.84	3.06	3.23	3.38	3.50	3.60	3.69	3.77	3.91	4.03	4.13	4.22	4.30
	0.10	1.97	2.26	2.46	2.62	2.74	2.85	2.94	3.03	3.10	3.22	3.33	3.42	3.50	3.57
	0.20	1.41	1.69	1.88	2.02	2.14	2.24	2.32	2.40	2.46	2.58	2.67	2.75	2.82	2.88
6	0.01	3.68	4.01	4.25	4.44	4.60	4.73	4.84	4.94	5.03	5.19	5.32	5.44	5.54	5.63
	0.05	2.42	2.70	2.89	3.05	3.18	3.28	3.37	3.46	3.53	3.65	3.76	3.85	3.93	4.00
	0.10	1.90	2.17	2.36	2.50	2.62	2.72	2.81	2.88	2.95	3.06	3.16	3.24	3.31	3.38
	0.20	1.38	1.65	1.83	1.97	2.08	2.17	2.25	2.32	2.38	2.49	2.57	2.65	2.72	2.77
7	0.01	3.48	3.77	3.98	4.15	4.29	4.40	4.50	4.59	4.67	4.81	4.92	5.02	5.11	5.19
	0.05	2.34	2.60	2.78	2.93	3.04	3.14	3.22	3.30	3.37	3.48	3.58	3.66	3.73	3.80
	0.10	1.86	2.11	2.29	2.43	2.54	2.63	2.71	2.78	2.84	2.95	3.04	3.12	3.19	3.25
	0.20	1.36	1.62	1.79	1.93	2.03	2.12	2.20	2.27	2.33	2.43	2.51	2.58	2.65	2.70
8	0.01	3.34	3.61	3.80	3.95	4.07	4.18	4.27	4.35	4.42	4.54	4.65	4.74	4.82	4.89
	0.05	2.28	2.53	2.70	2.84	2.95	3.04	3.12	3.19	3.25	3.36	3.45	3.53	3.59	3.65
	0.10	1.83	2.07	2.24	2.37	2.48	2.57	2.64	2.71	2.77	2.87	2.96	3.03	3.09	3.15
	0.20	1.35	1.60	1.77	1.90	2.00	2.09	2.16	2.23	2.28	2.38	2.46	2.53	2.59	2.65
9	0.01	3.24	3.49	3.66	3.80	3.91	4.01	4.09	4.17	4.23	4.35	4.44	4.53	4.60	4.67
	0.05	2.24	2.48	2.64	2.77	2.88	2.96	3.04	3.10	3.16	3.27	3.35	3.42	3.49	3.55
	0.10	1.81	2.04	2.21	2.33	2.43	2.52	2.59	2.66	2.71	2.81	2.89	2.96	3.03	3.08
	0.20	1.34	1.58	1.75	1.88	1.98	2.06	2.13	2.20	2.25	2.35	2.43	2.49	2.55	2.61
10	0.01	3.16	3.39	3.56	3.69	3.80	3.89	3.96	4.03	4.09	4.20	4.29	4.37	4.44	4.50
	0.05	2.21	2.44	2.60	2.72	2.82	2.90	2.98	3.04	3.10	3.19	3.28	3.35	3.41	3.46
	0.10	1.79	2.02	2.18	2.30	2.40	2.48	2.55	2.61	2.67	2.76	2.84	2.91	2.97	3.02
	0.20	1.33	1.57	1.73	1.86	1.96	2.04	2.11	2.17	2.23	2.32	2.40	2.46	2.52	2.57

Reprinted from Hochberg and Tamhane (1987), pp. 383–390, by courtesy of John Wiley & Sons.

continued

Table B.4. *Continued*

		\multicolumn{13}{c}{$\rho = 0.0$}												
ν	$\alpha\backslash p$	2	3	4	5	6	7	8	9	10	12	14	16	18	20
12	0.01	3.05	3.26	3.41	3.53	3.63	3.71	3.78	3.84	3.90	3.99	4.07	4.14	4.21	4.26
	0.05	2.16	2.38	2.53	2.65	2.74	2.82	2.89	2.95	3.00	3.09	3.17	3.23	3.29	3.34
	0.10	1.76	1.98	2.14	2.25	2.35	2.43	2.49	2.55	2.60	2.70	2.77	2.84	2.89	2.94
	0.20	1.31	1.55	1.71	1.83	1.93	2.01	2.08	2.14	2.19	2.28	2.35	2.42	2.47	2.52
16	0.01	2.92	3.11	3.24	3.35	3.43	3.50	3.56	3.62	3.67	3.75	3.82	3.88	3.94	3.98
	0.05	2.11	2.31	2.45	2.56	2.64	2.72	2.78	2.83	2.88	2.97	3.04	3.10	3.15	3.20
	0.10	1.73	1.94	2.08	2.20	2.28	2.36	2.42	2.48	2.53	2.61	2.68	2.74	2.80	2.84
	0.20	1.30	1.53	1.68	1.80	1.89	1.97	2.04	2.09	2.14	2.23	2.30	2.36	2.42	2.46
20	0.01	2.84	3.02	3.15	3.24	3.32	3.39	3.44	3.49	3.54	3.62	3.68	3.74	3.79	3.83
	0.05	2.08	2.27	2.41	2.51	2.59	2.66	2.72	2.77	2.82	2.89	2.96	3.02	3.07	3.11
	0.10	1.71	1.91	2.06	2.16	2.25	2.32	2.38	2.43	2.48	2.56	2.63	2.69	2.74	2.78
	0.20	1.29	1.51	1.67	1.78	1.87	1.95	2.01	2.07	2.12	2.20	2.27	2.33	2.38	2.43
24	0.01	2.79	2.97	3.09	3.18	3.25	3.31	3.37	3.41	3.46	3.53	3.59	3.64	3.69	3.73
	0.05	2.05	2.24	2.38	2.47	2.55	2.62	2.68	2.73	2.77	2.85	2.91	2.97	3.01	3.06
	0.10	1.69	1.90	2.04	2.14	2.22	2.29	2.35	2.41	2.45	2.53	2.60	2.65	2.70	2.75
	0.20	1.28	1.51	1.66	1.77	1.86	1.93	2.00	2.05	2.10	2.18	2.25	2.31	2.36	2.41
30	0.01	2.75	2.91	3.03	3.11	3.18	3.24	3.29	3.34	3.38	3.45	3.50	3.55	3.60	3.64
	0.05	2.03	2.22	2.35	2.44	2.52	2.58	2.64	2.69	2.73	2.80	2.86	2.92	2.96	3.00
	0.10	1.68	1.88	2.02	2.12	2.20	2.27	2.33	2.38	2.42	2.50	2.56	2.62	2.66	2.71
	0.20	1.27	1.50	1.65	1.76	1.85	1.92	1.98	2.03	2.08	2.16	2.23	2.29	2.34	2.38
40	0.01	2.70	2.86	2.97	3.05	3.12	3.17	3.22	3.26	3.30	3.37	3.42	3.47	3.51	3.55
	0.05	2.01	2.19	2.32	2.41	2.48	2.55	2.60	2.65	2.69	2.76	2.82	2.87	2.91	2.95
	0.10	1.67	1.86	2.00	2.10	2.18	2.24	2.30	2.35	2.39	2.47	2.53	2.58	2.63	2.67
	0.20	1.27	1.49	1.64	1.75	1.83	1.90	1.96	2.02	2.06	2.14	2.21	2.27	2.31	2.36
60	0.01	2.66	2.81	2.91	2.99	3.06	3.11	3.15	3.19	3.23	3.29	3.34	3.38	3.42	3.46
	0.05	1.99	2.17	2.29	2.38	2.45	2.51	2.56	2.61	2.65	2.71	2.77	2.82	2.86	2.90
	0.10	1.66	1.85	1.98	2.08	2.16	2.22	2.27	2.32	2.36	2.44	2.50	2.55	2.59	2.63
	0.20	1.26	1.48	1.63	1.73	1.82	1.89	1.95	2.00	2.05	2.13	2.19	2.24	2.29	2.33
120	0.01	2.62	2.76	2.86	2.93	2.99	3.04	3.09	3.12	3.16	3.22	3.26	3.30	3.34	3.37
	0.05	1.97	2.14	2.26	2.35	2.42	2.48	2.53	2.57	2.61	2.67	2.73	2.77	2.81	2.85
	0.10	1.64	1.83	1.96	2.06	2.13	2.20	2.25	2.30	2.34	2.41	2.46	2.51	2.56	2.59
	0.20	1.26	1.47	1.61	1.72	1.81	1.88	1.93	1.99	2.03	2.11	2.17	2.22	2.27	2.31
∞	0.01	2.57	2.71	2.81	2.88	2.93	2.98	3.02	3.06	3.09	3.14	3.19	3.23	3.26	3.29
	0.05	1.95	2.12	2.23	2.32	2.39	2.44	2.49	2.53	2.57	2.63	2.68	2.73	2.77	2.80
	0.10	1.63	1.82	1.94	2.04	2.11	2.17	2.22	2.27	2.31	2.38	2.43	2.48	2.52	2.56
	0.20	1.25	1.46	1.60	1.71	1.79	1.86	1.92	1.97	2.01	2.09	2.15	2.20	2.25	2.29

Table B.4. *Continued*

		$\rho = 0.1$												
ν	$\alpha\backslash p$	2	3	4	5	6	7	8	9	10	12	14	16	18	20
2	0.01	9.36	10.9	12.1	13.0	13.0	14.4	15.0	15.4	15.9	16.6	17.2	17.7	18.2	18.6
	0.05	4.03	4.75	5.28	5.70	6.04	6.32	6.57	6.79	6.98	7.31	7.58	7.82	8.02	8.20
	0.10	2.71	3.23	3.61	3.91	4.15	4.36	4.53	4.69	4.82	5.06	5.25	5.42	5.56	5.69
	0.20	1.71	2.10	2.38	2.59	2.77	2.92	3.04	3.15	3.25	3.42	3.55	3.67	3.77	3.86
3	0.01	5.68	6.40	6.93	7.35	7.69	7.98	8.23	8.46	8.65	9.00	9.28	9.53	9.75	9.94
	0.05	3.07	3.51	3.83	4.08	4.29	4.46	4.61	4.74	4.86	5.06	5.23	5.37	5.50	5.61
	0.10	2.24	2.61	2.87	3.08	3.24	3.38	3.50	3.61	3.70	3.87	4.00	4.12	4.22	4.30
	0.20	1.52	1.84	2.06	2.23	2.36	2.48	2.57	2.66	2.74	2.86	2.97	3.06	3.14	3.21
4	0.01	4.53	5.01	5.36	5.64	5.87	6.06	6.23	6.38	6.51	6.74	6.93	7.10	7.25	7.38
	0.05	2.71	3.05	3.30	3.49	3.65	3.79	3.90	4.00	4.09	4.25	4.38	4.49	4.59	4.68
	0.10	2.06	2.36	2.58	2.75	2.89	3.00	3.10	3.19	3.27	3.40	3.51	3.61	3.69	3.76
	0.20	1.44	1.72	1.92	2.07	2.19	2.29	2.38	2.45	2.52	2.63	2.73	2.81	2.88	2.94
5	0.01	3.99	4.36	4.64	4.85	5.03	5.18	5.31	5.43	5.53	5.71	5.86	5.99	6.10	6.20
	0.05	2.52	2.82	3.03	3.20	3.33	3.45	3.54	3.63	3.71	3.84	3.95	4.05	4.13	4.21
	0.10	1.96	2.23	2.43	2.58	2.70	2.80	2.89	2.97	3.03	3.15	3.25	3.34	3.41	3.47
	0.20	1.40	1.66	1.84	1.98	2.09	2.19	2.27	2.34	2.40	2.50	2.59	2.67	2.73	2.79
6	0.01	3.68	4.00	4.23	4.41	4.56	4.68	4.79	4.89	4.98	5.13	5.25	5.36	5.46	5.54
	0.05	2.41	2.68	2.87	3.02	3.14	3.24	3.33	3.41	3.47	3.59	3.69	3.78	3.85	3.92
	0.10	1.89	2.15	2.33	2.47	2.58	2.68	2.76	2.83	2.89	3.00	3.09	3.17	3.24	3.30
	0.20	1.37	1.62	1.80	1.93	2.03	2.12	2.20	2.26	2.32	2.42	2.50	2.58	2.64	2.69
7	0.01	3.48	3.76	3.97	4.13	4.26	4.37	4.46	4.55	4.62	4.76	4.87	4.96	5.05	5.12
	0.05	2.33	2.58	2.76	2.90	3.01	3.10	3.19	3.26	3.32	3.43	3.52	3.60	3.67	3.73
	0.10	1.85	2.10	2.27	2.40	2.50	2.59	2.67	2.74	2.80	2.90	2.98	3.06	3.12	3.18
	0.20	1.35	1.59	1.76	1.89	1.99	2.08	2.15	2.21	2.27	2.37	2.44	2.51	2.57	2.62
8	0.01	3.34	3.60	3.78	3.93	4.05	4.15	4.24	4.31	4.38	4.50	4.60	4.69	4.76	4.83
	0.05	2.28	2.52	2.68	2.81	2.92	3.01	3.08	3.15	3.21	3.31	3.40	3.47	3.54	3.60
	0.10	1.82	2.06	2.22	2.35	2.45	2.53	2.61	2.67	2.73	2.82	2.90	2.97	3.03	3.09
	0.20	1.33	1.57	1.74	1.86	1.96	2.05	2.12	2.18	2.23	2.32	2.40	2.47	2.52	2.58
9	0.01	3.23	3.48	3.65	3.79	3.90	3.99	4.07	4.14	4.20	4.31	4.41	4.49	4.56	4.62
	0.05	2.24	2.47	2.63	2.75	2.85	2.93	3.01	3.07	3.13	3.23	3.31	3.38	3.44	3.49
	0.10	1.80	2.03	2.19	2.31	2.40	2.49	2.56	2.62	2.67	2.77	2.84	2.91	2.97	3.02
	0.20	1.32	1.56	1.72	1.84	1.94	2.02	2.09	2.15	2.20	2.29	2.37	2.43	2.49	2.54
10	0.01	3.16	3.39	3.55	3.68	3.78	3.87	3.94	4.01	4.07	4.17	4.26	4.33	4.40	4.46
	0.05	2.20	2.43	2.58	2.70	2.80	2.88	2.95	3.01	3.06	3.16	3.24	3.30	3.36	3.42
	0.10	1.78	2.00	2.16	2.28	2.37	2.45	2.52	2.58	2.63	2.72	2.80	2.86	2.92	2.97
	0.20	1.31	1.55	1.71	1.83	1.92	2.00	2.07	2.13	2.18	2.27	2.34	2.40	2.46	2.51

continued

Table B.4. *Continued*

		$\rho = 0.1$													
ν	$\alpha\backslash p$	2	3	4	5	6	7	8	9	10	12	14	16	18	20
12	0.01	3.04	3.26	3.40	3.52	3.61	3.69	3.76	3.82	3.88	3.97	4.05	4.12	4.18	4.23
	0.05	2.16	2.37	2.52	2.63	2.72	2.80	2.86	2.92	2.97	3.06	3.13	3.20	3.25	3.30
	0.10	1.75	1.97	2.12	2.23	2.32	2.40	2.46	2.52	2.57	2.66	2.73	2.79	2.85	2.89
	0.20	1.30	1.53	1.69	1.80	1.89	1.97	2.04	2.09	2.15	2.23	2.30	2.36	2.42	2.46
16	0.01	2.91	3.10	3.24	3.34	3.42	3.49	3.55	3.61	3.65	3.74	3.80	3.86	3.92	3.96
	0.05	2.10	2.30	2.44	2.54	2.63	2.70	2.76	2.81	2.86	2.94	3.01	3.07	3.12	3.16
	0.10	1.72	1.93	2.07	2.18	2.26	2.33	2.40	2.45	2.50	2.58	2.65	2.71	2.76	2.80
	0.20	1.28	1.51	1.66	1.77	1.86	1.94	2.00	2.05	2.10	2.19	2.26	2.31	2.36	2.41
20	0.01	2.84	3.02	3.14	3.24	3.31	3.38	3.44	3.48	3.53	3.60	3.67	3.72	3.77	3.81
	0.05	2.07	2.26	2.39	2.49	2.58	2.64	2.70	2.75	2.80	2.87	2.94	2.99	3.04	3.08
	0.10	1.70	1.90	2.04	2.14	2.23	2.30	2.36	2.41	2.45	2.53	2.60	2.65	2.70	2.75
	0.20	1.28	1.50	1.64	1.75	1.84	1.92	1.98	2.03	2.08	2.16	2.23	2.28	2.33	2.38
24	0.01	2.79	2.96	3.08	3.17	3.25	3.31	3.36	3.41	3.45	3.52	3.58	3.63	3.68	3.72
	0.05	2.05	2.24	2.37	2.46	2.54	2.61	2.66	2.71	2.75	2.83	2.89	2.94	2.99	3.03
	0.10	1.69	1.89	2.02	2.12	2.21	2.27	2.33	2.38	2.43	2.50	2.57	2.62	2.67	2.71
	0.20	1.27	1.49	1.63	1.74	1.83	1.90	1.96	2.02	2.06	2.14	2.21	2.26	2.31	2.36
30	0.01	2.75	2.91	3.02	3.11	3.18	3.24	3.29	3.33	3.37	3.44	3.50	3.54	3.59	3.63
	0.05	2.03	2.21	2.34	2.43	2.51	2.57	2.62	2.67	2.71	2.79	2.85	2.90	2.94	2.98
	0.10	1.67	1.87	2.00	2.10	2.18	2.25	2.31	2.36	2.40	2.47	2.54	2.59	2.64	2.68
	0.20	1.26	1.48	1.62	1.73	1.82	1.89	1.95	2.00	2.05	2.13	2.19	2.25	2.29	2.34
40	0.01	2.70	2.86	2.96	3.05	3.11	3.17	3.22	3.26	3.30	3.36	3.41	3.46	3.50	3.54
	0.05	2.01	2.19	2.31	2.40	2.47	2.54	2.59	2.63	2.67	2.74	2.80	2.85	2.89	2.93
	0.10	1.66	1.85	1.98	2.08	2.16	2.23	2.28	2.33	2.37	2.44	2.51	2.56	2.60	2.64
	0.20	1.26	1.47	1.61	1.72	1.81	1.88	1.93	1.99	2.03	2.11	2.17	2.23	2.27	2.31
60	0.01	2.66	2.81	2.91	2.99	3.05	3.10	3.15	3.19	3.22	3.28	3.33	3.38	3.42	3.45
	0.05	1.99	2.16	2.28	2.37	2.44	2.50	2.55	2.60	2.63	2.70	2.76	2.80	2.85	2.88
	0.10	1.65	1.84	1.97	2.06	2.14	2.20	2.26	2.30	2.35	2.42	2.47	2.52	2.57	2.61
	0.20	1.25	1.46	1.60	1.71	1.79	1.86	1.92	1.97	2.02	2.09	2.15	2.21	2.25	2.29
120	0.01	2.62	2.76	2.86	2.93	2.99	3.04	3.08	3.12	3.15	3.21	3.26	3.30	3.34	3.37
	0.05	1.97	2.14	2.25	2.34	2.41	2.47	2.52	2.56	2.60	2.66	2.71	2.76	2.80	2.84
	0.10	1.64	1.82	1.95	2.04	2.12	2.18	2.23	2.28	2.32	2.39	2.44	2.49	2.54	2.57
	0.20	1.25	1.45	1.59	1.70	1.78	1.85	1.91	1.96	2.00	2.07	2.14	2.19	2.23	2.27
∞	0.01	2.57	2.71	2.80	2.87	2.93	2.98	3.02	3.06	3.09	3.14	3.18	3.22	3.26	3.29
	0.05	1.95	2.12	2.23	2.31	2.38	2.43	2.48	2.52	2.56	2.62	2.67	2.72	2.75	2.79
	0.10	1.63	1.81	1.93	2.02	2.10	2.16	2.21	2.25	2.29	2.36	2.41	2.46	2.50	2.54
	0.20	1.24	1.45	1.59	1.69	1.77	1.84	1.89	1.94	1.98	2.06	2.12	2.17	2.21	2.25

Table B.4. *Continued*

		$\rho = 0.3$													
ν	$\alpha\backslash p$	2	3	4	5	6	7	8	9	10	12	14	16	18	20
2	0.01	9.15	10.5	11.5	12.3	13.0	13.5	14.0	14.4	14.7	15.4	15.9	16.3	16.7	17.1
	0.05	3.93	4.56	5.02	5.37	5.66	5.91	6.12	6.30	6.46	6.74	6.97	7.17	7.35	7.50
	0.10	2.63	3.09	3.42	3.68	3.88	4.06	4.21	4.34	4.45	4.65	4.81	4.95	5.08	5.19
	0.20	1.64	1.99	2.23	2.41	2.56	2.69	2.80	2.89	2.97	3.11	3.23	3.33	3.42	3.49
3	0.01	5.60	6.24	6.71	7.08	7.38	7.63	7.84	8.04	8.21	8.50	8.74	8.95	9.14	9.30
	0.05	3.01	3.41	3.69	3.91	4.09	4.24	4.37	4.48	4.58	4.75	4.90	5.02	5.13	5.22
	0.10	2.20	2.52	2.75	2.93	3.08	3.20	3.30	3.39	3.47	3.61	3.73	3.82	3.91	3.99
	0.20	1.47	1.75	1.95	2.09	2.21	2.31	2.40	2.47	2.53	2.64	2.74	2.82	2.88	2.94
4	0.01	4.48	4.92	5.24	5.48	5.69	5.86	6.00	6.13	6.25	6.45	6.61	6.75	6.88	6.99
	0.05	2.67	2.98	3.20	3.38	3.51	3.63	3.73	3.82	3.90	4.03	4.15	4.24	4.33	4.40
	0.10	2.02	2.30	2.49	2.64	2.76	2.86	2.95	3.02	3.09	3.20	3.30	3.38	3.45	3.52
	0.20	1.40	1.65	1.83	1.96	2.06	2.15	2.23	2.29	2.35	2.45	2.53	2.60	2.66	2.71
5	0.01	3.95	4.30	4.55	4.75	4.91	5.04	5.15	5.26	5.35	5.50	5.63	5.75	5.84	5.93
	0.05	2.49	2.76	2.95	3.10	3.22	3.32	3.41	3.49	3.55	3.67	3.77	3.85	3.92	3.98
	0.10	1.92	2.17	2.35	2.48	2.59	2.68	2.76	2.83	2.89	2.99	3.07	3.15	3.21	3.27
	0.20	1.36	1.60	1.76	1.88	1.98	2.06	2.13	2.19	2.25	2.34	2.41	2.48	2.54	2.59
6	0.01	3.65	3.95	4.16	4.33	4.46	4.58	4.67	4.76	4.84	4.97	5.08	5.18	5.26	5.33
	0.05	2.38	2.63	2.80	2.94	3.05	3.14	3.22	3.28	3.34	3.45	3.54	3.61	3.68	3.73
	0.10	1.86	2.10	2.26	2.39	2.49	2.57	2.64	2.71	2.76	2.86	2.93	3.00	3.06	3.11
	0.20	1.33	1.56	1.72	1.83	1.93	2.01	2.07	2.13	2.18	2.27	2.34	2.40	2.46	2.50
7	0.01	3.45	3.72	3.91	4.06	4.18	4.28	4.37	4.44	4.51	4.63	4.73	4.81	4.89	4.95
	0.05	2.31	2.54	2.70	2.83	2.93	3.01	3.09	3.15	3.21	3.30	3.38	3.45	3.51	3.57
	0.10	1.82	2.05	2.20	2.32	2.42	2.50	2.56	2.62	2.68	2.77	2.84	2.90	2.96	3.01
	0.20	1.31	1.54	1.69	1.80	1.89	1.97	2.03	2.09	2.14	2.22	2.29	2.35	2.40	2.45
8	0.01	3.32	3.56	3.74	3.87	3.98	4.07	4.15	4.22	4.28	4.39	4.48	4.56	4.63	4.69
	0.05	2.25	2.48	2.63	2.75	2.85	2.93	2.99	3.05	3.11	3.20	3.27	3.34	3.40	3.45
	0.10	1.79	2.01	2.16	2.27	2.37	2.44	2.51	2.56	2.61	2.70	2.77	2.83	2.89	2.93
	0.20	1.30	1.52	1.67	1.78	1.87	1.94	2.00	2.06	2.11	2.19	2.26	2.31	2.36	2.41
9	0.01	3.22	3.45	3.61	3.74	3.84	3.92	4.00	4.06	4.12	4.22	4.30	4.37	4.44	4.49
	0.05	2.21	2.43	2.58	2.69	2.78	2.86	2.93	2.98	3.03	3.12	3.19	3.26	3.31	3.36
	0.10	1.77	1.98	2.13	2.24	2.33	2.40	2.46	2.52	2.57	2.65	2.72	2.78	2.83	2.88
	0.20	1.29	1.50	1.65	1.76	1.85	1.92	1.98	2.03	2.08	2.16	2.23	2.28	2.33	2.38
10	0.01	3.14	3.36	3.51	3.63	3.73	3.81	3.88	3.94	3.99	4.09	4.17	4.23	4.29	4.35
	0.05	2.18	2.39	2.54	2.65	2.73	2.81	2.87	2.93	2.98	3.06	3.13	3.19	3.24	3.29
	0.10	1.75	1.96	2.10	2.21	2.30	2.37	2.43	2.48	2.53	2.61	2.68	2.74	2.79	2.83
	0.20	1.28	1.49	1.64	1.75	1.83	1.90	1.96	2.02	2.06	2.14	2.21	2.26	2.31	2.35
12	0.01	3.03	3.23	3.37	3.48	3.57	3.65	3.71	3.77	3.82	3.90	3.97	4.03	4.09	4.14
	0.05	2.14	2.34	2.48	2.58	2.66	2.73	2.79	2.85	2.89	2.97	3.04	3.10	3.15	3.19
	0.10	1.73	1.93	2.07	2.17	2.25	2.32	2.38	2.43	2.48	2.56	2.62	2.68	2.73	2.77
	0.20	1.27	1.48	1.62	1.72	1.81	1.88	1.94	1.99	2.03	2.11	2.17	2.23	2.27	2.32

continued

Table B.4. *Continued*

		$\rho = 0.3$												
ν	$\alpha\backslash p$	2	3	4	5	6	7	8	9	10	12	14	16	18	20
16	0.01	2.90	3.09	3.21	3.31	3.39	3.46	3.51	3.56	3.61	3.68	3.75	3.80	3.85	3.89
	0.05	2.09	2.27	2.40	2.50	2.58	2.64	2.70	2.75	2.79	2.87	2.93	2.98	3.03	3.07
	0.10	1.69	1.89	2.02	2.12	2.20	2.27	2.32	2.37	2.42	2.49	2.55	2.60	2.65	2.69
	0.20	1.25	1.46	1.60	1.70	1.78	1.85	1.91	1.95	2.00	2.07	2.13	2.19	2.23	2.27
20	0.01	2.83	3.00	3.12	3.21	3.29	3.35	3.40	3.45	3.49	3.56	3.62	3.67	3.72	3.76
	0.05	2.05	2.24	2.36	2.45	2.53	2.59	2.65	2.69	2.74	2.81	2.87	2.92	2.96	3.00
	0.10	1.68	1.87	2.00	2.09	2.17	2.23	2.29	2.34	2.38	2.45	2.51	2.56	2.60	2.64
	0.20	1.24	1.45	1.58	1.68	1.76	1.83	1.89	1.94	1.98	2.05	2.11	2.16	2.21	2.25
24	0.01	2.78	2.95	3.06	3.15	3.22	3.28	3.33	3.37	3.41	3.48	3.54	3.59	3.63	3.67
	0.05	2.03	2.21	2.33	2.43	2.50	2.56	2.61	2.66	2.70	2.77	2.82	2.87	2.92	2.95
	0.10	1.66	1.85	1.98	2.07	2.15	2.21	2.27	2.31	2.35	2.42	2.48	2.53	2.57	2.61
	0.20	1.24	1.44	1.57	1.67	1.75	1.82	1.87	1.92	1.96	2.04	2.10	2.15	2.19	2.23
30	0.01	2.74	2.90	3.01	3.09	3.16	3.21	3.26	3.30	3.34	3.41	3.46	3.51	3.55	3.58
	0.05	2.01	2.19	2.31	2.40	2.47	2.53	2.58	2.62	2.66	2.73	2.78	2.83	2.87	2.91
	0.10	1.65	1.84	1.96	2.05	2.13	2.19	2.24	2.29	2.33	2.40	2.45	2.50	2.55	2.58
	0.20	1.23	1.43	1.56	1.66	1.74	1.81	1.86	1.91	1.95	2.02	2.08	2.13	2.17	2.21
40	0.01	2.69	2.85	2.95	3.03	3.09	3.15	3.19	3.23	3.27	3.33	3.38	3.43	3.47	3.50
	0.05	1.99	2.16	2.28	2.37	2.44	2.49	2.54	2.59	2.63	2.69	2.74	2.79	2.83	2.87
	0.10	1.64	1.82	1.94	2.04	2.11	2.17	2.22	2.27	2.31	2.37	2.43	2.48	2.52	2.55
	0.20	1.23	1.42	1.56	1.65	1.73	1.80	1.85	1.90	1.94	2.01	2.07	2.12	2.16	2.20
60	0.01	2.65	2.80	2.90	2.97	3.03	3.09	3.13	3.17	3.20	3.26	3.31	3.35	3.39	3.42
	0.05	1.98	2.14	2.25	2.34	2.41	2.46	2.51	2.55	2.59	2.65	2.71	2.75	2.79	2.82
	0.10	1.63	1.81	1.93	2.02	2.09	2.15	2.20	2.24	2.28	2.35	2.40	2.45	2.49	2.52
	0.20	1.22	1.42	1.55	1.64	1.72	1.78	1.84	1.88	1.92	1.99	2.05	2.10	2.14	2.18
120	0.01	2.61	2.75	2.85	2.92	2.98	3.03	3.07	3.10	3.14	3.19	3.24	3.28	3.31	3.34
	0.05	1.96	2.12	2.23	2.31	2.38	2.43	2.48	2.52	2.56	2.62	2.67	2.71	2.75	2.78
	0.10	1.62	1.79	1.91	2.00	2.07	2.13	2.18	2.22	2.26	2.32	2.38	2.42	2.46	2.50
	0.20	1.22	1.41	1.54	1.63	1.71	1.77	1.83	1.87	1.91	1.98	2.04	2.08	2.13	2.16
∞	0.01	2.57	2.70	2.80	2.86	2.92	2.97	3.01	3.04	3.07	3.12	3.17	3.20	3.24	3.27
	0.05	1.94	2.10	2.20	2.28	2.35	2.40	2.45	2.49	2.52	2.58	2.63	2.67	2.71	2.74
	0.10	1.61	1.78	1.90	1.98	2.05	2.11	2.16	2.20	2.24	2.30	2.35	2.40	2.43	2.47
	0.20	1.21	1.40	1.53	1.63	1.70	1.76	1.81	1.86	1.90	1.97	2.02	2.07	2.11	2.15

Table B.4. *Continued*

		\multicolumn{13}{c}{$\rho = 0.5$}
ν	$\alpha\backslash p$	2	3	4	5	6	7	8	9	10	12	14	16	18	20
---	---	---	---	---	---	---	---	---	---	---	---	---	---	---	---
2	0.01	8.88	10.0	10.9	11.5	12.0	12.5	12.9	13.2	13.5	14.0	14.4	14.7	15.1	15.3
	0.05	3.80	4.34	4.71	5.00	5.24	5.43	5.60	5.75	5.88	6.11	6.29	6.45	6.59	6.72
	0.10	2.54	2.92	3.20	3.40	3.57	3.71	3.83	3.94	4.03	4.19	4.32	4.44	4.54	4.62
	0.20	1.57	1.85	2.05	2.21	2.33	2.43	2.52	2.59	2.66	2.77	2.87	2.95	3.02	3.08
3	0.01	5.48	6.04	6.44	6.74	6.99	7.20	7.38	7.53	7.67	7.91	8.11	8.28	8.43	8.56
	0.05	2.94	3.28	3.52	3.70	3.85	3.97	4.08	4.17	4.25	4.39	4.51	4.61	4.70	4.78
	0.10	2.13	2.41	2.61	2.76	2.87	2.97	3.06	3.13	3.20	3.31	3.41	3.49	3.56	3.62
	0.20	1.41	1.65	1.81	1.94	2.03	2.12	2.19	2.25	2.30	2.39	2.47	2.53	2.59	2.64
4	0.01	4.41	4.80	5.07	5.28	5.45	5.59	5.72	5.82	5.92	6.08	6.22	6.34	6.44	6.53
	0.05	2.61	2.88	3.08	3.22	3.34	3.44	3.52	3.59	3.66	3.77	3.86	3.94	4.01	4.07
	0.10	1.96	2.20	2.37	2.50	2.60	2.68	2.75	2.82	2.87	2.97	3.05	3.11	3.17	3.22
	0.20	1.34	1.56	1.71	1.82	1.91	1.98	2.04	2.10	2.15	2.23	2.30	2.35	2.40	2.45
5	0.01	3.90	4.21	4.43	4.60	4.73	4.85	4.94	5.03	5.11	5.24	5.34	5.44	5.52	5.59
	0.05	2.44	2.68	2.85	2.98	3.08	3.16	3.24	3.30	3.36	3.45	3.53	3.60	3.66	3.71
	0.10	1.87	2.09	2.24	2.36	2.45	2.53	2.59	2.65	2.70	2.78	2.86	2.92	2.97	3.02
	0.20	1.30	1.51	1.65	1.76	1.84	1.91	1.97	2.02	2.06	2.14	2.20	2.26	2.30	2.34
6	0.01	3.61	3.88	4.06	4.21	4.32	4.42	4.51	4.58	4.64	4.76	4.85	4.93	5.00	5.06
	0.05	2.34	2.56	2.71	2.83	2.92	3.00	3.06	3.12	3.17	3.26	3.33	3.40	3.45	3.50
	0.10	1.82	2.02	2.17	2.27	2.36	2.43	2.49	2.54	2.59	2.67	2.74	2.79	2.84	2.89
	0.20	1.28	1.48	1.61	1.71	1.79	1.86	1.92	1.97	2.01	2.08	2.14	2.19	2.24	2.28
7	0.01	3.42	3.66	3.83	3.96	4.06	4.15	4.22	4.29	4.35	4.45	4.53	4.60	4.67	4.72
	0.05	2.27	2.48	2.82	2.73	2.81	2.89	2.95	3.00	3.05	3.13	3.20	3.26	3.31	3.36
	0.10	1.78	1.98	2.11	2.22	2.30	2.37	2.42	2.47	2.52	2.59	2.66	2.71	2.76	2.80
	0.20	1.26	1.46	1.59	1.69	1.76	1.83	1.88	1.93	1.97	2.04	2.10	2.15	2.19	2.23
8	0.01	3.29	3.51	3.66	3.78	3.88	3.96	4.03	4.09	4.14	4.23	4.31	4.38	4.43	4.49
	0.05	2.22	2.42	2.55	2.66	2.74	2.81	2.87	2.92	2.96	3.04	3.11	3.16	3.21	3.25
	0.10	1.75	1.94	2.08	2.17	2.25	2.32	2.38	2.42	2.47	2.54	2.60	2.65	2.70	2.74
	0.20	1.25	1.44	1.57	1.67	1.74	1.81	1.86	1.90	1.95	2.01	2.07	2.12	2.16	2.20
9	0.01	3.19	3.40	3.54	3.66	3.75	3.82	3.89	3.94	3.99	4.08	4.15	4.21	4.26	4.31
	0.05	2.18	2.37	2.50	2.60	2.68	2.75	2.81	2.86	2.90	2.97	3.04	3.09	3.14	3.18
	0.10	1.73	1.92	2.05	2.14	2.22	2.28	2.34	2.39	2.43	2.50	2.56	2.61	2.65	2.69
	0.20	1.24	1.43	1.56	1.65	1.73	1.79	1.84	1.89	1.93	1.99	2.05	2.10	2.14	2.17
10	0.01	3.11	3.31	3.45	3.56	3.64	3.72	3.78	3.83	3.88	3.96	4.03	4.08	4.14	4.18
	0.05	2.15	2.34	2.47	2.56	2.64	2.70	2.76	2.81	2.85	2.92	2.98	3.03	3.08	3.12
	0.10	1.71	1.90	2.02	2.12	2.19	2.26	2.31	2.35	2.40	2.46	2.52	2.57	2.61	2.65
	0.20	1.23	1.42	1.55	1.64	1.71	1.77	1.83	1.87	1.91	1.98	2.03	2.08	2.12	2.15
12	0.01	3.01	3.19	3.32	3.42	3.50	3.56	3.62	3.67	3.71	3.79	3.85	3.91	3.95	3.99
	0.05	2.11	2.29	2.41	2.50	2.58	2.64	2.69	2.74	2.78	2.84	2.90	2.95	2.99	3.03
	0.10	1.69	1.87	1.99	2.08	2.16	2.22	2.27	2.31	2.35	2.42	2.47	2.52	2.56	2.60
	0.20	1.22	1.41	1.53	1.62	1.69	1.75	1.80	1.85	1.89	1.95	2.01	2.05	2.09	2.13

continued

Table B.4. *Continued*

		$\rho = 0.5$												
ν	$\alpha\backslash p$	2	3	4	5	6	7	8	9	10	12	14	16	18	20
16	0.01	2.88	3.05	3.17	3.26	3.33	3.39	3.44	3.48	3.52	3.59	3.65	3.70	3.74	3.78
	0.05	2.06	2.23	2.34	2.43	2.50	2.56	2.61	2.65	2.69	2.75	2.81	2.85	2.89	2.93
	0.10	1.66	1.83	1.95	2.04	2.11	2.17	2.22	2.26	2.30	2.36	2.41	2.46	2.50	2.53
	0.20	1.21	1.39	1.51	1.60	1.67	1.73	1.78	1.82	1.86	1.92	1.97	2.02	2.06	2.09
20	0.01	2.81	2.97	3.08	3.17	3.23	3.29	3.34	3.38	3.42	3.48	3.53	3.58	3.62	3.65
	0.05	2.03	2.19	2.30	2.39	2.46	2.51	2.56	2.60	2.64	2.70	2.75	2.80	2.83	2.87
	0.10	1.64	1.81	1.93	2.01	2.08	2.14	2.19	2.23	2.26	2.33	2.38	2.42	2.46	2.49
	0.20	1.20	1.38	1.50	1.59	1.65	1.71	1.76	1.80	1.84	1.90	1.95	2.00	2.04	2.07
24	0.01	2.77	2.92	3.03	3.11	3.17	3.22	3.27	3.31	3.35	3.41	3.46	3.50	3.54	3.57
	0.05	2.01	2.17	2.28	2.36	2.43	2.48	2.53	2.57	2.60	2.66	2.72	2.76	2.80	2.83
	0.10	1.63	1.80	1.91	2.00	2.06	2.12	2.17	2.21	2.24	2.30	2.35	2.40	2.43	2.47
	0.20	1.19	1.37	1.49	1.58	1.65	1.70	1.75	1.79	1.83	1.89	1.94	1.99	2.02	2.06
30	0.01	2.72	2.87	2.97	3.05	3.11	3.16	3.21	3.25	3.28	3.34	3.39	3.43	3.46	3.50
	0.05	1.99	2.15	2.25	2.34	2.40	2.45	2.50	2.54	2.57	2.63	2.68	2.72	2.76	2.79
	0.10	1.62	1.79	1.90	1.98	2.05	2.10	2.15	2.19	2.22	2.28	2.33	2.37	2.41	2.44
	0.20	1.19	1.36	1.48	1.57	1.64	1.69	1.74	1.78	1.82	1.88	1.93	1.97	2.01	2.04
40	0.01	2.68	2.82	2.92	2.99	3.05	3.10	3.14	3.18	3.21	3.27	3.32	3.36	3.39	3.42
	0.05	1.97	2.13	2.23	2.31	2.37	2.42	2.47	2.51	2.54	2.60	2.65	2.69	2.72	2.75
	0.10	1.61	1.77	1.88	1.96	2.03	2.08	2.13	2.17	2.20	2.26	2.31	2.35	2.39	2.42
	0.20	1.18	1.36	1.47	1.56	1.63	1.68	1.73	1.77	1.81	1.87	1.92	1.96	2.00	2.03
60	0.01	2.64	2.78	2.87	2.94	3.00	3.04	3.08	3.12	3.15	3.20	3.25	3.29	3.32	3.35
	0.05	1.95	2.10	2.21	2.28	2.34	2.40	2.44	2.48	2.51	2.57	2.61	2.65	2.69	2.72
	0.10	1.60	1.76	1.87	1.95	2.01	2.06	2.11	2.15	2.18	2.24	2.29	2.33	2.36	2.39
	0.20	1.18	1.35	1.47	1.55	1.62	1.67	1.72	1.76	1.80	1.86	1.91	1.95	1.98	2.02
120	0.01	2.60	2.73	2.82	2.89	2.94	2.99	3.02	3.06	3.09	3.14	3.18	3.22	3.25	3.28
	0.05	1.93	2.08	2.18	2.26	2.32	2.37	2.41	2.45	2.48	2.53	2.58	2.62	2.65	2.68
	0.10	1.59	1.75	1.85	1.93	1.99	2.05	2.09	2.13	2.16	2.22	2.27	2.31	2.34	2.37
	0.20	1.17	1.35	1.46	1.54	1.61	1.66	1.71	1.75	1.79	1.84	1.89	1.93	1.97	2.00
∞	0.01	2.56	2.68	2.77	2.84	2.89	2.93	2.97	3.00	3.03	3.08	3.12	3.15	3.18	3.21
	0.05	1.92	2.06	2.16	2.23	2.29	2.34	2.38	2.42	2.45	2.50	2.55	2.58	2.62	2.64
	0.10	1.58	1.73	1.84	1.92	1.98	2.03	2.07	2.11	2.14	2.20	2.24	2.28	2.32	2.35
	0.20	1.17	1.34	1.45	1.54	1.60	1.65	1.70	1.74	1.77	1.83	1.88	1.92	1.96	1.99

Table B.4. *Continued*

		\multicolumn{13}{c	}{$\rho = 0.7$}
ν	$\alpha\backslash p$	2	3	4	5	6	7	8	9	10	12	14	16	18	20
	0.01	8.51	9.40	10.0	10.5	10.9	11.2	11.5	11.7	11.9	12.3	12.6	12.9	13.1	13.3
2	0.05	3.63	4.04	4.32	4.54	4.71	4.86	4.98	5.09	5.19	5.35	5.49	5.60	5.70	5.79
	0.10	2.41	2.71	2.91	3.07	3.19	3.29	3.38	3.46	3.53	3.65	3.74	3.82	3.90	3.96
	0.20	1.46	1.68	1.84	1.95	2.04	2.12	2.18	2.24	2.29	2.37	2.44	2.50	2.55	2.60
	0.01	5.32	5.76	6.07	6.30	6.49	6.65	6.78	6.90	7.00	7.18	7.32	7.45	7.56	7.66
3	0.05	2.83	3.10	3.29	3.43	3.54	3.63	3.71	3.78	3.84	3.95	4.03	4.11	4.17	4.23
	0.10	2.04	2.26	2.41	2.52	2.61	2.69	2.75	2.81	2.86	2.94	3.01	3.07	3.12	3.17
	0.20	1.33	1.51	1.64	1.73	1.81	1.87	1.92	1.97	2.01	2.08	2.13	2.18	2.22	2.26
	0.01	4.30	4.61	4.83	4.99	5.12	5.23	5.33	5.41	5.48	5.60	5.71	5.80	5.87	5.94
4	0.05	2.53	2.75	2.90	3.01	3.10	3.18	3.24	3.30	3.35	3.43	3.50	3.56	3.61	3.66
	0.10	1.89	2.08	2.21	2.31	2.38	2.45	2.50	2.55	2.59	2.66	2.72	2.77	2.82	2.86
	0.20	1.26	1.44	1.55	1.64	1.71	1.76	1.81	1.85	1.89	1.95	2.00	2.04	2.08	2.12
	0.01	3.82	4.07	4.25	4.38	4.48	4.57	4.65	4.71	4.77	4.87	4.96	5.03	5.09	5.14
5	0.05	2.37	2.56	2.70	2.80	2.88	2.94	3.00	3.05	3.09	3.16	3.23	3.28	3.32	3.36
	0.10	1.80	1.98	2.10	2.19	2.26	2.32	2.37	2.41	2.45	2.52	2.57	2.62	2.66	2.69
	0.20	1.23	1.39	1.50	1.59	1.65	1.70	1.75	1.79	1.82	1.88	1.93	1.97	2.01	2.04
	0.01	3.54	3.76	3.91	4.03	4.12	4.20	4.26	4.32	4.37	4.45	4.53	4.59	4.64	4.69
6	0.05	2.27	2.45	2.57	2.67	2.74	2.80	2.85	2.90	2.94	3.00	3.06	3.11	3.15	3.19
	0.10	1.75	1.92	2.03	2.12	2.18	2.24	2.29	2.33	2.36	2.42	2.47	2.52	2.56	2.59
	0.20	1.21	1.37	1.47	1.55	1.62	1.67	1.71	1.75	1.78	1.84	1.88	1.92	1.96	1.99
	0.01	3.36	3.56	3.70	3.80	3.88	3.95	4.01	4.06	4.11	4.19	4.25	4.31	4.35	4.40
7	0.05	2.21	2.38	2.49	2.58	2.65	2.71	2.75	2.80	2.83	2.90	2.95	2.99	3.03	3.07
	0.10	1.72	1.88	1.99	2.07	2.13	2.18	2.23	2.27	2.30	2.36	2.41	2.45	2.49	2.52
	0.20	1.19	1.35	1.45	1.53	1.59	1.64	1.68	1.72	1.75	1.81	1.85	1.89	1.92	1.95
	0.01	3.23	3.42	3.55	3.64	3.72	3.78	3.84	3.89	3.93	4.00	4.06	4.11	4.15	4.19
8	0.05	2.16	2.32	2.43	2.52	2.58	2.64	2.68	2.72	2.76	2.82	2.87	2.91	2.95	2.98
	0.10	1.69	1.85	1.95	2.03	2.09	2.15	2.19	2.23	2.26	2.32	2.36	2.40	2.44	2.47
	0.20	1.18	1.34	1.44	1.51	1.57	1.62	1.66	1.70	1.73	1.78	1.83	1.87	1.90	1.93
	0.01	3.14	3.31	3.43	3.53	3.60	3.66	3.71	3.76	3.79	3.86	3.92	3.97	4.01	4.05
9	0.05	2.13	2.28	2.39	2.47	2.53	2.59	2.63	2.67	2.71	2.76	2.81	2.85	2.89	2.92
	0.10	1.67	1.82	1.93	2.00	2.07	2.12	2.16	2.19	2.23	2.28	2.33	2.37	2.40	2.43
	0.20	1.17	1.32	1.43	1.50	1.56	1.61	1.65	1.68	1.71	1.77	1.81	1.85	1.88	1.91
	0.01	3.07	3.24	3.35	3.44	3.51	3.56	3.61	3.66	3.69	3.76	3.81	3.86	3.90	3.93
10	0.05	2.10	2.25	2.36	2.43	2.50	2.55	2.59	2.63	2.66	2.72	2.77	2.81	2.84	2.87
	0.10	1.66	1.81	1.91	1.98	2.04	2.09	2.13	2.17	2.20	2.26	2.30	2.34	2.37	2.40
	0.20	1.17	1.32	1.42	1.49	1.55	1.60	1.64	1.67	1.70	1.75	1.80	1.83	1.87	1.89
	0.01	2.97	3.12	3.23	3.31	3.37	3.43	3.47	3.51	3.55	3.61	3.66	3.70	3.74	3.77
12	0.05	2.06	2.21	2.31	2.38	2.44	2.49	2.53	2.57	2.60	2.65	2.70	2.74	2.77	2.80
	0.10	1.63	1.78	1.88	1.95	2.01	2.06	2.10	2.13	2.16	2.22	2.26	2.30	2.33	2.36
	0.20	1.16	1.30	1.40	1.48	1.53	1.58	1.62	1.65	1.68	1.73	1.78	1.81	1.84	1.87

continued

Table B.4. *Continued*

		$\rho = 0.7$												
ν	$\alpha\backslash p$	2	3	4	5	6	7	8	9	10	12	14	16	18	20
16	0.01	2.85	2.99	3.09	3.16	3.22	3.27	3.31	3.35	3.38	3.43	3.48	3.52	3.55	3.58
	0.05	2.01	2.15	2.25	2.32	2.37	2.42	2.46	2.50	2.53	2.58	2.62	2.66	2.69	2.72
	0.10	1.61	1.75	1.84	1.91	1.97	2.02	2.06	2.09	2.12	2.17	2.21	2.25	2.28	2.30
	0.20	1.14	1.29	1.39	1.46	1.51	1.56	1.60	1.63	1.66	1.71	1.75	1.79	1.82	1.84
20	0.01	2.78	2.91	3.01	3.08	3.13	3.18	3.22	3.25	3.28	3.34	3.38	3.42	3.45	3.48
	0.05	1.98	2.12	2.21	2.28	2.34	2.38	2.42	2.45	2.48	2.53	2.57	2.61	2.64	2.67
	0.10	1.59	1.73	1.82	1.89	1.95	1.99	2.03	2.06	2.09	2.14	2.18	2.22	2.25	2.27
	0.20	1.14	1.28	1.38	1.45	1.50	1.55	1.59	1.62	1.65	1.70	1.74	1.77	1.80	1.83
24	0.01	2.73	2.87	2.96	3.02	3.08	3.12	3.16	3.19	3.22	3.27	3.31	3.35	3.38	3.41
	0.05	1.96	2.10	2.19	2.26	2.31	2.36	2.39	2.43	2.45	2.50	2.54	2.58	2.61	2.63
	0.10	1.58	1.72	1.81	1.88	1.93	1.98	2.01	2.05	2.08	2.12	2.16	2.20	2.23	2.25
	0.20	1.13	1.27	1.37	1.44	1.49	1.54	1.58	1.61	1.64	1.69	1.73	1.76	1.79	1.82
30	0.01	2.69	2.82	2.91	2.97	3.02	3.07	3.10	3.13	3.16	3.21	3.25	3.28	3.31	3.34
	0.05	1.95	2.08	2.17	2.23	2.29	2.33	2.37	2.40	2.43	2.47	2.51	2.55	2.58	2.60
	0.10	1.57	1.71	1.80	1.86	1.92	1.96	2.00	2.03	2.06	2.11	2.15	2.18	2.21	2.23
	0.20	1.13	1.27	1.36	1.43	1.49	1.53	1.57	1.60	1.63	1.68	1.72	1.75	1.78	1.81
40	0.01	2.65	2.77	2.86	2.92	2.97	3.01	3.05	3.08	3.11	3.15	3.19	3.22	3.25	3.28
	0.05	1.93	2.06	2.15	2.21	2.26	2.30	2.34	2.37	2.40	2.45	2.49	2.52	2.55	2.57
	0.10	1.56	1.69	1.78	1.85	1.90	1.95	1.98	2.01	2.04	2.09	2.13	2.16	2.19	2.21
	0.20	1.12	1.26	1.36	1.42	1.48	1.52	1.56	1.59	1.62	1.67	1.71	1.74	1.77	1.80
60	0.01	2.61	2.73	2.81	2.87	2.92	2.96	2.99	3.02	3.05	3.09	3.13	3.16	3.19	3.21
	0.05	1.91	2.04	2.12	2.19	2.24	2.28	2.32	2.35	2.37	2.42	2.46	2.49	2.52	2.54
	0.10	1.55	1.68	1.77	1.84	1.89	1.93	1.97	2.00	2.03	2.07	2.11	2.14	2.17	2.20
	0.20	1.12	1.26	1.35	1.42	1.47	1.52	1.55	1.58	1.61	1.66	1.70	1.73	1.76	1.79
120	0.01	2.57	2.69	2.76	2.82	2.87	2.91	2.94	2.97	2.99	3.04	3.07	3.10	3.13	3.15
	0.05	1.89	2.02	2.10	2.17	2.22	2.26	2.29	2.32	2.35	2.39	2.43	2.46	2.49	2.51
	0.10	1.54	1.67	1.76	1.82	1.87	1.92	1.95	1.98	2.01	2.05	2.09	2.12	2.15	2.18
	0.20	1.11	1.25	1.34	1.41	1.46	1.51	1.54	1.58	1.60	1.65	1.69	1.72	1.75	1.78
∞	0.01	2.53	2.64	2.72	2.78	2.82	2.86	2.89	2.92	2.94	2.98	3.02	3.05	3.07	3.09
	0.05	1.88	2.00	2.08	2.14	2.19	2.23	2.27	2.30	2.32	2.37	2.40	2.43	2.46	2.48
	0.10	1.53	1.66	1.75	1.81	1.86	1.90	1.94	1.97	1.99	2.04	2.08	2.11	2.13	2.16
	0.20	1.11	1.25	1.34	1.40	1.46	1.50	1.54	1.57	1.60	1.64	1.68	1.71	1.74	1.77

Table B.5. Upper-α Equicoordinate Point $|T|^{(\alpha)}_{p,\nu,1/2}$ of the Absolute Value of the p-Variate t-Distribution with Unit Variances, Common Correlation 1/2, and ν Degrees of Freedom for $\alpha = 0.01, 0.05, 0.10, 0.25$, $p = 2(1)10(2)20$, $\nu = 2(1)10, 12(4)24, 30, 40, 60, 120, \infty$

		\multicolumn{12}{c}{p}
ν	α	2	3	4	5	6	7	8	9	10	12	14	16	18	20
2	0.01	12.4	13.8	14.8	15.6	16.2	16.7	17.1	17.5	17.8	18.4	18.9	19.2	19.6	19.9
	0.05	5.42	6.06	6.51	6.85	7.12	7.35	7.54	7.71	7.85	8.10	8.31	8.49	8.64	8.77
	0.10	3.72	4.18	4.50	4.74	4.93	5.09	5.23	5.34	5.45	5.62	5.77	5.89	6.00	6.09
	0.20	2.47	2.80	3.03	3.20	3.33	3.44	3.54	3.62	3.70	3.82	3.92	4.01	4.08	4.15
3	0.01	6.97	7.64	8.10	8.46	8.75	8.98	9.19	9.37	9.52	9.79	10.02	10.21	10.37	10.52
	0.05	3.87	4.26	4.54	4.75	4.92	5.06	5.18	5.28	5.37	5.53	5.66	5.77	5.87	5.95
	0.10	2.91	3.23	3.45	3.62	3.75	3.87	3.96	4.04	4.12	4.24	4.34	4.43	4.51	4.58
	0.20	2.10	2.36	2.54	2.68	2.78	2.87	2.95	3.01	3.07	3.17	3.25	3.32	3.38	3.43
4	0.01	5.36	5.81	6.12	6.36	6.55	6.72	6.85	6.98	7.08	7.27	7.42	7.55	7.66	7.77
	0.05	3.31	3.62	3.83	3.99	4.13	4.23	4.33	4.41	4.48	4.60	4.71	4.79	4.87	4.94
	0.10	2.60	2.86	3.05	3.18	3.30	3.39	3.47	3.54	3.60	3.70	3.79	3.86	3.92	3.98
	0.20	1.95	2.18	2.34	2.45	2.55	2.63	2.69	2.75	2.80	2.89	2.96	3.02	3.08	3.12
5	0.01	4.63	4.97	5.22	5.41	5.56	5.68	5.79	5.89	5.97	6.12	6.24	6.34	6.43	6.51
	0.05	3.03	3.29	3.48	3.62	3.73	3.82	3.90	3.97	4.03	4.14	4.23	4.30	4.37	4.42
	0.10	2.43	2.67	2.83	2.96	3.05	3.14	3.21	3.27	3.32	3.41	3.49	3.56	3.61	3.66
	0.20	1.86	2.08	2.22	2.33	2.42	2.49	2.56	2.61	2.66	2.74	2.80	2.86	2.91	2.95
6	0.01	4.21	4.51	4.71	4.87	5.00	5.10	5.20	5.28	5.35	5.47	5.57	5.66	5.74	5.80
	0.05	2.86	3.10	3.26	3.39	3.49	3.57	3.64	3.71	3.76	3.86	3.94	4.00	4.06	4.11
	0.10	2.33	2.55	2.70	2.81	2.91	2.98	3.05	3.10	3.15	3.24	3.31	3.37	3.42	3.47
	0.20	1.81	2.01	2.15	2.26	2.34	2.41	2.47	2.52	2.56	2.64	2.70	2.76	2.80	2.85
7	0.01	3.95	4.21	4.39	4.53	4.64	4.74	4.82	4.89	4.95	5.06	5.15	5.23	5.29	5.35
	0.05	2.75	2.97	3.12	3.24	3.33	3.41	3.47	3.53	3.58	3.67	3.74	3.81	3.86	3.91
	0.10	2.26	2.47	2.61	2.72	2.81	2.88	2.94	2.99	3.04	3.12	3.19	3.24	3.29	3.34
	0.20	1.77	1.97	2.10	2.20	2.28	2.35	2.41	2.46	2.50	2.57	2.63	2.68	2.73	2.77
8	0.01	3.77	4.00	4.17	4.29	4.40	4.48	4.56	4.62	4.68	4.78	4.86	4.93	4.99	5.05
	0.05	2.67	2.88	3.02	3.13	3.22	3.29	3.35	3.41	3.46	3.54	3.61	3.67	3.72	3.76
	0.10	2.22	2.41	2.55	2.65	2.73	2.80	2.86	2.91	2.96	3.03	3.10	3.15	3.20	3.24
	0.20	1.75	1.94	2.07	2.16	2.24	2.31	2.36	2.41	2.45	2.52	2.58	2.63	2.68	2.71
9	0.01	3.63	3.85	4.01	4.12	4.22	4.30	4.37	4.43	4.48	4.57	4.65	4.71	4.77	4.82
	0.05	2.61	2.81	2.95	3.05	3.14	3.20	3.26	3.32	3.36	3.44	3.51	3.56	3.61	3.65
	0.10	2.18	2.37	2.50	2.60	2.68	2.74	2.80	2.85	2.89	2.97	3.03	3.08	3.13	3.17
	0.20	1.73	1.91	2.04	2.14	2.21	2.27	2.33	2.37	2.41	2.48	2.54	2.59	2.63	2.67
10	0.01	3.53	3.74	3.88	3.99	4.08	4.16	4.22	4.28	4.33	4.42	4.49	4.55	4.60	4.65
	0.05	2.57	2.76	2.89	2.99	3.07	3.14	3.19	3.24	3.29	3.36	3.43	3.48	3.53	3.57
	0.10	2.15	2.34	2.46	2.56	2.64	2.70	2.75	2.80	2.84	2.92	2.98	3.03	3.07	3.11
	0.20	1.71	1.89	2.02	2.11	2.19	2.25	2.30	2.35	2.39	2.45	2.51	2.56	2.60	2.64
12	0.01	3.39	3.58	3.71	3.81	3.89	3.96	4.02	4.07	4.12	4.19	4.26	4.32	4.36	4.41
	0.05	2.50	2.68	2.81	2.90	2.98	3.04	3.09	3.14	3.18	3.25	3.31	3.36	3.41	3.45
	0.10	2.11	2.29	2.41	2.50	2.57	2.63	2.69	2.73	2.77	2.84	2.90	2.95	2.99	3.03
	0.20	1.69	1.87	1.99	2.08	2.15	2.21	2.26	2.30	2.34	2.41	2.46	2.51	2.55	2.59

Reprinted from Hochberg and Tamhane (1987), pp. 391–398, by courtesy of John Wiley & Sons.

continued

Table B.5. *Continued*

ν	α	\multicolumn{13}{c}{p}
		2	3	4	5	6	7	8	9	10	12	14	16	18	20
16	0.01	3.22	3.39	3.51	3.60	3.67	3.73	3.78	3.83	3.87	3.94	4.00	4.05	4.09	4.13
	0.05	2.42	2.59	2.71	2.80	2.87	2.92	2.97	3.02	3.06	3.12	3.18	3.22	3.26	3.30
	0.10	2.06	2.23	2.34	2.43	2.50	2.56	2.61	2.65	2.69	2.75	2.80	2.85	2.89	2.93
	0.20	1.66	1.83	1.95	2.04	2.11	2.16	2.21	2.25	2.29	2.36	2.41	2.45	2.49	2.53
20	0.01	3.13	3.29	3.40	3.48	3.55	3.60	3.65	3.69	3.73	3.80	3.85	3.90	3.94	3.97
	0.05	2.38	2.54	2.65	2.73	2.80	2.86	2.90	2.95	2.98	3.05	3.10	3.14	3.18	3.22
	0.10	2.03	2.19	2.30	2.39	2.46	2.51	2.56	2.60	2.64	2.70	2.75	2.79	2.83	2.87
	0.20	1.64	1.81	1.93	2.01	2.08	2.14	2.18	2.22	2.26	2.32	2.37	2.42	2.46	2.49
24	0.01	3.07	3.22	3.32	3.40	3.47	3.52	3.57	3.61	3.64	3.70	3.76	3.80	3.84	3.87
	0.05	2.35	2.51	2.61	2.70	2.76	2.81	2.86	2.90	2.94	3.00	3.05	3.09	3.13	3.16
	0.10	2.01	2.17	2.28	2.36	2.43	2.48	2.53	2.57	2.60	2.66	2.71	2.76	2.79	2.83
	0.20	1.63	1.80	1.91	1.99	2.06	2.12	2.16	2.20	2.24	2.30	2.35	2.39	2.43	2.47
30	0.01	3.01	3.15	3.25	3.33	3.39	3.44	3.49	3.52	3.56	3.62	3.66	3.71	3.74	3.77
	0.05	2.32	2.47	2.58	2.66	2.72	2.77	2.82	2.86	2.89	2.95	3.00	3.04	3.08	3.11
	0.10	1.99	2.15	2.25	2.33	2.40	2.45	2.50	2.54	2.57	2.63	2.68	2.72	2.76	2.79
	0.20	1.62	1.78	1.90	1.98	2.04	2.10	2.15	2.19	2.22	2.28	2.33	2.37	2.41	2.44
40	0.01	2.95	3.09	3.19	3.26	3.32	3.37	3.41	3.44	3.48	3.53	3.58	3.62	3.65	3.68
	0.05	2.29	2.44	2.54	2.62	2.68	2.73	2.77	2.81	2.85	2.90	2.95	2.99	3.02	3.05
	0.10	1.97	2.13	2.23	2.31	2.37	2.42	2.47	2.51	2.54	2.60	2.65	2.69	2.72	2.75
	0.20	1.61	1.77	1.88	1.96	2.03	2.08	2.13	2.17	2.20	2.26	2.31	2.35	2.39	2.42
60	0.01	2.90	3.03	3.12	3.19	3.25	3.29	3.33	3.37	3.40	3.45	3.49	3.53	3.56	3.59
	0.05	2.27	2.41	2.51	2.58	2.64	2.69	2.73	2.77	2.80	2.86	2.90	2.94	2.97	3.00
	0.10	1.95	2.10	2.21	2.28	2.34	2.40	2.44	2.48	2.51	2.57	2.61	2.65	2.69	2.72
	0.20	1.60	1.76	1.87	1.95	2.01	2.06	2.11	2.15	2.18	2.24	2.29	2.33	2.36	2.39
120	0.01	2.85	2.97	3.06	3.12	3.18	3.22	3.26	3.29	3.32	3.37	3.41	3.45	3.48	3.50
	0.05	2.24	2.38	2.47	2.55	2.60	2.65	2.69	2.73	2.76	2.81	2.86	2.89	2.93	2.95
	0.10	1.93	2.08	2.18	2.26	2.32	2.37	2.41	2.45	2.48	2.53	2.58	2.62	2.65	2.68
	0.20	1.59	1.75	1.85	1.93	1.99	2.05	2.09	2.13	2.16	2.22	2.27	2.30	2.34	2.37
∞	0.01	2.79	2.92	3.00	3.06	3.11	3.15	3.19	3.22	3.25	3.29	3.33	3.37	3.40	3.42
	0.05	2.21	2.35	2.44	2.51	2.57	2.61	2.65	2.69	2.72	2.77	2.81	2.85	2.88	2.91
	0.10	1.92	2.06	2.16	2.23	2.29	2.34	2.38	2.42	2.45	2.50	2.55	2.58	2.62	2.64
	0.20	1.58	1.73	1.84	1.92	1.98	2.03	2.07	2.11	2.14	2.20	2.24	2.28	2.32	2.35

Table B.6. Upper-α Point $Q_{p,\nu}^{(\alpha)}$ of the Studentized Range Distribution for $\alpha = 0.05, 0.10, 0.20$, $p = 3(1)15$, and $\nu = 1(1)20, 24, 30, 40, 60, 120, \infty$

	\multicolumn{13}{c}{$\alpha = 0.05$}
$\nu\backslash p$	3	4	5	6	7	8	9	10	11	12	13	14	15
1	27.0	32.8	37.1	40.4	43.1	45.4	47.4	49.1	50.6	52.0	53.2	54.3	55.4
2	8.33	9.80	10.9	11.7	12.4	13.0	13.5	14.0	14.4	14.8	15.1	15.4	15.7
3	5.91	6.83	7.50	8.04	8.48	8.85	9.18	9.46	9.72	9.95	10.2	10.4	10.5
4	5.04	5.76	6.29	6.71	7.05	7.35	7.60	7.83	8.03	8.21	8.37	8.53	8.66
5	4.60	5.22	5.67	6.03	6.33	6.58	6.80	7.00	7.17	7.32	7.47	7.60	7.72
6	4.34	4.90	5.31	5.63	5.90	6.12	6.32	6.49	6.65	6.79	6.92	7.03	7.14
7	4.17	4.68	5.06	5.36	5.61	5.82	6.00	6.16	6.30	6.43	6.55	6.66	6.76
8	4.04	4.53	4.89	5.17	5.40	5.60	5.77	5.92	6.05	6.18	6.29	6.39	6.48
9	3.95	4.42	4.76	5.02	5.24	5.43	5.60	5.74	5.87	5.98	6.09	6.19	6.28
10	3.88	4.33	4.65	4.91	5.12	5.31	5.46	5.60	5.72	5.83	5.94	6.03	6.11
11	3.82	4.26	4.57	4.82	5.03	5.20	5.35	5.49	5.61	5.71	5.81	5.90	5.98
12	3.77	4.20	4.51	4.75	4.95	5.12	5.27	5.40	5.51	5.62	5.71	5.80	5.88
13	3.74	4.15	4.45	4.69	4.89	5.05	5.19	5.32	5.43	5.53	5.63	5.71	5.79
14	3.70	4.11	4.41	4.64	4.83	4.99	5.13	5.25	5.36	5.46	5.55	5.64	5.71
15	3.67	4.08	4.37	4.60	4.78	4.94	5.08	5.20	5.31	5.40	5.49	5.57	5.65
16	3.65	4.05	4.33	4.56	4.74	4.90	5.03	5.15	5.26	5.35	5.44	5.52	5.59
17	3.63	4.02	4.30	4.52	4.71	4.86	4.99	5.11	5.21	5.31	5.39	5.47	5.54
18	3.61	4.00	4.28	4.50	4.67	4.82	4.96	5.07	5.17	5.27	5.35	5.43	5.50
19	3.59	3.98	4.25	4.47	4.65	4.79	4.92	5.04	5.14	5.23	5.32	5.39	5.46
20	3.58	3.96	4.23	4.45	4.62	4.77	4.90	5.01	5.11	5.20	5.28	5.36	5.43
24	3.53	3.90	4.17	4.37	4.54	4.68	4.81	4.92	5.01	5.10	5.18	5.25	5.32
30	3.49	3.85	4.10	4.30	4.46	4.60	4.72	4.82	4.92	5.00	5.08	5.15	5.21
40	3.44	3.79	4.04	4.23	4.39	4.52	4.64	4.74	4.82	4.90	4.98	5.04	5.11
60	3.40	3.74	3.98	4.16	4.31	4.44	4.55	4.65	4.73	4.81	4.88	4.94	5.00
120	3.36	3.69	3.92	4.10	4.24	4.36	4.47	4.56	4.64	4.71	4.78	4.84	4.90
∞	3.31	3.63	3.86	4.03	4.17	4.29	4.39	4.47	4.55	4.62	4.69	4.74	4.80

Reprinted from Hochberg and Tamhane (1987), pp. 407–409, by courtesy of John Wiley & Sons.

continued

Table B.6. *Continued*

	$\alpha = 0.10$												
$\nu \backslash p$	3	4	5	6	7	8	9	10	11	12	13	14	15
1	13.4	16.4	18.5	20.2	21.5	22.6	23.6	24.5	25.2	25.9	26.5	27.1	27.6
2	5.73	6.77	7.54	8.14	8.63	9.05	9.41	9.73	10.0	10.3	10.5	10.7	10.9
3	4.47	5.20	5.74	6.16	6.51	6.81	7.06	7.29	7.49	7.67	7.83	7.98	8.12
4	3.98	4.59	5.04	5.39	5.68	5.93	6.14	6.33	6.50	6.65	6.78	6.91	7.03
5	3.72	4.26	4.66	4.98	5.24	5.46	5.65	5.82	5.97	6.10	6.22	6.34	6.44
6	3.56	4.07	4.44	4.73	4.97	5.17	5.34	5.50	5.64	5.76	5.88	5.98	6.08
7	3.45	3.93	4.28	4.56	4.78	4.97	5.14	5.28	5.41	5.53	5.64	5.74	5.83
8	3.37	3.83	4.17	4.43	4.65	4.83	4.99	5.13	5.25	5.36	5.46	5.56	5.64
9	3.32	3.76	4.08	4.34	4.55	4.72	4.87	5.01	5.13	5.23	5.33	5.42	5.51
10	3.27	3.70	4.02	4.26	4.47	4.64	4.78	4.91	5.03	5.13	5.23	5.32	5.40
11	3.23	3.66	3.97	4.21	4.40	4.57	4.71	4.84	4.95	5.05	5.15	5.23	5.31
12	3.20	3.62	3.92	4.16	4.35	4.51	4.65	4.78	4.89	4.99	5.08	5.16	5.24
13	3.18	3.59	3.89	4.12	4.31	4.46	4.60	4.72	4.83	4.93	5.02	5.10	5.18
14	3.16	3.56	3.85	4.08	4.27	4.42	4.56	4.68	4.79	4.88	4.97	5.05	5.12
15	3.14	3.54	3.83	4.05	4.24	4.39	4.52	4.64	4.75	4.84	4.93	5.01	5.08
16	3.12	3.52	3.80	4.03	4.21	4.36	4.49	4.61	4.71	4.81	4.89	4.97	5.04
17	3.11	3.50	3.78	4.00	4.19	4.33	4.46	4.58	4.68	4.77	4.86	4.94	5.01
18	3.10	3.49	3.77	3.98	4.17	4.31	4.44	4.55	4.66	4.75	4.83	4.91	4.98
19	3.09	3.47	3.75	3.97	4.15	4.29	4.42	4.53	4.63	4.72	4.80	4.88	4.95
20	3.08	3.46	3.74	3.95	4.13	4.27	4.40	4.51	4.61	4.70	4.78	4.86	4.92
24	3.05	3.42	3.69	3.90	4.07	4.21	4.34	4.45	4.54	4.63	4.71	4.78	4.85
30	3.02	3.39	3.65	3.85	4.02	4.16	4.28	4.38	4.47	4.56	4.64	4.71	4.77
40	2.99	3.35	3.61	3.80	3.97	4.10	4.22	4.32	4.41	4.49	4.56	4.63	4.70
60	2.96	3.31	3.56	3.76	3.92	4.04	4.16	4.25	4.34	4.42	4.49	4.56	4.62
120	2.93	3.28	3.52	3.71	3.86	3.99	4.10	4.19	4.28	4.35	4.42	4.49	4.54
∞	2.90	3.24	3.48	3.66	3.81	3.93	4.04	4.13	4.21	4.29	4.35	4.41	4.47

Table B.6. *Continued*

						$\alpha = 0.20$							
$\nu\backslash p$	3	4	5	6	7	8	9	10	11	12	13	14	15
1	6.62	8.08	9.14	9.97	10.6	11.2	11.7	12.1	12.5	12.8	13.1	13.4	13.7
2	3.82	4.56	5.10	5.52	5.87	6.16	6.41	6.63	6.83	7.00	7.16	7.31	7.44
3	3.25	3.83	4.26	4.60	4.87	5.10	5.31	5.48	5.64	5.78	5.91	6.02	6.13
4	3.00	3.53	3.91	4.21	4.45	4.66	4.83	4.99	5.13	5.25	5.37	5.47	5.57
5	2.87	3.36	3.71	3.99	4.21	4.41	4.57	4.72	4.84	4.96	5.07	5.16	5.25
6	2.79	3.25	3.59	3.85	4.07	4.25	4.40	4.54	4.66	4.77	4.87	4.97	5.05
7	2.73	3.18	3.50	3.76	3.96	4.14	4.29	4.42	4.54	4.64	4.74	4.83	4.91
8	2.69	3.13	3.44	3.69	3.89	4.06	4.20	4.33	4.44	4.55	4.64	4.73	4.81
9	2.66	3.09	3.39	3.63	3.83	3.99	4.14	4.26	4.37	4.87	4.56	4.65	4.72
10	2.63	3.05	3.36	3.59	3.78	3.94	4.08	4.21	4.32	4.41	4.50	4.59	4.66
11	2.61	3.03	3.33	3.56	3.75	3.91	4.04	4.16	4.27	4.37	4.45	4.53	4.61
12	2.60	3.01	3.30	3.53	3.72	3.87	4.01	4.13	4.23	4.33	4.41	4.49	4.57
13	2.58	2.99	3.28	3.51	3.69	3.84	3.98	4.10	4.20	4.29	4.38	4.46	4.53
14	2.57	2.97	3.26	3.49	3.67	3.82	3.95	4.07	4.17	4.27	4.35	4.43	4.50
15	2.56	2.96	3.25	3.47	3.65	3.80	3.93	4.05	4.15	4.24	4.32	4.40	4.47
16	2.55	2.95	3.23	3.45	3.63	3.78	3.91	4.03	4.13	4.22	4.30	4.38	4.45
17	2.54	2.94	3.22	3.44	3.62	3.77	3.90	4.01	4.11	4.20	4.28	4.36	4.43
18	2.54	2.93	3.21	3.43	3.60	3.75	3.88	3.99	4.09	4.18	4.26	4.34	4.41
19	2.53	2.92	3.20	3.42	3.59	3.74	3.87	3.98	4.08	4.17	4.25	4.32	4.39
20	2.52	2.91	3.19	3.41	3.58	3.73	3.86	3.97	4.07	4.15	4.23	4.31	4.38
24	2.51	2.89	3.17	3.38	3.55	3.69	3.82	3.93	4.02	4.11	4.19	4.26	4.33
30	2.49	2.87	3.14	3.35	3.52	3.66	3.78	3.89	3.98	4.07	4.15	4.22	4.28
40	2.47	2.85	3.11	3.32	3.48	3.62	3.74	3.85	3.94	4.03	4.10	4.17	4.23
60	2.46	2.83	3.09	3.29	3.45	3.59	3.71	3.81	3.90	3.98	4.06	4.12	4.19
120	2.44	2.81	3.06	3.26	3.42	3.55	3.67	3.77	3.86	3.94	4.01	4.08	4.14
∞	2.42	2.78	3.04	3.23	3.39	3.52	3.63	3.73	3.82	3.90	3.97	4.03	4.09

Table B.7. Upper-α Point $|M|_{p,\nu}^{(\alpha)}$ of the Studentized Maximum Modulus Distribution Having Dimension p and ν Degrees of Freedom for $\alpha = 0.10, 0.05, 0.01$, $p = 2(1)10, 12, 14, 16(1)27, 29, 31$, and $\nu = 2(1)12(2)20, 24, 30, 40, 60, \infty$

ν	α	2	3	4	5	6	7	8	9	10	12	14	16
2	0.10	3.83	4.38	4.77	5.06	5.30	5.50	5.67	5.82	5.96	6.18	6.37	6.53
	0.05	5.57	6.34	6.89	7.31	7.65	7.93	8.17	8.38	8.57	8.89	9.16	9.39
	0.01	12.73	14.44	15.65	16.59	17.35	17.99	18.53	19.01	19.43	20.15	20.75	21.26
3	0.10	2.99	3.37	3.64	3.84	4.01	4.15	4.27	4.38	4.47	4.63	4.76	4.88
	0.05	3.96	4.43	4.76	5.02	5.23	5.41	5.56	5.69	5.81	6.01	6.18	6.33
	0.01	7.13	7.91	8.48	8.92	9.28	9.58	9.84	10.06	10.27	10.61	10.90	11.15
4	0.10	2.66	2.98	3.20	3.37	3.51	3.62	3.72	3.81	3.89	4.02	4.13	4.23
	0.05	3.38	3.74	4.00	4.20	4.37	4.50	4.62	4.72	4.82	4.98	5.11	5.22
	0.01	5.46	5.99	6.36	6.66	6.90	7.10	7.27	7.43	7.57	7.80	8.00	8.17
5	0.10	2.49	2.77	2.96	3.12	3.24	3.34	3.43	3.51	3.58	3.69	3.79	3.88
	0.05	3.09	3.40	3.62	3.79	3.93	4.04	4.14	4.23	4.31	4.45	4.56	4.66
	0.01	4.70	5.11	5.40	5.63	5.81	5.97	6.11	6.23	6.33	6.52	6.67	6.81
6	0.10	2.39	2.64	2.82	2.96	3.07	3.17	3.25	3.32	3.38	3.49	3.58	3.66
	0.05	2.92	3.19	3.39	3.54	3.66	3.77	3.86	3.94	4.01	4.13	4.23	4.32
	0.01	4.27	4.61	4.86	5.05	5.20	5.33	5.45	5.55	5.64	5.80	5.93	6.04
7	0.10	2.31	2.56	2.73	2.86	2.96	3.05	3.13	3.19	3.25	3.35	3.44	3.51
	0.05	2.80	3.06	3.24	3.38	3.49	3.59	3.67	3.74	3.80	3.92	4.01	4.09
	0.01	4.00	4.30	4.51	4.68	4.81	4.93	5.03	5.12	5.20	5.33	5.45	5.55
8	0.10	2.26	2.49	2.66	2.78	2.88	2.97	3.04	3.10	3.16	3.26	3.34	3.41
	0.05	2.72	2.96	3.13	3.26	3.36	3.45	3.53	3.60	3.66	3.76	3.85	3.93
	0.01	3.81	4.08	4.27	4.42	4.55	4.65	4.74	4.82	4.89	5.02	5.12	5.21
9	0.10	2.22	2.45	2.60	2.72	2.82	2.90	2.97	3.03	3.09	3.18	3.26	3.32
	0.05	2.66	2.89	3.05	3.17	3.27	3.36	3.43	3.49	3.55	3.65	3.73	3.80
	0.01	3.67	3.92	4.10	4.24	4.35	4.45	4.53	4.61	4.67	4.79	4.88	4.96
10	0.10	2.19	2.41	2.56	2.68	2.77	2.85	2.92	2.98	3.03	3.12	3.20	3.26
	0.05	2.61	2.83	2.98	3.10	3.20	3.28	3.35	3.41	3.47	3.56	3.64	3.71
	0.01	3.57	3.80	3.97	4.10	4.20	4.29	4.37	4.44	4.50	4.61	4.70	4.78
11	0.10	2.17	2.38	2.53	2.64	2.73	2.81	2.88	2.93	2.98	3.07	3.15	3.21
	0.05	2.57	2.78	2.93	3.05	3.14	3.22	3.29	3.35	3.40	3.49	3.57	3.63
	0.01	3.48	3.71	3.87	3.99	4.09	4.17	4.25	4.31	4.37	4.47	4.55	4.63
12	0.10	2.15	2.36	2.50	2.61	2.70	2.78	2.84	2.90	2.95	3.03	3.10	3.17
	0.05	2.54	2.75	2.89	3.00	3.09	3.17	3.24	3.29	3.34	3.43	3.51	3.57
	0.01	3.42	3.63	3.78	3.90	4.00	4.08	4.15	4.21	4.26	4.36	4.44	4.51
14	0.10	2.12	2.32	2.46	2.57	2.65	2.72	2.79	2.84	2.89	2.97	3.04	3.10
	0.05	2.49	2.69	2.83	2.94	3.02	3.09	3.16	3.21	3.26	3.34	3.41	3.48
	0.01	3.32	3.52	3.66	3.77	3.85	3.93	3.99	4.05	4.10	4.19	4.26	4.33
16	0.10	2.10	2.29	2.43	2.53	2.62	2.69	2.75	2.80	2.85	2.93	2.99	3.05
	0.05	2.46	2.65	2.78	2.89	2.97	3.04	3.10	3.15	3.20	3.28	3.35	3.40
	0.01	3.25	3.43	3.57	3.67	3.75	3.82	3.88	3.94	3.99	4.07	4.14	4.20
18	0.10	2.08	2.27	2.41	2.51	2.59	2.66	2.72	2.77	2.81	2.89	2.96	3.01
	0.05	2.43	2.62	2.75	2.85	2.93	3.00	3.05	3.11	3.15	3.23	3.29	3.35
	0.01	3.19	3.37	3.50	3.60	3.68	3.74	3.80	3.85	3.90	3.98	4.04	4.10
20	0.10	2.07	2.26	2.39	2.49	2.57	2.63	2.69	2.74	2.79	2.86	2.93	2.98
	0.05	2.41	2.59	2.72	2.82	2.90	2.96	3.02	3.07	3.11	3.19	3.25	3.31
	0.01	3.15	3.32	3.45	3.54	3.62	3.68	3.74	3.79	3.83	3.91	3.97	4.03
24	0.10	2.05	2.23	2.36	2.46	2.53	2.60	2.66	2.70	2.75	2.82	2.88	2.94
	0.05	2.38	2.56	2.68	2.77	2.85	2.91	2.97	3.02	3.06	3.13	3.19	3.25
	0.01	3.09	3.25	3.37	3.46	3.53	3.59	3.64	3.69	3.73	3.80	3.86	3.91
30	0.10	2.03	2.21	2.33	2.43	2.50	2.57	2.62	2.67	2.71	2.78	2.84	2.89
	0.05	2.35	2.52	2.64	2.73	2.80	2.87	2.92	2.96	3.00	3.07	3.13	3.18
	0.01	3.03	3.18	3.29	3.38	3.45	3.51	3.55	3.60	3.64	3.70	3.76	3.81
40	0.10	2.01	2.18	2.30	2.40	2.47	2.53	2.58	2.63	2.67	2.74	2.80	2.85
	0.05	2.32	2.49	2.60	2.69	2.76	2.82	2.87	2.91	2.95	3.02	3.08	3.12
	0.01	2.97	3.12	3.22	3.30	3.37	3.42	3.47	3.51	3.54	3.61	3.66	3.71
60	0.10	1.99	2.16	2.28	2.37	2.44	2.50	2.55	2.59	2.63	2.70	2.76	2.80
	0.05	2.29	2.45	2.56	2.65	2.72	2.77	2.82	2.86	2.90	2.96	3.02	3.06
	0.01	2.91	3.05	3.15	3.23	3.29	3.34	3.38	3.42	3.46	3.51	3.56	3.61
∞	0.10	1.95	2.11	2.23	2.31	2.38	2.43	2.48	2.52	2.56	2.62	2.67	2.72
	0.05	2.24	2.39	2.49	2.57	2.63	2.68	2.73	2.77	2.80	2.86	2.91	2.95
	0.01	2.81	2.93	3.02	3.09	3.14	3.19	3.23	3.26	3.29	3.34	3.38	3.42

Table B.7. *Continued*

ν	α	\\ p										
		17	18	19	20	21	22	23	24	25	27	29	31
2	0.10	6.60	6.67	6.74	6.80	6.85	6.91	6.96	7.01	7.05	7.14	7.22	7.30
	0.05	9.49	9.59	9.68	9.77	9.85	9.92	10.00	10.07	10.13	10.26	10.37	10.48
	0.01	21.49	21.71	21.91	22.11	22.29	22.46	22.63	22.78	22.93	23.21	23.47	23.71
3	0.10	4.93	4.98	5.02	5.07	5.11	5.15	5.18	5.22	5.25	5.31	5.37	5.42
	0.05	6.39	6.45	6.51	6.57	6.62	6.67	6.71	6.76	6.80	6.88	6.95	7.02
	0.01	11.27	11.37	11.47	11.56	11.65	11.74	11.82	11.89	11.97	12.11	12.23	12.35
4	0.10	4.27	4.31	4.35	4.38	4.42	4.45	4.48	4.51	4.54	4.59	4.64	4.68
	0.05	5.27	5.32	5.37	5.41	5.45	5.49	5.52	5.56	5.59	5.66	5.71	5.77
	0.01	8.25	8.32	8.39	8.45	8.51	8.57	8.63	8.68	8.73	8.83	8.92	9.00
5	0.10	3.92	3.95	3.99	4.02	4.05	4.08	4.10	4.13	4.16	4.20	4.25	4.29
	0.05	4.70	4.74	4.78	4.82	4.85	4.89	4.92	4.95	4.98	5.03	5.08	5.13
	0.01	6.87	6.93	6.98	7.03	7.08	7.13	7.17	7.21	7.25	7.33	7.40	7.46
6	0.10	3.70	3.73	3.76	3.79	3.82	3.84	3.87	3.89	3.92	3.96	4.00	4.04
	0.05	4.36	4.39	4.43	4.46	4.49	4.52	4.55	4.58	4.60	4.65	4.70	4.74
	0.01	6.09	6.14	6.18	6.23	6.27	6.31	6.34	6.38	6.41	6.48	6.54	6.59
7	0.10	3.55	3.58	3.61	3.63	3.66	3.69	3.71	3.73	3.75	3.80	3.83	3.87
	0.05	4.13	4.16	4.19	4.22	4.25	4.28	4.31	4.33	4.35	4.40	4.44	4.48
	0.01	5.59	5.64	5.68	5.72	5.75	5.79	5.82	5.85	5.88	5.94	5.99	6.04
8	0.10	3.44	3.47	3.50	3.52	3.55	3.57	3.59	3.61	3.64	3.67	3.71	3.74
	0.05	3.96	3.99	4.02	4.05	4.08	4.10	4.13	4.15	4.18	4.22	4.26	4.29
	0.01	5.25	5.29	5.33	5.36	5.39	5.43	5.45	5.48	5.51	5.56	5.61	5.65
9	0.10	3.35	3.38	3.41	3.44	3.46	3.48	3.50	3.53	3.55	3.58	3.62	3.65
	0.05	3.84	3.87	3.90	3.92	3.95	3.97	4.00	4.02	4.04	4.08	4.12	4.15
	0.01	5.00	5.04	5.07	5.10	5.13	5.16	5.19	5.21	5.24	5.29	5.33	5.37
10	0.10	3.29	3.32	3.34	3.37	3.39	3.41	3.43	3.45	3.47	3.51	3.54	3.57
	0.05	3.74	3.77	3.80	3.82	3.85	3.87	3.89	3.91	3.94	3.97	4.01	4.04
	0.01	4.81	4.84	4.88	4.91	4.93	4.96	4.99	5.01	5.03	5.08	5.12	5.15
11	0.10	3.24	3.26	3.29	3.31	3.34	3.36	3.38	3.40	3.42	3.45	3.48	3.51
	0.05	3.66	3.69	3.72	3.74	3.77	3.79	3.81	3.83	3.85	3.89	3.92	3.95
	0.01	4.66	4.69	4.72	4.75	4.78	4.80	4.83	4.85	4.87	4.91	4.95	4.99
12	0.10	3.19	3.22	3.24	3.27	3.29	3.31	3.33	3.35	3.37	3.40	3.43	3.46
	0.05	3.60	3.63	3.65	3.68	3.70	3.72	3.74	3.76	3.78	3.82	3.85	3.88
	0.01	4.54	4.57	4.60	4.63	4.65	4.67	4.70	4.72	4.74	4.78	4.82	4.85
14	0.10	3.13	3.15	3.18	3.20	3.22	3.24	3.26	3.28	3.29	3.33	3.36	3.39
	0.05	3.50	3.53	3.55	3.58	3.60	3.62	3.64	3.66	3.68	3.71	3.74	3.77
	0.01	4.36	4.39	4.41	4.44	4.46	4.48	4.50	4.52	4.54	4.58	4.61	4.65
16	0.10	3.08	3.10	3.12	3.15	3.17	3.19	3.20	3.22	3.24	3.27	3.30	3.33
	0.05	3.43	3.46	3.48	3.50	3.52	3.54	3.56	3.58	3.60	3.63	3.66	3.69
	0.01	4.23	4.25	4.28	4.30	4.32	4.34	4.36	4.38	4.40	4.43	4.47	4.50
18	0.10	3.04	3.06	3.08	3.11	3.13	3.14	3.16	3.18	3.20	3.23	3.26	3.28
	0.05	3.38	3.40	3.42	3.44	3.46	3.48	3.50	3.52	3.54	3.57	3.60	3.62
	0.01	4.13	4.15	4.18	4.20	4.22	4.24	4.26	4.27	4.29	4.33	4.36	4.38
20	0.10	3.01	3.03	3.05	3.07	3.09	3.11	3.13	3.15	3.16	3.19	3.22	3.25
	0.05	3.33	3.36	3.38	3.40	3.42	3.44	3.46	3.47	3.49	3.52	3.55	3.57
	0.01	4.05	4.07	4.10	4.12	4.14	4.16	4.17	4.19	4.21	4.24	4.27	4.30
24	0.10	2.96	2.98	3.01	3.03	3.04	3.06	3.08	3.10	3.11	3.14	3.17	3.19
	0.05	3.27	3.29	3.31	3.33	3.35	3.37	3.39	3.40	3.42	3.45	3.48	3.50
	0.01	3.94	3.96	3.98	4.00	4.02	4.04	4.05	4.07	4.09	4.12	4.14	4.17
30	0.10	2.92	2.94	2.96	2.98	3.00	3.01	3.03	3.05	3.06	3.09	3.12	3.14
	0.05	3.21	3.23	3.25	3.27	3.29	3.30	3.32	3.33	3.35	3.38	3.40	3.43
	0.01	3.83	3.85	3.87	3.89	3.91	3.92	3.94	3.95	3.97	4.00	4.02	4.04
40	0.10	2.87	2.89	2.91	2.93	2.95	2.97	2.98	3.00	3.01	3.04	3.06	3.09
	0.05	3.14	3.17	3.18	3.20	3.22	3.24	3.25	3.27	3.28	3.31	3.33	3.36
	0.01	3.73	3.74	3.76	3.78	3.80	3.81	3.83	3.84	3.85	3.88	3.90	3.93
60	0.10	2.83	2.85	2.87	2.88	2.90	2.92	2.93	2.95	2.96	2.99	3.01	3.04
	0.05	3.08	3.10	3.12	3.14	3.16	3.17	3.19	3.20	3.21	3.24	3.26	3.28
	0.01	3.63	3.64	3.66	3.68	3.69	3.71	3.72	3.73	3.75	3.77	3.79	3.81
∞	0.10	2.74	2.76	2.77	2.79	2.81	2.82	2.84	2.85	2.86	2.89	2.91	2.93
	0.05	2.97	2.98	3.00	3.02	3.03	3.04	3.06	3.07	3.08	3.11	3.13	3.15
	0.01	3.44	3.45	3.47	3.48	3.49	3.50	3.52	3.53	3.54	3.56	3.58	3.60

APPENDIX C

FORTRAN Programs

In addition to the programs described below, there is a variety of commercial and public domain software available for implementing the procedures described in this book. Perhaps the most widely available commercial software is that for computing Tukey simultaneous confidence intervals for all pairwise treatment means (Section 4.3) and Dunnett simultaneous confidence intervals for comparing t treatment means with a control mean (Section 5.4). More recently, several packages have added Hsu simultaneous confidence intervals for the difference between each treatment mean and the best of the other treatment means (Section 4.4). Examples of software that calculate Hsu intervals are MINITAB (as a subcommand of "ONEWAY" in Release 8 and higher) and JMP (in the "FIT Y BY X" platform in Version 2 and higher).

We also make special mention of the very useful public domain FORTRAN programs MVNPRD and MVTPRD in Dunnett (1989) that compute probabilities of rectangular regions for multivariate normal and multivariate t-distributions with product correlation structure, that is, correlations of the form $\lambda_i \times \lambda_j$ for $i \neq j$. These programs can be obtained via ftp from the statlib archive maintained at Carnegie Mellon University at the Internet address lib.stat.cmu.edu (128.2.241.142) or at the URL address http://lib.stat.cmu.edu/apstat/ using your favorite WWW reader. The MVNPRD [MVTPRD] program calculates

$$P\{a_i \le W_i \le b_i \quad (1 \le i \le p)\}$$

for given a_i and b_i when $\mathbf{W} = (W_1, \ldots, W_p)$ has the multivariate normal distribution [multivariate t-distribution with arbitrary degrees of freedom] with arbitrary mean vector, unit variances and product correlation structure. Dunnett's programs allow any of the endpoints a_i to be $-\infty$ and any of the endpoints b_i to be $+\infty$. (Note that using MVTPRD with the internal variable NDF = 0 invokes MVNPRD.)

As an example, consider the calculation of the critical point h in Equation (2.5.2) of Section 2.5.1 to implement the version of procedure $\mathcal{N}_B$ that selects the s best of t treatments. In terms of the notation of this Appendix, we are required to solve

$$P\{-\infty < W_i \le h/\sqrt{2} \quad (1 \le i \le t-s);$$
$$0 < W_i < \infty \quad (t-s+1 \le i < t)\} = P^\star/s \qquad \text{(C.1)}$$

for h, where $(W_1, \ldots, W_{t-1})$ has the multivariate normal distribution with mean vector zero, unit variances and common correlation $1/2$. We take $\lambda_i = 1/\sqrt{2}$ for $1 \le i < t$, $(a_i, b_i) = (-\infty, h/\sqrt{2})$ for $1 \le i \le t - s$ and $(a_i, b_i) = (0, +\infty)$ for $t - s + 1 \le i < t$ and use MVNPRD together with a bisection method or some other zero-finding algorithm to solve (C.1). In particular, the Dunnett programs can be used to determine $Z_{p,1/2}^{(1-P^*)}$ and $T_{p,\nu,1/2}^{(1-P^*)}$. However, because $Z_{p,1/2}^{(1-P^*)}$ is frequently occurring and simple to compute, we include the stand-alone program USENB to determine its value.

The main stand-alone programs in this Appendix are described in 1–7, below. Grouped by function, the programs USENB and UNEQNB evaluate quantities related to the Bechhofer procedure $\mathcal{N}_B$, NPMC uses Monte Carlo simulation to study performance characteristics of Paulson's procedure $\mathcal{N}_P$, RINOTT calculates constants necessary for procedure $\mathcal{N}_R$, EVALNG evaluates performance quantities for the Gupta procedure $\mathcal{N}_G$, USEGSA determines constants needed to implement procedure $\mathcal{N}_{GSa}$, and BPMC uses Monte Carlo simulation to study performance characteristics of Paulson's procedure $\mathcal{B}_P$. Several of the main programs use the FORTRAN 'INCLUDE file' statement, functions and subroutine given in A–I, below.

A number of other FORTRAN programs are available from the authors. These programs cover a range of topics involving exact and Monte Carlo analysis of various normal, Bernoulli and multinomial procedures.

These programs are provided in good faith. The software has been tested on a number of platforms and compilers to assure the reproducibility of results and its accuracy checked in a variety of ways. However, none of the authors, publishers, or distributors warrant their accuracy nor can be held accountable for the consequences of their use. We shall be unable to assist with platform or compiler dependent difficulties.

Supporting FORTRAN Common Input File, Functions and Subroutines

A. GLQUAD

Description: Input file containing weights and zeroes to compute 64-point Gauss-Laguerre quadrature.

B. ZCDF

Description: Function that approximates the $N(0, 1)$ c.d.f. using Equation (26.2.17) from Abramowitz and Stegun (1972).

C. FACTOR

Description: Function that calculates $n!/x!(n - x)!$ where n, x and $(n - x) \ge 0$.

D. MULTZ

Description: Function that calculates $Z_{p,1/2}^{(1-P^*)}$.

E. PCSNB

Description: Function that evaluates the $P\{CS\}$ for procedure $\mathcal{N}_B$ at the slippage configuration $\boldsymbol{\mu} = (0, \ldots, 0, \delta)$, for fixed t and σ and respective sample sizes $(n(1), \ldots, n(t))$.

FORTRAN PROGRAMS

F. PCSNGS

Description: Function that evaluates the $P\{CS\}$ for procedure $\mathcal{N}_{GSa}$ at the slippage configuration $\boldsymbol{\mu} = (0,\ldots,0,\delta)$, for fixed t and σ and specified q, n and yardstick h.

G. UNIF

Description: Function that generates $U(0,1)$ random numbers using code adapted from Bratley, Fox and Schrage (1987).

H. RNORML

Description: Function that generates standard normal random variates using code adapted from Bratley, Fox and Schrage (1987).

I. STATS

Description: Subroutine that calculates sample averages and variances.

Main FORTRAN Programs

1. USENB

Description: Calculates $P^\star$, $\delta^\star$ or n so that procedure $\mathcal{N}_B$ from Section 2.2 satisifies the probability requirement (2.1.1) for fixed t and σ.

2. UNEQNB

Description: Computes the $P\{CS \mid LF\}$ in (2.2.5) for procedure $\mathcal{N}_B$ for fixed t and σ and given $\delta^\star$ and $(n_{(1)},\ldots,n_{(t)})$.

3. NPMC

Description: Uses Monte Carlo simulation to study performance characteristics of Paulson's normal means procedure $\mathcal{N}_P$ from Section 2.3.3.

4. RINOTT

Description: Can be used to extend the Rinott tables in Section 2.8.

5. EVALNG

Description: Uses Monte Carlo simulation to to study performance characteristics of procedure $\mathcal{N}_G$ from Section 3.2.1.

6. USEGSA

Description: Calculates n, h or $\delta^\star$ so that procedure $\mathcal{N}_{GSa}$ from Section 3.4.3 satisfies the probability requirement (3.4.7) for fixed t and σ and given q and $P^\star$.

7. BPMC

Description: Uses Monte Carlo simulation to study performance characteristics of Paulson's Bernoulli procedure $\mathcal{B}_P$ from Section 7.4.2.

```
C      INCLUDE FILE GLQUAD
C
       DATA X/.44489365833267018419D-1,  .23452610951961853745,
      *  .57688462930188642649,          .10724487538178176330D1,
      *  .17224087764446454411D1,        .25283367064257948811D1,
      *  .34922132730219944896D1,        .46164567697497673878D1,
      *  .59039585041742439466D1,        .73581267331862411132D1,
      *  .89829409242125961034D1,        .10783018632539972068D2,
      *  .12763697986742725115D2,        .14931139755522557320D2,
      *  .17292454336715314789D2,        .19855860940336054740D2,
      *  .22630889013196774489D2,        .25628636022459247767D2,
      *  .28862101816323474744D2,        .32346629153964737003D2,
      *  .36100494805751973804D2,        .40145719771539441536D2,
      *  .44509207995754937976D2,        .49224394987308639177D2,
      *  .54333721333396907333D2,        .59892509162134018196D2,
      *  .65975377287935052797D2,        .72687628090662708639D2,
      *  .80187446977913523067D2,        .88735340417892398689D2,
      *  .98829542868283972559D2,        .11175139809793769521D3/
C
       DATA W/.10921834195238497114,  .21044310793881323294,
      *  .23521322966984800539,          .19590333597288104341,
      *  .12998378628607176061,          .70578623865717441560D-1,
      *  .31760912509175070306D-1,       .11918214834838557057D-1,
      *  .37388162946115247897D-2,       .98080330661495513223D-3,
      *  .21486491880136418802D-3,       .39203419679879472043D-4,
      *  .59345416128686328784D-5,       .74164045786675522191D-6,
      *  .76045678791207814811D-7,       .63506022266258067424D-8,
      *  .42813829710409288788D-9,       .23058994918913360793D-10,
      *  .97993792887270940633D-12,      .32378016577292664623D-13,
      *  ,81718234434207194332D-15,      .15421338333938233722D-16,
      *  .21197922901636186120D-18,      .20544296737880454267D-20,
      *  .13469825866373951558D-22,      .56612941303973593711D-25,
      *  .14185605454630369059D-27,      .19133754944542243094D-30,
      *  .11922487600982223565D-33,      .26715112192401369860D-37,
      *  .13386169421062562827D-41,      .45105361938989742322D-47/
C
       DO 10 I = 1,32
10     WEX(I) = W(I) * DEXP(X(I))
```

```fortran
      DOUBLE PRECISION FUNCTION ZCDF(UU)
C
C     APPROXIMATE N(0,1) CDF USING ABRAMOWITZ AND STEGUN (1972) - 26.2.17
C
      DOUBLE PRECISION UU,B(5),P,T,TEMP,IRT2PI
C
      IF (UU.GT.5.) THEN
              ZCDF = 1.0D0
              RETURN
      ENDIF
C
      P = 0.2316419
      DATA B/0.319381530, -0.356563782, 1.781477937, -1.821255978,
     *      1.330274429/
      IRT2PI = 0.3989422803
C
      T = 1.0D0/(1.0D0 + P*DABS(UU))
      TEMP = ((((B(5)*T + B(4))*T + B(3))*T + B(2))*T + B(1))*T
C
      ZCDF = 1.0D0 - TEMP*DEXP(-UU*UU*0.5D0)*IRT2PI
      IF (UU .LT. 0.0D0) ZCDF = 1.0D0 - ZCDF
      RETURN
      END




      DOUBLE PRECISION FUNCTION FACTOR(N,X)
C
C     CALCULATE N!/{X!(N-X)!} WHERE N, X AND (N-X) >=0
C
      DOUBLE PRECISION C1,C2,C3
      INTEGER I,N,X,NMX
C
      IF((N .GE. 0) .AND. (X .GE. 0) .AND. (N .GE. X)) THEN
              GOTO 100
      ELSE
              FACTOR = -1.0D0
              WRITE(*,*) ' ENTRY ERROR!!'
              RETURN
      END IF
C
100   C1 = 1.0
      IF(N .EQ. 0) GOTO 110
      DO 120 I = 1,N
120   C1 = C1*DFLOAT(I)
C
110   C2 = 1.0
      IF(X .EQ. 0) GOTO 130
      DO 140 I = 1,X
140   C2 = C2*DFLOAT(I)
C
130   C3 = 1.0
      IF(N .EQ. X) GOTO 150
      NMX = N - X
      DO 160 I = 1,NMX
160   C3 = C3*DFLOAT(I)
C
150   FACTOR = C1/(C2*C3)
      RETURN
      END
```

```fortran
      DOUBLE PRECISION FUNCTION MULTZ(PSTAR,P)
C
C     CALCULATES Z^{1-PSTAR}_{P,1/2}, THE ONE-SIDED UPPER (1-PSTAR)
C     EQUICOORDINATE POINT OF THE P-DIMENSIONAL MULTIVARIATE NORMAL
C     DISTRIBUTION WITH COMMON CORRELATION 1/2 AND UNIT VARIANCES
C
C  INPUT:  P      = DIMENSION OF THE MULTIVARIATE NORMAL DISTRIBUTION
C          1-PSTAR = UPPER TAIL PROBABILITY
C
C  OUTPUT: Z^{1-PSTAR}_{P,1/2}
C
      INTEGER MAXT
C     MAXT DEFINES THE MAXIMUM NUMBER OF TREATMENTS
      PARAMETER (MAXT=50)
      DOUBLE PRECISION PCSNB
      DOUBLE PRECISION PSTAR
      DOUBLE PRECISION ANS,H,LOWERH,TMP,UPPERH,IP
      INTEGER P,T,NTMP(MAXT),I
      EXTERNAL PCSNB
C
C     CHECK INPUT VALUES
      IP = 1.0D0/DFLOAT(MAX0(P,2))
      IF ((IP .LT. PSTAR) .AND. (PSTAR .LT. 1.0D0)) THEN
             GOTO 100
      ELSE
             MULTZ = -1.0D0
             WRITE(*,*) ' P^* ENTRY ERROR!!'
             RETURN
      END IF
C
100   T = P + 1
      DO 110 I = 1,T
110   NTMP(I) = 1
C
C     DETERMINE UPPER AND LOWER VALUES AT WHICH TO START BISECTION
C
      H = 0.0D0
      LOWERH = 0.0D0
C
C     EVALUATE THE P{CS} AT A TRIAL DELTA-STAR
C
120   TMP = H*DSQRT(2.0D0)
      ANS = PCSNB(T,NTMP,TMP,1.0D0)
C
      IF(ANS .GT. PSTAR) GOTO 130
      H = H + 1.0D0
      GOTO 120
130   UPPERH = H
      H = (UPPERH + LOWERH)/2.0D0
C
C     PERFORM BINARY SEARCH WITH A MAXIMUM OF 50 ITERATIONS
      DO 140 I = 1,50
      TMP = H*DSQRT(2.0D0)
      ANS = PCSNB(T,NTMP,TMP,1.0D0)
      IF(DABS(ANS-PSTAR).LE.0.000001) GOTO 150
      IF(ANS.GT.PSTAR) GOTO 160
      LOWERH = H
      H = (LOWERH + UPPERH)/2.
      GOTO 140
160   UPPERH = H
      H = (LOWERH + UPPERH)/2.
140   CONTINUE
150   MULTZ = H
      RETURN
      END
```

FORTRAN PROGRAMS

```fortran
      DOUBLE PRECISION FUNCTION PCSNB(T,N,DSTAR,SIGMA)
C
C     EVALUATE P{CS} USING PROCEDURE N_B WHEN THE TRUE MEANS ARE IN
C     THE SLIPPAGE CONFIG, THE MEASUREMENT ERROR FOR EACH TREATMENT IS
C     SPECIFIED BY SIGMA, AND THE SAMPLE SIZE OF THE TREATMENT
C     ASSOCIATED WITH THE ITH ORDERED MEAN IS N(I).
C
C     INPUT: T       = NUMBER OF TREATMENTS (T .LEQ. 50 - RESET BY MAXT)
C            N(I)    = SAMPLE SIZE OF THE TREATMENT ASSOCIATED WITH THE
C                      ITH ORDERED MEAN MU[I]
C            DSTAR   = DIFFERENCE TO BE DETECTED
C            SIGMA   = STANDARD DEVIATION OF THE MEASURE ERROR
C
C     OUTPUT: PROBABILITY OF CORRECT SELECTION USING N_B
C
      INTEGER MAXT
C     MAXT DEFINES THE MAXIMUM NUMBER OF TREATMENTS
      PARAMETER (MAXT=50)
      DOUBLE PRECISION ZCDF
      DOUBLE PRECISION DSTAR,SIGMA
      DOUBLE PRECISION DN(MAXT),X(32),W(32),WEX(32),DUMP,DUMN,ANS
      INTEGER T,N(MAXT),I,LK,TM1
      EXTERNAL ZCDF
      INCLUDE 'GLQUAD'
C
      DO 100 I = 1,T
  100 DN(I) = DSQRT(DFLOAT(N(I)))
      TM1 = T - 1
C
      ANS = 0.0D0
C
C     EVALUATE P{CS}
C
      DO 110 I = 1,32
      DUMP = 1.0D0
      DUMN = 1.0D0
      DO 120 LK = 1,TM1
      DUMP = DUMP * ZCDF(X(I) * DN(LK)/DN(T) + DN(LK) * DSTAR/SIGMA)
      DUMN = DUMN * ZCDF(-X(I) * DN(LK)/DN(T) + DN(LK) * DSTAR/SIGMA)
  120 CONTINUE
      ANS = ANS + WEX(I)*(DUMP+DUMN)*0.3989422803*DEXP(-X(I)*X(I)/2.0D0)
  110 CONTINUE
      PCSNB = ANS
      RETURN
      END
```

```
      DOUBLE PRECISION FUNCTION PCSNGS(T,Q,N,DSTAR,SIGMA,H)
C
C     EVALUATE P{CS} USING PROCEDURE N_{GSa} AT THE SLIPPAGE
C     CONFIGURATION (0,...,0,DSTAR) WHEN SIGMA IS STANDARD DEVIATION OF
C     THE MEASUREMENT ERROR FOR GIVEN T, Q, N AND YARDSTICK H
C
C INPUT: T        = NUMBER OF TREATMENTS
C        Q        = SIZE OF THE SUBSET
C        N        = COMMON SAMPLE SIZE
C        DSTAR    = DIFFERENCE TO BE DETECTED
C        SIGMA    = STANDARD DEVIATION OF THE MEASURE ERROR
C        H        = CRITICAL VALUE CHOSEN TO GUARANTEE PROBABILITY
C                   P-STAR OF DELTA-STAR CS FOR GIVEN (N, SIGMA)
C
C OUTPUT: PROBABILITY OF CORRECT SELECTION FOR N_{GSa}
C
      DOUBLE PRECISION ZCDF,FACTOR
      DOUBLE PRECISION DSTAR,SIGMA,H
      DOUBLE PRECISION DN,CONST,X(32),W(32),WEX(32),DUMP,DUMN,
     *ANS,ANS1
      INTEGER T,Q,N,I,J,TM1,TMQ
      EXTERNAL ZCDF,FACTOR
C
      INCLUDE 'GLQUAD'
C
      DN = DSQRT(DFLOAT(N))
      TM1 = T - 1
      TMQ = T - Q
      CONST = DN*DSTAR/SIGMA
C
      ANS = 0.0D0
C
C     EVALUATE P{CS}
C
      DO 100 I = TMQ,TM1
      ANS1 = 0.0D0
      DO 120 J = 1,32
      DUMP = ((ZCDF(X(J)+CONST))**I)*((ZCDF(X(J)+H*
     *DSQRT(DFLOAT(2))+CONST)-ZCDF(X(J)+CONST))**(TM1-I))
      DUMN = ((ZCDF(-X(J)+CONST))**I)*((ZCDF(-X(J)+H*
     *DSQRT(DFLOAT(2))+CONST)-ZCDF(-X(J)+CONST))**(TM1-I))
      ANS1 = ANS1+WEX(J)*(DUMP+DUMN)*0.3989422803D0*
     *DEXP(-X(J)*X(J)/2.0D0)
  120 CONTINUE
      ANS = ANS + FACTOR(TM1,I)*ANS1
  100 CONTINUE
      PCSNGS = ANS
C
      RETURN
      END
```

FORTRAN PROGRAMS

```
      FUNCTION UNIF(IX)
C
C     PORTABLE U(0,1) GENERATOR FROM BRATLEY, FOX AND SCHRAGE (1987)
C
C     INPUT:  IX = INTEGER > 0 BUT < 2147483647
C     OUTPUT: IX = NEW SEED, UNIF = U(0,1) DEVIATE
C
      K1 = IX/127773
      IX = 16807*(IX-K1*127773)-K1*2836
      IF(IX.LT.0)IX = IX+2147483647
      UNIF = IX*4.656612875E-10
      RETURN
      END




      FUNCTION RNORML(IX)
C
C     GIVES N(0,1) DEVIATE BY COMPOSITION METHOD OF AHRENS AND DIETER.
C     SEE BRATLEY, FOX, AND SCHRAGE (1987), P. 330.
C
C     INPUT:  IX = INTEGER > 0 BUT < 2147483647
C     OUTPUT: IX = NEW SEED, RNORML = N(0,1) DEVIATE
C
      U = UNIF(IX)
      U0 = UNIF(IX)
      IF(U.GE. 0.919544)GOTO 160
C
      RNORML = 2.40376*(U0+U*.825339)-2.11403
      RETURN
160   IF(U.LT. 0.965487)GOTO 210
C
180   RNORML = (4.46911-2*ALOG(UNIF(IX)))**0.5
      IF(RNORML*UNIF(IX).GT. 2.11403)GOTO 180
      GOTO 340
210   IF(U.LT. 0.949991)GOTO 260
C
230   RNORML = 1.8404+UNIF(IX)*0.273629
      IF(0.398942*EXP(-RNORML*RNORML/2)-0.443299+
     &  RNORML*0.209694 .LT. UNIF(IX)*4.27026E-02)GOTO 230
      GOTO 340
260   IF(U.LT. 0.925852)GOTO 310
C
280   RNORML = 0.28973+UNIF(IX)*1.55067
      IF(0.398942*EXP(-RNORML*RNORML/2)-0.443299+
     &  RNORML*.209694 .LT. UNIF(IX)*1.59745E-02)GOTO 280
      GOTO 340
310   RNORML = UNIF(IX)*0.28973
C
      IF(0.398942*EXP(-RNORML*RNORML/2)-0.382545
     &  .LT. UNIF(IX)*1.63977E-02)GOTO 310
340   IF(U0.GT. 0.5)GOTO 370
      RNORML = -RNORML
370   RETURN
      END
```

```
      SUBROUTINE STATS(DATASET,NUMRNS)
C
C     CALCULATES SAMPLE AVERAGES AND VARIANCES FOR THE OBSNS IN DATASET
C
C  INPUT: DATASET (UP TO SIZE 20,000)
C         NUMRNS (NUMBER OF ENTRIES IN DATASET)
C
      DOUBLE PRECISION DUM,EXPN,VARY,STNDEV,STNERR
      INTEGER DATASET(20000)
C
      VARY = 0.0
      EXPN = 0.0
C
      DO 100 I = 1,NUMRNS
100   EXPN = EXPN + DATASET(I)
      EXPN = EXPN/NUMRNS
C
      DO 150 I = 1,NUMRNS
      DUM = DATASET(I) - EXPN
150   VARY = VARY + DUM*DUM
      VARY = VARY/(NUMRNS - 1.)
C
      STNDEV = VARY**0.5
      STNERR = (VARY/NUMRNS)**0.5
C
      WRITE(6,200)EXPN
200   FORMAT(1X,'SAMPLE AVERAGE = ',F12.4)
      WRITE(6,205)STNERR
205   FORMAT(1X,'STANDARD ERROR = ',F12.4)
      WRITE(6,210)STNDEV
210   FORMAT(1X,'SAMPLE STANDARD DEVIATION = ',F12.4/)
      RETURN
      END
```

FORTRAN PROGRAMS

```
      PROGRAM USENB
C
C     INTERACTIVE CALCULATION OF N, P-STAR OR DELTA-STAR FOR THE
C     SINGLE-STAGE PROCEDURE N_B TO ACHIEVE PROB REQUIREMENT (2.1.1)
C
C  INPUT: PROGRAM INTERACTIVELY ASKS FOR
C         T       = NUMBER OF TREATMENTS  (T .LEQ. 50-RESET BY MAXT)
C         SIGMA   = STANDARD DEVIATION OF THE MEASURE ERROR
C  THEN PROMPTS FOR 1 OF 3 CHOICES FROM THE MENU
C
C           CHOICE           GIVEN              FIND
C             1             N, P-STAR          DELTA-STAR
C             2          DELTA-STAR, P-STAR        N
C             3          N, DELTA-STAR          P-STAR
C  WHERE
C       DELTA-STAR = DIFFERENCE TO BE DETECTED
C       P-STAR     = MINIMUM P{CS} OVER THE INDIFFERENCE-ZONE
C       N          = SMALLEST COMMON NUMBER OF OBSERVATIONS PER
C                    TREATMENT TO ATTAIN PROB REQUIREMENT (2.1.1)
C
C  OUTPUT: DELTA-STAR, N OR P-STAR
C
      INTEGER MAXT
C     MAXT = THE MAXIMUM NUMBER OF TREATMENTS
      PARAMETER (MAXT=50)
      DOUBLE PRECISION ZCDF,PCSNB
      DOUBLE PRECISION PSTAR,DSTAR,SIGMA,H
      DOUBLE PRECISION ANS,TMP,LOWERH,UPPERH
      INTEGER T,N,CHOICE1,NTMP(MAXT),I190
      CHARACTER CHOICE2
      EXTERNAL ZCDF,PCSNB
C
C     READ T AND SIGMA
      WRITE(6,*) 'ENTER T AND SIGMA'
      READ(5,*) T, SIGMA
C
100   WRITE(6,510)
110   WRITE(6,*) 'ENTER CHOICE (1, 2, OR 3) '
      READ(5,*) CHOICE1
      IF(CHOICE1 .EQ. 1) GOTO 120
      IF(CHOICE1 .EQ. 2) GOTO 130
      IF(CHOICE1 .EQ. 3) GOTO 140
      WRITE(6,*) 'ENTRY ERROR, TRY AGAIN '
      GOTO 110
120   WRITE(6,*) 'ENTER N AND P-STAR'
      READ(5,*) N, PSTAR
      GOTO 150
130   WRITE(6,*) 'ENTER DELTA-STAR AND P-STAR'
      READ(5,*) DSTAR, PSTAR
C
C     COMPUTES Z^{1-PSTAR}_{T-1,1/2} USING A BINARY SEARCH
C
150   DO 160 I = 1,T
160   NTMP(I) = 1
C
C     DETERMINE UPPER AND LOWER VALUES AT WHICH TO START BISECTION
C
      H = 0.0D0
      LOWERH = 0.0D0
      TMP=1.0D0
170   ANS = PCSNB(T,NTMP,H,TMP)
C
      IF(ANS .GT. PSTAR) GOTO 180
      H = H + 1.0D0
      GOTO 170
180   UPPERH = H
      H = (UPPERH + LOWERH)/2.0D0
```

```
C
C       PERFORM BINARY SEARCH WITH A MAXIMUM OF 50 ITERATIONS
C
        DO 190 I190 = 1,50
        ANS = PCSNB(T,NTMP,H,TMP)
        IF(DABS(ANS-PSTAR).LE.0.000001) GOTO 200
        IF(ANS.GT.PSTAR) GOTO 210
        LOWERH = H
        H = (LOWERH + UPPERH)/2.
        GOTO 190
210     UPPERH = H
        H = (LOWERH + UPPERH)/2.
190     CONTINUE
C
200     IF(CHOICE1 .EQ. 1) GOTO 220
        N = DINT((SIGMA * H / DSTAR)**2) + 1
        WRITE(6,520) T, DSTAR, PSTAR, SIGMA, N
C
        WRITE(6,*) 'DO YOU WANT TO CONTINUE? PLEASE ANSWER Y/N'
        READ(5,500) CHOICE2
        IF (CHOICE2 .EQ. 'Y' .OR. CHOICE2 .EQ. 'y') GOTO 100
        STOP
C
220     DSTAR = H * SIGMA / DSQRT(DFLOAT(N))
        WRITE(6,530) T, PSTAR, SIGMA, N, DSTAR
C
        WRITE(6,*) 'DO YOU WANT TO CONTINUE? PLEASE ANSWER Y/N'
        READ(5,500) CHOICE2
        IF (CHOICE2 .EQ. 'Y' .OR. CHOICE2 .EQ. 'y') GOTO 100
        STOP
C
C       CHOICE1 = 3 BRANCH
C
140     WRITE(6,*) 'ENTER N AND DELTA-STAR'
        READ(5,*) N, DSTAR
        DO 230 I =1,T
        NTMP(I) = N
230     CONTINUE
        ANS = PCSNB(T,NTMP,DSTAR,SIGMA)
        WRITE(6,540) T,SIGMA,N,DSTAR,ANS
C
        WRITE(6,*) 'DO YOU WANT TO CONTINUE? PLEASE ANSWER Y/N'
        READ(5,500) CHOICE2
        IF (CHOICE2 .EQ. 'Y' .OR. CHOICE2 .EQ. 'y') GOTO 100
C
500     FORMAT(A1)
510     FORMAT(1X,'            CHOICE         GIVEN              FIND'/
       *'              1          N, P-STAR        DELTA-STAR'/
       *'              2       DELTA-STAR, P-STAR      N'/,
       *'              3          N, DELTA-STAR       P-STAR')
520     FORMAT(1X,'FOR (T, DELTA*, P*, SIGMA) = ( ', I4,', ',
       * F5.2, ', ',F5.3, ', ',F5.2,')' /1X, 'THE REQUIRED '
       *'SAMPLE SIZE FOR PROCEDURE N_B IS N = ', I6/)
530     FORMAT(1X,'FOR (T, P*, SIGMA,N) = ( ', I4,', ',
       * F5.2, ', ',F5.3, ', ', I5,')' / ' THE MINIMUM ',
       *'DELTA STAR WHICH CAN BE DETECTED WHILE' / ' ACHIEVING (2.1.1) '
       *'IS DELTA-STAR = ', F7.3/)
540     FORMAT(1X,'FOR (T, SIGMA,N,DELTA*) = ( ', I4,', ',
       *F5.2,', ', I5,', ',F7.3,')' / ' THE ACHIEVED P-STAR IN '
       *'(2.1.1) IS P-STAR = ', F5.3/)
        STOP
        END
```

FORTRAN PROGRAMS

```fortran
      PROGRAM UNEQNB
C
C     INTERACTIVE CALCULATION OF P{CS} FOR PROCEDURE N_B FOR A GIVEN
C     ASSIGNMENT OF SAMPLE SIZES TO THE ORDERED MEANS
C
C  INPUT (PROGRAM INTERACTIVELY ASKS FOR EACH):
C        T          = NUMBER OF TREATMENTS (T .LEQ. 50-RESET BY MAXT)
C        DELTA-STAR = DIFFERENCE TO BE DETECTED
C        SIGMA      = STANDARD DEVIATION OF THE MEASURE ERROR
C        N(I)       = SAMPLE SIZE OF THE TREATMENT ASSOCIATED WITH THE
C                     ITH ORDERED MEAN MU[I]
C
C  OUTPUT: PROBABILITY OF CORRECT SELECTION
C
      INTEGER MAXT
C     MAXT = THE MAXIMUM NUMBER OF TREATMENTS
      PARAMETER (MAXT=50)
      DOUBLE PRECISION ZCDF,PCSNB
      DOUBLE PRECISION DSTAR,SIGMA
      DOUBLE PRECISION ANS
      INTEGER I,T,N(MAXT)
      EXTERNAL ZCDF, PCSNB
C
C     READ IN DATA
C
      WRITE(6,*) 'ENTER T, DELTA-STAR AND SIGMA '
      READ(5,*) T, DSTAR, SIGMA
      DO 100 I = 1,T
      WRITE(6,500) I,I
100   READ(5,*) N(I)
C
      ANS = PCSNB(T,N,DSTAR,SIGMA)
C
      WRITE(6,510) T, DSTAR, SIGMA
      DO 110 I = 1,T
110   WRITE(6,520) I, N(I)
      WRITE(6,530) ANS
C
500   FORMAT(1X,'ENTER N(',I2,'), THE SAMPLE SIZE ASSOCIATED WITH
     1ORDERED MEAN ',I2)
510   FORMAT(1X,'FOR (T, DELTA*, SIGMA) = ', I4, F5.2, 2X, F5.3)
520   FORMAT(1X,'N(',I2,') = ',I5)
530   FORMAT(1X,'THE PCS IN THE SLIPPAGE CONFIGURATION IS ',F7.4)
      STOP
      END
```

```
      PROGRAM NPMC
C
C     INTERACTIVE MONTE CARLO IMPLEMENTATION OF PAULSON'S SEQUENTIAL
C     NORMAL MEANS PROCEDURE N_P FOR THE CASE OF COMMON KNOWN VARIANCE
C     (SEE SECTION 2.3.3)
C
C  INPUT (PROGRAM INTERACTIVELY PROMPTS FOR):
C        T       = NUMBER OF TREATMENTS
C        SIGMA   = STANDARD DEVIATION OF THE MEASUREMENT ERROR
C        CONFIG  = 1, 2, 3 (LF-, EM-, ES-CONFIGURATIONS)
C        DELTA*  = DIFFERENCE TO BE DETECTED
C        P*      = MINIMUM P{CS} OVER THE INDIFFERENCE ZONE
C        C       = RUN-TIME CONSTANT IN [0,1] (PAULSON RECOMMENDS 0.85)
C        NUMRNS  = NUMBER OF INDEPENDENT MONTE CARLO REPLICATIONS
C        ISEED   = INTEGER RANDOM NUMBER SEED IN [1, 2^{31}-1]
C
C  OUTPUT: P{CS}, E[OBSERVATIONS], E[STAGES] (AND STANDARD ERRORS)
C
      DOUBLE PRECISION DELTA,PSTAR,C,SIGMA
      DOUBLE PRECISION SHFT(10),X(10,1000),XP(10,1000),Y(10)
      DOUBLE PRECISION CON1,CON2,CON3,DUM1,DUM2
      DOUBLE PRECISION PCS,SE
      INTEGER NUMOBS(10000),NUMSTG(10000),KALIVE(10),ETIME(10),MSTAGE
      INTEGER T,CONFIG,NSTG,NOBS,NALIVE,MAXSTG
      CHARACTER CHOICE2
      EXTERNAL RNORML,UNIF,STATS
C     KALIVE(I) = 1 IF TREATMENT I IS STILL IN CONTENTION, 0 O'WISE
C     ETIME(I) = TIME AT WHICH TREATMENT I IS ELIMINATED
C
C  READ IN PARAMETERS FOR THE RUN:
C
100   WRITE(6,*) 'ENTER T, SIGMA (2 <= T <= 10, SIGMA > 0)'
      READ(5,*) T,SIGMA
      WRITE(6,*) 'ENTER CONFIGURATION (1=LF, 2=EM, 3=ES)'
      READ(5,*) CONFIG
      WRITE(6,*) 'ENTER DELTA*, P* (DELTA* > 0, 1/T < P* < 1)'
      READ(5,*) DELTA,PSTAR
      WRITE(6,*) 'ENTER C (0 < C < 1, PAULSON RECOMMENDS 0.85)'
      READ(5,*) C
      WRITE(6,*) 'ENTER NUMBER OF RUNS (2 <= NUMRNS <= 10000)'
      READ(5,*) NUMRNS
      WRITE(6,*) 'ENTER INTEGER SEED (1 <= SEED <= 2^31 - 1)'
      READ(5,*) ISEED
C
      WRITE(6,*)' PAULSON NORMAL MEANS PROCEDURE MONTE CARLO RESULTS'
      WRITE(6,*)' (COMMON KNOWN VARIANCE)...'
      WRITE(6,*)' '
      WRITE(6,210)T,SIGMA,DELTA,PSTAR
210   FORMAT(1X,'T, SIGMA, DELTA*, P* = ',I3,3F7.2)
      WRITE(6,*)' '
      IF(CONFIG.EQ.1)WRITE(6,*)' LEAST-FAVORABLE CONFIGURATION...'
      IF(CONFIG.EQ.2)WRITE(6,*)' EQUAL-MEANS CONFIGURATION...'
      IF(CONFIG.EQ.3)WRITE(6,*)' EQUALLY-SPACED CONFIGURATION...'
      WRITE(6,*)' '
      WRITE(6,220)C
220   FORMAT(1X,'C = ',F8.5)
      WRITE(6,*)' '
      WRITE(6,*)' NUMRNS, ISEED = ',NUMRNS,ISEED
      WRITE(6,*)' '
C
      IF(CONFIG.EQ.2)GOTO 250
      IF(CONFIG.EQ.3)GOTO 260
C
C     SET-UP FOR LF CONFIG...
      DO 245 I = 1,T
245   SHFT(I) = 0.0D0
      SHFT(T) = DELTA
```

FORTRAN PROGRAMS

```
      GOTO 270
C
C     SET-UP FOR EM CONFIG...
250   DO 255 I = 1,T
255   SHFT(I) = 0.0D0
      GOTO 270
C
C     SET-UP FOR ES CONFIG...
260   DO 265 I = 1,T
265   SHFT(I) = DELTA*(I-1)
C
C  INITIALIZE SOME CONSTANTS THAT WILL BE USED FOR ALL NUMRNS REPS...
270   CON1 = DFLOAT(T-1)/(1.0D0-PSTAR)
      CON2 = C*(1.0D0-C)*DELTA*DELTA/(SIGMA*SIGMA)
      CON3 = C * DELTA/(SIGMA*SIGMA)
      MAXSTG = 1000
C
C     SET NUMBER OF CORRECT SELECTIONS TO ZERO...
      KPCS = 0
C
C  MAIN LOOP...
      DO 500 KKK = 1,NUMRNS
C
C  INITIALIZE SOME VARIABLES FOR EACH RUN...
      DO 300 I = 1,6
      KALIVE(I) = 1
      ETIME(I) = MAXSTG
      DO 300 J = 1,MAXSTG
      X(I,J) = 0.0D0
300   XP(I,J) = 0.0D0
C
C  INITIALIZE # STAGES, # OBSERVATIONS, # REMAINING TREATMENTS...
      NSTG = 0
      NOBS = 0
      NALIVE = T
C
      DO 400 MSTAGE = 1,MAXSTG
C
      NSTG = NSTG + 1
      NOBS = NOBS + NALIVE
C
      DO 310 I = 1,T
      IF(KALIVE(I).EQ.0)GOTO 310
      Y(I) = RNORML(ISEED) + SHFT(I)
      IF(MSTAGE.EQ.1)X(I,1) = Y(I)
      IF(MSTAGE.GE.2)X(I,MSTAGE) = X(I,MSTAGE-1) + Y(I)
310   CONTINUE
C
      IF(MSTAGE.EQ.1)GOTO 330
      DO 325 I = 1,T
      KDUM = 0
      DUM1 = -999999.999
      DO 320 J = 1,T
      IF((I.EQ.J).OR.(KALIVE(J).EQ.0))GOTO 320
      IF(X(J,MSTAGE-1).LE.DUM1)GOTO 320
         DUM1 = X(J,MSTAGE-1)
         KDUM = J
320   CONTINUE
      XP(I,MSTAGE) = XP(I,MSTAGE-1)+Y(KDUM)
325   CONTINUE
C
330   DO 350 I = 1,T
      IF(KALIVE(I).EQ.0)GOTO 350
      DUM1 = 0.0D0
      DO 340 J = 1,T
      JMIN = MIN0(MSTAGE,ETIME(J))
      DUM2 = CON3 * (X(J,JMIN)-X(I,MSTAGE)+XP(I,MSTAGE)-XP(I,JMIN))
```

```
340   DUM1 = DUM1 + DEXP(DUM2)
C
      DUM2 = CON1*DEXP(-DBLE(MSTAGE)*CON2)
      IF(DUM1.LE.DUM2)GOTO 350
      KALIVE(I) = 0
      NALIVE = NALIVE - 1
      ETIME(I) = MSTAGE
350   CONTINUE
C
      IF(NALIVE.EQ.1)GOTO 420
C
400   CONTINUE
C
420   IF(KALIVE(T).EQ.1)KPCS = KPCS+1
      NUMOBS(KKK) = NOBS
      NUMSTG(KKK) = NSTG
C
500   CONTINUE
C
      PCS = DFLOAT(KPCS)/NUMRNS
      SE = (PCS*(1.-PCS)/NUMRNS)**0.5
      WRITE(6,510)PCS,SE
510   FORMAT(1X,'ESTIMATED PCS AND STANDARD ERROR = ',2F12.6/)
      WRITE(6,*)'STATISTICS FOR OBSERVATIONS...'
      CALL STATS(NUMOBS,NUMRNS)
      WRITE(6,*)' '
      WRITE(6,*)'STATISTICS FOR STAGES...'
      CALL STATS(NUMSTG,NUMRNS)
      WRITE(6,*)' '
C
      WRITE(6,*) 'DO YOU WANT TO CONTINUE? PLEASE ANSWER Y/N'
      READ(5,600) CHOICE2
600   FORMAT(A1)
      IF (CHOICE2 .EQ. 'Y' .OR. CHOICE2 .EQ. 'y') GOTO 100
      STOP
      END
```

```fortran
      PROGRAM RINOTT
C
C     PROGRAM TO CALCULATE G(T,PSTAR,NU) OF RINOTT PROCEDURE
C
      DOUBLE PRECISION ZCDF,CHIPDF
      DOUBLE PRECISION PSTAR,H
      DOUBLE PRECISION X(32),W(32),WEX(32),DUMMY,LNGAM(50),
     *LOWERH,UPPERH,ANS,TMP
      INTEGER I,T,NU,LOOPH
      EXTERNAL ZCDF
C
      INCLUDE 'GLQUAD'
C
C     CALCULATE LNGAM(K) = LN(GAMMA(K/2))
C
      LNGAM(1) = 0.5723649429
      LNGAM(2) = 0.0
      DO 100 I = 2,25
      LNGAM(2*I - 1) = DLOG(DBLE(I)-1.50D0) + LNGAM(2*I - 3)
      LNGAM(2*I) = DLOG(DBLE(I) - 1.0D0) + LNGAM(2*I - 2)
100   CONTINUE
C
C     READ PARAMETERS FOR WHICH TO DETERMINE G(T,PSTAR,\NU)
C
110   WRITE(6,*) 'ENTER T,P-STAR AND NU'
      READ(5,*) T,PSTAR,NU
      IF (NU .LT. 5) THEN
        WRITE(6,520)
        STOP
      ELSE
           IF (T .LT. 2) THEN
              WRITE(6,530)
              STOP
           ENDIF
      ENDIF
      DUMMY = 1.0D0
C
C
      H = 4.0D0
      LOWERH = 0.0D0
      UPPERH = 20.00
C
      DO 120 LOOPH = 1,50
C
      ANS = 0.0D0
      DO 130 J = 1,32
      TMP = 0.0D0
      DO 140 I = 1,32
      TMP = TMP + WEX(I)
     *  * ZCDF(H/DSQRT(DBLE(NU)*(1.0D0/X(I) + 1.0D0/X(J))/DUMMY))
     *  * CHIPDF(NU,DUMMY*X(I),LNGAM) * DUMMY
140   CONTINUE
      TMP = TMP ** (T-1)
130   ANS = ANS + WEX(J) * TMP * CHIPDF(NU,DUMMY*X(J),LNGAM)*DUMMY
C
      IF(DABS(ANS-PSTAR) .GT. 0.000001D0) GOTO 150
C
C     IF NOT, THEN DONE SO WRITE G(T,PSTAR,\NU)
C
      WRITE(6,500) PSTAR,T,NU,H
      WRITE(6,*) 'DO YOU WANT TO CONTINUE? PLEASE ANSWER Y/N'
      READ(5,510) CHOICE2
      IF (CHOICE2 .EQ. 'Y' .OR. CHOICE2 .EQ. 'y') GOTO 110
      STOP
150   IF(ANS .GT. PSTAR) GOTO 160
      LOWERH = H
      H = (LOWERH + UPPERH)/2.0D0
```

```
          GOTO 120
160   UPPERH = H
      H = (LOWERH + UPPERH)/2.0D0
120   CONTINUE
C
500   FORMAT(1X,' G^{',F5.3,'}_{',I3,',',I3,'} = ',F6.3)
510   FORMAT(A1)
520   FORMAT(1X,'THIS PROGRAM WILL NOT COMPUTE   G    FOR LESS',/,
     *' THAN 5 DEGREES OF FREEDOM')
530   FORMAT(1X,' T MUST BE AN INTEGER GREATER THAN OR EQUAL TO 2')
C
      END
C
C
      DOUBLE PRECISION FUNCTION CHIPDF(N,C,LNGAM)
C
C     THIS GIVES THE PDF OF THE CHI^2 DISTN WITH N DF FOR N .LE. 50.
C     LNGAM(N) IS LN(GAMMA(N/2))
C
      DOUBLE PRECISION C,TMP,LNGAM(50),FLN2
      INTEGER N
      FLN2 = DBLE(N)/2.
      TMP = -FLN2*DLOG(2.0D0) - LNGAM(N) +
     *    (FLN2 - 1.0D0)*DLOG(C) - C/2.0D0
      CHIPDF = DEXP(TMP)
      RETURN
      END
```

FORTRAN PROGRAMS

```
      PROGRAM EVALNG
C
C     USES MONTE CARLO SIMULATION TO ESTIMATE THE C.D.F. AND THE
C     EXPECTED NUMBER OF TREATMENTS SELECTED BY PROCEDURE N_G WHEN THE
C     MEANS ARE IN THE SLIPPAGE (SC) AND EQUALLY-SPACED (ES) CONFIGS.
C
C  INPUT: PROGRAM INTERACTIVELY PROMPTS FOR
C        T         = NUMBER OF TREATMENTS    (T .LEQ. 50-RESET BY MAXT)
C        DELTA     = DIFFERENCE BETWEEN LARGEST AND REMAINING MEANS
C        P-STAR    = MINIMUM PROBABILITY OF CORRECT SELECTION
C        N         = NUMBER OF OBSERVATIONS PER TREATMENT
C        SIGMA     = STANDARD DEVIATION OF THE MEASURE ERROR
C        REPS      = NUMBER OF INDEPENDENT SIMULATION REPLICATIONS
C        ISEED     = INTEGER RANDOM NUMBER SEED
C
C  OUTPUT: EST'D P{S.GEQ.K | SC}, P{S.GEQ.K | ES} (1 .LEQ. K .LEQ. T),
C          E{S|SC} AND E{S|ES}, WHERE
C                 SC = (0,...,0,DELTA)
C          AND    ES = (0,DELTA,...,(T-1)*DELTA)
C
      INTEGER MAXT
C     MAXT DEFINES THE MAXIMUM NUMBER OF TREATMENTS
      PARAMETER (MAXT=50)
      DOUBLE PRECISION MULTZ
      DOUBLE PRECISION DELTA,PSTAR,SIGMA
      DOUBLE PRECISION CON1,CON2,H,CDFSC(MAXT),CDFES(MAXT),
     *      YBARSC(MAXT),YBARES(MAXT),YBMXSC,YBMXES,ESSC,ESES
      INTEGER ISEED,N,REPS,T
      INTEGER I,I400,SIZEES,SIZESC,TM1
      CHARACTER CHOICE2
      EXTERNAL MULTZ,UNIF,RNORML
C
C     READ PARAMETERS TO BE STUDIED
C
      WRITE(6,*) 'ENTER T AND SIGMA'
      READ(5,*) T,SIGMA
      WRITE(6,*)'ENTER NUMBER OF SIMULATION REPS AND INTEGER SEED'
      READ(5,*)REPS,ISEED
      WRITE(6,*) ' '
40    WRITE(6,*) 'ENTER N, DELTA AND P-STAR'
      READ(5,*) N,DELTA,PSTAR
C
      TM1 = T-1
      H = MULTZ(PSTAR,TM1)
C
      CON1 = SIGMA/N**0.5
      CON2 = H*SIGMA*DSQRT(2.0D0/N)
C
      DO 50 I = 1,T
      CDFSC(I) = 0.0D0
50    CDFES(I) = 0.0D0
      ESSC = 0.0D0
      ESES = 0.0D0
C
C     MAIN LOOP...
C
      DO 400 I400 = 1,REPS
C
C     GENERATE T SAMPLE MEANS (ASSUME TREATMENT T IS BEST)
C
      DO 110 I = 1,T
      YBARSC(I) = CON1 * RNORML(ISEED)
110   YBARES(I) = YBARSC(I) + (I-1)*DELTA
      YBARSC(T) = YBARSC(T) + DELTA
C
C     FIND MAXIMUM YBAR'S FOR SC AND ES CONFIG'S
      YBMXSC = YBARSC(1)
```

```
              YBMXES = YBARES(1)
              DO 120 I = 2,T
              IF(YBARSC(I).GT.YBMXSC)YBMXSC = YBARSC(I)
 120          IF(YBARES(I).GT.YBMXES)YBMXES = YBARES(I)
       C
       C      FIND SIZE OF SELECTED SUBSET
              SIZESC = 0
              SIZEES = 0
              DO 130 I = 1,T
              IF(YBARSC(I).GE.YBMXSC-CON2)SIZESC = SIZESC + 1
 130          IF(YBARES(I).GE.YBMXES-CON2)SIZEES = SIZEES + 1
       C
       C      UPDATE ESTIMATED C.D.F.'S
       C
              CDFSC(SIZESC) = CDFSC(SIZESC) + 1.0D0
              CDFES(SIZEES) = CDFES(SIZEES) + 1.0D0
 400          CONTINUE
       C
       C      GET P.M.F.'S AND EXPECTED SUBSET SIZES
       C
              WRITE(6,*)' '
              DO 410 I = 1,T
              CDFSC(I) = CDFSC(I)/REPS
              CDFES(I) = CDFES(I)/REPS
              ESSC = ESSC + I*CDFSC(I)
 410          ESES = ESES + I*CDFES(I)
       C
       C      GET COMPLEMENTS OF C.D.F.'S
       C
              DO 420 I = 2,T
              CDFSC(T+1-I) = CDFSC(T+1-I) + CDFSC(T+2-I)
 420          CDFES(T+1-I) = CDFES(T+1-I) + CDFES(T+2-I)
       C
              WRITE(6,430)
 430          FORMAT(1X,'   K    P{S.GEQ.K | SC}   P{S.GEQ.K | ES} ')
              DO 500 I = 1,T
              WRITE(6,440)I,CDFSC(I),CDFES(I)
 440          FORMAT(1X,I3,4X,F8.4,9X,F8.4)
 500          CONTINUE
              WRITE(6,510)ESSC,ESES
 510          FORMAT(1X,' E{S|SC} = ',F8.4,'   E{S|ES} = ',F8.4)
       C
              WRITE(6,*)' '
              WRITE(6,*) 'DO YOU WANT TO CONTINUE? PLEASE ANSWER Y/N'
              READ(5,600) CHOICE2
 600          FORMAT(A1)
              IF (CHOICE2 .EQ. 'Y' .OR. CHOICE2 .EQ. 'y') GOTO 40
              STOP
              END
```

FORTRAN PROGRAMS

```fortran
      PROGRAM USEGSA
C
C     INTERACTIVE CALCULATION OF N, H OR DELTA-STAR FOR THE
C     SINGLE-STAGE PROCEDURE N_{GSa} TO ACHIEVE THE PROBABILITY
C     REQUIREMENT OF SECTION 3.4.3
C
C  INPUT (PROGRAM INTERACTIVELY PROMPTS FOR):
C        T          = NUMBER OF TREATMENTS
C        Q          = SIZE OF THE SUBSET
C        SIGMA      = STANDARD DEVIATION OF THE MEASURE ERROR
C        P-STAR     = MINIMUM P{CS} OVER THE INDIFFERENCE-ZONE
C  THEN PROMPTS FOR 1 OF 3 CHOICES FROM THE MENU
C
C             CHOICE          GIVEN              FIND
C                1             N, H             DELTA-STAR
C                2          DELTA-STAR, H          N
C                3          DELTA-STAR, N          H
C  WHERE
C        DELTA-STAR = DIFFERENCE TO BE DETECTED
C        N          = SMALLEST COMMON OF OBSERVATIONS PER TREATMENT TO
C                     ATTAIN THE PROBABILITY REQUIREMENT (3.4.7)
C        H          = LENGTH OF THE YARDSTICK USED BY N_{GSa}
C
C  OUTPUT: DELTA-STAR, N OR H
C
      DOUBLE PRECISION ZCDF,PCSNGS,FACTOR
      DOUBLE PRECISION PSTAR,DSTAR,SIGMA
      DOUBLE PRECISION LOWERC,UPPERC,C,LOWERH,UPPERH,H,ANS,
     *TMP,ANSLO,ANSHI
      INTEGER T,Q,N,CHOICE1,NTMP,I190,I260
      CHARACTER CHOICE2
      EXTERNAL ZCDF,PCSNGS,FACTOR
C
C     READ IN PARAMETERS FIXED FOR THE RUN: T,Q,SIGMA AND P-STAR
C
100   WRITE(6,*) 'ENTER T,Q,SIGMA AND P-STAR'
      READ(5,*) T,Q,SIGMA,PSTAR
      IF(Q .GE. T) THEN
      WRITE(6,*) 'Q MUST BE STRICTLY LESS THAN T, TRY AGAIN '
      GOTO 100
      ENDIF
C
110   WRITE(6,510)
120   WRITE(6,*) 'ENTER CHOICE (1, 2, OR 3) '
      READ(5,*) CHOICE1
      IF(CHOICE1 .EQ. 1) GOTO 130
      IF(CHOICE1 .EQ. 2) GOTO 140
      IF(CHOICE1 .EQ. 3) GOTO 150
      WRITE(6,*) 'ENTRY ERROR, TRY AGAIN '
      GOTO 120
130   WRITE(6,*) 'ENTER N AND H'
      READ(5,*) N,H
      GOTO 160
140   WRITE(6,*) 'ENTER DELTA-STAR AND H'
      READ(5,*) DSTAR,H
160   TMP=1.0D0
      NTMP=1
      C = 4.0D0
      LOWERC = 0.0D0
C
C     DETERMINE UPPER AND LOWER VALUES AT WHICH TO START BISECTION
C
170   ANS = PCSNGS(T,Q,NTMP,C,TMP,H)
C
      IF(ANS .GT. PSTAR) GOTO 180
      C = C + 1
      GOTO 170
```

```
      180  UPPERC = C
           C = (UPPERC + LOWERC)/2.0D0
      C
      C    PERFORM BINARY SEARCH WITH A MAXIMUM OF 50 ITERATIONS
      C
           DO 190 I190 = 1,50
      C
      C    EVALUATE THE P{CS} AT A TRIAL DELTA-STAR
      C
           ANS = PCSNGS(T,Q,NTMP,C,TMP,H)
      C
           IF(DABS(ANS-PSTAR).LE.0.000001) GOTO 200
      210  IF(ANS.GT.PSTAR) GOTO 220
           LOWERC = C
           C = (LOWERC + UPPERC)/2.
           GOTO 190
      220  UPPERC = C
           C = (LOWERC + UPPERC)/2.
      190  CONTINUE
      C
      200  IF(CHOICE1 .EQ. 1) GOTO 230
           N = DINT((SIGMA * C / DSTAR)**2) + 1
           WRITE(6,520) T, Q, DSTAR, PSTAR, SIGMA, H, N
      C
           WRITE(6,*) 'DO YOU WANT TO CONTINUE? PLEASE ANSWER Y/N'
           READ(5,500) CHOICE2
           IF (CHOICE2 .EQ. 'Y' .OR. CHOICE2 .EQ. 'y') GOTO 110
           STOP
      C
      230  DSTAR = C * SIGMA / DSQRT(DFLOAT(N))
           WRITE(6,530) T, Q, PSTAR, SIGMA, H, N, DSTAR
      C
           WRITE(6,*) 'DO YOU WANT TO CONTINUE? PLEASE ANSWER Y/N'
           READ(5,500) CHOICE2
           IF (CHOICE2 .EQ. 'Y' .OR. CHOICE2 .EQ. 'y') GOTO 110
           STOP
      C
      150  WRITE(6,*) 'ENTER DELTA-STAR AND N'
           READ(5,*) DSTAR,N
           H = 0.0D0
           ANSLO =  PCSNGS(T,Q,N,DSTAR,SIGMA,H)
           H = 1000.0D0
           ANSHI =  PCSNGS(T,Q,N,DSTAR,SIGMA,H)
           IF ((PSTAR .LE. ANSLO) .OR. (PSTAR .GE. ANSHI)) GOTO 290
           H = 0.0D0
           LOWERH = 0.0D0
      240  ANS = PCSNGS(T,Q,N,DSTAR,SIGMA,H)
      C
           IF(ANS .GT. PSTAR) GOTO 250
           H = H + 1
           GOTO 240
      250  UPPERH = H
           H = (UPPERH + LOWERH)/2.0D0
      C
      C    PERFORM BINARY SEARCH WITH A MAXIMUM OF 50 ITERATIONS
      C
           DO 260 I260 = 1,50
      C
      C    EVALUATE THE P{CS} AT A TRIAL H
      C
           ANS = PCSNGS(T,Q,N,DSTAR,SIGMA,H)
      C
           IF(DABS(ANS-PSTAR).LE.0.000001) GOTO 270
           IF(ANS.GT.PSTAR) GOTO 280
           LOWERH = H
           H = (LOWERH + UPPERH)/2.
           GOTO 260
```

FORTRAN PROGRAMS

```
280     UPPERH = H
        H = (LOWERH + UPPERH)/2.
260     CONTINUE
270     WRITE(6,540) T,Q,SIGMA,N,DSTAR,H
C
        WRITE(6,*) 'DO YOU WANT TO CONTINUE? PLEASE ANSWER Y/N'
        READ(5,500) CHOICE2
        IF (CHOICE2 .EQ. 'Y' .OR. CHOICE2 .EQ. 'y') GOTO 110
        STOP
290     WRITE(5,550) ANSLO, ANSHI,PSTAR
        GOTO 150
C
500     FORMAT(A1)
510     FORMAT(1X,'         CHOICE          GIVEN                  FIND'/
       1'           1                N, H                DELTA-STAR'/,
       2'           2             DELTA-STAR, H           N'/,
       3'           3             DELTA-STAR, N           H')
520     FORMAT(1X,' FOR (T, Q, DELTA*, P*, SIGMA, H) = ( ', I3, ',',
       *I3, ',',F5.3, ', ',F5.3, ', ',F5.3, ', ',F5.3,')' /
       *'   THE REQUIRED SAMPLE SIZE FOR PROCEDURE N_{GSa} IS ', I6/)
530     FORMAT(1X,' FOR (T, Q, P*, SIGMA, H, N) = ( ', I3,',',I3,
       *',',F5.3, ', ',F5.3, ', ',F5.3,', ', I4,')' / ' THE '
       *'MINIMUM DELTA STAR WHICH CAN BE DETECTED WHILE' / ' ACHIEVING'
       *' (3.4.6) IS ', F7.3/)
540     FORMAT(1X,' FOR (T, Q, SIGMA, N, DELTA*) = ( ', I4,', ',
       * I4,', ',F5.3,', ', I5,', ',F7.3,')' / ' THE CRITICAL VALUE H
       * IS ', F6.3/)
550     FORMAT(1X,'ONLY P-STAR BETWEEN 'F7.3,' AND ',F7.3,
       1' HAVE  H  VALUES IN [0,\INFTY)'/,1X,'THAT ACHIEVE P-STAR---',
       2'YOUR P-STAR = ',F7.3,'. MODIFY DSTAR AND/OR N'/)
        END
```

```
      PROGRAM BPMC
C
C     INTERACTIVE MONTE CARLO IMPLEMENTATION OF PAULSON'S SEQUENTIAL
C     BERNOULLI PROCEDURE B_P (SEE SECTION 7.4.2)
C
C  INPUT (PROGRAM INTERACTIVELY PROMPTS FOR):
C       T       = NUMBER OF TREATMENTS
C       CONFIG  = 1, 2 (LF-, EP-CONFIGURATIONS)
C       P_[T]   = LARGEST P-VALUE
C       P*      = MINIMUM P{CS} OVER THE INDIFFERENCE-ZONE
C       THETA*  = ODDS RATIO TO BE DETECTED
C       NUMRNS  = NUMBER OF INDEPENDENT MONTE CARLO REPLICATIONS
C       ISEED   = INTEGER RANDOM NUMBER SEED IN [1, 2^{31}-1]
C
C  OUTPUT: P{CS}, E[OBSERVATIONS], E[STAGES] (AND STANDARD ERRORS)
C
      DOUBLE PRECISION PSTAR,THETA,PCS,SE,CON,P(20),PT
      DOUBLE PRECISION TPOWER(200),GIM
      INTEGER NUMOBS(10000),NUMSTG(10000)
      INTEGER KALIVE(10),ETIME(10),Y(10,1000),BERN,T,TM1,CONFIG
      INTEGER KPCS,NSTG,NOBS,NALIVE,MSTAGE,MAXSTG,KDUM,KDIFF,I,J,KKK
      EXTERNAL UNIF,BERN,STATS
C     KALIVE(I) = 1 IF TREATMENT I IS STILL IN CONTENTION, 0 O'WISE
C     ETIME(I)  = TIME AT WHICH TREATMENT I IS ELIMINATED
C
C  READ IN PARAMETERS FOR THE RUN:
C
100   WRITE(6,*) 'ENTER T (2 <= T <= 10)'
      READ(5,*) T
      WRITE(6,*) 'ENTER CONFIGURATION (1=LF, 2=EP)'
      READ(5,*) CONFIG
      WRITE(6,*) 'ENTER P_[T], THE LARGEST P-VALUE (0 < P_[T] < 1)'
      READ(5,*) PT
      WRITE(6,*) 'ENTER THETA*, P* (THETA* > 0, 1/T < P* < 1)'
      READ(5,*) THETA,PSTAR
      WRITE(6,*) 'ENTER NUMBER OF RUNS (2 <= NUMRNS <= 10000)'
      READ(5,*) NUMRNS
      WRITE(6,*) 'ENTER INTEGER SEED (1 <= SEED <= 2^31 - 1)'
      READ(5,*) ISEED
C
C  FIX SOME CONSTANTS...
      TM1 = T - 1
      CON = DFLOAT(TM1)/(1.-PSTAR)
      MAXSTG = 1000
C
      WRITE(6,*)' PAULSON BERNOULLI PROCEDURE MONTE CARLO RESULTS'
      WRITE(6,*)' '
      WRITE(6,200)T,THETA,PSTAR
200   FORMAT(1X,'T, THETA*, P* = ',I3,2F7.2)
      WRITE(6,*)' '
      IF(CONFIG.EQ.1)WRITE(6,*)' LEAST-FAVORABLE CONFIGURATION...'
      IF(CONFIG.EQ.2)WRITE(6,*)' EQUAL-PROBABILITY CONFIGURATION...'
      WRITE(6,*)' '
C
C  CALCULATE COMPONENTS OF P-VECTOR...
      DO 210 I = 1,TM1
      IF(CONFIG.EQ.1)P(I) = 1./(1. + THETA*(1.-PT)/PT)
210   IF(CONFIG.EQ.2)P(I) = PT
      P(T) = PT
C
      DO 230 I = 1,T
      WRITE(6,220)I,P(I)
220   FORMAT(1X,'P(',I2,') = ',F8.6)
230   CONTINUE
      WRITE(6,*)' '
      WRITE(6,*)' NUMRNS, ISEED = ',NUMRNS,ISEED
      WRITE(6,*)' '
```

FORTRAN PROGRAMS

```
C
C   CALCULATE POWERS OF THETA (TO SAVE ON COMPUTATION TIME LATER)...
      TPOWER(1) = THETA
      DO 240 I = 2,200
240   TPOWER(I) = TPOWER(I-1)*THETA
C
C   SET NUMBER OF CORRECT SELECTIONS TO ZERO...
      KPCS = 0
C
C   MAIN LOOP...
      DO 500 KKK = 1,NUMRNS
C
C   INITIALIZE SOME VARIABLES FOR EACH RUN...
      DO 250 I = 1,T
      KALIVE(I) = 1
      ETIME(I) = MAXSTG
      DO 250 J = 1,MAXSTG
250   Y(I,J) = 0
C
C   INITIALIZE # STAGES, # OBSERVATIONS, # REMAINING TREATMENTS...
      NSTG = 0
      NOBS = 0
      NALIVE = T
C
      DO 400 MSTAGE = 1,MAXSTG
C
      NSTG = NSTG + 1
      NOBS = NOBS + NALIVE
C
      DO 310 I = 1,T
      IF(KALIVE(I).EQ.0)GOTO 310
      IF(MSTAGE.EQ.1)Y(I,1) = BERN(P(I),ISEED)
      IF(MSTAGE.GE.2)Y(I,MSTAGE) = Y(I,MSTAGE-1) + BERN(P(I),ISEED)
310   CONTINUE
C
C   LOOP TO SEE IF ANYTHING ELSE GETS ELIMINATED...
      DO 350 I = 1,T
      IF(KALIVE(I).EQ.0)GOTO 350
         GIM = 0.0D0
         DO 340 J = 1,T
            IF(I.EQ.J)GOTO 340
            KDUM = MIN(ETIME(J),MSTAGE)
            KDIFF = Y(J,KDUM) - Y(I,KDUM)
            IF(KDIFF.EQ.0)GIM = GIM + 1.0d0
            IF(KDIFF.GT.0)GIM = GIM + TPOWER(KDIFF)
            IF(KDIFF.LT.0)GIM = GIM + 1./TPOWER(-KDIFF)
340      CONTINUE
      IF(GIM.LE.CON)GOTO 350
C   OTHERWISE, ELIMINATE TREATMENT I...
      KALIVE(I) = 0
      NALIVE = NALIVE - 1
      ETIME(I) = MSTAGE
350   CONTINUE
C
      IF(NALIVE.EQ.1)GOTO 420
C
400   CONTINUE
C
420   IF(KALIVE(T).EQ.1)KPCS = KPCS+1
      NUMOBS(KKK) = NOBS
      NUMSTG(KKK) = NSTG
C
500   CONTINUE
C
      PCS = DFLOAT(KPCS)/NUMRNS
      SE = (PCS*(1.-PCS)/NUMRNS)**0.5
      WRITE(6,510)PCS,SE
```

```
510     FORMAT(1X,'ESTIMATED PCS AND STANDARD ERROR = ',2F12.6/)
        WRITE(6,*)'STATISTICS FOR OBSERVATIONS...'
        CALL STATS(NUMOBS,NUMRNS)
        WRITE(6,*)' '
        WRITE(6,*)'STATISTICS FOR STAGES...'
        CALL STATS(NUMSTG,NUMRNS)
        WRITE(6,*)' '
C
        WRITE(6,*) 'DO YOU WANT TO CONTINUE? PLEASE ANSWER Y/N'
        READ(5,600) CHOICE2
600     FORMAT(A1)
        IF (CHOICE2 .EQ. 'Y' .OR. CHOICE2 .EQ. 'y') GOTO 100
C
        STOP
        END
C
C
        INTEGER FUNCTION BERN(A,IX)
C
C       INPUT:  IX = INTEGER > 0 BUT < 2147483647
C       OUTPUT: IX = NEW SEED, BERN = BERN(A) DEVIATE
C
        DOUBLE PRECISION A
        BERN = 0
        IF(UNIF(IX).LT.A)BERN = 1
        RETURN
        END
```

References

Abramowitz, M., and Stegun, I. A. (1972). *Handbook of Mathematical Functions*. New York: Dover Publications.

Alam, K. (1970). A two-sample procedure for selecting the population with the largest mean from k normal populations. *Ann. Inst. Stat. Math.* **22**, 127–136.

Alam, K. (1971). On selecting the most probable category. *Technometrics* **13**, 843–850.

Alam, K., and Thompson, J. R. (1972). On selecting the least probable multinomial event. *Ann. Math. Stat.* **43**, 1981–1990.

Anderson, V., and McLean, R. (1974). *Design of Experiments: A Realistic Approach*. New York: Marcel Dekker.

Anscombe, F. J. (1963). Sequential medical trials. *J. Am. Stat. Assoc.* **58**, 365–383.

Armitage, P. (1960, 1975). *Sequential Medical Trials*. New York: John Wiley & Sons.

Aubuchon, J. C., Gupta, S. S., and Hsu, J. C. (1986). PROC RSMCB: A procedure for ranking, selection, and multiple comparisons with the best. *Proc. SAS Users Group Int. Conf.* **11**, 761–765.

Bahadur, R. R. (1950). On a problem in the theory of k populations. *Ann. Math. Stat.* **21**, 362–375.

Barlow, R. E., and Gupta, S. S. (1969). Selection procedures for restricted families of probability distributions. *Ann. Math. Stat.* **40**, 905–917.

Bartlett, N. S., and Govindarajulu, Z. (1968). Some distribution-free statistics and their application to the selection problem. *Ann. Inst. Stat. Math.* **20**, 79–97.

Bawa, V. S. (1972). Asymptotic efficiency of one R-factor experiment relative to R one-factor experiments for selecting the best normal population. *J. Am. Stat. Assoc.* **67**, 660–661.

Bechhofer, R. E. (1954). A single-sample multiple decision procedure for ranking means of normal populations with known variances. *Ann. Math. Stat.* **25**, 16–39.

Bechhofer, R. E. (1969). Optimal allocation of observations when comparing several treatments with a control. In: *Multivariate Analysis, II* (P. R. Krishhnaiah, ed.), 463–473. New York: Academic Press.

Bechhofer, R. E., and Chen, P. (1991). A note on a curtailed sequential procedure for subset selection of multinomial cells. *Am. J. Math. Mgmt. Sci.* **11**, 309–324.

Bechhofer, R. E., and Dunnett, C. W. (1982). Multiple comparisons for orthogonal contrasts: examples and tables. *Technometrics* **24**, 213–222.

Bechhofer, R. E., and Dunnett, C. W. (1986). Two-stage selection of the best factor-level combination in multi-factor experiments: common unknown variance. In: *Statistical Design: Theory and Practice. A Conference in Honor of Walter T. Federer* (C. E. McCullogh, S. J. Schwager, G. Casella and S. R. Searle, eds.), 3–16. Biometrics Unit, Cornell University, Ithaca, NY.

Bechhofer, R. E., and Dunnett, C. W. (1987). Subset selection for normal means in multi-factor experiments. *Commun. Stat.—Theory and Methods* **A16**, 2277–2286.

Bechhofer, R. E., and Dunnett, C. W. (1988). Percentage points of multivariate Student t distributions. In: *Selected Tables in Mathematical Statistics*, Vol. 11. Providence, RI: American Mathematical Society.

Bechhofer, R. E., Dunnett, C. W., Goldsman, D. M., and Hartmann, M. (1990). A comparison of the performances of procedures for selecting the normal population having the largest mean when the populations have a common unknown variance. *Commun. Stat.—Simul. and Comput.* **B19**, 971–1006.

Bechhofer, R. E., Dunnett, C. W., and Sobel, M. (1954). A two-sample multiple-decision procedure for ranking means of normal populations with a common unknown variance. *Biometrika* **41**, 170–176.

Bechhofer, R. E., Elmaghraby, S., and Morse, N. (1959). A single-sample multiple decision procedure for selecting the multinomial event which has the highest probability. *Ann. Math. Stat.* **30**, 102–119.

Bechhofer, R. E., and Frisardi, T. (1983). A Monte Carlo study of the performance of a closed adaptive sequential procedure for selecting the best Bernoulli population. *J. Stat. Comput. Simul.* **18**, 179–213.

Bechhofer, R. E., and Goldsman, D. M. (1985). Truncation of the Bechhofer–Kiefer–Sobel sequential procedure for selecting the multinomial event which has the largest probability. *Commun. Stat.—Simul. and Comput.* **B14**, 283–315.

Bechhofer, R. E., and Goldsman, D. M. (1986). Truncation of the Bechhofer–Kiefer–Sobel sequential procedure for selecting the multinomial event which has the largest probability (II): extended tables and an improved procedure. *Commun. Stat.—Simul. and Comput.* **B15**, 829–851.

Bechhofer, R. E., and Goldsman, D. M. (1987). Truncation of the Bechhofer–Kiefer–Sobel sequential procedure for selecting the normal population which has the largest mean. *Commun. Stat.—Simul. and Comput.* **B16**, 1067–1092.

Bechhofer, R. E., and Goldsman, D. M. (1988a). Sequential selection procedures for multi-factor experiments involving Koopman–Darmois populations with additivity. In: *Statistical Decision Theory and Related Topics, IV* (S. S. Gupta and J. O. Berger, eds.), Vol. 2, 3–21. New York: Springer-Verlag.

Bechhofer, R. E., and Goldsman, D. M. (1988b). Truncation of the Bechhofer–Kiefer–Sobel sequential procedure for selecting the normal population which has the largest mean (II): 2-factor experiments with no interaction. *Commun. Stat.—Simul. and Comput.* **B17**, 103–125.

Bechhofer, R. E., and Goldsman, D. M. (1989a). Truncation of the Bechhofer–Kiefer–Sobel sequential procedure for selecting the normal population which has the largest mean (III): supplementary truncation numbers and resulting performance characteristics. *Commun. Stat.—Simul. and Comput.* **B18**, 63–81.

Bechhofer, R. E., and Goldsman, D. M. (1989b). A comparison of the performances of procedures for selecting the normal population having the largest mean when the variances are known and equal. In: *Contributions to Probability and Statistics—Essays in Honor of Ingram Olkin* (L. J. Gleser, M. D. Perlman, S. J. Press and A. R. Sampson, eds.), 303–317. New York: Springer-Verlag.

Bechhofer, R. E., Goldsman, D. M., and Hartmann, M. (1993). Performances of selection procedures for 2-factor additive normal populations with common known variance. In: *Multiple Comparisons, Selection, and Applications in Biometry* (F. M. Hoppe, ed.), 209–224. New York: Marcel Dekker.

Bechhofer, R. E., Goldsman, D. M., and Jennison, C. (1989). A single-stage selection procedure for multi-factor multinomial experiments with multiplicativity. *Commun. Stat.—Simul. and Comput.* **B18**, 31–61.

Bechhofer, R. E., Hayter, A. J., and Tamhane, A. C. (1991). Designing experiments for selecting the largest normal mean when the variances are known and unequal: optimal sample size allocation. *J. Stat. Plan. Infer.* **28**, 271–289.

Bechhofer, R. E., Kiefer, J., and Sobel, M. (1968). *Sequential Identification and Ranking Procedures (with Special Reference to Koopman–Darmois Populations)*. Chicago: University of Chicago Press.

Bechhofer, R. E., and Kulkarni, R. V. (1982). Closed adaptive sequential procedures for selecting the best of $k \geq 2$ Bernoulli populations. In: *Statistical Decision Theory and Related Topics, III* (S. S. Gupta and J. O. Berger, eds.), Vol. 1, 61–108. New York: Academic Press.

Bechhofer, R. E., and Kulkarni, R. V. (1984). Closed sequential procedures for selecting the multinomial events which have the largest probabilities. *Commun. Stat.—Theory and Methods* **A13**, 2997–3031.

Bechhofer, R. E., and Nocterne, D. J. (1972). Optimal allocation of observations when comparing several treatments with a control, II: 2-sided comparisons. *Technometrics* **14**, 423–436.

Bechhofer, R. E., Santner, T. J., and Turnbull, B. W. (1977). Selecting the largest interaction in a two-factor experiment. In: *Statistical Decision Theory and Related Topics, II* (S. S. Gupta and D. S. Moore, eds.), Vol. 2, 1–18. New York: Academic Press.

Bechhofer, R. E., and Sobel, M. (1958). Non-parametric multiple-decision procedures for selecting that one of k populations which has the highest probability of yielding the largest observation (preliminary report). Abstract. *Ann. Math. Stat.* **29**, 325.

Bechhofer, R. E., and Tamhane, A. C. (1981). Incomplete block designs for comparing treatments with a control: general theory. *Technometrics* **23**, 45–57.

Bechhofer, R. E., and Tamhane, A. C. (1983a). Design of experiments for comparing treatments with a control: tables of optimal allocations of observations. *Technometrics* **25**, 87–95.

Bechhofer, R. E., and Tamhane, A. C. (1983b). Tables of admissible and optimal balanced treatment incomplete block (BTIB) designs for comparing treatments with a control. In: *Selected Tables in Mathematical Sciences*, Vol. 8. Providence, RI: American Mathematical Society.

Bechhofer, R. E., and Turnbull, B. (1978). Two $(k + 1)$-decision selection procedures for comparing k normal means with a specified standard. *J. Am. Stat. Assoc.* **73**, 385–392.

Beeson, J. R. (1965). *A Simulator for Evaluating Prosthetic Cardiac Valves*. Unpublished M.S. Thesis, Purdue University Library, West Lafayette, IN.

Berger, J. O., and Deely, J. J. (1988). A Bayesian approach to ranking and selection of related means with alternatives to analysis-of-variance methodology. *J. Am. Stat. Assoc.* **83**, 364–373.

Berger, R. L. (1979). Minimax subset selection for loss measured by subset size. *Ann. Stat.* **7**, 1333–1338.

Berger, R. L. (1980). Minimax subset selection for the multinomial distribution. *J. Stat. Plan. Infer.* **4**, 391–402.

Berger, R. L. (1982). A minimax and admissible subset selection rule for the least probable multinomial cell. In: *Statistical Decision Theory and Related Topics, III* (S. S. Gupta and J. O. Berger, eds.), Vol. 1, 143–156. New York: Academic Press.

Berger, R. L., and Gupta, S. S. (1980). Minimax subset selection rules with applications to unequal variance (unequal sample size) problems. *Scand. J. Stat.* **7**, 21–26.

Berry, D. A. (1972). A Bernoulli two-armed bandit. *Ann. Math. Stat.* **43**, 871–897.

Berry, D. A. (1978). Modified two-armed bandit strategies for certain clinical trials. *J. Am. Stat. Assoc.* **73**, 339–345.

Bhapkar, V. P., and Gore, A. (1971). Some selection problems based on U-statistics for the location and scale problems. *Ann. Inst. Stat. Math.* **23**, 375–386.

Bhapkar, V. P., and Somes, G. W. (1976). Multiple comparisons of matched proportions. *Commun. Stat.—Theory and Methods* **A5**, 17–25.

Bickel, P. J., and Yahav, J. A. (1977). On selecting a subset of good populations. In: *Statistical Decision Theory and Related Topics, II* (S. S. Gupta and D. S. Moore, eds.), 37–55. New York: Academic Press.

Bickel, P. J., and Yahav, J. A. (1982). Asymptotic theory of selection procedures and optimality of Gupta's rules. In: *Statistics and Probability: Essays in Honor of C. R. Rao* (G. Kallianpur, P. R. Krishhnaiah and J. K. Ghosh, eds.), 109–124. Amsterdam: North Holland.

Bofinger, E., and Lewis, G. J. (1992). Simultaneous comparisons with a control and with the best: two stage procedures (with Discussion). *The Frontiers of Modern Statistical Inference Procedures: Proceedings of the IPASRAS-II Conference* (E. Bofinger, E. J. Dudewicz, G. J. Lewis, and K. Mengersen, eds.), 25–45. Columbus, OH: American Sciences Press.

Bofinger, E., and Mengersen, K. (1986). Subset selection of the t best populations. *Commun. Stat.—Theory and Methods* **A15**, 3145–3161.

Borowiak, D.S., and De Los Reyes, J. P. (1992). Selection of the best in 2×2 factorial designs. *Commun. Stat.—Theory and Methods* **A21**, 2493–2500.

Box, G. E. P., Hunter, W. G., and Hunter, J. S. (1978). *Statistics for Experimenters*. New York: John Wiley & Sons.

Bradt, R. N., Johnson, S. M., and Karlin, S. (1956). On sequential designs for maximizing the sum of n observations. *Ann. Math. Stat.* **27**, 1060–1074.

Bratcher, T. L., and Bland, R. P. (1975). On comparing binomial populations from a Bayesian viewpoint. *Commun. Stat.—Theory and Methods* **A4**, 975–985.

Bratley, P., Fox, B. L., and Schrage, L. E. (1987). *A Guide to Simulation*, 2nd edition. New York: Springer-Verlag.

Bromaghm, J. (1993). Sample size determination for interval estimation of multinomial probabilities. *Am. Stat.* **47**, 203–206.

Brownlee, K. A. (1965). *Statistical Theory and Methodology in Science and Engineering*, 2nd edition. New York: John Wiley & Sons.

Büringer, H., Martin, H., and Schriever, K. H. (1980). *Nonparametric Sequential Selection Procedures*. Boston: Birkhauser.

REFERENCES

Cacoullos, T., and Sobel, M. (1966). An inverse-sampling procedure for selecting the most probable event in a multinomial distribution. In: *Multivariate Analysis* (P. R. Krishhnaiah, ed.), 423–455. New York: Academic Press.

Canner, P. L. (1970). Selecting one of two treatments when the responses are dichotomous. *J. Am. Stat. Assoc.* **65**, 293–306.

Carroll, R. J., Gupta, S. S., and Huang, D.-Y. (1975). On selection procedures for the t best populations and some related problems. *Commun. Stat.—Theory and Methods* **A4**, 987–1008.

Chen, H. J., Dudewicz, E. J., and Lee, Y. J. (1976). Subset selection procedures for normal means under unequal sample sizes. *Sankhyā* **B38**, 249–255.

Chen, P. (1985). Subset selection for the least probable multinomial cell. *Ann. Inst. Stat. Math.* **37**, 303–314.

Chen, P. (1988). Selecting the best multinomial cell—provided it is better than a control. *Biometrical J.* **8**, 985–992.

Chen, P. (1992). Truncated selection procedures for the most probable event and the least probable event. *Ann. Inst. Stat. Math.* **B44**, 613–622.

Chen, P., and Hsu, L. (1991). A composite stopping rule for multinomial subset selection. *Br. J. of Math. Stat. Psychol.* **44**, 403–411.

Chernoff, H., and Yahav, J. (1977). A subset selection problem employing a new criterion. In: *Statistical Decision Theory and Related Topics, II* (S. S. Gupta and D. S. Moore, eds.), 93–119. New York: Academic Press.

Chiu, W. K. (1974). The ranking of means of normal populations for a generalized selection goal. *Biometrika* **61**, 579–584.

Chiu, W. K. (1977). On correct selection for a ranking problem. *Ann. Inst. Stat. Math.* **29**, 59–66.

Cochran, W. G., and Cox, G. M. (1957). *Experimental Design*, 2nd edition. New York: John Wiley & Sons.

Coe, P. R., and Tamhane, A. C. (1993a). Exact repeated confidence intervals for Bernoulli parameters in a group sequential clinical trial. *Controlled Clin. Trials* **14**, 19–29.

Coe, P. R., and Tamhane, A. C. (1993b). Small sample confidence intervals for the difference, ratio and odds ratio of two success probabilities. *Commun. Stat.— Simul. and Comput.* **B22**, 925–938.

Cohen, D. S. (1959). *A Two-Sample Decision Procedure for Ranking Means of Normal Populations with a Common Known Variance*. Unpublished M.S. Thesis, Department of Operations Research, Cornell University, Ithaca, NY.

Colton, T. (1963). A model for selecting one of two medical treatments. *J. Am. Stat. Assoc.* **58**, 388–400.

Cornfield, J., Halperin, M., and Greenhouse, S. W. (1969). An adaptive procedure for sequential clinical trials. *J. Am. Stat. Assoc.* **64**, 759–770.

Cox, D. R. (1958). *Planning of Experiments*. New York: John Wiley & Sons.

Dantzig, G. B. (1940). On the non-existence of tests of "Student's" hypotheses having power functions independent of σ. *Ann. Math. Stat.* **11**, 186.

Davies, O. L. (1967). *Design and Analysis of Industrial Experiments*, 2nd edition. London: Oliver and Boyd.

Deely, J. J. (1965). *Multiple Decision Procedures from an Empirical Bayes Approach*. Ph.D. Dissertation, Department of Statistics, Purdue University, West Lafayette, IN.

Deely, J. J., and Gupta, S. S. (1988). Hierarchical Bayesian selection procedures for the best binomial population. Technical Report 88-21C, Department of Statistics, Purdue University, West Lafayette, IN.

Desu, M. M. (1970). A selection problem. *Ann. Math. Stat.* **41**, 1596–1603.

Desu, M. M., and Sobel, M. (1968). A fixed subset-size approach to a selection problem. *Biometrika* **55**, 401–410. Corrections and amendments: **63**, 685.

Dodge, H. F., and Romig, H. G. (1944). *Sampling Inspection Tables*. New York: John Wiley & Sons.

Domröse, H., and Rasch, D. (1987). Robustness of selection procedures. *Biometrical J.* **29**, 541–553.

Dourleijn, C. J. (1993). *On Statistical Selection in Plant Breeding*. Ph.D. Dissertation, Agricultural University, Wageningen, The Netherlands.

Driessen, S. G. (1991). Multiple comparisons with and selection of the best treatment in (incomplete) block designs. *Commun. Stat.—Theory and Methods* **A20**, 179–218.

Driessen, S. G. (1992). *Statistical Selection: Multiple Comparison Approach*. Ph.D. Dissertation, University of Technology, Eindhoven, The Netherlands.

Driessen, S. G. (1993). A note on selection procedures and selection constants. In: *Multiple Comparisons, Selection, and Applications in Biometry* (F. M. Hoppe, ed.), 367–380. New York: Marcel Dekker.

Driessen, S. G., van der Laan, P., and van Putten, B. (1990). Robustness of the probability of correct selection against deviations from the assumption of a common known variance. *Biometrical J.* **32**, 131–142.

Dudewicz, E. J. (1971). Non-existence of a single-sample selection procedure whose $P\{CS\}$ is independent of the variances. *S. Afr. Stat. J.* **5**, 37–39.

Dudewicz, E. J., and Chen, H. J. (1992). Subset selection with expected subset size control (with Discussion). In: *The Frontiers of Modern Statistical Inference Procedures, II. Proceedings of the IPASRAS-II Conference* (E. Bofinger, E. J. Dudewicz, G. J. Lewis, and K. Mengersen, eds.), 177–222. Columbus, OH: American Sciences Press.

Dudewicz, E. J., and Dalal, S. R. (1975). Allocation of observations in ranking and selection with unequal variances. *Sankhyā* **B37**, 28–78.

Dudewicz, E. J., and Koo, J. O. (1982). *The Complete Categorized Guide to Statistical Selection and Ranking Procedures*. Columbus, OH: American Sciences Press.

Dudewicz, E. J., and Mishra, S. N. (1984). The robustness of Bechhofer's normal means selection procedure. In *Design of Experiments: Ranking and Selection—Essays in Honor of Robert E. Bechhofer* (T. J. Santner and A. C. Tamhane, eds.), 35–45. New York: Marcel Dekker.

Dudewicz, E. J., and Zaino, N. A., Jr. (1977). Allowance for correlation in setting simulation run-length via ranking-and-selection procedures. *TIMS Studies in the Management Sciences* **7**, 51–61.

Dunnett, C. W. (1955). A multiple comparison procedure for comparing several treatments with a control. *J. Am. Stat. Assoc.* **50**, 1096–1121.

Dunnett, C. W. (1960). On selecting the largest of k normal population means (with Discussion). *J. R. Stat. Soc.* **B22**, 1–40.

Dunnett, C. W. (1964). New tables for multiple comparisons with a control. *Biometrics* **20**, 482–491.

Dunnett, C.W. (1982). Robust multiple comparisons. *Commun. Stat.—Theory and Methods* **A11**, 2611–2629.

Dunnett, C. W. (1984). Selection of the best treatment in comparison to a control with an application to a medical trial. In *Design of Experiments: Ranking and Selection—Essays in Honor of Robert E. Bechhofer* (T. J. Santner and A. C. Tamhane, eds.), 47–66. New York: Marcel Dekker.

Dunnett, C. W. (1989). Multivariate normal probability integrals with product correlation structure. *Applied Stat.* **38**, 564–579. Correction: **42**, 709.

Du Preez, J. P., Swanepoel, J. W. H., Venter, J. H., and Somerville, P. N. (1985). Some properties of Somerville's multiple range subset selection procedure for three populations. *South Af. Stat. J.* **19**, 45–72.

Eaton, M. L. (1967). Some optimum properties of ranking procedures. *Ann. Math. Stat.* **38**, 124–137.

Eaton, M. L., and Gleser, L. J. (1989). Some results on convolutions and a statistical application. *Contributions to Probability and Statistics—Essays in Honor of Ingram Olkin* (L. J. Gleser, M. D. Perlman, S. J. Press and A. R. Sampson, eds.), 75–90. New York: Springer-Verlag.

Edwards, D. G. (1987). Extended-Paulson sequential selection. *Ann. Stat.* **15**, 449–455.

Edwards, D. G., and Hsu, J. C. (1983). Multiple comparisons with the best treatment. *J. Am. Stat. Assoc.* **78**, 965–971. Corrigendum: **79**, 965.

Fabian, V. (1962). On multiple decision methods for ranking population means. *Ann. Math. Stat.* **33**, 248–254.

Fabian, V. (1974). Note on Anderson's sequential procedures with triangular boundary. *Ann. Stat.* **2**, 170–176; Acknowledgement of priority to "Note on Anderson's sequential procedures with triangular boundary." *Ann. Stat.* **2**, 1063.

Fabian, V. (1991). On the problem of interactions in the analysis of variance (with Discussion). *J. Am. Stat. Assoc.* **86**, 362–375.

Fairweather, W. R. (1968). Some extensions of Somerville's procedure for ranking means of normal populations. *Biometrika* **55**, 411–418.

Faltin, F. (1980). A quantile unbiased estimation of the probability of correct selection achieved by Bechhofer's single-stage procedure for the two population normal means problem. Abstract 80t-60, *IMS Bull.* **9**, 180–181.

Faltin, F., and McCulloch, C. E. (1983). On the small sample properties of the Olkin–Sobel–Tong estimator of the probability of correct selection. *J. Am. Stat. Assoc.* **78**, 464–467.

Federer, W. T., and McCulloch, C. E. (1984). Multiple comparison procedures for some split plot and split block designs. In: *Design of Experiments: Ranking and Selection—Essays in Honor of Robert E. Bechhofer* (T. J. Santner and A. C. Tamhane, eds.), 7–22. New York: Marcel Dekker.

Feigin, P. D., and Weissman, I. (1981). On the indifference zone approach to selection—a consistency result. *Ann. Inst. Stat. Math.* **33**, 471–474.

Finner, H., and Giani, G. (1994). Closed subset selection procedures for selecting good populations. *J. Stat. Plan. Infer.* **38**, 179–200.

Fong, D. K. H. (1992a). Ranking and estimation of related means in the presence of a covariate—a Bayesian approach. *J. Am. Stat. Assoc.* **87**, 1128–1136.

Fong, D. K. H. (1992b). A Bayesian approach to the estimation of the largest normal mean. *J. Stat. Comput. Simul.* **40**, 119–133.

Fong, D. K. H., and Chow, M. (1991). Computation of posterior ranking probabilities of related means. Technical Report, Department of Management Science, Pennsylvania State University, University Park, PA.

Gabriel, K. R. (1969). Simultaneous test procedures—some theory of multiple comparisons. *Ann. Math. Stat.* **40**, 224–250.

Ghosh, M. (1973). Nonparametric selection procedure for symmetric location parameter populations. *Ann. Stat.* **1**, 773–779.

Gibbons, J. D., Olkin, I., and Sobel, M. (1977). *Selecting and Ordering Populations: A New Statistical Methodology.* New York: John Wiley & Sons.

Gnanadesikan, M. (1966). *Some Selection and Ranking Procedures for Multivariate Normal Populations.* Ph.D. Dissertation, Department of Statistics, Purdue University, West Lafayette, IN.

Goel, P. K., and Rubin, H. (1977). On selecting a subset containing the best population—a Bayesian approach. *Ann. Stat.* **5**, 969–983.

Goldsman, D. M., and Petit, T. (1994). On the robustness of subset selection procedures. Technical Report, School of Industrial and Systems Engineering, Georgia Institute of Technology, Atlanta, GA.

Goldsman, D. M., Nelson, B. L., and Schmeiser, B. W. (1991). Methods for selecting the best system. *Proceedings of the 1991 Winter Simulation Conference* (B. L. Nelson, W. D. Kelton and G. M. Clark, eds.), 177–186. Piscataway, NJ: Institute of Electrical and Electronics Engineers.

Gupta, S. S. (1956). *On a Decision Rule for a Problem in Ranking Means.* Ph.D. Dissertation (Mimeo. Ser. No. 150). Institute of Statistics, Univ. of North Carolina, Chapel Hill, NC.

Gupta, S. S. (1963). On a selection and ranking procedure for gamma populations. *Ann. Inst. Stat. Math.* **14**, 199–216.

Gupta, S. S. (1965). On some multiple decision (selection and ranking) rules. *Technometrics* **7**, 225–245.

Gupta, S. S., and Han, S. (1991). An elimination type two-stage procedure for selecting the population with the largest mean from k logistic populations. *Am. J. Math. Mgmt. Sci.* **11**, 351–370.

Gupta, S. S., and Hsiao, P. (1983). Empirical Bayes rules for selecting good populations. *J. Stat. Plan. Infer.* **8**, 87–101.

Gupta, S. S., and Hsu, J. C. (1978). On the performance of some subset selection procedures. *Commun. Stat.—Simul. and Comput.* **B7**, 561–591.

Gupta, S. S., and Hsu, J. C. (1980). Subset selection procedures with application to motor vehicle fatality data in a two-way layout. *Technometrics* **22**, 543–546.

Gupta, S. S., and Hsu, J. C. (1984). A computer package for ranking, selection, and multiple comparisons with the best. *Proceedings of the 1984 Winter Simulation Conference* (S. Sheppard, U. Pooch and C. D. Pegden, eds.), 251–257. Piscataway, NJ: Institute of Electrical and Electronics Engineers.

Gupta, S. S., and Hsu, J. C. (1985). RS-MCB: ranking, selection, and multiple comparisons with the best. *Am. Stat.* **39**, 313–314.

Gupta, S. S., and Huang, D.-Y. (1976). Subset selection procedures for the means and variances of normal populations: unequal sample sizes case. *Sankhyā* **B38**, 112–128.

Gupta, S. S., and Huang, D.-Y. (1980). A note on optimal subset selection procedures. *Ann. Stat.* **8**, 1164–1167.

Gupta, S.S., and Huang, D.-Y. (1981). *Multiple Decision Theory: Recent Developments.* Lecture Notes in Statistics, Vol. 6, New York: Springer-Verlag.

Gupta, S.S., Huang, D.-Y., and Huang, W.-T. (1976). On ranking and selection procedures and tests of homogeneity for binomial populations. In: *Essays in Probability and Statistics* (S. Ikeda et al., eds.), 501–533. Tokyo: Shinko Tsusho Co. Ltd.

Gupta, S. S., and Huang, W.-T. (1974). A note on selecting a subset of normal populations with unequal sample sizes. *Sankhyā* **A36**, 389–396.

Gupta, S. S., and Kim, W.-C. (1980). γ-Minimax and minimax decision rules for comparison of treatments with a control. In: *Recent Developments in Statistical Inference and Data Analysis* (K. Matusita, ed.), 55–71. Amsterdam: North-Holland.

Gupta, S. S., and Kim, W.-C. (1984). A two-stage elimination-type procedure for selecting the largest of several normal means with a common unknown variance. In: *Design of Experiments: Ranking and Selection—Essays in Honor of Robert E. Bechhofer* (T. J. Santner and A. C. Tamhane, eds.), 77–94. New York: Marcel Dekker.

Gupta, S. S., Leu, L.-Y., and Liang, T.-C. (1990). On lower confidence bounds for PCS in truncated location parameter models. *Commun. Stat.—Theory and Methods* **A19**, 527–546.

Gupta, S. S., and Liang, T.-C. (1986). Empirical Bayes rules for selecting good binomial populations. In: *Adaptive Statistical Procedures and Related Topics* (J. Van Ryzin, ed.), 110–128. Hayword, CA: Institute of Mathematical Statistics.

Gupta, S. S., and Liang, T.-C. (1989a). Selecting the best binomial population: parametric empirical Bayes approach. *J. Stat. Plan. and Infer.* **23**, 21–31.

Gupta, S. S., and Liang, T.-C. (1989b). Parametric empirical Bayes rules for selecting the most probable multinomial event. In: *Contributions to Probability and Statistics—Essays in Honor of Ingram Olkin* (L. J. Gleser, M. D. Perlman, S. J. Press and A. R. Sampson, eds.), 318–328. New York: Springer-Verlag.

Gupta, S. S., and Liang, T.-C. (1989c). On Bayes and empirical Bayes two-stage allocation procedures for selection problems. In: *Statistical Data Analysis and Inference* (Y. Dodge, ed.), 61–70. Elsevier Science Publishers B.V. (North-Holland).

Gupta, S. S., and Liang, T.-C. (1991a). On a lower confidence bound for the probability of a correct selection: analytical and simulation studies. In: *The Frontiers of Statistical Computation, Simulation, and Modeling: Proceedings of the First International Conference on Statistical Computing* (P. R. Nelson, E. J. Dudewicz, A. Öztürk and E. C. van der Meulen, eds.), Vol. II, 77–95. Columbus, OH: American Sciences Press.

Gupta, S. S., and Liang, T.-C. (1991b). On restricted subset selection rules for selecting the best population. Technical Report No. 91-27C, Dept. of Statistics, Purdue University, West Lafayette, IN.

Gupta, S.S., Liang, T.-C., and Rau, R. B. (1994). Empirical Bayes two-stage procedures for selecting best Bernoulli populations compared with a control. In: *Statistical Decision Theory and Related Topics, V* (S. S. Gupta and J. O. Berger, eds.), 277–292, New York: Springer-Verlag.

Gupta, S. S., and Lu, M.-W. (1979). Subset selection procedures for restricted families of probability distributions. *Ann. Inst. Stat. Math.* **31**, 235–252.

Gupta, S. S., and McDonald, G. C. (1970). Some selection procedures with applications to reliability problems. In: *Nonparametric Techniques in Statistical Inference* (M. Puri, ed.), 491–514. Cambridge, England: Cambridge University Press.

Gupta, S. S., and McDonald, G. C. (1980). Nonparametric procedures in multiple decisions (ranking and selection procedures). *Colloq. Math. Soc. Janos Bolyai* **32**, 361–389.

Gupta, S. S., and McDonald, G. C. (1986). A statistical selection approach to binomial models. *J. Qual. Tech.* **18**, 103–115.

Gupta, S. S., Miao, B., and Sun, D. (1993). A two-stage procedure for selecting the population with the largest mean when the common variance is unknown. Technical Report #93-3C, Dept. of Statistics, Purdue University, West Lafayette, IN.

Gupta, S. S., and Miescke, K. J. (1981). Optimality of subset selection procedures for ranking means of three normal populations. *Sankhyā* **B43**, 1–17.

Gupta, S. S., and Miescke, K. J. (1984). On two-stage Bayes selection procedures. *Sankhyā* **B46**, 123–134.

Gupta, S. S., and Nagel, K. (1967). On selection and ranking procedures and order statistics from the multinomial distribution. *Sankhyā* **B29**, 1–17.

Gupta, S. S., Nagel, K., and Panchapakesan, S. (1973). On the order statistics from equally correlated normal random variables. *Biometrika* **60**, 403–413.

Gupta, S. S., and Panchapakesan, S. (1979). *Multiple Decision Procedures: Theory and Methodology of Selecting and Ranking Populations*. New York: John Wiley & Sons.

Gupta, S. S., and Panchapakesan, S. (1993). Selection and screening procedures in multivariate analysis. In: *Multivariate Analysis: Future Directions* (C. R. Rao, ed.), 233–262. Elsevier Science Publishers B.V. (North-Holland).

Gupta, S. S., and Santner, T. J. (1973). On selection and ranking procedures—a restricted subset selection rule. In: *Proceedings of the 39th Session of the International Statistical Institute*, 409–417. Vienna, Austria.

Gupta, S. S., and Sobel, M. (1958). On selecting a subset which contains all populations better than a standard. *Ann. Math. Stat.* **29**, 235–244.

Gupta, S. S., and Sobel, M. (1960). Selecting a subset containing the best of several binomial populations. In: *Contributions to Probability and Statistics—Essays in Honor of Harold Hotelling* (I. Olkin, S. G. Ghurye, H. Hoeffding, W. G. Madow and H. B. Mann, eds.), 224–248. Stanford, CA: Stanford University Press.

Gupta, S. S., and Sobel, M. (1962). On selecting a subset containing the population with the smallest variance. *Biometrika* **49**, 495–507.

Gupta, S. S., and Wong, W.-Y. (1982). Subset selection procedures for the means of normal populations with unequal variances: unequal sample sizes case. *Selecta Stat. Can.* **6**, 109–150.

Guttman, I., and Tiao, G. (1964). A Bayesian approach to some best population problems. *Ann. Math. Stat.* **35**, 825–835.

Hall, W. J. (1959). The most economical character of Bechhofer and Sobel decision rules. *Ann. Math. Stat.* **30**, 964–969.

Harter, L. (1960). Tables of range and Studentized range. *Ann. Math. Stat.* **31**, 1122–1147.

Harter, L. (1969). *Order Statistics and Their Use in Testing and Estimation, Vol. 1: Tests Based on Range and Studentized Range of Samples from a Normal Population*, Aerospace Research Laboratories, Office of Aerospace Research, U.S. Air Force.

Hartmann, M. (1991). An improvement on Paulson's procedure for selecting the population with the largest mean from k normal populations with a common unknown variance. *Sequential Anal.* **10**, 1–16.

Hartmann, M. (1993). Multi-factor extensions of Paulson's procedures for selecting the best normal population. In: *Multiple Comparisons, Selection, and Applications in Biometry* (F. M. Hoppe, ed.), 225–245. New York: Marcel Dekker.

Hayter, A. J. (1984). A proof of the conjecture that the Tukey–Kramer multiple comparisons procedure is conservative. *Ann. Stat.* **12**, 61–75.

Hayter, A. J. (1989). Selecting the normal population with the largest mean when the variances are bounded. *Commun. Stat.—Theory and Methods* **A18**, 1455–1467.

Healy, W. C., Jr. (1956). Two-sample procedures in simultaneous estimation. *Ann. Math. Stat.* **27**, 687–702.

Heyl, P. R. (1930). A redetermination of the constant of gravitation. *J. Res. Bur. Stand.* **5**, 1243–1250.

Hochberg, Y., and Lachenbruch, P. (1976). Two stage multiple comparison procedures based on the studentized range. *Commun. Stat.—Theory and Methods* **A5**, 1447–1454.

Hochberg, Y., and Tamhane, A. C. (1987). *Multiple Comparison Procedures*. New York: John Wiley & Sons.

Hoel, D. G., Sobel, M., and Weiss, G. H. (1975). A survey of adaptive sampling for clinical trials. *Perspect. Biometry* **I**, 29–61.

Hooper, J. H., and Santner, T. J. (1979). Design of experiments for selection from ordered families of distributions. *Ann. Stat.* **7**, 615–643.

Hoover, D. R. (1991). Simultaneous comparisons of multiple treatments to two (or more) controls. *Biometrical J.* **33**, 913–921.

Hsu, J. C. (1981). Simultaneous confidence intervals for all distances from the 'best'. *Ann. Stat.* **9**, 1026–1034.

Hsu, J. C. (1982). Simultaneous inference with respect to the best treatment in block designs. *J. Am. Stat. Assoc.* **77**, 461–467.

Hsu, J. C. (1984a). Constrained simultaneous intervals for multiple comparisons with the best. *Ann. Stat.* **12**, 1136–1144.

Hsu, J. C. (1984b). Ranking and selection and multiple comparisons with the best. *Design of Experiments: Ranking and Selection—Essays in Honor of Robert E. Bechhofer* (T. J. Santner and A. C. Tamhane, eds.), 23–33. New York: Marcel Dekker.

Hsu, J. C. (1988). Simultaneous confidence intervals in the general linear model. In: *Computer Science and Statistics: Proceedings of the 20th Symposium on the Interface* (E. J. Wegman, ed.), 453–457. Alexandria, VA: American Statistical Association.

Hsu, J. C. (1989). Multiple comparisons in the general linear model. In *Computer Science and Statistics: Proceedings of the 21st Symposium on the Interface* (K. Berk and L. Malone, eds.), 398–401. Alexandria, VA: American Statistical Association.

Hsu, J. C. (1992). The factor analytic approach to simultaneous inference in the general linear model. *J. Graph. Comput. Stat.* **1**, 151–168.

Hsu, J. C. (1994). *Multiple Comparisons: Theory and Methods*, forthcoming.

Hsu, J. C., and Edwards, D. G. (1983). Sequential multiple comparisons with the best. *J. Am. Stat. Assoc.* **78**, 958–964.

Huang, D.-Y. (1976). On some selection procedures for normal means problems with unequal sample sizes. *Commun. Stat.—Theory and Methods* **A5**, 1489–1499.

Hustý, J. (1981). Ranking and selection procedures for location parameter case based on L-estimates. *Aplikace Mat.* **26**, 377–388.

Hustý, J. (1984). Subset selection of the largest location parameter based on L-estimates. *Aplikace Mat.* **29**, 397–410.

Jennison, C. (1983). Equal probability of correct selection for Bernoulli selection procedures. *Commun. Stat.—Theory and Methods* **A12**, 2887–2896.

Kesten, H., and Morse, N. (1959). A property of the multinomial distribution. *Ann. Math. Stat.* **30**, 120–127.

Kim, W.-C. (1986). A lower confidence bound on the probability of a correct selection. *J. Am. Stat. Assoc.* **81**, 1012–1017.

Kulkarni, R. V., and Jennison, C. (1984). Optimal procedures for selecting the best s out of k Bernoulli populations. In: *Design of Experiments: Ranking and Selection—Essays in Honor of Robert E. Bechhofer* (T. J. Santner and A. C. Tamhanem, eds.), 113–125. New York: Marcel Dekker.

Kulkarni, R. V., and Kulkarni, V. (1987). Optimal Bayes procedures for selecting the better of two Bernoulli populations. *J. Stat. Plan. Infer.* **24**, 311–330.

Lam, K. (1986). A new procedure for selecting good populations. *Biometrika* **73**, 201–206.

Law, A. M., and Kelton, W. D. (1991). *Simulation Modeling and Analysis.* New York: McGraw-Hill.

Lehmann, E. L. (1963). A class of selection procedures based on ranks. *Math. Ann.* **150**, 268–275.

Liang, T.-C. (1984). *Some Contributions to Empirical Bayes, Sequential and Locally Optimal Subset Selection Rules.* Ph.D. Dissertation, Department of Statistics, Purdue University, West Lafayette, IN.

Liang, T.-C. (1990). Empirical Bayes estimation of binomial parameter with symmetric priors. *Commun. Stat.—Theory and Methods* **A19**, 1671–1683.

Liang, T.-C. (1991). Empirical Bayes selection for the highest success probability in Bernoulli processes with negative binomial sampling. *Stat. Decisions* **9**, 213–234.

Lin, L.-Y., and Jen, B.-C. (1987). A procedure for the selection of populations with means fall within normal limits. *Soochow J. Math.* **13**, 179–195.

Listing, J., and Rasch, D. (1993). *Robustness of selection procedures.* Preprint.

Lowe, B. (1935). Data from Iowa Agricultural Experiment Station.

Mahamunulu, D. M. (1967). Some fixed-sample ranking and selection problems. *Ann. Math. Stat.* **38**, 1079–1091.

Matejcik, F., and Nelson, B. L. (1993). Two-stage multiple comparisons with the best for computer simulation. *Oper. Res.*

McDonald, G. C. (1969). *On Some Distribution-Free Ranking and Selection Procedures.* Ph.D. Dissertation, Department of Statistics, Purdue University, West Lafayette, IN.

McDonald, G. C. (1979). Nonparametric selection procedures applied to state traffic fatality rates. *Technometrics* **21**, 515–523.

McDonald, G. C. (1985). Characteristics of block design selection procedures and a counterexample. *Sankhyā* **B47**, 47–55.

Mengersen, K. (1992). Robustness to normality of a selection rule. *Commun. Stat.—Simul. and Comput.* **B21**, 35–56.

Mengersen, K., and Bofinger, E. (1988). Confidence bounds and selection of the t best populations. *Commun. Stat.—Simul. and Comput.* **B17**, 927–945.

Mengersen, K., and Bofinger, E. (1989). Selecting the t best populations: scale parameter case. *Aust. J. Stat.* **31**, 297–307. Correction: **32**, 127.

Miescke, K. J. (1979). Bayesian subset selection for additive and linear loss functions. *Commun. Stat.—Theory and Methods* **A8**, 1205–1226.

Milliken, G., and Johnson, D. (1984). *Analysis of Messy Data, Vol. 1: Designed Experiments.* New York: Van Nostrand Reinhold.

Milton, R. C. (1970). *Rank Order Probabilities: Two-Sample Normal Shift Alternatives.* New York: John Wiley & Sons.

Moberg, T. F., Ramberg, J. S., and Randles, R. H. (1978). An adaptive M-estimator and its application to a selection problem. *Technometrics* **20**, 255–263.

Moshman, J. (1958). A method for selecting the size of the initial sample in Stein's 2-sample procedure. *Ann. Math. Stat.* **29**, 1271–1275.

Mukhopadhyay, N. (1979). Some comments on two-stage selection procedures. *Commun. Stat.—Theory and Methods* **A8**, 671–684.

Mukhopadhyay, N., and Solanky, T. K. S. (1994). *Multistage Selection and Ranking Procedures: Second-Order Asymptotics.* New York: Marcel Dekker.

Naik, U. D. (1977). Some subset selection problems. *Commun. Stat.—Theory and Methods* **A6**, 955–966.

Nelson, B. L., and Matejcik, F. J. (1993). Using common random numbers for indifference-zone selection and multiple comparisons in simulation. Technical Report, Department of ISE, The Ohio State University, Columbus, OH.

Norell, L. (1990). Partitioning a set of normal populations by their locations with respect to a number of controls. *Scand. J. Stat.* **17**, 217–233.

Olkin, I., Sobel, M., and Tong, Y. L. (1982). Bounds for a k-fold integral for location and scale parameter models with application to statistical ranking and selection problems. In: *Statistical Decision Theory and Related Topics, III* (S. S. Gupta and J. O. Berger, eds.), Vol. 2, 193–212. New York: Academic Press.

Pan, G., and Santner, T. J. (1994). Indifference Zone Selection in additive two-factor experiments using randomization restricted designs. In preparation.

Panchapakesan, S. (1971). On a subset selection procedure for the most probable event in a multinomial distribution. In: *Statistical Decision Theory and Related Topics* (S. S. Gupta and Y. Yackel, eds.), 275–298. New York: Academic Press.

Panchapakesan, S., and Santner, T. J. (1977). Subset selection for Δ_p-superior populations. *Commun. Stat.—Theory and Methods* **A6**, 1081–1090.

Parnes, M., and Srinivasan, R. (1986). Some inconsistencies and errors in the indifference zone formulation of selection. *Sankhyā* **A48**, 86–97.

Paulson, E. (1949). A multiple comparison procedure for certain problems in the analysis of variance. *Ann. Math. Stat.* **20**, 95–98.

Paulson, E. (1952). On the comparison of several experimental categories with a control. *Ann. Math. Stat.* **23**, 239–246.

Paulson, E. (1962). A sequential procedure for comparing several experimental categories with a standard or control. *Ann. Math. Stat.* **33**, 438–443.

Paulson, E. (1964). A sequential procedure for selecting the population with the largest mean from k normal populations. *Ann. Math. Stat.* **35**, 174–180.

Paulson, E. (1969). A new sequential procedure for selecting the best one of k binomial populations [Abstract]. *Ann. Math. Stat.* **40**, 1865–1866.

Paulson, E. (1993). Personal Communication.

Percus, O. E., and Percus, J. K. (1984). Modified Bayes technique in sequential clinical trials. *Comput. Biol. Med.* **14**, 127–134.

Raiffa, H., and Schlaiffer, R. (1961). *Applied Statistical Decision Theory*. Boston: Harvard University.

Ramberg, J. S. (1972). Selection sample size approximations. *Ann. Math. Stat.* **43**, 1977–1980.

Ramey, J. T., Jr., and Alam, K. (1979). A sequential procedure for selecting the most probable multinomial event. *Biometrika* **66**, 171–173.

Randles, R. H. (1970). Some robust selection procedures. *Ann. Math. Stat.* **41**, 1640–1645.

Randles, R. H., and Hollander, M. (1971). γ-Minimax selection procedures in treatments versus control problems. *Ann. Math. Stat.* **42**, 330–341.

Randles, R. H., Ramberg, J. S., and Hogg, R. V. (1973). An adaptive procedure for selecting the population with the largest location parameter. *Technometrics* **15**, 769–778.

Rasch, G. (1960). *Probabilistic Models for Some Intelligence and Attainment Tests*. Copenhagen: Danmarks Paedagogiske Institut.

Rasch, G. (1980). *Probabilistic Models for Some Intelligence and Attainment Tests*. Chicago: University of Chicago Press.

Rinott, Y. (1978). On two-stage procedures and related probability-inequalities. *Commun. Stat.—Theory and Methods* **A8**, 799–811.

Rizvi, M. H., and Woodworth, G. G. (1970). On selection procedures based on ranks: counterexamples concerning the least favorable configurations. *Ann. Math. Stat.* **41**, 1942–1951.

Robbins, H. (1952). Some aspects of the sequential design of experiments. *Bull. Am. Math. Soc.* **58**, 527–535.

Robbins, H. (1956). A sequential decision problem with a finite memory. *Proc. Natl. Acad. Sci. USA* **42**, 920–923.

Robbins, H. (1964). The empirical Bayes approach to statistical decision problems. *Ann. Math. Stat.* **35**, 1–20.

Rodman, L. (1978). On the many-armed bandit problem. *Ann. Prob.* **6**, 491–498.

Roth, A. J. (1978). A new procedure for selecting a subset containing the best normal population. *J. Am. Stat. Assoc.* **73**, 613–617.

Sanchez, S. M. (1987a). Small-sample performance of a modified least-failures sampling procedure for Bernoulli subset selection. *Commun. Stat.—Theory and Methods* **A16**, 1051–1065.

Sanchez, S. M. (1987b). A modified least-failures sampling procedure for Bernoulli subset selection. *Commun. Stat.—Theory and Methods* **A16**, 3609–3629.

Sanchez, S. M. (1989). Unbiased estimation following selection for Bernoulli populations. *Commun. Stat.—Theory and Methods* **A18**, 4275–4301.

Sanchez, S. M., and Higle, J. L. (1992). Observational studies of rare events: a subset selection approach. *J. Am. Stat. Assoc.* **87**, 878–883.

Santner, T. J. (1975). A restricted subset selection approach to ranking and selection problems. *Ann. Stat.* **3**, 334–349.

Santner, T. J. (1976). A two-stage procedure for selection of $\delta^{\star}$-optimal means in the normal case. *Commun. Stat.—Theory and Methods* **A5**, 283–292.

Santner, T. J. (1981). Designing two-factor experiments for selecting interactions. *J. Stat. Plan. Infer.* **5**, 45–55.

Santner, T. J., and Behaxeteguy, M. (1992). A two-stage procedure for selecting the best normal population whose first stage selects a bounded random number of populations. *J. Stat. Plan. Infer.* **31**, 147–148.

Santner, T. J., and Hayter, A. J. (1993). The least favorable configuration for a two-stage procedure for selecting the largest normal mean. In: *Multiple Comparisons, Selection, and Applications in Biometry* (F. M. Hoppe, ed.), 247–265. New York: Marcel Dekker.

Santner, T. J., and Pan, G. (1994). Subset selection in two-factor experiments using randomized restricted designs. Submitted.

Santner, T. J., and Snell, M. (1980). Small sample confidence limits for $p_1 - p_2$ and p_1/p_2 in 2×2 contingency tables. *J. Am. Stat. Assoc.* **75**, 386–394.

Santner, T. J., and Yamagami, S. (1993). Invariant small sample confidence intervals for the difference of two success probabilities. *Commun. Stat.—Simul. and Comput.* **B22**, 33–59.

Schafer, R. E. (1977). On selecting which of k populations exceed a standard. In: *The Theory and Applications of Reliability* (C. P. Tsokos and I. N. Shimi, eds.), 449–473. New York: Academic Press.

Scheffé, H. (1959). *The Analysis of Variance.* New York: John Wiley & Sons.

Seal, K. C. (1955). On a class of decision procedures for ranking means of normal populations. *Ann. Math. Stat.* **26**, 387–398.

Sehr, J. (1988). On a conjecture concerning the least favorable configuration of a two-stage selection procedure. *Commun. Stat.—Theory and Methods* **A17**, 3221–3233.

Smith, H. (1969). The analysis of a designed experiment. *J. Qual. Technol.* **1**, 259–263.

Snedecor, G. W., and Cochran, W. G. (1967). *Statistical Methods*, 6th edition. Ames, Iowa: Iowa State University Press.

Sobel, M., and Huyett, M. J. (1957). Selecting the best one of several binomial populations. *Bell Syst. Tech. J.* **36**, 537–576.

Sobel, M., and Weiss, G. H. (1972). Recent results on using the play-the-winner sampling rule with binomial selection problems. In: *Proceedings of the Sixth Berkeley Symposium on Mathematical Statistics and Probability* (L. LeCam, J. Neyman and E. L. Scott, eds.), 717–736. Berkeley: University of California Press.

Somerville, P. N. (1954). Some problems of optimum sampling. *Biometrika* **41**, 420–429.

Somerville, P. N. (1984). A multiple range subset selection procedure. *J. Stat. Comput. and Simul.* **19**, 215–226.

Stein, C. (1945). A two-sample test for a linear hypothesis whose power is independent of the variance. *Ann. Math. Stat.* **24**, 669–673.

Sullivan, D. W., and Wilson, J. R. (1984). Restricted subset selection for normal populations with unknown and unequal variances. In: *Proceedings of the 1984 Winter Simulation Conference* (S. Sheppard, U. Pooch and C. D. Pegden, eds.), 123–128. Piscataway, NJ: Institute of Electrical and Electronics Engineers.

Sullivan, D. W., and Wilson, J. R. (1989). Restricted subset selection procedures for simulation. *Oper. Res.* **37**, 52–71.

Taheri, H., and Young, D. H. (1974). A comparison of sequential sampling procedures for selecting the better of two binomial populations. *Biometrika* **61**, 585–592.

Tamhane, A. C. (1980). Selecting the better Bernoulli treatment using a matched samples design. *J. R. Stat. Soc.* **B42**, 26–30.

Tamhane, A. C. (1985). Some sequential procedures for selecting the better Bernoulli treatment by using a matched samples design. *J. Am. Stat. Assoc.* **80**, 455–460.

Tamhane, A. C., and Bechhofer, R. E. (1977). A two-stage minimax procedure with screening for selecting the largest normal mean. *Commun. Stat.—Theory and Methods* **A6**, 1003–1033.

Tamhane, A. C., and Bechhofer, R. E. (1979). A two-stage minimax procedure with screening for selecting the largest normal mean (ii): an improved PCS lower bound and associated tables. *Commun. Stat.—Theory and Methods* **A8**, 337–358.

Taneja, B. K. (1986). Selection of the best normal mean in complete factorial experiments with interaction and with common unknown variance. *J. Jpn. Stat. Soc.* **16**, 55–65.

Thompson, S. K. (1987). Sample size for estimating multinomial proportions. *Am. Stat.* **41**, 42–46.

Tiao, G. C., and Afonja, B. (1976). Some Bayesian considerations of the choice of design for ranking, selection and estimation. *Ann. Inst. Stat. Math.* **28**, 167–186.

Tong, Y. L. (1969). On partitioning a set of normal populations by their locations with respect to a control. *Ann. Math. Stat.* **40**, 1300–1324.

Tortora, R. D. (1978). A note on sample size estimation for multinomial populations. *Am. Stat.* **32**, 100–101.

Tukey, J. W. (1953). *The Problem of Multiple Comparisons*. Unpublished Notes, Princeton University, Princeton, NJ.

Turnbull, B. W., Kaspi, H., and Smith, R. L. (1978). Adaptive sequential procedures for selecting the best of several normal populations. *J. Stat. Comput. Simul.* **7**, 133–150.

van der Laan, P. (1992a). Subset selection: robustness and imprecise selection. Memorandum COSOR 92-10, Eindhoven University of Technology, The Netherlands.

van der Laan, P. (1992b). Subset selection of an almost best treatment. *Biometrical J.* **34**, 647–656.

van der Laan, P., and van Putten, B. (1988). The robustness of the probability of correct selection against various distributional assumptions. Technical Report, Departmemt of Statistics, Agricultural University, Wageningen, The Netherlands.

Wilcox, R. R. (1984). Selecting the best population, provided it is better than a standard: the unequal variance case. *J. Am. Stat. Assoc.* **79**, 887–891.

Wu, K. H., and Cheung, S. H. (1994). Subset selection for normal means in a two-way design. *Biometrical J.* **36**, 165–175.

Zelen, M. (1969). Play-the-winner rule and the controlled clinical trial. *J. Am. Stat. Assoc.* **64**, 131–146.

Author Index

Abramowitz, M., 278
Afonja, B., 66
Alam, K., 32, 240, 245
Anderson, V., 159, 171
Anscombe, F. J., 177
Armitage, P., 177, 211
Aubuchon, J. C., 115

Büringer, H., 3, 211, 212
Bahadur, R. R., xi
Barlow, R. E., 98
Bartlett, N. S., 68
Bawa, V. S., 149
Bechhofer, R. E., 3, 18, 31, 32, 35, 42, 43, 50, 52, 54, 56, 57, 61, 70, 74, 78, 118, 123, 135, 138, 141, 146, 148, 149, 151, 152, 155, 162, 164, 165, 167, 169, 172, 175, 187, 189, 192, 212, 217, 219, 228, 229, 232, 233, 238–240, 243–245
Beeson, J. R., 159, 171
Behaxeteguy, M., 35, 66, 70, 74
Berger, J. O., 66, 213
Berger, R. L., 97, 245
Berry, D. A., 212
Bhapkar, V. P., 68, 212
Bickel, P. J., 97
Bland, R. P., 213
Bofinger, E., 86, 91, 98, 141
Borowiak, D. S., 143
Box, G. E. P., 26, 78, 83
Bradt, R. N., 212
Bratcher, T. L., 213
Bratley, P., 279
Bromaghm, J., 246
Brownlee, K. A., 102

Cacoullos, T., 245
Canner, P. L., 211
Carroll, R. J., 86
Chen, H., 97
Chen, P., 238–240, 244, 245
Chernoff, H., 97
Cheung, S. H., 175
Chiu, W. K., 67
Chow, M., 66
Cochran, W. G., 26, 75, 106, 108
Coe, P. R., 212
Cohen, D. S., 32
Colton, T., 211
Cornfield, J., 211
Cox, D. R., 25
Cox, G. M., 26

Dalal, S. R., 60, 97
Dantzig, G. B., 52
Davies, O. L., 109
De Los Reyes, J. P., 143
Deely, J. J., 66, 96–97, 213
Desu, M. M., 98
Dodge, H. F., 177
Domröse, H., 28
Dourleijn, C. J., 3, 28
Driessen, S. G., 3, 28, 29, 79, 97, 115
du Preez, J. P., 97
Dudewicz, E. J., 4, 28, 30, 52, 60, 97, 234
Dunnett, C. W., 25, 44, 52, 54, 56, 57, 61, 66, 72, 77, 78, 87, 91, 114, 125, 130, 132, 134, 135, 139, 141, 149, 162–165, 167, 169, 172, 174, 209, 253, 277

319

Eaton, M. L., 24, 31, 184
Edwards, D. G., 67, 114, 115
Elmaghraby, S., 217

Fabian, V., 67, 92, 143, 153
Fairweather, W. R., 66
Faltin, F., 66
Federer, W. T., 175
Finner, H., 97, 99
Fong, D. K. H., 66
Fox, B. L., 279

Gabriel, K. R., 111
Ghosh, M., 68
Giani, G., 97, 99
Gibbons, J. D., 3, 46, 217
Gleser, L. J., 184
Gnanadesikan, M., 97, 98
Goel, P. K., 97
Goldsman, D. M., 31, 35, 42, 54, 56, 57, 61, 79, 146, 151, 152, 155, 212, 229, 243, 244
Gore, A., 68
Govindarajulu, Z., 68
Greenhouse, S. W., 211
Gupta, S. S., 3, 6, 32, 54, 66, 67, 71, 76, 86, 94, 96–98, 115, 127, 141, 167, 176, 181, 201, 211–213, 236–238, 240, 245, 253
Guttman, I., 66

Hall, W. J., 24, 31, 185
Halperin, M., 211
Han, S., 67
Harter, L., 253
Hartmann, M., 52, 54, 56, 57, 153, 155, 158
Hayter, A. J., 50, 60, 66, 111
Healy, W. C., 133, 137
Heyl, P. R., 102
Higle, J. L., 211
Hochberg, Y., 3, 100, 111, 114, 212, 253
Hoel, D. G., 67
Hogg, R. V., 68
Hollander, M., 141
Hooper, J. H., 94, 98
Hoover, D. R., 139
Hsiao, P., 141
Hsu, J. C., 6, 67, 97, 100, 113–115, 176

Hsu, L., 240, 244
Huang, D.-Y., 3, 76, 86, 97
Huang, W.-T., 97
Hunter, J., 26, 78, 83
Hunter, W., 26, 78, 83
Hustý, J., 29, 98
Huyett, M. J., 180, 182

Jen, B.-C., 98, 99
Jennison, C., 189, 243
Johnson, D., 160
Johnson, S. M., 212

Karlin, S., 212
Kaspi, H., 67
Kelton, W. D., 30
Kiefer, J., 3, 31, 192, 229, 232, 244
Kim, W.-C., 54, 66, 141
Koo, J. O., 4
Kulkarni, R. V., 187, 189, 213, 219, 228, 245
Kulkarni, V., 213

Lachenbruch, P., 112
Lam, K., 98
Law, A. M., 30
Lee, Y. J., 97
Lehmann, E. L., 68
Leu, L.-Y., 66
Lewis, G. J., 141
Liang, T.-C., 66, 98, 212, 213, 245
Lin, L.-Y., 98, 99
Listing, J., 79, 98
Lowe, B., 75
Lu, M.-W., 98

Mahamunulu, D. M., 45, 98
Martin, H., 3, 211, 212
Matejcik, F., 115
Matejcik, F. J., 115
McCulloch, C. E., 66, 175
McDonald, G. C., 6, 98, 211
McLean, R., 159, 172
Mengersen, K., 79, 86, 91, 98
Miao, B., 67
Miescke, K. J., 67, 96, 97
Milliken, G., 160
Milton, R. C., 45
Mishra, S. N., 28

Moberg, T. F., 68
Morse, N., 217
Mukhopadhyay, N., 3, 67

Nagel, K., 236–238, 240, 245, 253
Naik, U. D., 98, 99
Nelson, B. L., 61, 115
Nocterne, D. J., 135
Norell, L., 141

Olkin, I., 3, 46, 66, 217

Pan, G., 98, 161, 162, 167, 170, 173, 175, 176
Panchapakesan, S., 3, 93, 97–99, 240, 245, 253
Parnes, M., 67
Paulson, E., xi, 31, 38, 52, 54, 67, 125, 141, 153, 177
Percus, J., 189
Percus, O., 189
Petit, T., 79

Raiffa, H., 66
Ramberg, J. S., 22, 68
Ramey, J. T., 240, 245
Randles, R. H., 68, 141
Rasch, D., 28, 79, 98
Rasch, G., 195
Rinott, Y., 60, 61, 67, 141
Rizvi, M. H., 68
Robbins, H., 212, 213
Rodman, L., 212
Romig, H. G., 177
Roth, A. J., 92
Rubin, H., 97

Sanchez, S. M., 208, 211, 212
Santner, T. J., 35, 66, 70, 74, 93, 94, 96, 98, 99, 161, 162, 167, 170, 173, 175, 176, 212
Schafer, R. E., 141
Scheffé, H., 29
Schlaiffer, R., 66
Schmeiser, B. W., 61
Schrage, L. E., 279
Schriever, K. H., 3, 211, 212
Seal, K. C., 96

Sehr, J., 66
Smith, H., 165, 169
Smith, R. L., 67
Snedecor, G. W., 75, 106, 108
Snell, M., 212
Sobel, M., 3, 31, 46, 52, 61, 66, 67, 98, 127, 180–182, 192, 201, 211, 217, 229, 232, 233, 244, 245
Solanky, T. K. S., 3
Somerville, P. N., 66, 97
Somes, G. W., 212
Srinivasan, R., 67
Stegun, I. A., 278
Stein, C., 52, 112, 123, 137
Sullivan, D. W., 96
Sun, D., 67
Swanepoel, J. W. H., 97

Taheri, H., 211
Tamhane, A. C., 3, 31, 32, 50, 70, 74, 100, 114, 135, 138, 212, 253
Taneja, B. K., 143
Thompson, J. R., 245
Thompson, S. K., 246
Tiao, G. C., 66
Tong, Y. L., 66, 141
Tortora, R. D., 246
Tukey, J. W., 110
Turnbull, B. W., 67, 118, 123, 141, 175, 245

van der Laan, P., 29, 79, 92
van Putten, B., 29, 79
Venter, J. H., 97

Weiss, G. H., 67, 211
Wilcox, R. R., 61, 141
Wilson, J. R., 96
Wong, W.-Y., 97
Woodworth, G. G., 68
Wu, K. H., 175

Yahav, J. A., 97
Yamagami, S., 212
Young, D. H., 211

Zaino, N. A., 30
Zelen, M., 211

Subject Index

Adaptive sampling, 186, 211
Alternative formulations, 184, 211, 212, 245
   Confidence statements, 22, 35, 150, 166, 185
   IZ selection of s best with regard to order, 46
   IZ selection of s best without regard to order, 43, 44
   Restricted SS, 13, 94
   SS selection of δ^*-near-best, 91, 94
   SS selection of s best without regard to order, 86
Autoregressive model, 29, 79

Bernoulli distribution; *see* Distributions
Bernoulli sequential procedures
   Closed adaptive, 188
   Open multi-stage, 192, 195
Bernoulli single-stage procedures
   Indifference-zone, 182, 184
   Subset selection, 201
Blocks; *see* Experimental designs

Comparison of procedures, 40, 56, 96, 154, 196
Configurations
   Equal means, 21, 41, 57
   Equally spaced, 41
   Least-favorable, 20, 37, 41, 44, 57, 65
   Slippage, 21
Control treatment, 13, 116
   Confidence intervals with respect to two controls, 138
   Selecting all treatments better than, 129
   Selecting an experimental treatment better than, 125
Critical point; *see* Quantiles
Curtailment, 185, 219, 237, 240
   Strong, 187
   Weak, 186

Distributions
   Bernoulli, 11, 177
   Multinomial, 214
   Normal, 15, 69, 100, 109, 116, 142

Eliminating procedure; *see* Normal sequential procedures and Bernoulli sequential procedure
Equicoordinate upper-α point
   Multivariate t, 53, 72, 77, 247
   Multivariate normal, 18, 72, 77, 248
Experimental designs, 5, 6
   Completely randomized
      Balanced factorial, 103, 106
      Balanced incomplete block, 25, 81, 84
      Balanced single-factor experiments, 2, 71, 109, 112, 126, 129, 134, 201
      Balanced treatment incomplete blocks, 138
      Balanced two-factor experiments, 10, 144, 163, 167
      Cross-over, 27
      Latin square, 27
      Unbalanced experiments, 22, 76, 78, 111, 114, 128, 133, 134, 209

Cross-classified multinomial; *see* Experimental designs and Discrete bivariate
Discrete bivariate, 241
Discrete univariate, 216
Randomized complete block, 24, 40, 80, 130, 150, 194
Split-plot, 160, 166, 169, 175

Hyper-rectangle normal probability, 20, 44, 65, 209, 277

Indifference-zones, 8
　Bernoulli
　　In terms of differences, 178, 182
　　In terms of odds ratios, 11, 179, 192
　　In terms of relative risks, 179
　Multinomial
　　In terms of ratios, 12, 216
　Normal, 8

Multi-factor response; *see* Experimental designs
Multinomial sequential procedures
　Closed nonadaptive
　　Without elimination, 229
　Open multi-stage, 232
Multinomial single-stage procedures, 216, 236
　Indifference-zone, 216, 217, 243
　Subset selection, 236
Multiple comparisons; *see* Simultaneous confidence intervals

Noneliminating procedure; *see* Multinomial sequential procedures, Bernoulli sequential procedures, and Normal sequential procedures
Normal distribution; *see* Distributions
Normal sequential procedures, 31
　Closed nonadaptive
　　With elimination, 32, 38, 74, 152
　　Without elimination, 35, 151
　Open multi-stage, 54, 56
　Open two-stage, 52, 54
Normal single-stage procedures, 15
　Indifference-zone, 18, 25, 26, 48, 146, 166
　Subset selection, 167, 170

Normal variance assumptions
　Bounded but unknown, 60
　Common known, 18, 40, 42, 71, 76, 81, 87, 118, 125, 144, 168, 171
　Common unknown, 72, 77, 81, 87, 123, 134, 163, 168, 171
　Known but unequal, 48
　Unknown, 60

Odds ratio; *see* Indifference-zones
Optimality property, 24, 49, 97, 111, 134, 149, 185, 189

Percentile; *see* Quantiles
Probability requirement, 8–10

Qualitative factors, 4
Quantiles
　z^α, 6, 248
　$|M|_{p,\nu}^{(\alpha)}$, 101, 247
　$Q_{p,\nu}^\alpha$, 110, 247
　$T_{p,\nu,\rho}^{(\alpha)}$, 53, 247
　$|T|_{p,\nu,\rho}^{(\alpha)}$, 132, 248
　$Z_{p,\rho}^{(\alpha)}$, 18, 248
　$|Z|_{p,\rho}^{(\alpha)}$, 134, 248
Quantitative factors, 5

Randomization, 5, 6
Relative risk; *see* Indifference-zones
Replication, 6
Robustness
　Of $\mathcal{N}_B$, 28
　Of $\mathcal{N}_G$, 79

Screening; *see* Subset selection
Sequential procedure; *see* Normal sequential procedures, Bernoulli sequential procedures, and Multinomial sequential procedures
Simultaneous confidence intervals, 9, 100
　For orthogonal contrasts, 11, 101
　With a control treatment, 134
　With a given standard, 131
　With all pairwise differences, 109
　With the best, 112
　With the best of the rest, 113, 175
　With two control treatments, 138

Single-stage procedure; *see* Bernoulli sequential procedure, Normal sequential procedure, and Multinomial sequential procedures
Standard, 13, 116
  Selecting all treatments better than, 126
  Selecting an experimental treatment better than, 118
Subset selection, 9, 69
  Bernoulli, 200
  normal, 9, 71, 80, 86

Two-factor experiments; *see* Experimental designs completely randomized
Two-stage procedure; *see* Normal sequential procedures

Unbalanced experiments; *see* Normal single-stage procedures and Bernoulli single-stage procedures

Variance; *see* Normal variance assumptions

WILEY SERIES IN PROBABILITY AND STATISTICS

ESTABLISHED BY WALTER A. SHEWHART AND SAMUEL S. WILKS

Editors
*Vic Barnett, Ralph A. Bradley, Nicholas I. Fisher, J. Stuart Hunter,
J. B. Kadane, David G. Kendall, David W. Scott, Adrian F. M. Smith,
Jozef L. Teugels, Geoffrey S. Watson*

Probability and Statistics
  ANDERSON · An Introduction to Multivariate Statistical Analysis, *Second Edition*
 *ANDERSON · The Statistical Analysis of Time Series
  ARNOLD, BALAKRISHNAN, and NAGARAJA · A First Course in Order Statistics
  BACCELLI, COHEN, OLSDER, and QUADRAT · Synchronization and Linearity:
    An Algebra for Discrete Event Systems
  BARNETT · Comparative Statistical Inference, *Second Edition*
  BERNARDO and SMITH · Bayesian Statistical Concepts and Theory
  BHATTACHARYYA and JOHNSON · Statistical Concepts and Methods
  BILLINGSLEY · Convergence of Probability Measures
  BILLINGSLEY · Probability and Measure, *Second Edition*
  BOROVKOV · Asymptotic Methods in Queuing Theory
  BRANDT, FRANKEN, and LISEK · Stationary Stochastic Models
  CAINES · Linear Stochastic Systems
  CAIROLI and DALANG · Sequential Stochastic Optimization
  CHEN · Recursive Estimation and Control for Stochastic Systems
  CONSTANTINE · Combinatorial Theory and Statistical Design
  COOK and WEISBERG · An Introduction to Regression Graphics
  COVER and THOMAS · Elements of Information Theory
 *DOOB · Stochastic Processes
  DUDEWICZ and MISHRA · Modern Mathematical Statistics
  ETHIER and KURTZ · Markov Processes: Characterization and Convergence
  FELLER · An Introduction to Probability Theory and Its Applications, Volume 1,
    Third Edition, Revised; Volume II, *Second Edition*
  FULLER · Introduction to Statistical Time Series
  FULLER · Measurement Error Models
  GIFI · Nonlinear Multivariate Analysis
  GUTTORP · Statistical Inference for Branching Processes
  HALD · A History of Probability and Statistics and Their Applications before 1750
  HALL · Introduction to the Theory of Coverage Processes
  HANNAN and DEISTLER · The Statistical Theory of Linear Systems
  HEDAYAT and SINHA · Design and Inference in Finite Population Sampling
  HOEL · Introduction to Mathematical Statistics, *Fifth Edition*
  HUBER · Robust Statistics
  IMAN and CONOVER · A Modern Approach to Statistics
  JUREK and MASON · Operator-Limit Distributions in Probability Theory
  KAUFMAN and ROUSSEEUW · Finding Groups in Data: An Introduction to Cluster
    Analysis
  LARSON · Introduction to Probability Theory and Statistical Inference, *Third Edition*
  LESSLER and KALSBEEK · Nonsampling Error in Surveys
  LINDVALL · Lectures on the Coupling Method
  MANTON, WOODBURY, and TOLLEY · Statistical Applications Using Fuzzy Sets
  MORGENTHALER and TUKEY · Configural Polysampling: A Route to Practical
    Robustness
  MUIRHEAD · Aspects of Multivariate Statistical Theory

*Now available in a lower priced paperback edition in the Wiley Classics Library.

Probability and Statistics (Continued)
    OLIVER and SMITH · Influence Diagrams, Belief Nets and Decision Analysis
   *PARZEN · Modern Probability Theory and Its Applications
    PILZ · Bayesian Estimation and Experimental Design in Linear Regression Models
    PRESS · Bayesian Statistics: Principles, Models, and Applications
    PUKELSHEIM · Optimal Experimental Design
    PURI and SEN · Nonparametric Methods in General Linear Models
    PURI, VILAPLANA, and WERTZ · New Perspectives in Theoretical and Applied Statistics
    RAO · Asymptotic Theory of Statistical Inference
    RAO · Linear Statistical Inference and Its Applications, *Second Edition*
    RENCHER · Methods of Multivariate Analysis
    ROBERTSON, WRIGHT, and DYKSTRA · Order Restricted Statistical Inference
    ROGERS and WILLIAMS · Diffusions, Markov Processes, and Martingales, Volume II: Îto Calculus
    ROHATGI · An Introduction to Probability Theory and Mathematical Statistics
    ROSS · Stochastic Processes
    RUBINSTEIN · Simulation and the Monte Carlo Method
    RUZSA and SZEKELY · Algebraic Probability Theory
    SCHEFFE · The Analysis of Variance
    SEBER · Linear Regression Analysis
    SEBER · Multivariate Observations
    SEBER and WILD · Nonlinear Regression
    SERFLING · Approximation Theorems of Mathematical Statistics
    SHORACK and WELLNER · Empirical Processes with Applications to Statistics
    SMALL and McLEISH · Hilbert Space Methods in Probability and Statistical Inference
    STAPLETON · Linear Statistical Models
    STAUDTE and SHEATHER · Robust Estimation and Testing
    STOYANOV · Counterexamples in Probability
    STYAN · The Collected Papers of T. W. Anderson: 1943–1985
    WHITTAKER · Graphical Models in Applied Multivariate Statistics
    YANG · The Construction Theory of Denumerable Markov Processes

Applied Probability and Statistics
    ABRAHAM and LEDOLTER · Statistical Methods for Forecasting
    AGRESTI · Analysis of Ordinal Categorical Data
    AGRESTI · Categorical Data Analysis
    ANDERSON and LOYNES · The Teaching of Practical Statistics
    ANDERSON, AUQUIER, HAUCK, OAKES, VANDAELE, and WEISBERG · Statistical Methods for Comparative Studies
  *ARTHANARI and DODGE · Mathematical Programming in Statistics
    ASMUSSEN · Applied Probability and Queues
  *BAILEY · The Elements of Stochastic Processes with Applications to the Natural Sciences
    BARNETT · Interpreting Multivariate Data
    BARNETT and LEWIS · Outliers in Statistical Data, *Second Edition*
    BARTHOLOMEW, FORBES, and McLEAN · Statistical Techniques for Manpower Planning, *Second Edition*
    BATES and WATTS · Nonlinear Regression Analysis and Its Applications
    BECHHOFER, SANTNER, and GOLDSMAN · Design and Analysis of Experiments for Statistical Selection, Screening, and Multiple Comparisons
    BELSLEY · Conditioning Diagnostics: Collinearity and Weak Data in Regression
    BELSLEY, KUH, and WELSCH · Regression Diagnostics: Identifying Influential Data and Sources of Collinearity
    BHAT · Elements of Applied Stochastic Processes, *Second Edition*

*Now available in a lower priced paperback edition in the Wiley Classics Library.

Applied Probability and Statistics (Continued)
BHATTACHARYA and WAYMIRE · Stochastic Processes with Applications
BIEMER, GROVES, LYBERG, MATHIOWETZ, and SUDMAN · Measurement Errors in Surveys
BIRKES and DODGE · Alternative Methods of Regression
BLOOMFIELD · Fourier Analysis of Time Series: An Introduction
BOLLEN · Structural Equations with Latent Variables
BOULEAU · Numerical Methods for Stochastic Processes
BOX · R. A. Fisher, the Life of a Scientist
BOX and DRAPER · Empirical Model-Building and Response Surfaces
BOX and DRAPER · Evolutionary Operation: A Statistical Method for Process Improvement
BOX, HUNTER, and HUNTER · Statistics for Experimenters: An Introduction to Design, Data Analysis, and Model Building
BROWN and HOLLANDER · Statistics: A Biomedical Introduction
BUCKLEW · Large Deviation Techniques in Decision, Simulation, and Estimation
BUNKE and BUNKE · Nonlinear Regression, Functional Relations and Robust Methods: Statistical Methods of Model Building
CHATTERJEE and HADI · Sensitivity Analysis in Linear Regression
CHATTERJEE and PRICE · Regression Analysis by Example, *Second Edition*
CLARKE and DISNEY · Probability and Random Processes: A First Course with Applications, *Second Edition*
COCHRAN · Sampling Techniques, *Third Edition*
*COCHRAN and COX · Experimental Designs, *Second Edition*
CONOVER · Practical Nonparametric Statistics, *Second Edition*
CONOVER and IMAN · Introduction to Modern Business Statistics
CORNELL · Experiments with Mixtures, Designs, Models, and the Analysis of Mixture Data, *Second Edition*
COX · A Handbook of Introductory Statistical Methods
*COX · Planning of Experiments
COX, BINDER, CHINNAPPA, CHRISTIANSON, COLLEDGE, and KOTT · Business Survey Methods
CRESSIE · Statistics for Spatial Data, *Revised Edition*
DANIEL · Applications of Statistics to Industrial Experimentation
DANIEL · Biostatistics: A Foundation for Analysis in the Health Sciences, *Sixth Edition*
DAVID · Order Statistics, *Second Edition*
*DEGROOT, FIENBERG, and KADANE · Statistics and the Law
*DEMING · Sample Design in Business Research
DILLON and GOLDSTEIN · Multivariate Analysis: Methods and Applications
DODGE and ROMIG · Sampling Inspection Tables, *Second Edition*
DOWDY and WEARDEN · Statistics for Research, *Second Edition*
DRAPER and SMITH · Applied Regression Analysis, *Second Edition*
DUNN · Basic Statistics: A Primer for the Biomedical Sciences, *Second Edition*
DUNN and CLARK · Applied Statistics: Analysis of Variance and Regression, *Second Edition*
ELANDT-JOHNSON and JOHNSON · Survival Models and Data Analysis
EVANS, PEACOCK, and HASTINGS · Statistical Distributions, *Second Edition*
FISHER and VAN BELLE · Biostatistics: A Methodology for the Health Sciences
FLEISS · The Design and Analysis of Clinical Experiments
FLEISS · Statistical Methods for Rates and Proportions, *Second Edition*
FLEMING and HARRINGTON · Counting Processes and Survival Analysis
FLURY · Common Principal Components and Related Multivariate Models
GALLANT · Nonlinear Statistical Models
GLASSERMAN and YAO · Monotone Structure in Discrete-Event Systems
GROSS and HARRIS · Fundamentals of Queueing Theory, *Second Edition*

*Now available in a lower priced paperback edition in the Wiley Classics Library.

Applied Probability and Statistics (Continued)

GROVES · Survey Errors and Survey Costs
GROVES, BIEMER, LYBERG, MASSEY, NICHOLLS, and WAKSBERG ·
  Telephone Survey Methodology
HAHN and MEEKER · Statistical Intervals: A Guide for Practitioners
HAND · Discrimination and Classification
*HANSEN, HURWITZ, and MADOW · Sample Survey Methods and Theory,
  Volume 1: Methods and Applications
*HANSEN, HURWITZ, and MADOW · Sample Survey Methods and Theory,
  Volume II: Theory
HEIBERGER · Computation for the Analysis of Designed Experiments
HELLER · MACSYMA for Statisticians
HINKELMAN and KEMPTHORNE: · Design and Analysis of Experiments, Volume 1:
  Introduction to Experimental Design
HOAGLIN, MOSTELLER, and TUKEY · Exploratory Approach to Analysis of Variance
HOAGLIN, MOSTELLER, and TUKEY · Exploring Data Tables, Trends and Shapes
HOAGLIN, MOSTELLER, and TUKEY · Understanding Robust and Exploratory
  Data Analysis
HOCHBERG and TAMHANE · Multiple Comparison Procedures
HOEL · Elementary Statistics, *Fifth Edition*
HOGG and KLUGMAN · Loss Distributions
HOLLANDER and WOLFE · Nonparametric Statistical Methods
HOSMER and LEMESHOW · Applied Logistic Regression
HØYLAND and RAUSAND · System Reliability Theory: Models and Statistical Methods
HUBERTY · Applied Discriminant Analysis
IMAN and CONOVER · Modern Business Statistics
JACKSON · A User's Guide to Principle Components
JOHN · Statistical Methods in Engineering and Quality Assurance
JOHNSON · Multivariate Statistical Simulation
JOHNSON and KOTZ · Distributions in Statistics
        Continuous Univariate Distributions—2
        Continuous Multivariate Distributions
JOHNSON, KOTZ, and BALAKRISHNAN · Continuous Univariate Distributions,
  Volume 1, *Second Edition*
JOHNSON, KOTZ, and KEMP · Univariate Discrete Distributions, *Second Edition*
JUDGE, GRIFFITHS, HILL, LÜTKEPOHL, and LEE · The Theory and Practice of
  Econometrics, *Second Edition*
JUDGE, HILL, GRIFFITHS, LÜTKEPOHL, and LEE · Introduction to the Theory and
  Practice of Econometrics, *Second Edition*
KALBFLEISCH and PRENTICE · The Statistical Analysis of Failure Time Data
KASPRZYK, DUNCAN, KALTON, and SINGH · Panel Surveys
KISH · Statistical Design for Research
*KISH · Survey Sampling
LANGE, RYAN, BILLARD, BRILLINGER, CONQUEST, and GREENHOUSE ·
  Case Studies in Biometry
LAWLESS · Statistical Models and Methods for Lifetime Data
LEBART, MORINEAU., and WARWICK · Multivariate Descriptive Statistical
  Analysis: Correspondence Analysis and Related Techniques for Large Matrices
LEE · Statistical Methods for Survival Data Analysis, *Second Edition*
LePAGE and BILLARD · Exploring the Limits of Bootstrap
LEVY and LEMESHOW · Sampling of Populations: Methods and Applications
LINHART and ZUCCHINI · Model Selection
LITTLE and RUBIN · Statistical Analysis with Missing Data
MAGNUS and NEUDECKER · Matrix Differential Calculus with Applications in
  Statistics and Econometrics

*Now available in a lower priced paperback edition in the Wiley Classics Library.

Applied Probability and Statistics (Continued)

MAINDONALD · Statistical Computation
MALLOWS · Design, Data, and Analysis by Some Friends of Cuthbert Daniel
MANN, SCHAFER, and SINGPURWALLA · Methods for Statistical Analysis of Reliability and Life Data
MASON, GUNST, and HESS · Statistical Design and Analysis of Experiments with Applications to Engineering and Science
McLACHLAN · Discriminant Analysis and Statistical Pattern Recognition
MILLER · Survival Analysis
MONTGOMERY and MYERS · Response Surface Methodology: Process and Product in Optimization Using Designed Experiments
MONTGOMERY and PECK · Introduction to Linear Regression Analysis, *Second Edition*
NELSON · Accelerated Testing, Statistical Models, Test Plans, and Data Analyses
NELSON · Applied Life Data Analysis
OCHI · Applied Probability and Stochastic Processes in Engineering and Physical Sciences
OKABE, BOOTS, and SUGIHARA · Spatial Tesselations: Concepts and Applications of Voronoi Diagrams
OSBORNE · Finite Algorithms in Optimization and Data Analysis
PANKRATZ · Forecasting with Dynamic Regression Models
PANKRATZ · Forecasting with Univariate Box-Jenkins Models: Concepts and Cases
PORT · Theoretical Probability for Applications
PUTERMAN · Markov Decision Processes: Discrete Stochastic Dynamic Programming
RACHEV · Probability Metrics and the Stability of Stochastic Models
RÉNYI · A Diary on Information Theory
RIPLEY · Spatial Statistics
RIPLEY · Stochastic Simulation
ROSS · Introduction to Probability and Statistics for Engineers and Scientists
ROUSSEEUW and LEROY · Robust Regression and Outlier Detection
RUBIN · Multiple Imputation for Nonresponse in Surveys
RYAN · Statistical Methods for Quality Improvement
SCHUSS - Theory and Applications of Stochastic Differential Equations
SCOTT · Multivariate Density Estimation: Theory, Practice, and Visualization
SEARLE · Linear Models
SEARLE · Linear Models for Unbalanced Data
SEARLE · Matrix Algebra Useful for Statistics
SEARLE, CASELLA, and McCULLOCH · Variance Components
SKINNER, HOLT, and SMITH · Analysis of Complex Surveys
STOYAN · Comparison Methods for Queues and Other Stochastic Models
STOYAN, KENDALL, and MECKE · Stochastic Geometry and Its Applications
STOYAN and STOYAN · Fractals, Random Shapes and Point Fields
THOMPSON · Empirical Model Building
THOMPSON · Sampling
TIERNEY · LISP-STAT: An Object-Oriented Environment for Statistical Computing and Dynamic Graphics
TIJMS · Stochastic Modeling and Analysis: A Computational Approach
TITTERINGTON, SMITH, and MAKOV · Statistical Analysis of Finite Mixture Distributions
UPTON and FINGLETON · Spatial Data Analysis by Example, Volume 1: Point Pattern and Quantitative Data
UPTON and FINGLETON · Spatial Data Analysis by Example, Volume II: Categorical and Directional Data
VAN RIJCKEVORSEL and DE LEEUW · Component and Correspondence Analysis
WEISBERG · Applied Linear Regression, *Second Edition*
WESTFALL and YOUNG · Resampling-Based Multiple Testing: Examples and Methods for *p*-Value Adjustment

Applied Probability and Statistics (Continued)
  WHITTLE · Optimization Over Time: Dynamic Programming and Stochastic Control, Volume I and Volume II
  WHITTLE · Systems in Stochastic Equilibrium
  WONNACOTT and WONNACOTT · Econometrics, *Second Edition*
  WONNACOTT and WONNACOTT · Introductory Statistics, *Fifth Edition*
  WONNACOTT and WONNACOTT · Introductory Statistics for Business and Economics, *Fourth Edition*
  WOODING · Planning Pharmaceutical Clinical Trials: Basic Statistical Principles
  WOOLSON · Statistical Methods for the Analysis of Biomedical Data

Tracts on Probability and Statistics
  BILLINGSLEY · Convergence of Probability Measures
  TOUTENBURG · Prior Information in Linear Models